hi-pass

Expressway Design Manual

Korea Expressway Corporation

목 차

Table of Contents

제5장 교통 분석 및 경제성 검토

Chapter 5. Traffic Analysis and Economic Efficiency Evaluation

Chapter 6. Route Location and Alignment Design

Chapter 7. Access Facilities

제8장 토 공

Chapter 8. Earth Work

제 9 장 배수공

Chapter 9. Drainage

제 10 장 구조물공

Chapter 10. Structure Construction

제 11 장 포장공

Chapter 11. Pavement

제 12 장 교통안전시설공

Chapter 12. Traffic Safety Facilities

제 13 장 부대공

Chapter 13. Subsidiary Facilities

제 14 장 터널공

Chapter 14. Tunnelling

제 15 장 영향평가

Chapter 15. Effect Assessment

제 16 장 도로구역 결정

제 17 장 총사업비

Chapter 16. Road Area Determination

Chapter 17. Total Cost

Chapter 1

General Provision

1. 목 적

- 설계란 건설공사의 시공 직전 단계에서 이루어지는 것으로서, 궁극적으로는 시공에 필요한 설계도면 및 시방서 등의 설계도서를 작성하는 것을 목표로 한다.

- 아울러 각 설계요소들의 적용치 결정을 위한 기술적 분석 및 합리적 의사결정을 위해 법적으로 정한 각종 절차를 수행하는 것 뿐만 아니라 도로사업과 연관된 농지, 산지, 국방, 하천 등에 대한 각종 법령에서 정한 관련 기관과의 협의 및 인·허가 등을 수행하는 것도 설계 업무에 포함된다.

- 이렇게 단순한 과정으로 정리하기 힘든 실시설계의 복잡성과 더불어 설계 준비단계에서부터 공사 발주 시까지 2년 내외의 기간이 필요한 설계의 긴 주기 때문에 실시설계 경험이 있는 설계자라 할지라도 각 단계별로 추진해야할 설계업무를 사전에 파악하는 것은 쉬운 일이 아니다.

- 본 매뉴얼은 설계의 준비단계에서부터 설계준공 후 인·허가 및 총사업비 등 공사 발주 시까지 필요한 주요 업무를 정리하여 제시함으로써 설계감독 및 설계용역사의 참여기술자가 설계의 일련의 과정을 숙지하고 각자의 맡은바 임무를 사전에 준비하여 짜임새 있는 설계업무 수행을 돕는 데 그 목적이 있다.

2. 적용범위

- 본 매뉴얼은 한국도로공사에서 시행하는 고속도로 건설공사의 설계 용역을 기준으로 작성되었다.

- 참고로 설계용역의 단계 구분은 "타당성조사→기본설계→실시설계" 이며, 공사규모 및 기타 필요에 따라 일부 설계단계를 생략 또는 통합하여 시행할 수 있다.

- 다음은 발주청에서 시행 가능한 설계용역의 단계별 종류이다.
  - 타당성조사 → 기본설계 → 실시설계
  - 타당성조사 → [기본 및] 실시설계　　　　* [ ] : 생략 가능한 단계
  - 타당성조사 [및 기본설계] → 실시설계
  - (타당성조사 생략) → 기본설계 → 실시설계
  - (타당성조사 생략) → [기본 및] 실시설계
  - 타당성조사 → (기본설계 생략) → 실시설계(Turn Key 또는 기술공모)

1. Purpose

- The design is required at the phase right before the construction, and the ultimate goal is to design drawing, specification, and design documents required for executing construction.

- Design work includes technical analysis for determining application values for each design component, executing various legal procedures for rational decision making, and negotiation and approval/authorization process with relevant organizations determined by various regulations on farming land, mountain area, national defense, and river in relation to road construction.

- As it requires around two years of long lead time from the preparation stage to the construction order along with the complexity of execution design, which cannot be simplified easily, it is very challenging to understand the design work to be done on each step beforehand even for the engineer who has rich experience on execution design.

- As this manual summarizes and presents the major work required from the preparation stage of design to the construction order including approval/ authorization process and total cost calculation after the completion of design, the purpose is to enable design director and participating engineers of design contractors to be familiar with the series of process and prepare the duties in advance to execute the structured design work.

2. Scope

- This manual has been prepared based on the design service for highway construction work by Korea Expressway Corporation.

- For reference, the work flow of design service is "Feasibility Study → Basic Design → Detaied Design", and certain steps can be skipped or combined during the execution depending on the size of construction and other needs.

- The following procedures show the executable work flow of design service by order-placing department.
  • Feasibility Study → Basic Design → Detailed Design
  • Feasibility Study → [Basic and] Detailed Design   * [ ] : can be skipped
  • Feasibility Study [and Basic Design] → Detailed Design
  • (Skip Feasibility Study) → Basic Design → Detailed Design
  • (Skip Feasibility Study) → [Basic and] Detailed Design
  • Feasibility Study → (Skip Basic Design) → Detailed Design (Turn Key or public offering of technology)

1. 매뉴얼의 구성 및 특징

- 실시설계 업무 성격에 따라 다음의 항목들로 구성되어 있으며, 목차 순서는 가급적 업무 순서를 따르도록 하였다.

  ① 설계행정 : 발주, 용역업자 선정, 계약 등

  ② 기본계획 : 노선선정, 설계기준 및 방침 설정, 관계기관 협의 등

  ③ 상세설계 : 공종별 설계(기술적 검토)

  ④ 인·허가 : 영향평가 등, 도로구역, 총사업비 등

- 내용의 깊이는 설계행정 관련 내용에 대해서는 주로 관련 법령 및 규정 등을 인용하거나 참고 가능토록 근거를 제시하였고, 기술적 검토에 해당하는 부분은 설계의 기본개념 설명과 설계하는 방법 정도만 제시하였다.

- 따라서, 도로설계자로서 상식수준의 내용이나 계산방법 등 설계의 세부사항은 도로의 구조 및 시설 기준에 관한 규칙과 도로설계요령 등의 기술서적을 참고하도록 한다.

3. Configuration and Features of Manual

- This manual comprises of the following items according to the needs of the design works, and the sequence has been designed to follow the work flow as much as possible.

  ① Design Administration : Placing order, selecting service provider, contract, etc

  ② Basic Planning : Route selection, Setting design guideline and policy, negotiation with relevant organization, etc

  ③ Detailed Design : Design for each construction type (Technical review)

  ④ Approval/Authorization : Evaluation of effect, road zone, total cost of the project, etc

- Contents related to design administration have provided relevant laws or regulations as a basis for citation or reference, and technical review parts explains only to the level of basic concept and procedures of the design.

- Therefore, road designers shall refer to the regulations on the standard of road structure and facilities and other technical documents including road design guide for the details of designs like calculation rules and other common sense-level of knowledge.

Chapter 2

Design Preparation

1. 용역업체의 선정

1-1 개　요

- 설계용역 수행을 위한 용역업체의 선정은 건설기술관리법 시행령 제38조1항에 따라 실시토록 되어 있는 사업수행능력평가(PQ 평가)를 실시하여 입찰에 참가할 자를 선정한 후 입찰을 실시하여 최종적으로 선정한다.

1-2 발　주

- 한국도로공사에서 국토해양부로 사업계획을 상신하여 기획재정부와의 협의를 통하여 설계하고자 하는 노선의 설계예산이 배정되면 사업이 확정된다.

- 국토해양부에서 한국도로공사로 예산배정이 완료되면 사업시행부서(설계처)와 예산주무부서(기획조정실)의 협의를 통하여 사업규모, 설계기간, 설계금액, 착수시기 등에 대한 방침을 확정한다.

- PQ 평가 및 입찰을 실시하기 위해서는 다음과 같은 각종 설계지침 및 안내문 등을 준비하고, 입찰공고를 실시한다.
  - 입찰공고문
  - 실시설계 추진개요
  - 사업수행능력 평가자료 작성지침
  - 일반과업지시서
  - 특별과업지시서
  - 사전자격심사신청서작성지침
  - 사업수행능력평가기준

- 기타 입찰 및 계약 등에 관한 기준은 계약팀에서 전자조달시스템에 등록한 자료를 참조하도록 한다.

1. Selecting Service Provider

1-1 Overview

- Final selection of service provider shall be made through conducting the bidding after selecting candidates for the bidding by executing the project execution capability evaluation (PQ evaluation) which shall be done in accordance with the Clause 38 Article 1 of the Enforcement Decree of the Construction Technology Management Act.

1-2 Placing Order

- The project will be confirmed when the road design budget is allocated to the route to be designed through the negotiation with the Ministry of Strategy and Finance after reporting the project proposal to the MLTM (Ministry of Land, Transport, and Maritime Affairs).

- When the budget is allocated to our corporation (Korea Expressway Corporation) from the MLTM, project plans regarding scale of project, design period, design cost, and time point of commencement will be finalized through the negotiation with Project Execution Department (Design Department) and Budget Handling Department (Planning and Coordination Department).

- In order to implement PQ evaluation and bidding, the design instructions and guidelines as followings shall be prepared and notice of tender shall be posted.
  - Notice of tender
  - Design project overview
  - Instructions on preparation for PQ evaluation data
  - General Tasks Instructions
  - Special Tasks Instructions
  - Instructions on preparing application for pre-qualification
  - Criteria for PQ evaluation

- Refer to the documents registered in the electronic procurement system by the contract team for the details about contract and bidding.

1-3 사업수행능력 평가

- 사업수행능력평가(PQ 평가)는 건설기술자의 능력, 사업의 수행실적, 신용도 등을 평가하는 것으로서 본 평가를 실시하여 관련 기준을 만족하는 업체에 한하여 입찰에 참가할 자격을 부여한다.

- 사업수행능력 평가에 관한 사항 및 입찰에 관한 사항을 공고한 후, 입찰에 참가를 원하는 용역업자를 대상으로 사업수행능력평가 기준 및 서류 작성에 관한 사항에 대하여 현장설명을 실시한다.

- 공고문에서 정한 기한까지 제출된 업체별 PQ 신청서류를 입찰공고시 공개한 사업수행능력평가기준에 의거하여 평가하여 업체별 점수를 산출한다.
  (사업수행능력평가기준은 국토해양부에서 고시하는 평가기준을 한국도로공사 자체 실정에 맞도록 정한 것이다.)

- 입찰에 참가할 자격은 사업수행능력평가 결과점수가 80점 이상인 용역업체에 부여한다.

1.4 입찰 및 업체선정

- 사업수행능력평가를 통과하여 입찰참가요청을 받은 용역업자는 한국도로공사 용역입찰유의서에 따라 입찰을 실시한다.

- 낙찰자 결정은 입찰자 중 예정가격 이하로서 최저가격으로 입찰한 자 순으로 적격심사를 실시하여 종합평점이 85점 이상인 자를 낙찰자로 결정한다.

- 적격심사 점수는 사업수행능력평가 점수를 용역금액에 따라 다음표에서 정한 점수로 환산하고 입찰가격에 의한 점수를 합산하여 산출한다.

1-3 PQ Evaluation

- Project execution capability evaluation (PQ evaluation) is for the assessment of the capability of construction engineering, accomplishments, and credit rating, and the companies that meet the relevant requirements through PQ evaluation are qualified for bidding.

- After the public notification of details on PQ evaluation and bidding, there will be a presentation on construction site regarding PQ evaluation criteria and document preparation.

- Scores of each applicant shall be calculated for the PQ evaluation applications received by the due date in accordance with the PQ evaluation criteria published at notice of tender.
  (PQ evaluation is based on the evaluation criteria of the MLTM (the Ministry of Land, Transport, and Maritime Affairs) and is customized to the situation of Korea Expressway Corporation)

- Candidates who get more than 80 points from the PQ evaluation are qualified to apply for the bidding.

1-4 Bidding and Selecting Service Provider

- Candidates who passed the PQ evaluation and gained the bidding request can participate in the bidding according to the bidding guideline of Korea Expressway Corporation.

- Among the tenders who proposed the bidding price lower than expected price, screening committee will review the proposal from each of the tender in the order of lower bidding price, and a tender who will have total score of over 85 points will be decided as a successful bidder.

- The score of screening is sum of the PQ evaluation points converted in accordance with the score stated in the following table by the service price and points determined by bidding price.

용역금액	배 점	
	사업수행능력 평가결과	입찰가격 평점
10억 원 이상	70	30
5억 원 이상 ~ 10억 원 미만	50	50
1.9억 원 이상 ~ 5억 원 미만	30	70

- 적격심사에 관한 세부사항은 한국도로공사『용역적격심사 세부기준』에 따른다.

2. 계 약

- 낙찰통지를 받은 용역업체는 한국도로공사와 10일 이내에 계약을 체결하여야 한다.

- 계약에 관한 세부사항은 한국도로공사 용역계약일반조건, 용역계약특수조건, 공동계약운용기준에 따른다.

3. 설계 착수

- 한국도로공사와 계약을 완료한 용역업체는 계약 후 5일 이내에 착수계를 제출하여야 한다.

- 착수계에 포함할 내용
  - 책임 기술자 선임계
  - 책임기술자 사용인감계
  - 책임기술자 이력서 및 자격증 사본
  - 예정공종표
  - 내역서
  - 보안각서 및 청렴이행 준수각서 등

Service Price	Points	
	PQ Evaluation Result	Bidding Price Point
Over one billion one	70	30
Over 500 million ~ Below one billion one	50	50
Over 190 Million ~ Below 500 Million	30	70

- The details on the screening shall be in accordance with the 『Detailed Criteria of Service Provider Screening』 of Korea Expressway Corporation.

2. Contract

- Service provider who won the bidding shall make the contract with Korea Expressway Corporation within 10 days from the notice.

- The details on the contract shall follow General Conditions of Contract, Special Conditions of Contract, and Common Contracts Operation Standard of Korea Expressway Corporation.

3. Commencement of Design

- Service provider who signed the contract shall submit the commencement plan within five days from the date of contract.

- The commencement plan shall include the followings
  - Technical department manager assignment plan
  - Technical department manager's legal seal plan
  - Resume and copy of certificate of technical department manager
  - Process time table plan
  - Cost details,
  - Security memorandum, integrity memorandum, etc.

4. 과업수행보고

4-1 착수보고

◎ 수급인은 현장여건, 사업내용 등을 검토한 후 용역 착수 후 15일 이내에 착수보고를 하여야 하며, 설계계획서를 작성·제출하여 감독원의 승인을 받아야 한다.

◎ 설계계획서에 포함할 내용

- 설계 업무의 범위
- 설계 공정표
- 설계 자원 투입 계획 · 인력 · 프로그램 · 장비 등
- 설계 분야 간 조직적 · 기술적 연계사항
- 설계 성과품 : 성과품 종류, 부수, 제출 시기 등

4-2 월간진도보고

◎ 수급인은 과업수행 기간 중 다음사항을 포함한 월간진도보고를 매월 말일을 기준으로 다음달 (5일)까지 용역감독원을 경유하여 서면으로 보고하여야한다.

◎ 보고내용

- 과업추진내용 및 공정현황
- 과업수행상 중요 문제점 및 대책
- 다음달 과업수행계획
- 각종 문서수발 현황
- 참여기술자 현황

4-3 중간보고

◎ 다음 각각의 경우에는 관련자료를 제출하고 담당 분야별 책임기술자로 하여금 설명토록 하여야 한다.

- 주요 단계별 과업 종료 시
- 주요계획 및 방침의 설정과 변경 시
- 감독원이 용역수행상 필요하여 요구 시

4-4 용역 준공 전 종합보고

◎ 수급인은 용역준공 30일전까지 용역수행 결과를 종합하여 용역 준공 전 종합보고를 하여야 하며, 사업책임기술자가 직접 보고하여야 한다.

4. Project Execution Report

4.1 Commencement Report

- The contractor shall make commencement report within 15 days from the starting the contract after reviewing the site conditions and project details, and shall prepare and submit design plans to the supervisor to get the approval.

- Design plans shall include following
  - Scope of design work
  - Design process table
  - Design resource plan – human resource, program, equipments, etc.
  - Organization and technical coorperation between design area
  - Output of design : Type of output, number of copies, submit dates, etc.

4.2 Monthly Progress Report

- The contactor shall submit written monthly progress report including the following items through contract supervisor for each month by the fifth of next month.

- Contents of the report
  - Details of project progress and process status
  - Various exchanged documents status
  - Major problems and measures during the project
  - Status of participating engineers
  - Project plan for the next month

4-3 Interim Report

- For the following cases, the contractor shall submit the related documents and technical manager of each department shall explain the details.
  - When major milestones are achieved
  - When major plans and policies are set or changed
  - When the supervisor requests in need for contract execution.

4-4 Comprehensive report prior to the completion of service

- The contractor shall submit the comprehensive report on the result of contract execution by 30 days before the completion of contract, and the project technical manger in charge shall report in person.

Chapter 3

Basic Design Review

1. 개 요

- 기본설계 당시와 상위계획, 현지여건, 교통 및 환경 등 설계조건의 변화여부를 조사하여 기본설계 내용을 재검토한 후 실시설계를 수행한다.

- 단, 기본설계를 시행하지 않은 확장(선형개량 포함)공사 실시설계는 타당성 조사에 준한 기본계획을 검토하여야 한다.

- 또한, 관계기관 협의 및 민원처리 현황도 확인하여 기본설계 시 미흡한 사항은 검토 후 추가 협의하여야 한다.

2. 주요검토사항

항 목	검 토 내 용
조 사	• 기본설계 시 조사하여 설계에 반영된 국가 및 지역 개발계획과 지장물 등 현지조사 결과는 그간의 시간경과 및 외부여건 변화로 실시설계 시 재조사를 통하여 기본설계 내용을 보완하여야 한다. • 특히 기본설계 완료 후 실시설계 착수 시까지 시간 간격이 큰 경우에는 보다 면밀히 조사를 실시하여야 한다.
교통재분석	• 교통량 및 교통용량은 차로 수 및 경제적 타당성을 결정하는 직접적 요소이다. 따라서, 기본설계 시 교통분석 수행 후 시간 경과로 인하여 계획노선의 교통량에 변동 요인이 발생하였다면 교통재분석을 실시하여 교통수요를 재예측하여 차로 수 산정 등 관련계획을 재검토하여야 한다.
노선 검토	• 기본설계 시 결정된 노선을 실시설계 시 재검토하여 보완한다. • 기본설계 당시와 비교하여 시간경과로 인한 현장여건 변화·관련계획 조사 시 추가로 발생한 주변 개발계획을 반영하여 기본설계 노선의 적합성 여부를 재검토하고, 다수의 비교 안 검토 및 자문 등의 절차를 통하여 최적의 노선을 결정한다.

1. Overview

- Perform detailed design after reviewing the basic design by considering changes in the upper level plan, site conditions, and transportation and environment compared to the situation at the time of basic design.

- However, for detailed designs for expansion construction (including path improvement) which didn't require basic design, the basic plan corresponding to the facility study shall be reviewed.

- Also, items not fully addressed during the basic design shall be agreed additionally after reviewing through the agreement with related institutes and the civil claims status.

2. Major Review Items

Item	Contents for Review
Investigation	• Site investigation including national & local development plan and obstacles reflected in the basic design shall be examined again during detailed design stage to complement the basic design considering the changes in time and external environment. • Especially when there is long time gap between the completion of basic design and the beginning of detailed design, thorough investigation shall be done.
Traffic Re-analysis	• Traffic and traffic capacity are factors directly affecting the number of lanes and economic validity. Therefore, traffic has to be analyzed again and the plan including the prediction of traffic demand for calculating number of lanes has to reviewed again if there are any changes in traffic for the planned route because of time changes after the traffic analysis for the basic design.
Route Review	• Review and complement the route determined at the time of basic design during the detailed design. • Review the suitability of the route of basic design reflecting the site changes over the time and additional development plan around the sites since the initial investigation compared to the time of basic design, and the best route has to be determined through the procedure of comparison of multiple proposals and consulting.

주요시설 검토 (출입시설, 영업소, 휴게소)	• 출입시설의 위치 및 형식은 기본설계에서 제안된 것을 우선적으로 고려하지만 교통량의 변화 및 주변여건의 변동 등을 고려하여 위치, 규모, 형식 추가설치 등을 재검토하여 설계한다. • 단, 추가로 출입시설을 설치할 경우 인접시설과의 간격, 교통패턴, 용지취득 여부, 지역의 특성, 공사비 등을 제시하여 결정하여야 한다. • 영업소, 휴게소 설치 필요 시 규모 및 위치는 계획도로의 선형과의 관계를 고려하고, 전·후 구간의 휴게소 위치 및 규모 등을 고려하여 적정위치에 설치될 수 있도록 검토되어야 한다.
영향평가 재협의	• 기본설계 시 실시 완료 또는 실시 중인 교통 및 환경영향평가 협의내용이 관련법에서 정한 재협의의 대상 여부인지를 검토하여 필요 시 재협의를 실시토록 한다. • 재협의 조건은 교통영향평가는 I.C 형식 및 접속부의 변경 등이며 환경영향평가는 일정규모 이상의 선형 변경이다.

※ 각 항목에 대한 세부사항은 본 매뉴얼 관련 장을 참고

Major facilities review (Access facilities, toll gate, rest area)	• Location and type of access facilities proposed in the basic design shall be considered first, but the location, size, and additional type of the facilities shall be reviewed and designed considering the changes in traffic and environment conditions. • However, where there additional access facility is to be constructed, decision has to be made by considering the distance from the adjacent facility, traffic pattern, land acquisition status, local characteristics, and construction cost. • The size and location of toll gate and rest area when there are needs to construct them shall be reviewed for installation at optimum location considering the relationship with path of planned road and location & size of the previous and next rest areas.
Re-negotiation of effects evaluation	• When the completed or on-going agreement on traffic and environmental effects shall be reviewed whether they need a further negotiation in accordance with the relevant regulations, and follow the procedure of agreement again if necessary. • The conditions of re-agreement are IC format and modification of connection parts for traffic effects evaluation and path modification over a certain scale for environmental effects evaluation.

※ Refer to the corresponding chapter of this manual for the details on each item.

Chapter 4

Investigation

1. 관련계획 조사

1-1 개 요

- 고속도로 건설사업은 국토종합계획 등 상위계획에 근거하여 위치 및 사업시행 시기가 결정되며, 노선대 및 주요 출입시설 계획은 해당 지역의 도로망 계획 · 주요 개발계획을 감안하여 실시한다. 따라서 고속도로 계획 전반에 영향을 미칠 수 있는 관련계획을 조사하여 실시설계에 반영하여야 한다.

1-2 상위계획

- 국가 차원의 중·장기 계획 등을 조사·반영

관 련 계 획	시 행 기 관 (자료수집기관)	주 요 분 석 대 상
국토종합계획	국토해양부 (교통정책실)	• 전국 도로망 구축 계획 (고속 교통망 및 지역 간 도로망) • 교통투자 우선순위 • 지역별 개발 방향 및 목표
권역별 종합개발계획	국토해양부 (국토정책국)	• 동남권 종합개발계획 • 중부권 종합개발계획 • 서남권 종합개발계획 • 수도권 정비 기본계획 • 서해안 개발사업 사전조사
도로정비기본계획(수정)	국토해양부, 지자체 (도로정책관, 도로과)	• 도로망 구축방향 및 총량적 지표 제시
국도건설5개년계획	국토해양부 (도로정책관)	• 2006~2010 국도확충계획

1. Related Plan Investigation

1-1 Overview

- Location and execution time of expressway construction projects are determined based upon the National Land Comprehensive Development Plan and other upper level plans, and the plans for the route and major access facilities shall be executed considering the road network plan of corresponding area and major development plan. Therefore, the related plans that could effect the overall planning of expressway shall be investigated and reflected upon the detail engineering design.

1-2 Upper Level Plan

- Investigation and reflection of the national-level mid-long term plan

Related Plan	Execution Organization (Data Collection Department)	Major Analysis Items
National Land Comprehensive Development Plan	MLTM (The Ministry of Land, Transport and Maritime Affairs (Traffic Policy Department)	• Nationwide road network building plan (Expressway network and local road network) • Priority of traffic investment • Development direction and goal by region
Comprehensive Development Plan by Regional groups	MLTM (National Land Policy Department)	• Comprehensive development plan of East-South area • Comprehensive development plan of central part of Korea • Comprehensive development plan of West-South area • Basic plan for improvement of Seoul metropolitan area • Pre-investigation of west coast development project
Road Maintenance Basic Plan (modified)	MLTM, local government (Road Policy Office, Department of Road)	• Provision of road network construction direction and index in total quantity
Five-Year Plan for National Road	MTLM (Road Policy Office)	• 2006~2010 National road and local road expansion plan

1-3 지역관련 계획

◎ 계획도로의 지역(지자체) 내 중·장기계획 및 시행중인 개발계획 등을 조사·반영

관 련 계 획	자 료 수 집 및 협 의 대 상 기 관	주 요 분 석 대 상
광역권 개발 계획	관할 지방자치단체	•부산·경남권 •아산만권 •대구·포항권 •광주·목포권 •군산·장항권 •대전·청주권 등
시·도 종합개발계획	관할 지방자치단체	•개발 방향 및 목표 •교통 관련 개발 계획
지역 및 도시계획	관할 지방자치단체	•장기발전 계획 및 목표 •도시 기본 계획 •교통 관련 계획
단지개발 및 조성계획	국토해양부 한국토지주택공사 관할 지방자치단체	•개발 방향 및 목표 •물동량 예측 •유통 경로 계획
기 타	관할 지방자치단체	•위락 시설 계획 •대단위 공장시설 부지 계획

1-4 교통관련 계획

◎ 교통관련 국가 차원의 교통망 계획, 지역 내 개별 지자체 교통·도로정비 기본계획 등 계획도로에 영향을 미칠 수 있는 사업계획을 조사·반영

관 련 계 획	자 료 수 집 및 협의 대상 기관	주 요 분 석 대 상
전국 도로망 체계 재정비 계획	국토해양부 국토연구원	•국도 기능의 재정립 및 개편방향 •국도 노선 조정 계획 •교통수요 분석결과 •도로 개선사업 우선순위 분석
도로사업계획 등 관련계획	국토해양부 해당 지방국토관리청 관할 지방자치단체	•국도 확장 및 포장 사업 •도시 간 우회도로 건설계획 •국가지원 지방도 건설계획
시·도·군 교통관련계획	관할 시·군 행정기관	•지방도 확장 및 포장 계획 •도로 개량 사업
공항·항만 관련계획	국토해양부 해당 지방해양수산청 및 항공청	•항만 건설 계획 •공항 건설 계획 •수송 수단간 연계 계획
기 타	국토해양부 한국철도시설공단 및 고속철도공단	•장·단기 철도건설 및 복선화 계획 •고속철도 건설계획

1-3 Plan Related to Local Development

- Investigation and reflection of local (inside local government) mid-long term plan and development plan under execution for the planned road.

Related Plan	Organization for Data Collection & Negotiation	Major analysis items
Metropolitan Development Plan	Jurisdictional Local Government	• Busan. Gyoengnam Area • Asan Bay area • Daegu. Pohang Area • Gawangju. Mokpo area • Gunsan. Janghan Area • Daejeon. Cheongju Area, etc.
City/Province Comprehensive Development Plan	Jurisdictional Local Government	• Development direction and goal • Transportation-related development plan
Local and Urban Plan	Jurisdictional Local Government	• Long-term development plan and goal • Urban basic plan • Transportation-related plan
Complex Development and Composition Plan	MLTM Korea Land Corporation Jurisdictional Local Government	• Development direction and goal • Prediction of traffic volume • Distribution route plan
Others	Jurisdictional Local Government	• Leisure facility plan • Large-scale plant facilities site plan

1-4 Transportation-Related Plan

- Investigation and reflection of project plans that might affect the planned routes such as nationwide road network plan and local transportation/road maintenance basic plan.

Related Plan	Organization for Data Collection & Negotiation	Major analysis items
Nationwide Road Network System Maintenance Plan	MLTM Korea Research Institute for Human Settlements (KRIHS)	• Re-establishment and reorganization plan for national road function • National road route adjustment plan • Traffic demand analysis result • Road improvement project priority analysis
Road Project Plan and Other Related Plan	MLTM Corresponding Regional Construction Management Administration Jurisdictional Local Government	• National road expansion and pavement plan • Inter-city detour road construction plan • Government-funded local road construction plan
Si/Do/Gun Transportation-Related Plan	Jurisdictional Si/Gun Administration Organization	• Local road expansion and pavement plan • Road improvement plan
Airport / Port Related Plan	MLTM Corresponding Regional Maritime Affairs & Port Office and Regional Aviation Adminstration	• Port construction plan • Airport construction plan • Connection plan between different types of transportation
Others	MLTM Korea Rail Network Authority and Korea Train Express	• Long/short term railway construction and dual-lane plan • Express railway construction plan

1-5 기타 기간시설 관련 계획

관 련 계 획		자료 수집 및 협의 대상 기관	주요 분석대상
기 간 시 설 계 획	통신	KT	• 통신 시설 현황 및 계획
	전기	한국전력공사	• 송전시설 현황 및 계획
	가스	한국가스공사	• 가스관 매설 현황 및 계획
	송유관	대한송유관공사	• 송유관 매설 현황 및 계획
	용수로	한국농촌공사	• 농업용수로 현황 및 계획
	광역상수도	한국수자원공사	• 상수관로 현황 및 계획
	군사시설	국방부 및 해당 군부대	• 군사시설 현황 및 계획
	항공시설	한국공항공사	• 항공운항현황 및 비행공역 등
농지 관련 계획		농림부 및 농어촌공사	• 대단위 농지정리 사업계획 • 농업진흥지역의 확인
산림 관련 계획		산림청 및 지방산림청	• 녹지자연도 등급별 현황 • 특정 보호림 지정 현황 • 국유림 실태조사
하천 관련 계획		국토해양부 수자원국 지방국토관리청 하천국 관할 지방자치단체	• 하천 정비 기본계획 수립 여부 • 수리 수문 자료 • 하천 관리 현황
공원 관련 계획		환경부, 관할 지자체 및 관리공단	• 국립 · 도립 · 시립공원 등 현황
광업 관련 계획		산업자원부 광업권등록사무소	• 광업권 및 광산 자료

2. 현지조사

- 지역 내의 기존도로망, 토지이용 현황, 지형 및 지질, 기상, 취락지 현황, 농업진흥구역 현황 등을 문헌조사, 통계조사 및 현지조사를 통하여 실시
- 주요 조사항목

항 목	주요 조사 내용
현지조사 및 답사	• 토지이용현황 • 각종 시설물 조사 • 문화재(민속문화, 무형문화 등) 조사 • 기존 및 계획노선 주변 접속시설물 조사 • 인근 공사현장 답사 및 자료 분석 • 농경지 구간의 경작현황 및 주민 이동경로(동선) 조사 • 농업용수 이용현황 조사 • 교육시설 및 통학경로 조사 • 응급·구호시설 현황조사 • 기타 지역주민의견 및 사전민원발생 요인 조사

1-5 Other Infrastructure-Related Plan

Related Plan		Organization for Data Collection & Negotiation	Major analysis items
Infrastructure Plan	Communication	KT	• Communication facility status and plan
	Electricity	KEPCO (Korea Electric Power Corporation)	• Electric power transmission facility status and plan
	Gas	Korea Gas Corporation	• Gas pipe burial status and plan
	Oil pipeline	DOPCO (Daehan Oil Pipeline Corporation)	• Oil pipeline burial status and plan
	Irrigation channel	Korea Rural Community and Agriculture Corporation	• Agricultural irrigation channel status and plan
	Wide area waterworks	KOWACO (Korea Water Resources Corporation)	• Water supply channel status and plan
	Military facilities	the Ministry of National Defense and Corresponding military camp site	• Military facilities status and plan
	Airport facilities	KAC (Korea Airports Corporation)	• Airplane flight status and airspace
Farming Land Related Plan		the Ministry of Food, Agriculture, Forestry & Fisheries and Korea Rural Community and Agriculture Corporation	• Large-scale farming land clearing project plan • Confirmation of agriculture promotion area
Forest related plan		Korea Forest Service and Local Forest Service	• Green area degree of naturalness - status by class • Special protected forest designation status • National forest status investigation
River related plan		MLTM Water Resource Department, Regional Construction Administration Office - River Department, Jurisdictional Local Government	• Whether or not a river maintenance basic plan is established • Irrigation hydrological data • River management status
Park related plan		the Ministry of Environment Republic of Korea, Jurisdictional Local Government and Management Corporation	• Status of national, provincial, city park, etc.
Mining related plan		the Ministry of Commerce, Industry & Energy Mine Registration Office	• Mining right and mine data

2. Site Investigation

- Investigation of the existing road network, land usage status, topography and geological features, weather, settlement area status, and agriculture promotion area status through document research, statistical research and site investigation.
- Major Investigation items

Items	Major Investigation Items
Site Investigation and Survey	• Land usage status • Various facilities investigation • Cultural assets investigation(folk culture, intangible culture, etc.) • Investigation of access facilities around existing and planned route • Investigation and data analysis of adjacent construction site • Cultivation status of farming area and activity path of residents • Agricultural water usage status • Educational facilities and commute routes investigation • Emergency·first aid facility status investigation • Investigation of other residents' opinion and potential civil appeal factors in advance

항 목	주요 조사 내용
교통량 및 시설조사	• 교통량 조사 • 교통시설 조사 • 교통량 특성 파악 • 교통사고 및 병목구간 현황 조사
수리·수문조사	• 하천 상태 및 자료 수집 • 하천정비계획 조사 • 저수지, 소류지 등 수리현황 조사 • 관정, 펌프장, 배수지 등 조사
기상 및 해상조사	• 기상 및 해상 자료 조사 • 안개, 결빙, 직광 등 교통안전과 관련된 기후 조사 • 최근 기상변화(태풍, 집중호우, 지진, 해일 등)에 따른 재해현황 조사
선박운항조사	• 통행 선박 및 부두시설 관련 조사
환경영향조사	• 주민생활에 미치는 영향 조사 • 동·식물에 미치는 영향 조사 • 습지, 늪 및 양서류, 파충류 등에 미치는 영향 조사 • 지하수 영향 조사 • 문화재 파악 및 지표 조사
지표 및 지질조사	• 지표 및 지반 조사
지장물조사	• 지상 및 지하 지장물 조사 • 폐광 및 광업권 조사
재료원조사	• 재료원, 토취장, 사토장 등 조사

3. 측 량

3-1 현황측량(축척 1:1,200)

- 현황측량은 기본설계 시 실시한 항공측량 자료를 활용하도록 하며, 기본설계 시 현황측량을 실시하지 않은 경우에는 별도의 용역을 시행하여 측량전문업체가 측량을 실시하도록 한다.

- 용역시행 중 불가피하게 노선변경이 필요하여 현황이 없는 부분이 발생 시는 별도의 현황측량을 실시하여야 한다.

Items	Major Investigation Items
Investigation of Traffic Volume and Facilities	• Traffic volume analysis • Traffic facilities investigation • Identification of traffic volume characteristics • Traffic accidents and bottle neck area status investigation
Irrigation·Water Gate Investigation	• River status and data collection • River maintenance plan investigation • Reservoir, pond, and other irrigation status investigation • Well, pump facility, water drain facility investigation
Weather and Marine Investigation	• Weather and maritime data collection • Investigation of weather that effect traffic safety such as fog, ice formation, direct sunlight, etc. • Investigation of natural disaster status resulting from weather change(typhoon, localized heavy rain, earthquake, tidal waves, etc.)
Ship Operation Status	• Passing ship and port facility investigation
Environmental Effects Investigation	• Investigation of effects on residents' lives • Investigation of effects on animal and plants • Investigation of effect on wet area, swamp, amphibian, and reptiles • Investigation of effects on underground water • Identification of cultural assets and investigation of ground surface
Surface and Geological Feature Investigation	• Surface and ground investigation
Obstacles Investigation	• Ground and underground obstacles investigation • Abandoned mine and mining right investigation
Material Source Investigation	• Material source, borrow pit, spoil bank, etc. investigation

3. Survey

3-1 Topographical Survey (Scale 1:1,200)

- Aerial survey data taken from the basic design stage will be used for topographical survey, and specialized surveying company shall be contracted for measurement if the topographical survey was not done during the basic design stage.

- If there is no topographical data because of unavoidable changes in route, additional topographical survey shall be implemented

3-2 중심선 측량

- 중심선은 1:1,200 현황도상에서 결정하고 측점간격은 20m로 하며, 지형상 종·횡단의 변화가 있는 지점, 구조물 설치점, 곡선의 시·종점(완화곡선의 시·종점)등 필요한 지점은 추가로 측량한다.

3-3 종·횡단 측량

- 종단측량 : 종단측량은 중심선을 따라 매 측점의 지반고를 측정하여야 하며, 반드시 왕복 실시하여 오차의 한계를 넘지 않도록 하여야 한다.

- 횡단측량 : 횡단측량은 토적산출 및 확장공사 종단계획의 기초가 되는 것이므로 중심선에 따른 측점을 포함한 각 측점과 지형이 급변하는 지점 또는 구조물 설치지점 등을 포함하여 종방향 중심선에서 직각 방향의 좌·우 충분한 폭으로 세밀히 측정하여야 한다.

4. 토질조사 및 시험

4-1 개 요

- 토질조사 및 시험은 보링, 오가보링, 시굴(test pit), 표준관입시험, 실내 CBR시험 및 재료원 조사 시험 등이 있으며 그 결과로부터 다음사항을 결정 또는 설계하여야 한다.

- 흙쌓기 재료의 적부 판단
- 흙쌓기부 자연지반의 성상 및 연약지반 설계
- 땅깎기부 비탈면 경사 선정과 토공량의 산정
- 교량기초 및 형식
- 터널 위치의 타당성과 터널 단면 해석 및 설계
- 포장 두께 설계
- 골재원 및 토취장의 사용 가능량 및 재료의 적정성 판단
- 기타 설계에 필요한 자료 확보

3-2 Center Line Survey

- The center line shall be determined according to the land usage map with the scale of 1:1,200, and the distance between measurement points 20 m. Additional survey shall be added for such area as point where vertical/cross-sectional changes occurred, where structure was installed, and beginning/end point of curve (beginning and end of transition curve).

3-3 Vertical/Cross-section Survey

- Vertical section survey : Ground height along the center line shall be measured at each measuring point and deviation shall be lower than the deviation limit through executing round trip measurement.

- Cross section survey : As traversing measurement is a foundation of longitudinal plan for mass calculation and expansion construction, precise measurement shall be done with enough width for both left and right 90 degree from the center line for various points including measuring points along the center line, points where sudden changes in topography, or structure installation points.

4. Soil Investigation and Test

4-1 Overview

- Soil investigation and test include boring, auger boring, test pit, standard penetration test, indoor CBR test, and material source survey test, and the followings shall be determined or designed based on the results of soil investigation and tests.
  - Whether soil material is adequate for earth work
  - Characteristics of natural ground under embankment
  - Selection of cutting slope inclination and estimation of earth work amount
  - Bridge foundation and type
  - Feasibility of tunnel location and analysis & design of tunnel section
  - Pavement design for estimation of thickness
  - Available amount of aggregate source & borrow pit and whether soil is appropriate
  - Collecting additional data required for the rest designs

4-2 흙쌓기부 및 땅깎기부

- 계획노선에 따라 시추조사, 핸드오거 혹은 시험굴조사를 원칙으로 하고, 필요에 따라 원위치 시험을 시행

- 땅깎기 구간에서 시험굴조사는 200m 간격으로, 시추조사는 땅깎기부 개소 당 1개소 이상 실시를 원칙으로 하며, 현지여건에 따라 조정하여 실시

- 절리면 전단시험, 일축압축 강도시험, 시추공 화상정보시험 및 안정성 분석 등을 통하여 절리·단층 및 풍화발달 정도를 조사·분석하고, 이에 대한 땅깎기 법면 경사선정 및 대책 공법을 수립

- 흙쌓기 구간에서 핸드오거는 300m 간격 기준으로 현지 여건을 감안하여 시행

- 시험굴조사, 시추조사 및 핸드오거 등 제반 토질시험은 체계적으로 정리하여 조사위치 및 주상도를 도면화하고, 토질시험 결과 등에 대한 토질조사보고서를 작성

- 모든 토질시료는 통일분류방식으로 분류하고 이러한 토질분류에 입각하여 노선을 동일 토질 구간으로 나누어야 하며, 각 대표구간에 대한 시료를 채취하여 다음과 같은 시험을 시행하여 포장설계의 기초자료로 활용
  - 토질분류 시험
  - 다짐시험
  - CBR 시험(수정)

- 절취 비탈면 높이 10m 이상을 원칙으로 하되, 다음의 경우는 10m 이하에서도 현장조건에 따라 안정성 검토를 시행
  - 구성지반이 붕적층 또는 퇴적층인 경우
  - 현재까지 붕괴이력이 있고 불안정한 상태에 있는 지반의 경우
  - 지하수위가 높고 용수가 많은 곳
  - 주변의 기존구조물(철탑 등)에 나쁜 영향을 미칠 것으로 예상되는 경우

4-2 Banking Section and Cutting Section

- Boring, hand auger, or test pit shall be carried out as a principal rule, and in-situs test shall be made if necessary.

- Test pit shall be carried out at approximately 200 m intervals in cutting section, and boring shall be done per each cutting section as a principal rule. The actual test shall be adjusted depending on the site conditions.

- Investigate and analyze the joint, dislocation, and development level of weathering through dislocation surface sheer test, unconfined compression strength test, bore-hole image information test, and stability test, and determine the slope of cutting surface and control method.

- Hand auger shall be carried out at approximately 300 meters intervals in banking section considering the site conditions.

- Soil test such as test pit, boring, and hand auger shall be systematically summarized, and investigation location and boring log shall be presented in drawing. Also, soil investigation report shall be prepared based on the soil investigation results.

- All soil samples shall be classified according to an unified classification method, and construction route shall be divided by the same soil type based on this soil analysis, and the following tests shall be done for samples from each representative block for utilizing as basic data for pavement.
  - Soil type classification test
  - Compaction test
  - Modified CBR test

- The height of cutting slope surface shall be over 10 meters as a principal rule for the stability analysis. the stability could be carried out for the following cases even when below 10 meters depending on the site conditions.
  - When the ground is colluvial deposit or sedimentary layer
  - When the ground has a history of collapse and is in an unstable condition
  - When the level of underground water is high and where there is excess water
  - When undesirable effects on the nearby existing structures (steel tower, etc) are expected

- 연약지반조사 : 핸드오거 실시결과 연약지반이 발견되면 연약지반 연장에 따라 피에조콘관입시험·간극수압소산시험 등을 시행하여 연약지반 분포, 토질분류 등을 파악 후 그 결과를 이용하여 시추조사를 피에조콘관입시험 위치에 자연시료를 채취하고, 채취한 시료로 비중·함수량·삼축압축·압밀 시험 등의 토성시험 및 역학시험을 시행하여 상호검증을 통하여 지반의 안정검토 및 처리공법 등의 대책을 강구하여야 한다. 또한 현장 베인시험을 추가로 시행하여 연약지반의 비배수강도에 대한 신뢰도를 확보하여야 한다.

4-3 교량기초 및 터널부

- 주요 교량 및 터널에 대해서는 시추, 표준관입시험 등 설계에 필요한 시험을 실시하여야 한다. 교량기초 하부에 공동이 예상되는 지역, 시추조사가 현장여건 등으로 불가한 경우 및 직접기초로 예상되는 지역의 지지층 깊이 변화가 큰 경우에는 물리 탐사를 이용하여 공동 및 지지층의 분포를 파악하여야 한다.

- 교량기초가 경사가 급한 비탈면에 위치하여 터파기로 인하여 비탈면 붕괴가 우려되는 경우 비탈면 안정검토를 할 수 있도록 시추공 영상촬영시험을 수행하여야 한다.

- 터널부 시추조사는 전기비저항탐사를 수행하여 그 결과을 이용하여 저토피, 파쇄대, 터널 내 용출수 산정에 필요한 적정 조사위치 및 공수를 검토 후 시추를 실시한다. 또한, 터널구조 및 내진해석을 위한 각종 시험을 실시하여야 하고, 단면형식을 결정하기 위하여 Cross hole Test에 의한 탄성파 탐사를 터널 상·하행 입·출구부에 시행하여야 한다.

- Soft Ground Investigation: If soft ground area is identified with hand auger test, piezo cone penetration test, pore water pressure dissipation test shall be carried out depending on the length of the soft ground to grasp the distribution and soil type. Natural samples shall be taken with the boring hole at the location of piezo cone penetration test and soil property test such as specific gravity, triaxial compression test, compaction test and mechanical test shall be performed. Safety review of the ground and other measures such as process method shall be devised along with mutual verification. Also, reliability of undrained strength shall be secured by carrying out an additional field vane test.

4-3 Bridge Foundation and Tunnel

- Boring, standard penetration test and other tests required for design shall be carried out for major bridges and tunnels. Geophysical exploration shall be done for the area to find the distribution of cavity and bearing stratum for the places where cavity is expected under the bridge, where boring is not possible because of field conditions, or when there is big variation in the depth of bearing stratum in the places where spread foundation is expected.

- When there is a possibility of slope collapse by excavation because the bridge foundation is located on steep slope, Borehole Image Processing System(BIPS) should be carried out stability analysis.

- Drilling for tunnel area should be carried out after reviewing the appropriate location and quantity required for the estimation of water in the tunnel, shallow soil cover, fault zone using the result of Electrical prospecting. And also, various tests for tunnel structure and seismic analysis shall be done, and seismic prospecting by cross hole test shall be done at the entrance and exit of both directions of tunnel in order to determine the section type.

Chapter 5

Traffic Analysis & Economic Efficiency Evaluation

1. 개 요

- 도로설계에서의 교통분석은 목표년도의 장래 교통수요를 예측하여 본선 및 연결로의 차로수를 결정, 출입시설의 배치 및 형식 결정·경제성 분석 등을 위하여 실시

- 교통수요분석을 위하여 사용되는 대표적인 기본자료는 O/D 및 Network 자료가 있으며, 이 자료는 국가교통DB에서 공식적으로 제공하는 자료를 이용하는 것을 원칙으로 한다. 다만, 매년 발표되는 사회·경제지표를 확인하여 국가교통 DB상의 장래 교통수요의 패턴과 큰 차이가 있는 경우 적절한 값을 추정 활용할 수 있음

- 교통분석에 관한 세부 지침 중 교통수요예측과 경제적 타당성 분석에 관한 사항은 국토해양부에서 발간한『교통시설 투자평가 지침(2007. 12)』을 참고하도록 하고, 차로 수 산정과 도로용량 및 서비스 수준 분석에 관한 사항은 국토해양부『도로용량편람(2004)』을 참고

- 보통 기본설계 시 사업도로의 차로 수, 용량 및 서비스 수준 등의 교통분석을 수행하며, 실시설계 시 기본설계 당시와의 교통관련 여건 변화로 인하여 기본설계 교통분석 결과를 그대로 활용할 수 없는 경우 교통재분석을 수행

- 교통재분석 수행 여부 결정을 위하여 변동사항을 조사하여야 하는 항목
  - 기본설계의 기준년도 경과에 따른 사회적·경제적 여건
  - 계획노선 주변의 대규모 교통유발 시설 계획
  - 출입시설 설치가 필요한 주요 도로계획
  - 기본설계 시 정한 목표년도의 변경 필요성

1. Overview

- Traffic analysis in road design is required for determining the number of main lane and connection road, the placement and type of access facilities, and cost analysis by predicting the future traffic demand of the goal year.

- The typical data for traffic demand analysis are O/D and Network data, and There are principal rule to use the data officially provided by National Traffic DB. However, if there is a major difference between the prediction of future traffic demand pattern in National Traffic DB and social/economic indexes that are published every year, the appropriate estimated values can be used.

- 『Transportation Facilities Investment Evaluation Instructions(2007. 12)』 published by MLTM (Ministry of Land, Transportation, and Maritime Affairs) shall be used as a reference for the traffic demand prediction and economic feasibility analysis among the detailed instructions on traffic analysis. 『Road Traffic Volume Manual(2004)』 shall be used as a reference for determining number of lanes, Traffic Volume, and service level analysis.

- Generally, traffic analysis such as the number of lanes, traffic volume and service level shall be executed during the basic design phase, and traffic re-analysis shall be executed traffic analysis result from the basic design cannot used without modification due to the change in traffic environment during the detailed design phase

- The following items shall be examined in order to determine the need of traffic re-analysis.
  - Changes of social and economic environment
  - Plan of large-scale traffic triggering facilities around the planned route
  - Plan of major roads required construction of access facilities
  - The need for changing target year determined in the basic design phase

2. 기초자료 수집

- 교통현황 조사 및 분석은 주변 변화에 대응하여 장래 교통수요 추정과 최적노선 선정을 위한 기초자료로 활용하는데 목적이 있으며, 자료의 분석을 통하여 노선의 타당성 및 경제성 분석의 기초자료로 활용

- 과업노선축 영향권의 외부 유출입지점, 주요 접속부, 기존 국도 및 지방도의 구간교통량 등 교통특성을 정확히 파악·분석하기 위하여 공신력 있는 관련 통계자료를 통하여 과업노선 및 주변지역의 도로시설 및 교통량 등의 현황 조사를 실시

- 또한, 통행의 유발 및 네트워크 구성을 위하여 상위 계획 및 과업구간 주변 개발계획을 조사하여 분석에 반영하여야 하는데, 이러한 관련계획 검토에 관한 내용은 본 매뉴얼 "조사"편을 참고

3. 시설규모 산정

3-1 계획목표년도의 설정

- 계획목표년도는 설계 및 공사기간을 감안하여 예측된 도로 개통년도로부터 20년 후로 설정

3-2 영향권 설정

- 영향권은 사업시행으로 인하여 현저한 교통패턴의 변화가 발생하여 사업의 타당성을 분석하는데 포함되어야 할 공간적 범위를 의미

- 영향권은 그 범위에 따라 크게 직접 영향권과 간접 영향권으로 구분할 수 있는데, 직접 영향권은 사업시행지역에 지리적으로 인접한 지역으로서 사업 시행 효과를 세밀하게 분석하기 위하여 상세한 O/D 및 네트워크의 구축이 필요한 공간적 범위를 의미하며, 대상 사업의 시행 구간을 모두 포함

- 간접 영향권은 사업시행으로 인한 통행패턴의 변화가 발생하여 편익 산정의 범위에 포함되어야 하는 공간적 범위 가운데 직접 영향권을 제외한 지역을 의미한다.

2. Basic Data Collection

- Traffic status investigation and analysis have a object to use basic data for future traffic demand estimation and optimum route selection in response to the changes in environment, and the data will be used for the basis of feasibility study and economical effieciency review.

- Current status investigation including project route, nearby road facilities and traffic volume shall be executed through the related statistical data from the credible institution in order to accurately identify and analyze the traffic characteristics such as traffic in external entrance/exit points within the sphere of influence at project routes, main junctions, interval traffic volume of existing national road and local road.

- Also, the upper level plan and nearby project route development plan shall be investigated and reflected to the analysis for traffic triggering and network organization, and refer to the "Investigation" chapter of this manual for the details on review of related plan.

3. Facility Scale Calculation

3-1 Setting Up Planned Goal Year

- Planned goal year shall be set as 20 years after expected road openiing year considering the design and construction period.

3-2 Setting Up Sphere Of Influence

- Sphere of influence means a spatial range that should be included in the feasibility analysis of the project when noticeable changes in traffic pattern occurs resulting from project execution.

- Sphere of influence can be divided into direct and indirect depending on the range. Sphere of direct influence is the area geographically close to the projection execution area, and detailed O/D and network establishment will be required in order to analyze the effects of project execution in detail. All project execution section shall be included.

- Sphere of indirect influence is the spatial range of area that shall be included for benefit estimation containing changes in traffic pattern due to the execution of the projects and excluding direct influence sphere.

- 영향권 설정은 O/D 기준 통행량 비율(PV), 사업시행으로 인한 교통량 변화율(RV) 등이 현저하게 변화하는 지역으로 결정

$$PV_{ij} = \frac{V_{ij}}{\sum_{j=1}^{n} V_{ij}} \times 100(\%), \qquad RV^k = \frac{V^k_{시행} - V^k_{미시행}}{V^k_{미시행}} \times 100(\%)$$

여기서, PV_{ij} : 존 i의 발생교통량 가운데 존 j의 도착교통량이 차지하는 비율(%)
V_{ij} : 교통량
RV^k : 사업시행 시 구간 k의 교통량 변화율(%)
$V^k_{미시행}$: 사업 미시행 시 구간 k의 교통량
$V^k_{시행}$: 사업 시행 시 구간 k의 교통량

- PV값은 2~3%, RV값 3~5% 이상 변화하는 지역을 영향권으로 설정

3-3 분석기준년도 O/D 및 네트워크 구축

- 국가교통DB의 O/D 및 네트워크 자료를 토대로 하여 현지 도로·교통 현황, 관련계획, 사회경제지표 등의 조사를 통하여 국가교통DB를 보정하여 적용

- 구축된 O/D 및 네트워크 자료는 현재 교통량 분석을 실시하고, 실제 관측 교통량과의 오차율이 20% 이내가 될 때까지 보정을 실시하여야 함

3-4 교통수요 예측

- EMME/2와 같은 교통수요 예측 프로그램을 활용하여 과업노선 및 주변 국도의 목표년도 교통량을 예측

3-5 도로용량 분석

- 도로용량은 예측된 장래교통량을 수용하기 위한 계획노선의 차로 수 산정 및 서비스수준 판단에 활용

- 고속도로 기본구간에서의 서비스 용량산정 공식

$$SF_i = MSF_i \times N \times f_W \times f_{HV}$$
$$= C_j \times (V/C)_i \times N \times f_W \times f_{HV}$$

여기서,

SF_i : 서비스수준 i에서 주어진 도로 및 교통 조건에 대한 서비스 교통량(vph)
N : 한 방향 차로 수
f_W : 차로폭 및 측방여유폭 보정계수
f_{HV} : 중차량 보정계수
MSF_i : 서비스수준 i에서 차로 당 최대 서비스 교통량(승용차/시/차로, pcphpl)
C_j : j 설계 속도의 용량(pcphpl)
$(V/C)_i$: 서비스수준 i에서 교통량 대 용량 비

- Setting up sphere of influence is determined by the area to change O/D reference traffic volume ratio(PV) and rate of change of traffic volume(RV) due to project execution significantly.

$$PV_{ij} = \frac{V_{ij}}{\sum_{j=1}^{n} V_{ij}} \times 100(\%), \qquad RV^k = \frac{V^k_{execution} - V^k_{noexecution}}{V^k_{noexecution}} \times 100(\%)$$

where, PV_{ij} : ratio of arrival traffic at zone j to the incurred traffic at zone i(%)
V_{ij} : traffic volume
RV^k : rate of change of traffic volume at zone k when executing project
$V^k_{no\ execution}$: traffic volume at zone k when not executing project
$V^k_{execution}$: traffic volume of zone k when executing project

- The area with PV over 2~3% and RV over 3~5% is set as sphere of influence

3-3 Analysis Reference Year O/D and Network Establishment

- Nationa traffic DB is applied with correction Through the investigation of local road, traffic status, related plan, social and economic index based on the O/D and network data of national traffic DB.

- Established O/D and network data require continuos correction until the error rate of the actual measured traffic rate reaches to less than 20% through current traffic volume analysis.

3-4 Traffic Demand Prediction

- Traffic volume of the goal year on project route and nearby national road shall be estimated using the traffic demand prediction program like EMME/2.

3-5 Road Capacity Analysis

- Road capacity is utilized for estimating the number of lanes of planned route and service level in order to accommodate the predicted future traffic volume.

- Formula to calculate the service capacity at unit zone of express way

$$\begin{aligned} SF_i &= MSF_i \times N \times f_W \times f_{HV} \\ &= C_j \times (V/C)_i \times N \times f_W \times f_{HV} \end{aligned}$$

Where, SF_i : service traffic volume for given road and traffic condition at service level i(vph)
N : number of lanes on one direction
f_W : road width and side margin compensation coefficient
f_{HV} : heavy vehicle compensation coefficient
MSF_i : maximum service traffic volume per lane at service level I (passenger car/hour/lane, pcphpl)
C_j : capacity of j design velocity (pcphpl)
$(V/C)_i$: ratio of traffic volume to capacity at service level I

3-6 차로 수 산정

- 차로 수의 산정은 설계 서비스수준, 장래 교통수요 및 교통특성 등을 고려하여 적절하게 검토되어야 하며, 일반적으로 최종목표년도의 교통량으로 도로의 차로수를 산정

- 고속도로의 차로 수 산정은 크게 본선구간과 연결로의 차로 수 산정으로 구분되며, 차로 수 산정 시 적용되는 각종 지표들은 용량분석 단계에서 산정

- 차로 수 산정식

$$
\begin{aligned}
N &= PDDHV / SF_i \\
&\quad (DDHV / PHF) / SF_i \\
&= [(DHV \times D) / PHF] / SF_i \\
&\quad [(AADT \times K \times D) / PHF] / SF_i
\end{aligned}
$$

여기서, $AADT$: 일교통량(대/일)
K : 설계시간계수
D : 중방향계수
PHF : 첨두시간계수
SF_i : 설계서비스교통량(대/시)

4. 경제성 검토

4-1 개　요

- 공공투자 사업의 경제성 분석은 계획 또는 논의되고 있는 대안들의 비용과 효과를 분석하여 경제적 효율성과 공공투자의 타당성을 검토하는 것으로서, 이는 사업의 효율성을 극대화하여 투자재원을 최적 배분할 수 있도록 투자의 적합성을 사전에 분석하는데 그 목적이 있음

- 수송부문 시설 투자사업의 평가에 일반적으로 사용되는 방법은 비용·편익 분석법이며, 분석과정을 세부적으로 살펴보면 다음 그림과 같음

3-6 Calculation of Number of Lane

- Number of lanes shall be properly reviewed considering design service level, future traffic demand, and traffic characteristics, and number of lanes is calculated with the traffic volume of the final goal year in general.

- Calculation of number of lanes of expressway is divided into main road and connection road, and various indexes applied for calculation of number of lanes shall be calculated during capacity analysis.

- Formula to calculate number of lanes

$$N = PDDHV / SF_i$$

$$(DDHV / PHF) / SF_i$$

$$= [(DHV \times D) / PHF] / SF_i$$

$$[(AADT \times K \times D) / PHF] / SF_i$$

Where, $AADT$: Daily traffic volume (car/day)
K : Design hour coefficient
D : Dual direction coefficient
PHF : Peak hour coefficient
SF_i : Design service traffic volume (car/hour)

4. Economic Efficiency Review

4-1 Overview

- Economic analysis of public investment project is the evaluation of economic efficiency and feasibility of public investment by analyzing the cost and effects of various proposals under discussion and planning, and the purpose is to analyze the appropriateness in advance for maximum efficiency of the project and optimum distribution of the investment resources.

- The general method of evaluating investment project in transportation segment facilities is cost-benefit analysis method, and the following picture shows the analysis process in detail.

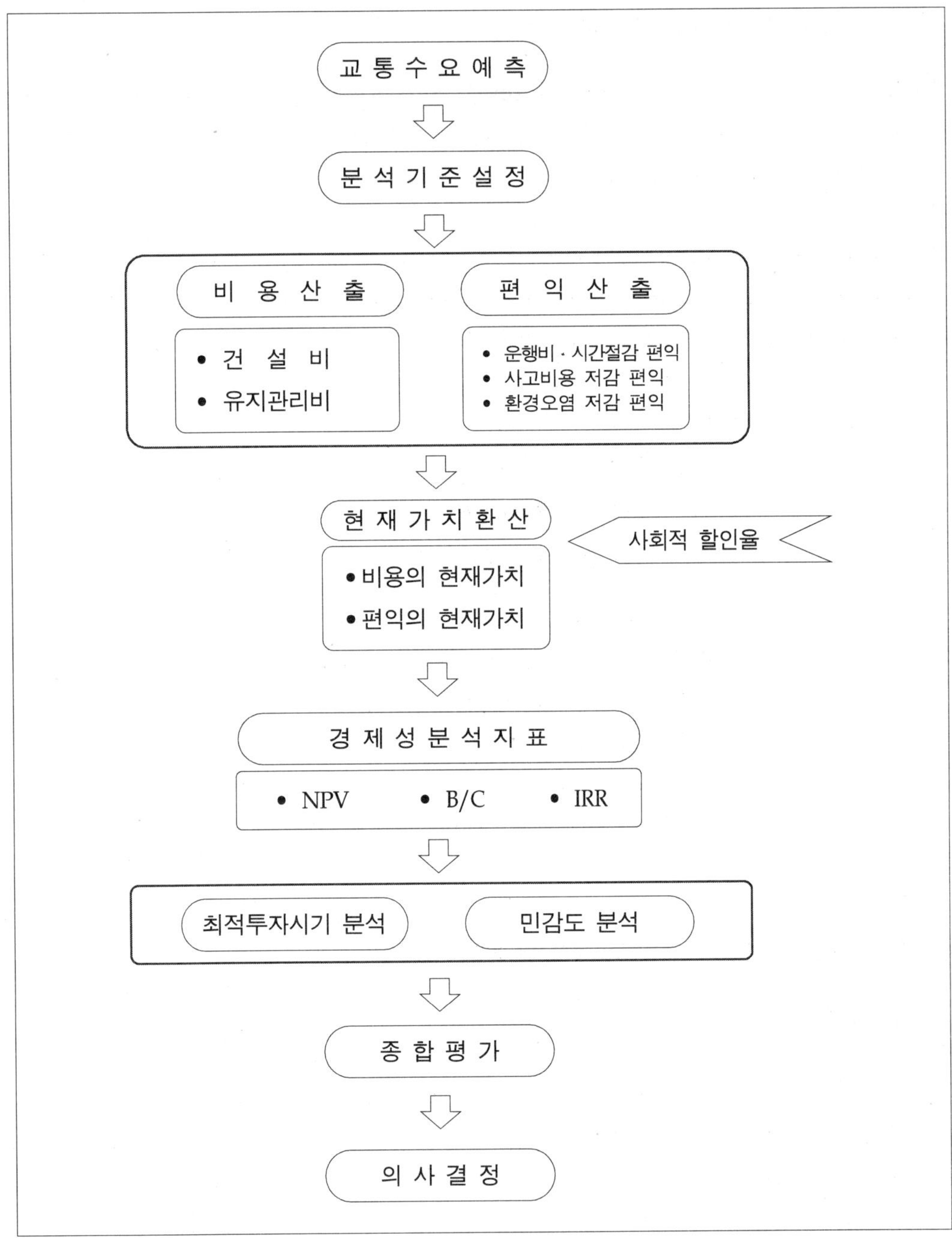
교 통 수 요 예 측
분 석 기 준 설 정
비 용 산 출
• 건 설 비
• 유지관리비
편 익 산 출
• 운행비 · 시간절감 편익
• 사고비용 저감 편익
• 환경오염 저감 편익
현 재 가 치 환 산
• 비용의 현재가치
• 편익의 현재가치
사회적 할인율
경 제 성 분 석 지 표
• NPV
• B/C
• IRR
최적투자시기 분석
민감도 분석
종 합 평 가
의 사 결 정

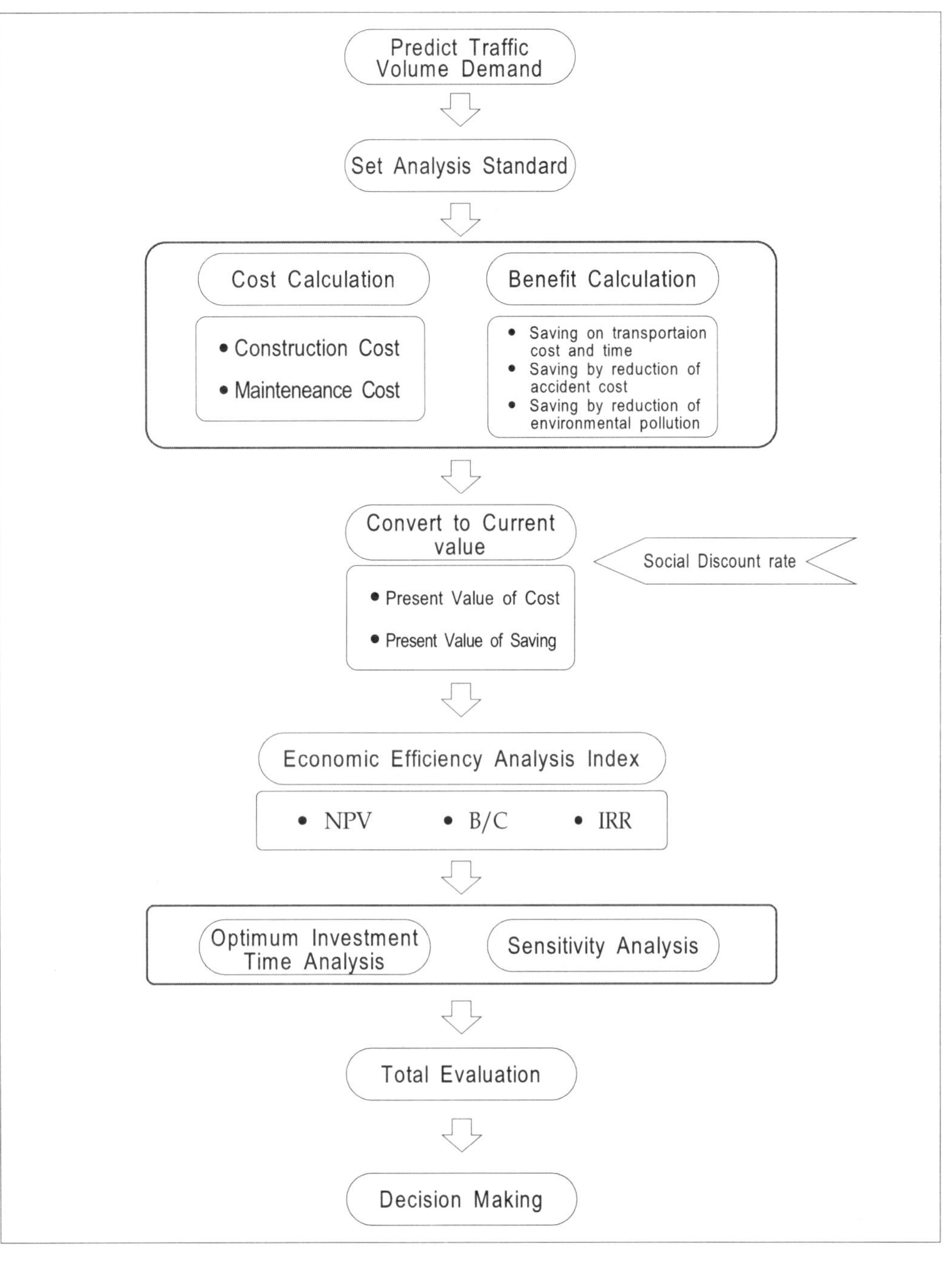
Predict Traffic Volume Demand
Set Analysis Standard
Cost Calculation
• Construction Cost
• Mainteneance Cost
Benefit Calculation
• Saving on transportaion cost and time
• Saving by reduction of accident cost
• Saving by reduction of environmental pollution
Convert to Current value
• Present Value of Cost
• Present Value of Saving
Social Discount rate
Economic Efficiency Analysis Index
• NPV • B/C • IRR
Optimum Investment Time Analysis
Sensitivity Analysis
Total Evaluation
Decision Making

4-2 분석기준

- 경제성 분석을 위한 각종 기준 설정은 「교통시설 투자평가지침, 2007.12, 국토해양부」에서 제시한 내용을 적용하되, 편익 산정의 기초가 되는 각종 원단위는 조사시점의 가격기준으로 변환하여 사용하여야 하므로 2007년을 기준으로 환산한 값을 적용

4-2-1 분석시점과 기준년도의 설정

- 경제성분석의 주요 항목인 비용, 편익 추정의 기준이 되는 분석 기준년도는 조사 착수 전 년도로 설정하는 것을 원칙으로 한다. 다만 주요 통계자료가 미비한 상황에서는 기준년도를 조사 착수 2년 전 말로 설정할 수 있으며, 할인율을 적용하는 기준년도도 동일

- 공사시간의 시작은 1월1일을 기준으로 하고, 공사의 완료시점은 12월 31일을 기준으로 함

4-2-2 분석기간 설정

- 투자사업의 평가기간은 보통 시설물의 수명과 미래의 여러 가지 변동요인들을 감안하여 정하게 되는데, 교통시설 투자평가지침에 따르면 도로사업의 경우 설계 및 시공기간 등 사업계획 기간과 개통 후 30년 기간을 포함하는 기간을 평가기간으로 설정하도록 하고 있음

4-2-3 할인율의 결정

- 경제성 분석을 위해서는 실제 자본의 기회비용(Opportunity Cost)을 적절히 반영하여 서로 다른 기간 중에 발생하는 비용과 편익을 비교할 수 있도록 적절한 할인율이 채택되어야 함

- 보통 교통시설 투자평가지침에서 제시하는 사회적 할인율을 적용

4-2 Analysis Standard

- The standards provided in 「Instruction of the evaluation of traffic facilities investment, 2007. 12, the Ministry of Transportation, Land, and Maritime Affairs」 shall be applied to set various standards for the analysis of economic efficiency, and various currency units shall be converted to the cost of the year 2007 since the cost shall be converted to the price standard at the time of investigation.

4-2-1 Analysis Time and Reference Year Setting

- The analysis reference year, which is a reference point for estimating cost and saving as major items in economic efficiency analysis, shall be set to the year one year earlier than beginning of the investigation. However, if there is not sufficient amount of major statistical data, reference year can be set to the end of two years prior to the investigation time, and the reference year for applying discount rate shall be same.

- January 1st is the reference date for the beginning of the construction, and December 31st is the reference date for the completion of the construction.

4-2-2 Analysis Period Setting

- Generally, evaluation period of the investment project is determined by considering various future variation factors including lifetime of the facilities, and the evaluation period shall be set to project planning period such as design and construction period and 30 years after opening to public according to the traffic facilities investment evaluation instructions.

4-2-3 Determination of Discount Rate

- Appropriate depreciation rate shall be applied in for the economic efficiency analysis by properly reflecting the opportunity cost of actual capital so that the cost and savings generated from other periods can be compared.

- In general, social discount rate provided in the traffic facility investment evaluation instructions is applied.

4-2-4 매몰비용의 처리

- 설계비는 기본설계 또는 실시설계가 이미 완료된 경우, 해당 사업의 추진을 중단하더라도 해당 비용을 회수할 수 없으므로 경제적으로 매몰비용으로 처리하는 것으로 원칙

- 공사비는 시공단계에서 타당성 재검증을 수행하는 경우, 이미 완료된 공사부분은 매몰비용으로 처리하고 해체비용은 무시함

4-3 분석방법

- 투자사업과 관련된 비용과 편익은 다른 시간대에 나타나므로 이들을 비교·분석하기 위한 지표가 필요

- 현재 가장 보편적으로 사용되는 것으로 사업의 할인된 총 비용과 총 편익의 차이인 순현재가치를 이용하는 방법, 비용과 편익의 비율을 이용하는 방법 등이 있음

4-3-1 편익/비용 비율(Benefit/Cost Ratio : B/C)

- 할인된 총 편익과 총 비용의 비율이며 B/C값이 "1"보다 크면 경제성이 있다고 평가되지만 비용과 편익 항목의 구분 여하에 따라 비율이 다르게 나타날 수 있음

- 편익/비용 비율(B/C) 산정식

$$B/C = \sum_{t=0}^{n} \frac{B_t}{(1+r)^t} \Big/ \sum_{t=0}^{n} \frac{C_t}{(1+r)^t}$$

여기서, B_t : 편익의 현재가치
C_t : 비용의 현재가치
r : 할인율(이자율)
n : 교통사업의 내구년도 (분석기간)

4-2-4 Processing the Sunk Cost

- If the basic design or detail design is completed, the design cost cannot be returned even when the project is cancelled immediately, therefore the cost shall be processed as sunk cost.

- When feasibility is re-evaluated during the construction, the expense spent for the completed part shall be processed as sunk cost, and the dismantle cost is ignored.

4-3 Analysis Method

- As the cost and saving related to the investment project happen in different time, indexes are required to compare and analyze them.

- The most commonly used one is net present value, which is the difference between total discounted cost of the project and total savings, and other methods like the ratio of cost to saving are used as well.

4-3-1 Benefit/Cost Ratio (B/C)

- This is the ratio of total discounted benefit to total cost, and it is considered to have economic efficiency if the value of "B/C" is bigger than 1. However, the ratio can be presented in different ways depending on the classification of the items on cost and benefit.

- Equation of computation for benefit/cost ratio

$$B/C = \sum_{t=0}^{n} \frac{B_t}{(1+r)^t} \Big/ \sum_{t=0}^{n} \frac{C_t}{(1+r)^t}$$

Where, B_t : present value of benefit
C_t : present value of cost
r : discount rate(interest rate)
n : durable year of traffic project (analysis period)

4-3-2 순현재가치(Net Present Value : NPV)

- 사업에 수반된 모든 비용과 편익을 기준년도의 현재가치로 할인된 총 편익과 총 비용의 차로 표시

- 순현재가치가 양(Positive)인 사업은 자본비용을 회수하고도 잉여가 발생한다는 것을 의미하므로 투자 타당성이 있는 것으로 판단할 수 있음

$$\text{순현재가치(NPV)} = \sum_{t=0}^{T} \frac{\sum_{i=n}^{N} B - \sum_{j=m}^{M} C_{jt}}{(1+r)^t}$$

여기서, B_{it} : i항목의 t연도 편익　　C_{jt} : j항목의 t연도 비용
n, N : 편익 항목의 종류　　m, M : 비용 항목의 종류
T : 기준년차로부터 평가대상기간 최종년차까지의 년수
t : 기준년차를 0으로 하는 년차

4-3-3 내부수익률(Internal Rate of Return, IRR)

- 결정된 평가기간 내에 총 편익이 투자된 총 비용과 일치되는 이자율을 의미

- 이 값이 사회적 할인율보다 크면 경제성이 있는 것으로 판단할 수 있음

- 일반적으로 교통시설사업의 경우 비용이 평가대상 기간의 초기에 집중하고 편익이 시설 완성 후와 평가대상 기간 내에 지속적으로 발생

- 경제적 내부수익률이 크다는 것은 미래의 편익을 낮게 평가함에도 불구하고 사업의 효과가 있다는 것을 의미하는 것이므로 경제적 내부수익률이 큰 교통시설사업은 우량한 사업이라 할 수 있음

$$\text{내부수익률(IRR)} = \sum_{t=0}^{T} \frac{\sum_{i=n}^{N} B - \sum_{j=m}^{M} C_{jt}}{(1+ir)^t} = 0$$

여기서, B_{it} : i항목의 t년도 편익
C_{jt} : j항목의 t년도 비용
n, N : 편익항목의 종류
m, M : 비용항목의 종류
T : 기준년차로부터 평가대상기간 최종년차까지의 년수
t : 기준년차를 0으로 하는 년차
ir : 내부수익률

4-3-2 Net Present Value (NPV)

- All cost and benefit associated with the project are shown as total benefit and total cost discounted by the present value of the reference year.

- Positive NPV means that the investment is returned with margin, so the investment is considered to be valid.

$$\text{Net present value(NPV)} = \sum_{t=0}^{T} \frac{\sum_{i=n}^{N} B - \sum_{j=m}^{M} C_{jt}}{(1+r)^t}$$

Where, B_{it} : benefit of item i at year t C_{jt} : cost of item j at year t
n, N : type of benefit items m, M : type of cost items
T : total number of years from reference year to the final year of the evaluation period
t : year index starting from 0 at the reference year

4-3-3 Internal Rate of Return (IRR)

- Internal rate of return means the interest rate that makes the total invested cost equivalent as total benefit during the determined evaluation period.

- It is considered to have economic efficient if this value is bigger than social discount rate.

- Generally, in case of traffic facility projects, the cost is concentrated on the intial period of the entire evaluation period, and benefit is generated continuously after the completion throughout the evaluation period.

- Bigger IRR means that the effects of project is bigger even when the future benefit is evaluated lower, therefore traffic facility project with bigger IRR can be referred to as a good project.

$$\text{Internal rate of return(IRR)} = \sum_{t=0}^{T} \frac{\sum_{i=n}^{N} B - \sum_{j=m}^{M} C_{jt}}{(1+ir)^t} = 0$$

Where, B_{it} : benefit of item i at year t C_{jt} : cost of item j at year t
n, N : type of benefit items m, M : type of cost items
T : total number of years from reference year to the final year of the evaluation period
t : year index starting from 0 at the reference year
ir : internal return

4-3-4 분석방법의 특징

- 편익/비용 비율, 순현재가치, 내부수익률에 의한 타당성 유무 판단이 항상 동일한 것은 아님
- 순현재가치는 순편익의 흐름을 사업개시년도의 가치로 평가하였지만 사업규모에 대하여 표준화되어 있지 않기 때문에 사업간 비교에는 적당하지 않다는 단점이 있음
- 내부수익률은 사업의 규모에 의존하지 않는다는 장점은 있으나 수익성이 극히 낮거나 높은 사업의 경우는 계산되지 않는 단점이 있음
- 편익/비용 비율은 특정 항목을 편익 혹은 비용으로 처리하는가에 따라 값이 달라진다는 단점이 있으나 일반적으로 투자심사기준으로 사용

4-4 비용 및 편익의 산출

4-4-1 비용 산출

- 비용은 공사비, 용지보상비, 부대비 등으로 구성되는 사업비와 유지관리비를 포함하고 제세공과금이 제외된 투자비로 크게 구분

구 분			내 용
비용	건설비 (사업비)	공사비	• 도로건설에 필요한 건설 사업비
		부대비	• 설계용역비(실시), 감리비 그리고 측량 및 조사비
		용 지 보상비	• 도로건설을 위한 토지소유권의 취득으로 도로 사업에 실질적으로 지출되는 비용 및 도로 사업으로 부득이하게 발생하는 손실에 대하여 지불하는 비용이며, 일반건설비용 및 전업에 따른 보상 등을 포함
		예비비	• 타당성조사 단계에서 발생될 수 있는 사업비 산출의 오차 및 이에 따른 영향을 최소화하기 위하여 책정된 비용
	유지관리비		• 도로 개통 후에 도로의 구조 및 교통 기능의 보존 또는 유지 등에 소요되는 비용

- 공사비는 해당 사업의 초기에 투자되는 건설 사업비로 토공구간·교량구간·터널구간 등으로 구분하며, 부대비는 설계·감리 및 이에 수반되는 측량·조사비를 말하며, 용지보상비는 사업 지역의 토지 매입 및 보상비를 말함
- 유지관리비는 사업의 초기 투자비용 뿐만 아니라 생애주기비용(Life Cycle Cost)까지 고려하기 위하여 추가하는 경상운영비 등을 말하며, 예비비는 사업비 산출오차에 따른 영향을 최소화하기 위하여 책정된 비용

4-3-4 Characteristics of Analysis Method

- The results of feasibility studies based on cost/benefit rate, net present value, and IRR are not necessarily same.
- Net present value evaluates the flow of net benefit with the value converted to the value at the year of project start, but it is not standardized by the scale of project, so the disadvantage of this method is that it is not suitable for comparing different projects.
- IRR method is independent of the scale of the project, which is an advantage, but disadvantage is that projects with extremely high or extremely low profit are not calculated properly.
- Cost/benefit ratio has a disadvantage of variation of the result depending on whether certain items are processed as benefit or cost, but this is the most commonly used investment judgment criterion.

4-4 Cost and Benefit Calculation

4-4-1 Cost Calculation

- Cost is classified into the project cost including construction cost, land acquisition cost, and other supplementary cost and the investment cost including maintenance and control cost excluding taxes due.

Classification			Contents
Cost	Building cost (project cost)	Construction cost	• Building and project cost required for constructing roads
		Supplementary cost	• Design service cos t(execution), supervision cost, measurement and investigation cost
		Land acquisition cost	• This is the expense paid for purchasing land property for road construction and expense paid for the unavoidable loss by road construction project, and it includes general construction expense and compensation for changing business.
		Reserve expense	• Expense allocated to minimize the errors in the calculation of project cost and the effects of errors, which was done during the feasibility analysis stage.
	Maintenance and control cost		• Expense required to preserve or maintain the road structure and traffic function after opening to public.

- Construction cost is a part of building project cost invested at the earlier stage of the corresponding project comprising of earth work section, bridge section, and tunnel section. Supplementary cost includes the expense for design, supervision and associated measurement and investigation expense, and land acquisition cost is the expense for land purchasing and compensation for the construction site.
- Maintenance and control cost is regular operation cost to consider the life cycle cost in addition to the initial investment cost, and reserve expense is the cost allocated to minimize the effects of errors in project cost calculation.

4-4-2 편익 산출

- 도로사업 시행 시 교통시설 이용자들에게 발생하는 편익으로서, 차량운행비용 절감·통행시간 절감·교통사고 감소·쾌적성 증가·정시성 향상·안정성 향상 등이 있음

- 차량운행비용 절감·통행시간 절감·교통사고 감소 등의 편익을 화폐가치화 작업은 비교적 용이하나, 교통의 쾌적성·정시성·안정성 향상 등의 효과는 개인별 주관적 만족도에 따라 가치가 달라질 수 있기 때문에 화폐가치화 하는데 어려움이 수반

- 간접편익은 도로사업 시행 시 교통시설 이용에 관계없이 모든 사람에게 발생하는 파급 효과임

- 간접편익 가운데 환경비용 절감 편익의 경우 대기오염, 소음 등 일부 항목을 계량화하는 연구 성과가 축적되어 비용편익분석에 반영할 수 있음

- 그러나, 지역개발·시장권 확대·산업구조 개편 효과 등이 실현되기 위해서는 도로사업 이외의 분야에 대한 투자가 병행되어야 하기 때문에 계량화에 어려움이 따르고, 투자의 구축효과(crowding out effects) 등으로 비용편익분석의 편익으로 직접 산정하는 데는 논란의 여지가 있어 편익항목에는 포함하지 않음

- 따라서, 통상적으로 편익 항목으로는 직·간접적인 편익 중 계량화가 가능한 차량운행비 절감과 통행시간 절감, 교통사고 저감 편익 및 환경비용 절감을 고려

① 차량운행비 절감편익

- 차량운행비는 우선 링크 상에 부하된 교통량에 따른 교통량 대 용량비(V/C)를 구하고, 교통량·속도의 곡선에 의하여 각각의 링크와 차종별로 평균 운행속도를 결정, 속도에 따른 차종별 차량운행비를 계산하여 이의 총합으로 산출

- 궁극적으로 과업노선에 대한 차량운행비 절감분은 기존 네트워크를 중심으로 산출된 총 차량운행비와의 비교를 통하여 산출

4-4-2 Benefit Calculation

- Benefits are generated to the users of the traffic facility including reduction of car operation expense, reduction in driving time, reduction in traffic accident, more pleasant driving, improving braking performance, and safety enhancement.

- It is relatively easy to convert some benefits into values in currency such as reduction of car operation expense, reduction in driving time, and reduction in traffic accident. however, it is difficult to covert other benefits such as more pleasant driving, improving braking performance, and safety enhancement into values in currency because of the values that might change depending on the individual subjective satisfaction level.

- Indirect benefit is a spreading effect to everyone regardless of using traffic facilities.

- Among indirect benefits, benefit of reduction in environmental cost can be reflected into the cost and benefit analysis due to the accumulation of the research output like quantification of some items such as air pollution and noise.

- However, as realization of the local development, expansion of market area, and improvement of industry structure requires parallel investment beside road construction project, it is difficult to quantify the benefit, and it is controversial to count them as benefit like crowding out effect, so those items shall not be included in the benefit items.

- Therefore, general benefit items of reduction in car operation cost, reduction in driving time, reduction in accident, and reduction in environmental cost, which can be quantified among the direct and indirect benefit items, are considered.

① Benefit of Reduction in Vehicle Operation Expense

- Vehicle operation expense is calculated by firstly calculating the ratio of traffic volume to capacity (V/C) depending on the traffic volume given to the link, and average operating velocity of each type of vehicles determined by the traffic volume/velocity curve, vehicle operation cost by vehicle type and velocities, and total sum will be calculated.

- Eventually, the saving on the vehicle operation cost for the project route shall be calculated based on the comparison with the vehicle operation cost calculated based on the existing network.

- 일반적으로 차량운행비는 고정비와 변동비로 구분되는데 변동비(Variable Cost)에는 연료비 · 엔진오일비 · 타이어 마모비 · 차량 유지 수선비 등이 포함되고, 고정비(Fixed Cost)에는 차량의 감가상각비 · 운전원 및 보조원의 임금 · 보험료 · 차량검사료 등이 포함되며, 도로조건과 교통여건에 따라 달라지므로 실제 차량운행비는 도로 및 교통조건에 부합되도록 각각 계산되어야 함

- 차량운행비는 세분된 비용 항목별로 최근의 공신력 있는 자료를 조사한 후 먼저 이상적인 교통상황 하에서의 차종별 · 속도별 차량운행비를 산출하고, 다음에 도로 및 교통조건에 따른 다양한 교통상황 하에서의 차종별 · 속도별 차량운행비를 ΔL 방법으로 계산

- 속도에 따른 차량운행비의 속도별 원단위는 교통시설 투자평가지침의 원단위 자료를 토대로 분석년도의 값으로 보정하여 사용
  - 차량운행비 산출 모형

모형식 : $VOC_k = a + b\frac{1}{V_k} + cV_k^2$

여기서, VOC_k : 차종 k의 속도 V에서의 운행비용(원/대-km)

V_k : 차종 k의 속도(km/h)

a, b, c : 계수(상수)

【차량운행비 산출모형의 계수】

차종(k)	a	b	c	R^2
승 용 차	162.729	1,981.054	-0.0017847	0.955
소형버스	116.380	1,847.230	0.0004326	0.950
대형버스	205.940	4,101.316	0.0022131	0.913
소형트럭	159.339	2,300.458	0.0040666	0.931
중형트럭	282.656	2,786.795	0.0018376	0.927
대형트럭	288.809	4,821.633	0.0090175	0.867

- In general, vehicle operation cost consists of fixed cost and variable cost. Variable cost includes fuel, engine oil, tire wear, vehicle maintenance and repair, and fixed cost includes depreciation of vehicle, salary of the driver or assistants, insurance, and vehicle inspection fee. As these cost vary depending on the road and traffic condition, the calculation of actual vehicle operation cost shall be estimated complying with the road and traffic conditions.

- When calculating vehicle operation cost, publically trusted data shall be searched first for each of detailed cost items, vehicle operation cost by vehicle type and velocity under ideal traffic conditions, and vehicle operation cost for each type and velocity shall be calculated in ΔL method under various traffic situations like road conditions and traffic conditions.

- Basic unit data of vehicle operation expense from the traffic facility investment evaluation instructions shall be compensated for the analysis period to use for the basic unit for velocity of vehicle operation expense.

  - Vehicle operation cost calculation model

    Model equation : $VOC_k = a + b\frac{1}{V_k} + cV_k^2$

    Where, VOC_k : operation cost of car type k at velocity V (won/car-km)

    V_k : velocity of car type k(km/h)

    a, b, c : coefficients (constants)

【Vehicle operation expense calculation model coefficients】

Vehicle type(k)	a	b	c	R^2
Passenger car	162.729	1,981.054	-0.0017847	0.955
Small bus	116.380	1,847.230	0.0004326	0.950
Large bus	205.940	4,101.316	0.0022131	0.913
Small truck	159.339	2,300.458	0.0040666	0.931
Mid-size truck	282.656	2,786.795	0.0018376	0.927
Large truck	288.809	4,821.633	0.0090175	0.867

- 차량운행비 절감 편익을 산출하기 위하여 장래 사업 미시행 시와 사업 시행 시의 도로망 체계에서의 운행비 총액을 구하고, 이 값들의 차를 구하여 차량운행비 절감 편익으로 적용

$$VOCS = VOC_{\text{사업 미시행 시}} - VOC_{\text{사업 시행 시}}$$

여기서, $VOCS$: 연간 총 차량운행비 절감편익

VOC : 도로 미건설 시(건설 시) 연간 총 차량운행비

$$VOC = \sum_{l}\sum_{k=1}^{6}(D_{kl} \times VT_k \times 365)$$

D_{kl} : 링크 l의 차종별 대·km

VT_k : 해당속도에 따른 차종별 차량운행비

k : 차종

② 통행시간가치 절감편익

- 도로사업의 편익 항목 중 통행시간 단축으로 인한 편익은 경제성 평가에 있어 차량운행비 절감 편익과 함께 중요한 항목임
- 통행시간가치 절감 편익은 통행자나 화물이 개선된 시설을 이용함으로써 단축되는 시간을 화폐단위로 계량화
- 통상적인 교통분석에서는 업무통행과 비업무통행으로 구분하여 각 통행목적에 따라 상이한 시간치를 적용
- 시간가치의 적용은 교통시설 투자평가지침에서 제시한 값을 분석년도의 값으로 보정하여 적용
- 통행시간가치 절감액은 사업 미시행 시 및 시행 시의 차종별 총 운행시간에 차종별 여객 및 화물시간가치를 적용하여 총 통행시간가치 비용을 산출한 후 사업 미시행 시와 사업 시행 시의 차이로부터 통행시간가치 절감액을 산출

$$VOTS = VOT_{\text{사업 미시행 시}} - VOT_{\text{사업 시행 시}}$$

여기서, $VOTS$: 연간 총 통행시간비용 절감편익

VOT : 도로 미건설 시(건설 시) 연간 총 통행시간비용

$$VOT = \left\{\sum_{l}\sum_{k=1}^{3}(T_{kl} \times P_k \times Q_{kl})\right\} \times 365$$

T_{kl} : 링크 l의 차종별 통행시간

P_k : 차종별 시간가치

Q_{kl} : 링크 l의 차종별 통행량

k : 차종(1 : 승용차, 2 : 버스, 3 : 화물차)

- In order to calculate the benefit of saving on vehicle operation cost, calculate the total cost of operation under the road network system for the case when project is executed and for the case when project is not executed, and the difference between two cases shall be applied to the benefit of saving on vehicle operation cost.

$$VOCS = VOC_{when\ project\ is\ not\ executed} - VOC_{when\ project\ is\ not\ executed}$$

Where, $VOCS$: Annual total benefit of saving on vehicle operation cost

VOC : Annual total vehicle operation cost when project is (not) executed

$$VOC = \sum_{l}\sum_{k=1}^{6}(D_{kl} \times VT_k \times 365)$$

D_{kl} : Number of vehicles by type at link l · km

VT_k : Vehicle operation cost by vehicle type for corresponding velocity

k : Vehicle type

② Benefit of Saving on Travel Time Value

- Benefit by saving in travel time is one of the most important items as well as vehicle operation cost saving benefit among the benefit items of road construction project.

- Reduction of time in moving passengers or cargo by using the improved facilities is quantified and converted into economic value to produce travel time value saving benefit.

- Generally, traffic analysis are divided into work traffic and non-work traffic, and different weights are applied depending on the travel purpose.

- Time value provided from the traffic facility investment instructions shall be corrected to the value of analysis period for the application of time value.

- Time value of passenger and cargo by vehicle type shall be applied to the operation time by vehicle type for the case when project is executed and project is not executed to calculate the total travel time value cost, and travel time value saving amount is calculated from the difference between two cases.

$$VOTS = VOT_{when\ project\ is\ not\ executed} - VOT_{when\ project\ is\ not\ executed}$$

Where, $VOTS$: Annual total benefit of saving on travel time cost

VOT : Annual travel time cost when project is (not) executed

$$VOT = \left\{\sum_{l}\sum_{k=1}^{3}(T_{kl} \times P_k \times Q_{kl})\right\} \times 365$$

T_{kl} : Travel time by vehicle type at the link l

P_k : Time value by vehicle type

Q_{kl} : Traffic volume by vehicle type at the link l

k : Vehicle type (1 : passenger car, 2 : bus, 3 : truck)

- 유료도로의 경우 통행배정과정에서 통행료를 시간으로 환산한 가치의 변화가 통행시간의 변화로 반영된다는 점에서 사회적 관점에서 이전 지출에 불과한 유료도로 통행료의 변화를 통행시간 절감 편익에서 제거

② 교통사고 절감 편익

- 일반적으로 도로의 시설이 개선되면 사고율이 낮아지며, 교통사고 발생건수는 도로형태·도로별 형태·교통특성 등에 따라 변함

- 따라서 도로사업 시행 전·후의 이 변수들의 변화가 반영되도록 하여 교통사고 절감 편익 산출

- 교통사고 자료는 도로교통공단에서 제시한 도로 유형별 교통사고 사상자수를 기준으로 분석하고, 사고 건당 및 사상자당 교통사고 비용은 한국교통연구원 등에서 교통사고 비용 추정자료를 활용

- 분석대상 도로의 교통사고 절감 편익의 화폐가치화 산출방법

$$VICS = VIC_{사업\ 미시행\ 시} - VIC_{사업\ 시행\ 시}$$

여기서, $VICS$: 연간 총 교통사고 감소 편익
VIC : 도로 미건설 시(건설 시) 연간 총 교통사고비용

$$VIC = \sum_{t}\sum_{s=1}^{2}(A_{ts} \times P_{s} \times VL_{ts})$$

A_{ts} : 도로유형별·사고유형별 1억대·km당 교통사고 사상자수
P_s : 사고유형별 사고 비용
VL_t : 연간 도로유형별 억대·km
t : 도로유형(고속도로, 국도, 지방도)
s : 사고유형(사망, 부상)

- 구축된 도로망의 링크 통행배정결과를 이용하여 고속도로 링크, 국도 링크 및 지방도 링크를 구분하여 사업시행과 사업미시행의 억대-km를 산출하여 그 차이를 활용하여 교통사고 감소편익을 산출

- For toll road, as the changes of the value converted from toll fee to time during the traffic allocation process are reflected as the changes of toll road fee, which is a merely transfer payment from the social view point, the changes of toll road fee shall be removed from the travel time saving benefit.

② Benefit of Reduction in Traffic Accident

- In general, improvement of road facilities reduce the accidents, and the number of accidents varies depending on the road shape, type of each road, and traffic characteristics.

- Therefore, reflect the changes of these variables before and after the execution of the project, and calculate the traffic accident reduction benefit.

- Traffic accidents shall be analyzed with the number of traffic accidents victims by road type provided from Road Traffic Authority, and the traffic accidents cost estimate data available from Korea Transport Institute and other organizations shall be used for the cost of traffic accidents per accident and per person.

- Calculation method of conversion to economic value of traffic accidents reduction benefit for the target analysis road

$$VICS = VIC_{when\ project\ is\ not\ executed} - VIC_{when\ project\ is\ not\ executed}$$

Where, $VICS$: Annual total benefit of reduction in traffic accident
VIC : Annual traffic accident cost when project is (not) executed

$$VIC = \sum_{t} \sum_{s=1}^{2} (A_{ts} \times P_s \times VL_{ts})$$

A_{ts} : Road type · 100 million cars by accident type · number of traffic accident victims per km
P_s : Accident cost by accident type
VL_t : Annual 100 million cars by road type · km
t : Road type(expressway, national road, local road)
s : Accident type (death, injury)

- Calculate the 100 million cars-km for the case when project is executed and when project is not executed by dividing the link into expressway link, national road link, and local road link using the link traffic allocation result of the existing road network, and use the difference to calculate the traffic accident reduction benefit.

④ 환경비용 절감 편익

- 교통부문 사업 시행에 따른 환경영향으로는 대기오염, 수질오염, 소음, 진동, 지반침하, 식물 및 동물 등 생태계 영향, 경관변화, 지구온난화 등을 들 수 있음

- 그러나 이러한 다양한 환경 변화요소에 대하여 모두 환경가치를 추정하는 것은 용이한 작업이 아님. 그 영향의 정도 자체를 파악하는 것이 힘든 경우도 있을 뿐만 아니라 설사 엄밀하게 영향을 파악한다 하여도 이를 경제적 가치로 환산하는 것은 보다 많은 불확실성을 내포하기 때문임

- 따라서, 이중에서도 비교적 영향의 정도가 크고 환경영향에 대한 평가 및 가치화가 용이한 대기오염과 소음발생에 초점을 맞추어 환경 비용을 추정

- 각각의 산정방법은 아래와 같으며, 대기오염배출계수와 차종별 대기오염비용 원단위를 적용한 차종별 속도별 대기오염 비용, 그리고 소음가치의 원단위 등은 교통시설 투자평가지침의 원단위 자료를 참고

1) 대기오염 절감편익 산정

- 교통시설사업 시행의 환경비용은 기본적으로 다음 3단계를 거쳐서 추정된다.
  - 1단계 : 자동차 주행으로 인하여 발생하는 오염물질별 배출계수를 산정한다.
  - 2단계 : 오염물질 단위당 환경피해비용을 산출한다.
  - 3단계 : 사업시행으로 인한 오염원별 오염물질의 변화를 산정하고, 이를 오염물질별 환경피해비용 원단위를 곱하여 화폐가치화 한다.

2) 소음 절감편익 산정

- 소음 절감편익 산정은 사업 미시행 시와 시행 시의 발생 소음도 차이를 구한 후, 유지비용법을 적용하여 단위 소음 당(1dB) 저감을 위하여 필요한 유지관리비용(방음벽 설치비용) 원단위를 곱하여 사업시행으로 인한 소음영향을 화폐가치화 한다.

- 도로소음 예측식은 일반도로(국도, 지방도 등) 와 고속도로에 대한 예측식이 별도로 제시되어 이를 적용하였으며, 소음가치의 평균 원단위 비용은 다음과 같다.
  - 소음비용의 추정식 EVN.

$$EVN = \sum_i \sum_j P\, l_{ij}\, L_{ij}$$

여기서, P: 소음가치의 원단위 　　l_{ij}: 대상노선 연장길이
L_{ij}: 예측소음도 　　i: 도로 및 철도 구분
j: 영향권 내 개별링크

④ Benefit of Reduction in Environmental Cost

- The environmental effects by the execution of traffic projects include air pollution, water pollution, noise, vibration, ground subsidence, effects to the ecology system of plant and animals, changes in scene, and global warming.

- However, it is not an easy task to estimate the environmental values for the various environment changing factors. This is not only because it is hard to estimate the degree of effects, but converting the effects into economic value contains more uncertainties even if the effects can be accurately identified.

- Therefore, environmental value is estimated focusing on the air pollution and noise generation, which have relatively big effects, and evaluation of effects to environment and converting to economic values are easier.

- Calculation method for each case is as followings. Air pollution cost by vehicle type and by velocity that applies the basic unit of air pollution cost by vehicle type and air pollution emission coefficient and basic unit of noise value shall use the basic unit data from traffic facility investment evaluation instructions.

1) Calculation of Air Pollution Reduction Benefit

- Environment cost of traffic facility project execution is estimated basically through the following three steps
  - Step1 : Calculate the emission coefficient for each pollution material generated by driving vehicles
  - Step2 : Calculate the environment damage cost per unit pollution material
  - Step3 : Calculate the change of pollution material for each pollution source due to the execution of projects, and multiply this by environment damage cost for each pollution material to convert into economic value.

2) Noise Reduction Benefit Calculation

- Calculate the difference of noise generation between the case when project is executed and the case when the project is not executed, apply the maintenance cost method to multiply the basic unit for maintenance and control expense (cost to install sound-absorbing wall) required to reduce the unit noise (1dB), and convert the effects of noise due to the execution of projects to economic value.

- Road noise prediction formula for general road (national road, local road, etc) and expressway are available and applied separately, and average basic unit of noise value is as followings.

- The following is noise cost estimate formula, EVN.

$$EVN = \sum_{i}\sum_{j} P\, l_{ij}\, L_{ij}$$

where, P : basic unit of noise value l_{ij} : expansion length of target route
L_{ij} : predicted noise level i : road or railroad
j : separate link unit in the sphere of influence

4-5 최적개통시기 결정

- 경제성 분석 결과 사업에 대한 경제적 타당성이 인정되더라도, 사업의 투자효율성을 높이기 위해서는 사업의 개통시기를 적절히 조정하는 과정이 필요

- 최적투자시기의 분석에 있어서는 시차적 분석 방법(Incremental Time Analysis)과 초기년도 수익율(First Year Rate of Return)법이 이용

- 시차적 분석 방법은 사업의 시행을 1년 단위로 연기시켜 순현재가치(NPV)가 최대가 되는 시기를 찾아내는 방법으로 수익이 안정적이거나 일정수준을 유지할 경우에 적용

- 초기년도 수익율법(FYRR)은 사업시행을 1년씩 연기시켜 사업완료 첫 해의 편익/비용 비가 할인율을 넘어서는 연도를 개통년도로 결정하는 방법

구 분	시차적 분석방법 (Incremental Time Analysis)	초기년도수익율법 (First Year Rate of Return)
분석방법	• 사업의 시행을 1년씩 연기시켜가며 NPV가 최대가 되는 연도를 찾아내는 방법	• 사업시행을 1년씩 늦추어 가면서 사업완료 첫 해의 수익률(First Year Rate of Return)이 적용 할인율을 초과하는 연도를 찾아내는 방법
적 용	• 편익이 안정적이거나 일정수준을 유지할 경우 • 최대이윤보장 시기를 찾을 경우	• 편익이 지속적으로 증가할 경우 • 이익 보장시점을 찾을 경우
특 성	• 편익이 지속적으로 늘어날 경우 FYRR은 높으나 NPV가 최대가 되는 시기가 계속 연기됨	• 투자 수익률이 사회적 할인율을 보장하는 시기를 찾기 때문에 시차적 분석방법에서의 개통시기보다 앞년도로 당겨질 경우가 많음

4-5 Determine Optimum Opening Time

- Even when the economic feasibility is accepted from the result of economic efficiency analysis, it is necessary to adjust the opening time properly to maximize the investment efficiency.

- Incremental time analysis and first year rate of return (FYRR) are used for optimum investment time analysis.

- Incremental analysis method is to delay the start of the project by one year unit to find the year with the maximum net present value (NPV), and this method is used when the profit is stable or maintains over certain level.

- FYRR method is to delay the project start year by one year unit to determine the year that the B/C ratio of the first year of project completion exceeds the discount rate as the opening year.

Classification	Incremental Time Analysis	First Year Rate of Return
Analysis method	• To find the year with maximum NPV by delaying project start by one year unit	• To find the first year rate of return exceeding the discount rate by delaying the project start year by one year unit.
Application	• When benefit is stable or maintains over certain level • When identifying the time that guarantees the maximum profit	• When benefit grows continuously • When identifying the point when profit is guaranteed
Characteristics	• When benefit grows continuously, FYRR is high, but the year with maximum NPV will keep getting delayed	• There are more chances to pull in the opening time to earlier year compared to the incremental time analysis due to finding the time that investment return rates guarantee the social discount rate

Chapter 6

Route Location and Alignment Design

1. 개 요

- 도로의 노선 및 선형은 사회적·기술적·경제적·환경적 측면 등을 고려한 평가를 통하여 이루어지며, 타당성조사－기본설계－실시설계의 단계를 거치면서 각 단계에 적합한 수준의 검토를 통하여 최적 선형을 결정한다.
- 실시설계 단계에서는 기본설계 시 도출된 노선의 재검토를 통하여 최종 노선을 결정하고, 세부 선형설계를 실시한다.

2. 노선선정

2-1 노선대 선정

- 타당성조사 단계에서 1:25,000 ～ 1:50,000의 지형도를 활용하여 다수의 후보노선대를 선정한 후 각 대안의 비교검토를 통하여 최적 노선대를 결정
- "노선대"란 노선계획의 범위를 의미하며, 최적 노선대를 정하는 이유는 설정된 노선대의 공간적 범위 내에서만 집중적으로 노선 검토를 실시하기 위함
- 최적 노선대 선정을 위하여 우선 기·종점을 연결 가능한 후보노선대를 정한 후, 다음의 항목들을 고려한 평가를 실시하여 최적 노선대를 결정

평가항목	평 가 내 용
접근성	사업 시행에 따른 영향권 지역의 접근성을 평가
시공성	건설 시 터널, 교량, 지장물 등 공사의 난이도를 평가
지역개발 효과	대안별 지역개발 수준을 영향권의 대소에 따라 평가하여 지자체의 의견을 반영
기존계획과의 관련성	도로 및 철도 등 교통망 확충계획과 도시기본계획 및 도시재정비계획 등 관련 계획과 적합성 평가
교통흐름	기존 도로의 통행량 전환 등 소통 개선 효과를 평가
환경영향	자연환경 및 생활환경에 대한 주요 영향의 조사 평가
공사비	개략 공사비를 산정하여 대안별 비교

1. Overview

- Route location and alignment design shall be done through overall evaluations considering social, technical and environmental factors. An optimal alignment is normally determined by reviews to the level appropriate for each design phase throughout feasibility study, basic design and detailed design.
- During the phase of detailed design, the route determined during the phase of basic design shall be re-evaluated in order to select the final route, and detailed alignment design shall be conducted accordingly.

2. Route location

2-1 Route Range Selection

- During the feasibility study, a number of candidate route ranges shall be selected, using topographic maps of 1:25,000 ~ 1:50,000 scale, and the final route range shall be determined by reviewing and comparing each one of the candidates.

- The term, "Route Range" signifies a range of route plan, and the reason for determining an optimum route range is to limit the scope of review within the spatial range of selected route range for intensive rout reviews.

- In order to select an optimum route range, a number of candidate route ranges connectable from the starting to the arrival points shall be selected first, and the following factors shall be considered and evaluated evaluations in order to determine an optimum route range.

Evaluation Criteria	Descriptions
Accessibility	To evaluate the accessibility of the area affected by the construction project.
Constructability	To evaluate the difficulty of various construction works including tunnels, bridges, and other obstacles.
Local development effects	To reflect the opinions of local government through evaluations of the level of local development by the significance of the affected range for each candidate plan.
Relevance to existing plan	To evaluate the alignment with related plans such as road/rail road & traffic network expansion plan, basic city plan, and city re-establishment plan, etc.
Traffic flow	To evaluate the effects of traffic improvement such as traffic-volume distribution from existing roads.
Environment impact	To survey and evaluate main impacts to natural and living environments
Construction cost	To compare the estimated cost of each alternative proposal

2-2 최적노선 선정

- 최적노선은 기본설계 단계에서 결정되는 것으로서, 타당성조사 시 1:50,000 지형도상에서 선정된 최적노선대의 범위 내에서 도로 등급에 부응할 수 있는 설계기준을 설정하고, 현지답사를 거쳐 1:5,000 지형도상에 비교 노선 대안을 도출

- 비교노선 대안에 대하여 사회적·기술적·경제적·환경적 측면의 고려 요소들을 비교 평가하여 1:1,200 지형도상에서서 최적 노선을 결정

- 노선선정 시 고려하여야 할 사항은 다음과 같으며, 계획노선의 특성에 따른 중요도를 판단하여 각 고려 요소들의 정성적·정량적·상호비교 분석 등을 통하여 최적 노선을 선정

구 분	항 목
사회적 측면	• 상위관련계획, 지역개발계획, 도시계획 등과의 관계 • 도시, 집단주거지, 항만, 공항, 산업단지, 관광지 등과의 관계 • 유적, 매장문화재 등과의 관계 • 학교, 병원, 종교시설, 공동묘지 등 지장물과의 관계 • 지역주민 의견수렴 및 민원 사전 해소가 가능한 노선
기술적 측면	• 주요 경유지 통과 및 인접 도로 노선과의 중복 고려 • 채택된 설계속도에 적합한 평면선형과 종단선형의 적용 및 평면 및 종단선형의 조화 • 적설이 많은 지역은 가능한 남향 노선을 선정 • 가급적 연약지반, 지질적으로 불안정한 절토사면을 유발지역 등은 우회 • 접속도로의 위치 등을 고려하여 출입시설의 설치가 용이한 노선
경제적 측면	• VE / LCC 기법을 활용하여 경제적인 노선 선정 • 교차로 설치위치와 시설규모에 따른 경제성 분석
환경적 측면	• 자연환경 및 생활환경 영향 측면을 고려 • 농경지 잠식 및 자투리 농지를 최소화 하는 노선 선정 • 보존가치가 있는 지형 및 지질유산과 해당지역의 경관자원을 파악하여 보존할 수 있는 노선선정 • 경관이 우수한 지역을 통과하는 경우 완속차로와 고속차로의 분리로 경관자원을 활용하는 노선선정 •「환경친화적인 도로건설 지침」 내용 준수

2-2 Selecting Optimum Route

- Optimum route is determined during the basic design phase. The design standards applicable to the road class within the route range selected from the topographical map of 1:50,000 scale during the feasibility study shall be set, and the candidate routes for comparison on the topographical map of 1:5,000 scale through the field survey shall be determined.

- An optimum route shall be determined on the topographical map of 1:1,200 scale through comparative evaluations, in consideration of social, technical, economical, and environmental factors.

- The following items must be considered in route location, and an optimum route shall be selected through qualitative, quantitative, comparative analysis of various factors from evaluations on 'priority & importance', reflecting the characteristics of planned routes.

Factors	Items
Social factors	• Relation with upper level related plans, local development plan, city plan, etc • Relation with city, dense dwelling area, ports, airport, industrial area, tourism area, etc • Relation with historic sites, buried cultural assets, etc • Relation with obstacles such as schools, hospitals, religious facilities, cemeteries, etc. • Collection of local residents and Route that can resolve the conflict with civil affairs in advance
Technical factors	• Consideration of passing major intermediate points and overlapping with nearby roads routes. • Application of horizontal alignment and longitudinal alignment suitable for the selected design speed and balance between horizontal alignment and longitudinal alignment. • Road facing south shall be selected as much as possible for the heavy snowfall area • Detour soft foundation and area of frequent geological unstable cutting edge surfaces as much as possible. • Routes with ease of construction of access facility considering the location of the connection road
Economical factors	• Selection of economical route with using VE / LCC techniques. • Analysis of economical efficiency for the location of interchanges and scale of facilities.
Environmental factors	• To consider the effects of natural environment and living environment. • To select the routes that minimize the encroachment of farming land and remnant farming land. • To select the route that can preserve the topography, geological heritage, and scenic resources that are worthwhile to preserve. • To select the route to utilize the scenic resources by dividing high speed lanes and low speed lanes when the routes pass the scenic area. • Comply with the details in 「Instructions of environmental friendly road construction」

2-3 선형설계

- 최적노선 결정 시 차량주행의 안전성 확보를 위한 선형검토를 동시에 실시하여야 한다. 도로의 선형은 실시설계 단계에서 최종 확정되며, 실시설계 단계 결정된 선형은 시공 시 수정하기 곤란하므로 면밀한 검토가 필요

- 선형설계에 관한 세부 사항은 "3. 선형설계"에서 상세히 설명

2-4 자문 등 의견 수렴

- 최적노선(출입시설 포함) 선정 시 비교 안에 대하여 다양한 분야의 전문가로 구성된 위원회를 통하여 기술자문을 실시하고, 자문의견을 면밀히 검토하여 원안의 문제점을 보완

- 자문은 우리 공사 기술심사부서에 요청하여 실시하며, 전문가로 구성된 설계자문위원회를 통한 외부자문과 기술심사부서에서 자체적으로 실시하는 경제성 검토(VE)를 동시에 수행

- 노선자문 실시 후 관계기관 협의를 실시하며, 관계기관 요구사항은 타당성 여부를 검토 후 반영

- 설계 준공 후에는 설계의 적정성을 평가하기 위하여 심의를 실시한다. 심의도 자문과 같이 기술심사부서를 통하여 전문가 의견을 수렴하지만 심의에서는 설계결과의 원안 채택 또는 재심의 여부를 결정

2-3 Alignment Design

- Alignment design securing the safety of driving vehicles has to be reviewed simultaneously when determining an optimum route. Alignment design is finalized at the detailed design phase, and the design has to be thoroughly reviewed since the determined alignment during the detailed design cannot be modified easily during the construction.

- The details on alignment design is described in the section, "3. Alignment Design".

2-4 Collecting Consulting Opinions

- When determining optimum routes (including access facilities), technical consultations shall be obtained from committees comprising specialists of various fields for the few candidate proposals, and the problems of original draft shall be complemented through scrupulous reviews of consulting opinions.

- Consultation shall be requested via Technical Assessment Department of Korea Expressway Corporation, and Value Engineering (VE), which is internally executed by Technical Assessment Department, is conducted in parallel with external consultations via Design Consultation Committee comprising a group of specialists.

- After acquiring consultations for route selection, feasibility reviews on the requests from related organizations shall be done while discussing the matter with the related organizations.

- Deliberation shall be followed after the design completion, in order to evaluate the design adequacy. In the deliberation process, the opinions of specialists are also collected via Technical Assessment Department, as in consultation pocess. However, the decisive difference lies on the fact that the original design draft are selected or re-deliberation are decided during this deliberation process.

3. 선형설계

3-1 선형의 구성 요소

- 도로의 선형은 다음과 같은 종류의 선형들의 조합으로 이루어짐

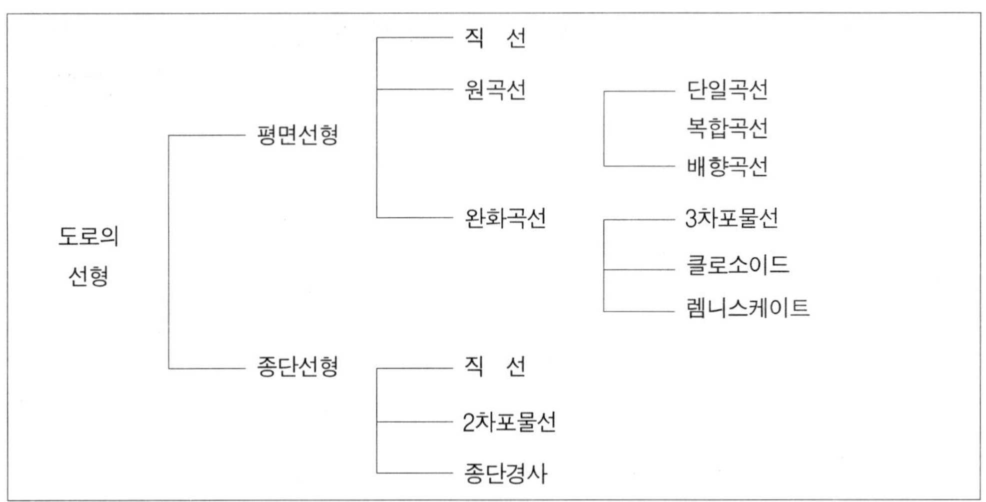

3-2 평면선형

3-2-1 평면곡선(원곡선) 반지름

- 평면 곡선부를 주행하는 자동차가 원운동을 하기 위해서는 구심력이 필요하며, 그에 반하여 평면곡선반지름과 속도에 따라 다음과 같은 원심력이 작용함

$$F=\frac{W}{g}\times\frac{v^2}{R}$$

여기서, F : 원심력(kg) W: 자동차의 총중량(kg)
g : 중력가속도(≒9.8m/s2) v : 자동차의 속도(m/s)
R : 평면곡선반지름(m)

- 윗 식을 노면마찰력 공식을 적용하고, 평면곡선반지름(R)에 대하여 정리하면,

$$R = \frac{v^2}{127(i+f)}$$

여기서, f: 횡방향 미끄럼 마찰계수 i : 편경사(%)

3. Alignment Design

3-1 Elements of Road Alignment

- Road Alignment is composed of the following elements :

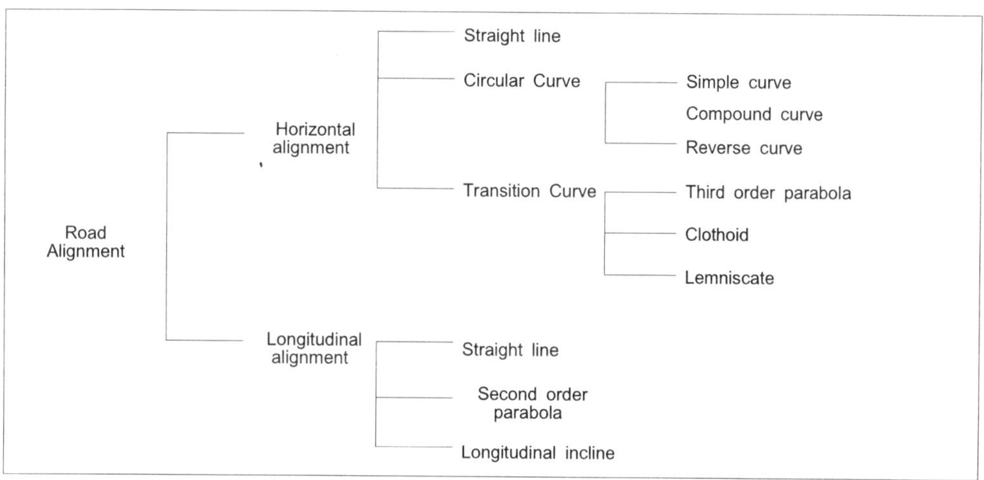

3-2 Horizontal Alignment

3-2-1 Radius of Horizontal Curve (Circular Curve)

- Vehicles moving through horizontal curve require centripetal force to make a circular move. On the other hand, the following centrifugal force is applied depending on the radius of horizontal curve and velocity.

$$F = \frac{W}{g} \times \frac{v^2}{R}$$

Where, F : Centrifugal force(kg)
W : Total weight of a vehicle (kg)
g : Acceleration of gravity (≒9.8m/s2)
v : Velocity of a vehicle (m/s)
R : Radius of horizontal curve (m)

- Applying and combing the surface friction formula with the above, the radius of horizontal curve (R) can be summarized as follows :

$$R = \frac{v^2}{127(i+f)}$$

Where, f : Coefficient of friction in lateral sliding direction
i : Bank ratio (%)

- 차량이 곡선부에서 미끄러지지 않고 쾌적한 주행을 하기 위해서는 원곡선을 차량의 속도(설계속도), 노면상태(마찰계수), 편경사 조건 하에서 곡선반지름이 윗 식에 의한 계산값보다 크도록 적용하여야 하며, 이때의 R값을 최소곡선반지름(Rmin)이라고 정의

- 최소곡선반지름은 설계속도에 따라 다음과 같음

설 계 속 도 (km/h)	횡방향 미끄럼 마찰계수	최 소 평 면 곡 선 반 지 름 (m)					
		최대 편경사 6%		최대 편경사 7%		최대 편경사 8%	
		계산값	적용값	계산값	적용값	계산값	적용값
120	0.10	709	710	667	670	630	630
110	0.10	596	600	560	560	529	530
100	0.11	463	460	437	440	414	420
90	0.11	375	380	354	360	336	340
80	0.12	280	280	265	265	252	250
70	0.13	203	200	193	190	184	180
60	0.14	142	140	135	135	129	130
50	0.16	89	90	86	85	82	80
40	0.16	57	60	55	55	52	50
30	0.16	32	30	31	30	30	30
20	0.16	14	15	14	15	13	15

- 횡방향 미끄럼 마찰계수의 특징
  - 속도가 증가하면 횡방향 미끄럼 마찰계수의 값은 감소
  - 습윤, 빙설상태의 포장면에서 횡방향 미끄럼 마찰계수 값은 감소
  - 타이어의 마모 정도에 따라 횡방향 미끄럼 마찰계수 값은 감소

3-2-2 평면곡선(원곡선) 길이

- 원곡선이 최소평면곡선의 조건을 만족하더라도 곡선구간의 길이가 짧으면 운전자의 핸들조작이 곤란하며, 도로 교각이 작은 경우에는 평면곡선반지름이 실제의 크기보다 작게 보이는 착각을 초래하므로 곡선구간은 일정 길이 이상을 확보하여 주어야 함

- 최소곡선길이는 차량이 곡선구간에서 4초간 주행하는 길이로 하며, 계산방법은 다음 식과 같음

- In order for vehicles to move along the curve without skidding, the value of curve radius higher than the above-mentioned calculation one shall be applied to circular curve under the conditions of velocity (Design Speed), surface state (coefficient of friction), and banked road. The R value is called 'Minimum Curve Radius' (Rmin) in this case.

- Values of Minimum Curve Radius are shown below, in relation to design speed

Design Speed (km/h)	Coefficients of friction in lateral skidding direction	Minimum horizontal curve radius (m)					
		Max bank 6%		Max bank 7%		Max bank 8%	
		Calculated	Applied	Calculated	Applied	Calculated	Applied
120	0.10	709	710	667	670	630	630
110	0.10	596	600	560	560	529	530
100	0.11	463	460	437	440	414	420
90	0.11	375	380	354	360	336	340
80	0.12	280	280	265	265	252	250
70	0.13	203	200	193	190	184	180
60	0.14	142	140	135	135	129	130
50	0.16	89	90	86	85	82	80
40	0.16	57	60	55	55	52	50
30	0.16	32	30	31	30	30	30
20	0.16	14	15	14	15	13	15

- Characteristics of coefficient of friction in lateral skidding direction
  - Coefficient of friction in lateral skidding direction decreases as the velocity increases
  - Coefficient of friction in lateral skidding direction decreases on wet or icy roads.
  - Coefficient of friction in lateral skidding direction decreases depending on the abrasion degree of tires.

3-2-2 Length of Horizontal Curve (Circular Curve)

- Even if the circular curve meets the conditions of minimum horizontal curve, maneuvering a vehicle is still difficult in case the length of the curve is short, and it might also create illusions of the radius of horizontal curve appearing smaller than actual when road intersection angle is small. Therefore, the length of horizontal curve has to be longer than certain value.

- The minimum curve length is calculated to the length of which vehicles move along to curve for four seconds, and the it is calculated as followings.

$$L = v \cdot t = \frac{V}{3.6} \cdot t$$

여기서, L : 평면곡선길이
t : 주행시간(4초)
v,V : 자동차 속도(m/s, ㎞/h)

◎ 위 계산방법을 통하여 설계속도별로 적용하는 곡선의 길이는 다음 표와 같음

설계속도(km/h)	최소평면곡선길이 (m)	
	계산값	적용값
120	133.3	140.0
110	122.2	130.0
100	111.1	110.0
90	100.0	100.0
80	88.9	90.0
70	77.8	80.0
60	66.7	70.0
50	55.6	60.0
40	44.4	50.0
30	33.3	40.0
20	22.2	30.0

3-2-3 완화곡선 및 완화구간

◎ 평면선형상 직선부에서 원곡선으로 진입하기 위해서는 운전자가 곡률의 차이만큼 급한 핸들 조작이 필요하며, 이를 해소하기 위하여 완화곡선(Transition Curve)을 직선과 원곡선 사이에 설치

◎ 완화곡선의 기능

- 두 개의 원곡선 사이나 직선과 원곡선 사이에서 선형의 변화에 대한 원심 가속도의 원활한 감소를 위한 유도 기능

◎ 편경사의 바람직한 변화를 위한 변이구간의 역할

- 일관성 있는 선형에 있어 곡률의 점진적인 변화와 일정한 속도의 주행이 가능하게 함.

- 선형의 시각적 외관에 대한 만족을 창출

$$L = v \cdot t = \frac{V}{3.6} \cdot t$$

Where, L : Length of horizontal curve
t : Running time (4 seconds)
v,V : Speed (m/s, ㎞/h)

- The curve length applied for each design speed are as shown in the following table

Design Speed (km/h)	Minimum length of horizontal curve segment (m)	
	Calculated	Applied Value
120	133.3	140.0
110	122.2	130.0
100	111.1	110.0
90	100.0	100.0
80	88.9	90.0
70	77.8	80.0
60	66.7	70.0
50	55.6	60.0
40	44.4	50.0
30	33.3	40.0
20	22.2	30.0

3-2-3 Transition Curve and Transition Segment

- Drivers will have to apply sudden steering operations by the difference of curvature when entering curved segment from linear segment in horizontal alignment, and transition curve shall be installed between linear and circular curve segments in order to resolve this issue.
- Function of transition curve
  - Transition function for smooth reduction of centrifugal acceleration according to the changes in the curvature between two circular curves or between linear alignment and circular curve.
- Function as a transition segment for the desirable changes of the bank.
  - Transition curve allows progressive changes of curvature in consistent alignments and consistent driving speed.
  - Satisfaction on the visual appearance of the alignments

- 완화곡선의 종류는 여러 가지가 있지만 도로설계에서는 클로소이드 곡선을 사용한다.

- 클로소이드란 두 개의 원곡선 사이 또는 직선과 원곡선 사이에 등속도로 주행하고 있는 자동차의 핸들을 등가속도로 회전시켰을 때의 주행궤적으로 일종의 나선형 곡선을 말한다.

- 클로소이드 설계요소의 산정식과 기하학적 형태는 다음과 같다.
  - 클로소이드 산정식

$$A = \sqrt{RL}$$

여기서, A : 클로소이드 파레메타

R : 곡선반지름(m)

L : 클로소이드 길이(m)

  - 클로소이드 기하 형태

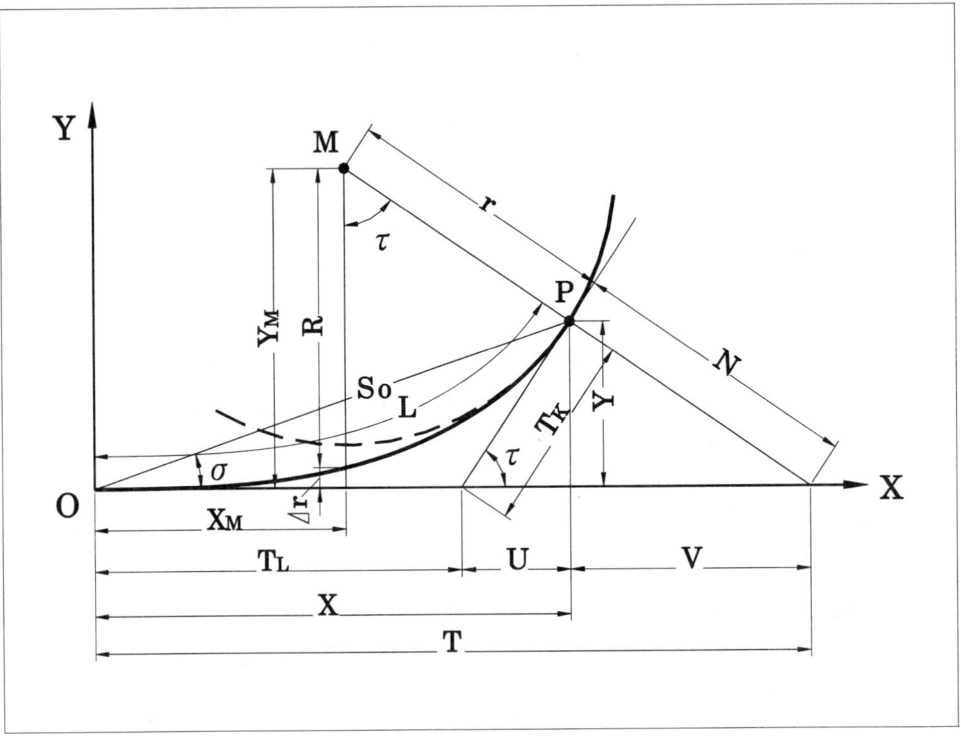

- 완화구간이란 자동차가 원활하게 주행하도록 하기 위하여 주행궤적의 변화에 따라 운전자가 쉽게 적응할 수 있도록 편경사의 변화 또는 확폭량을 설치하는 변이구간

- There are many types of transition curve, and clothoid is used in road design.

- Clothoid is a trace of vehicle, a type of spiral curve, when the steering wheel of a vehicle is turned in a constance acceleration when the vehicle moves in constant velocity between two circular curves or between linear segment and circular curve.

- The following equation is to calculate clothoid and geometrical shape.
  - Clothoid calculation equation

$$A = \sqrt{RL}$$

Where, A : Clothoid parameter
R : Radius of curvature (m)
L : Length of clothoid (m)

  - Geometrical shape of clothoid

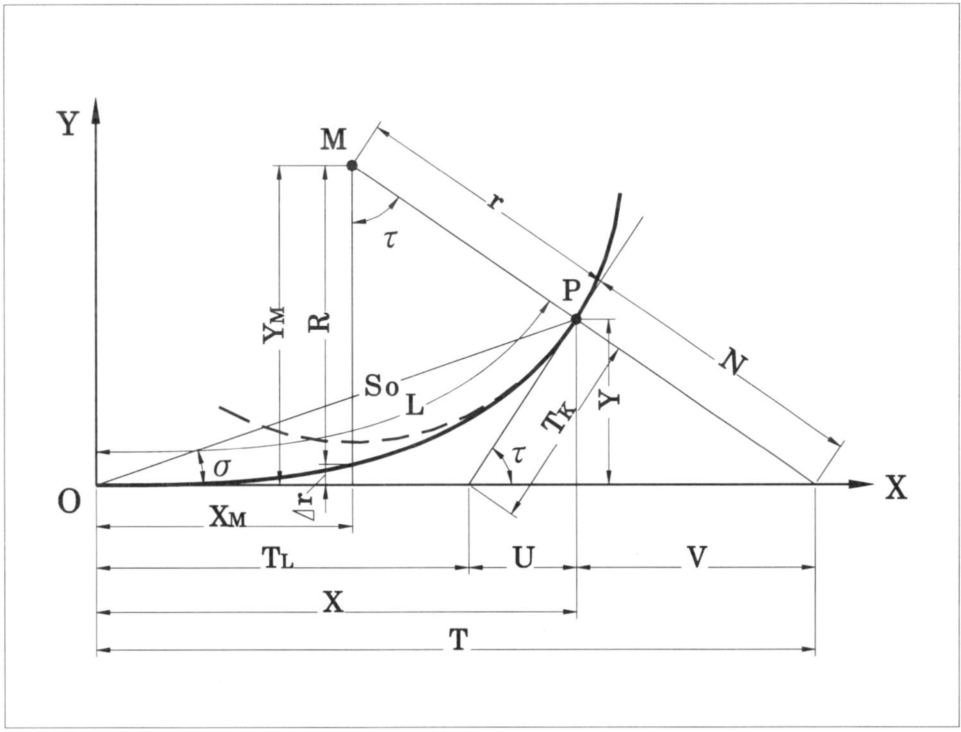

- Easement section is transition section for installation of variations in bank ratio or amount of flare so that drivers can easily adapt to the changes in driving trace for smooth movement of vehicles.

◎ 일반적으로 설계속도 60㎞/h 이상의 도로에서는 완화곡선을 설치하며, 설계속도 60㎞/h 미만의 도로에서는 완화구간만을 둠

• 완화곡선의 길이 및 완화구간 길이 산정식

$$L = v \cdot t = \frac{V}{3.6} \cdot t$$

여기서, L : 완화곡선길이(완화구간 길이)
t : 주행시간(t=2초)
v, V : 주행속도(m/s, km/h)

• 완화곡선 및 완화구간의 길이

설계속도(km/h)	계산값(m)	규정값(m)
120	66.7	70
100	55.6	60
80	44.4	50
70	38.9	40
60	33.3	35
50	27.8	30
40	22.2	25
30	16.7	20
20	11.1	15

◎ 직선과 접하는 원곡선이 일정 크기 이상 큰 경우 완화곡선을 생략할 수 있다. 다만 시각적으로나 주행상, 운전자의 쾌적성을 고려하여 다음 계산값의 3배 정도까지는 완화곡선을 설차하는 것이 바람직

• 완화곡선을 생략할 수 있는 곡선반지름의 산정
- R = 0.064V2

• 완화곡선을 생략할 수 있는 곡선반지름

설계속도(km/h)	계산값(m)	적용값(m)
120	921.6	3,000
100	640.0	2,000
80	409.6	1,300
70	313.6	1,000
60	230.4	700

- In general, transition curve shall be installed for the design speed of 60 km/h or higher, and only transition section will be installed for the design speed of 60 km/h or lower.
  - Equation to calculate the length of transition curve and transition section

$$L = v \cdot t = \frac{V}{3.6} \cdot t$$

Where, L : Length of transition curve (Length of tranbsition section)
t : Running time (t=2 seconds)
v, V : Running speed (m/s, km/h)

  - Length of transition curve and transition section

Design speed (km/h)	Calculated value (m)	Specified value (m)
120	66.7	70
100	55.6	60
80	44.4	50
70	38.9	40
60	33.3	35
50	27.8	30
40	22.2	25
30	16.7	20
20	11.1	15

- Transition curve can be omitted when the size of circular curve connecting to linear section is bigger than certain value. However, it is desirable to install transition curve at least three times bigger than following calculated values considering the visual satisfaction and the driving comfort & convenience.

  - Calculation of radius of curvature that transition curve can be omitted
    - R = 0.064V2

  - Radius of curvature that transition can be omitted

Design Speed (km/h)	Calculated Values (m)	Applied Values (m)
120	921.6	3,000
100	640.0	2,000
80	409.6	1,300
70	313.6	1,000
60	230.4	700

3-2-4 직 선

- 직선은 두 지점을 최단 거리로 연결할 수 있고 곡선부에 비하여 주행 안전성이 높은 편이나 지장물 때문에 긴 직선부로 설계하기가 곤란한 경우가 많으며, 긴 직선부로 설계가 가능할지라도 긴 직선부는 운전자에게 지루함을 주어 졸음운전을 유발시킬 수 있게 되므로 적정 길이를 초과하지 않는 직선구간으로 설계하는 것이 바람직

- 또한, 곡선부 사이의 짧은 직선부는 운전자의 핸들 조작에 불편함을 유발하고, 시각적으로 원활하지 못한 선형이 되므로 가급적 피하여야 함

- 일반적으로 설계에 적용되는 설치위치별 직선길이의 최대·최소값
  - 동일 종단경사 내의 직선 : 설계속도의 20배 이하
  - 배향곡선 사이의 직선 : 설계속도의 2배 이상
  - 동일방향 복합곡선 사이의 직선 : 설계속도의 6배 이상

3-3 종단선형

3-3-1 종단경사

- 종단경사는 동일한 주행상태를 유지될 수 있도록 설계속도에 따라 일정하게 정하여지는 도로의 기하구조 요소 중 하나로서, 설계속도 외에 지형여건 및 오르막 구간에서 가장 영향을 받는 자동차인 트럭의 오르막 능력을 감안하여 결정되며, 적용값은 다음 표와 같음

설계속도 (킬로미터/시간)	최대 종단경사(퍼센트)							
	고속도로		간선도로		집산도로 및 연결로		국지도로	
	평지	산지	평지	산지	평지	산지	평지	산지
120	3	4						
110	3	5						
100	3	5	3	6				
90	4	6	4	6				
80	4	6	4	7	6	9		
70			5	7	7	10		
60			5	8	7	10	7	13
50			5	8	7	10	7	14
40			6	9	7	11	7	15
30					7	12	8	16
20							8	16

3-2-4 Straight Line

- Straight line enables to connect two points in shortest distance, and driving stability is higher than in curved section, but quite often it is not possible to design long linear section due to obstacles. And also, it is desirable to keep the length of linear section within a certain value because long linear section makes the drivers bored and could cause dozing off while driving.

- And also the short linear section between curved sections causes inconvenience of drivers' steering and it does not look smooth visually, so this needs to be avoided if possible.

- In general, the following maximum and minimum lengths of linear section are applied for each installation locations.
  - Straight line within the same incline : lower than one twentieth of design speed
  - Straight line between reverse curve: higher than twice of design speed.
  - Straight line between complex curve in same direction: over six times of design speed

3-3 Longitudinal Alignment

3-3-1 Longitudinal Incline

- Longitudinal incline is one of the geometric components of road determined by the design speed for maintaining the same driving condition, and it is determined by the climb performance of trucks that are affected by climbing section and the lie of the road in addition to the design speed. The following table shows the applied values :

Maximum longitudinal incline (percent)								
Design Speed (km/h)	Expressway		Main roads		Collective road and connecting road		Local road	
	Flat area	Mountain area	Flat area	Mountain area	Flat area	Mountain area	Flat area	Mountain area
120	3	4						
110	3	5						
100	3	5	3	6				
90	4	6	4	6				
80	4	6	4	7	6	9		
70			5	7	7	10		
60			5	8	7	10	7	13
50			5	8	7	10	7	14
40			6	9	7	11	7	15
30					7	12	8	16
20							8	16

3-3-2 종단곡선

- 두 개의 다른 종단경사가 접속될 때는 접속지점을 통과하는 자동차의 운동량 변화에 따른 충격의 완화와 정지시거를 확보할 수 있도록 서로 다른 두 종단경사를 적당한 변화율로 접속시켜야 하며, 이러한 종단곡선은 그 형태에 따라 볼록형과 오목형으로 구분

- 종단곡선반지름(길이)과 종단곡선변화비율

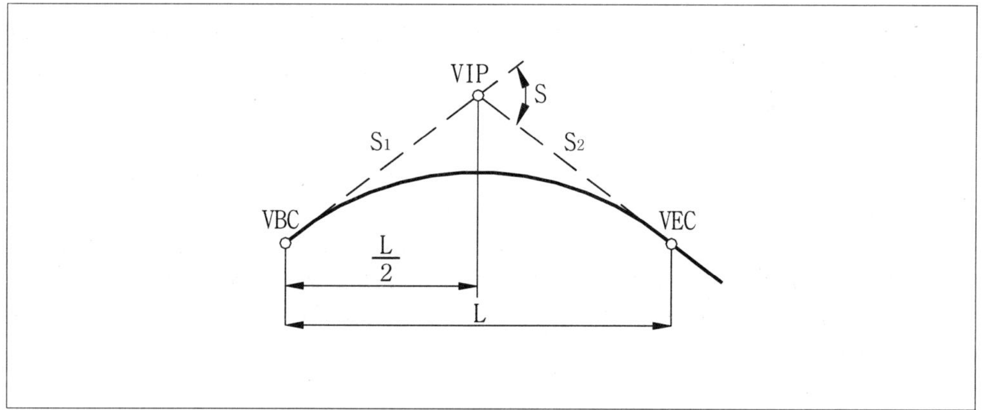

$$R_v = \frac{L}{s_2 - s_1}$$

여기서 R_v : 종단곡선반지름

S_1, S_2 : 종단경사1, 2

$$K = \frac{K_r(=R_v)}{100} = \frac{L}{(s_2 - s_1) \times 100} = \frac{L}{S}$$

여기서 K : 종단곡선 변화비율

L : 종단곡선 길이

S : 종단경사의 대수차($|S_1 - S_2|$) (%)

- 종단곡선길이 설계 적용 시 고려사항
  - 다른 두 경사구간에서 접하는 지점에서 주행하는 자동차의 운동량 변화로 인한 충격을 완화할 수 있도록 충분한 길이를 확보

3-3-2 Longitudinal Curve

- When two sections with different longitudinal incline are connected, two inclines shall be connected with appropriate rate of changes in order to secure the non-passing sight distance and softening of the shock from the changes of momentum of vehicles passing the connection point, and there are two types of longitudinal curves in terms of shape - convex and concave.

- The ratio of radius of longitudinal curve (length) and change of longitudinal curve.

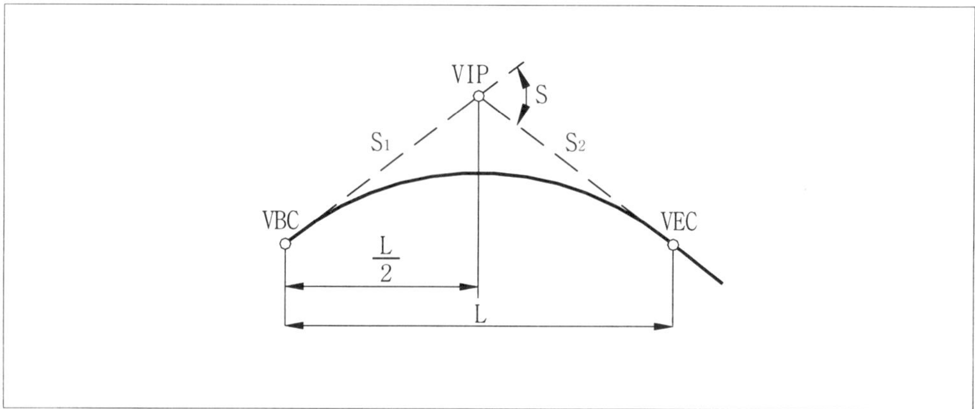

$$R_v = \frac{L}{s_2 - s_1}$$

Where R_V : Radius of longitudinal curvature

S_1, S_2 : Longitudinal incline (Grade) 1, 2

$$K = \frac{K_r(=R_v)}{100} = \frac{L}{(s_2 - s_1) \times 100} = \frac{L}{S}$$

Where K : Ratio of change of longitudinal curve

L : Length of longitudinal curve

S : Absolute difference between longitudinal inclines($|S_1 - S_2|$) (%)

- Design considerations for the length of longitudinal curve
  - To secure enough length to absorb the shock from the changes of the momentum of vehicles passing through the points where two different curves are connected.

- 정지시거를 확보할 수 있는 종단곡선 길이를 확보하여야 함. 통상 볼록형 종단곡선구간에 해당

- 전조등에 의한 야간 투시를 통하여 정지시거의 확보가 가능하여야 함

◎ 설계속도별 최소 종단곡선 변화비율(K)

설계속도 (km/h)	최 소 정지시거 (m)	볼록형 종단곡선의 종단곡선 변화비율(m/%)			오목형 종단곡선의 종단곡선 변화비율(m/%)		
		충격 완화를 위한 K값	정지시거 확보를 위한 K값	적용 K값	충격 완화를 위한 K값	전조등에 의한 정지시거 확보를 위한 K값	적용 K값
120	280	40.0	203.6	200.0	40.0	71.3	70.0
110	250	33.6	162.3	160.0	33.6	62.8	60.0
100	200	27.8	103.8	100.0	27.8	48.8	50.0
90	170	22.5	75.1	75.0	22.5	40.4	40.0
80	140	17.8	50.9	50.0	17.8	32.1	35.0
70	110	13.6	31.4	30.0	13.6	24.0	25.0
60	85	10.0	18.7	20.0	10.0	17.3	20.0
50	65	6.9	10.9	10.0	6.9	12.2	12.0
40	45	4.4	5.2	5.0	4.4	7.3	7.0
30	30	2.5	2.3	3.0	2.5	4.0	4.0
20	20	1.1	1.0	1.0	1.1	2.1	2.0

3-4 선형설계 방법

3-4-1 평면선형의 설계

◎ 평면선형 설계는 기본설계 시 선정된 기본선형에 근거를 두고 정밀도가 높은 1/1,200 지형도에 지형지물에 대한 선형의 적합 여부, 기준점(control point)에 대한 대책 등을 기본적으로 검토하고, 도로 중심선을 최종 결정하는 작업

◎ 평면선형은 주변 지형과 조화를 이루어야 함

◎ 도로 교각이 작은 경우에는 곡선반지름이 실제보다 작게 보이는 착각을 방지하기 위하여 충분한 곡선반지름을 확보

◎ 평면곡선 사이의 짧은 직선은 가급적 두 평면곡선반지름을 크게 하여 짧은 직선부를 삭제하거나 충분한 직선길이를 확보

- The length of longitudinal curve to secure stopping sight distance shall be secured. It is typically applied to convex type longitudinal curve section.

- Stopping sight distance shall be secured through the seeing-through with head lights during night time.

The ratio of changes of minimum longitudinal curve (K) for each design speed.

Design Speed (km/h)	Minimum stopping sight distance (m)	The ratio of changes of longitudinal curve convex type longitudinal curve (m/%)			The ratio of changes of longitudinal curve concave type longitudinal curve (m/%)		
		K value to absorb the shock	K value to secure the stopping sight distance	Applied K value	K value to absorb the shock	K value to secure the stopping sight distance by head lights	Applied K values
120	280	40.0	203.6	200.0	40.0	71.3	70.0
110	250	33.6	162.3	160.0	33.6	62.8	60.0
100	200	27.8	103.8	100.0	27.8	48.8	50.0
90	170	22.5	75.1	75.0	22.5	40.4	40.0
80	140	17.8	50.9	50.0	17.8	32.1	35.0
70	110	13.6	31.4	30.0	13.6	24.0	25.0
60	85	10.0	18.7	20.0	10.0	17.3	20.0
50	65	6.9	10.9	10.0	6.9	12.2	12.0
40	45	4.4	5.2	5.0	4.4	7.3	7.0
30	30	2.5	2.3	3.0	2.5	4.0	4.0
20	20	1.1	1.0	1.0	1.1	2.1	2.0

3-4 Alignment Design Method

3-4-1 Horizontal Alignment Design

Horizontal alignment design is a process of reviewing the adequacy of alignment to regional characteristics with high-precision topographical map of 1/1,200 scale based on the basic alignment selected during the basic design, to review the basic measure on control point, and to determine final road center line.

Horizontal alignment has to be balanced with the surrounding topographical conditions.

When the angle of intersection is small, enough radius of curvature must be secured to prevent the illusion of radius of curvature appearing smaller than actual.

Short linear section between horizontal curve must be either removed by extending the radius of two curvatures, or by well secured length of linear section.

- 두 원곡선을 같은 방향으로 복합시킬 경우에는 큰 원의 반지름이 작은 원의 반지름에 1.5배 이하가 되도록 하는 것이 바람직

- 직선과 원곡선 사이에 클로소이드를 삽입할 때, 클로소이드의 파라미터와 원곡선 반지름과의 관계를 가능한 한 R≥A≥R/3이 되도록 함

- 두 클로소이드가 반대방향으로 접속된 선형인 경우에는 두 클로소이드의 파라미터는 가급적 같도록 하며, 같지 않을 때에는 큰 파라미터가 작은 파라미터의 두 배 이하가 되도록 하여야 함

- 직선을 낀 두 평면곡선부가 반대방향으로 설치될 경우에 부득이 짧은 직선부를 두 곡선사이에 설치하여야 할 때에는 직선의 길이는 다음 조건을 만족하도록 하여야 함

$$\ell \leq \frac{A_1 + A_2}{40}$$

여기서 ℓ : 두 평면곡선 사이의 직선길이(m)

A_1, A_2 : 클로소이드의 파라미터

- 장대교 및 긴 고가 구조물 구간과 같이 건설비가 고가인 구조물이 연결되는 경우에는 시공의 용이성을 고려해서 직선으로 하는 것이 바람직

- 구조물 설계 · 시공의 용이성 및 주행의 안전성을 위하여 교량 및 터널구간 내에서는 직선과 곡선이 복합되지 않도록 함

- When two circular curved sections are connected, it is desirable that the larger circle radius is no bigger than 150% of smaller circle radius.

- When clothoid is inserted between linear section and circular curve, the relationship between clothoid parameter and radius of circular curve shall be intended to meet R≥A≥R/3 as much as possible.

- When two clothoids in opposite direction are connected together, their parameters shall be the same if possible, and the bigger parameter shall be smaller than twice of smaller parameter if they cannot be identical.

- When two horizontal curves with linear section in the middle are installed in opposite direction, and if linear section has to be installed in between, then the length of linear section shall meet the following condition :

$$\ell \leq \frac{A_1+A_2}{40}$$

Where ℓ : Length of linear section between two horizontal curves (m)
A_1, A_2 : Clothoid parameters

- It is desirable to apply linear section for the purpose of constructability on case of connecting high-cost structures such as long bridge and long elevated section.

- For the ease of design and construction of structures and also for the driving safety, lines and curves shall not be mixed in the bridge and tunnel.

3-4-2 종단선형의 설계

- 선형은 지형에 적합하고 원활한 것이어야 함

- 짧은 구간에서 오르내림(凹凸)이 많은 선형은 바람직하지 않음

- 운전자 관점에서 중간이 푹 패여 도로가 시각적으로 끊겨 보이는 선형은 피하여야 함

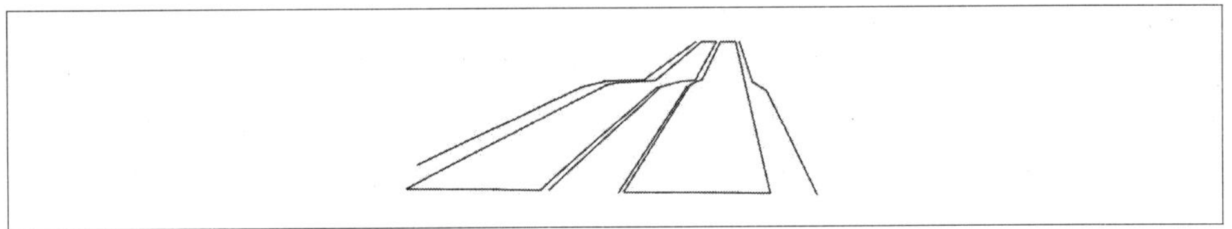

- 오목(凹)부분의 종단곡선길이가 짧은 선형은 도로가 급격하게 꺾여보이므로 충분히 긴 종단곡선을 설치하여 시각적으로 원활한 선형을 얻을 수 있도록 하여야 함

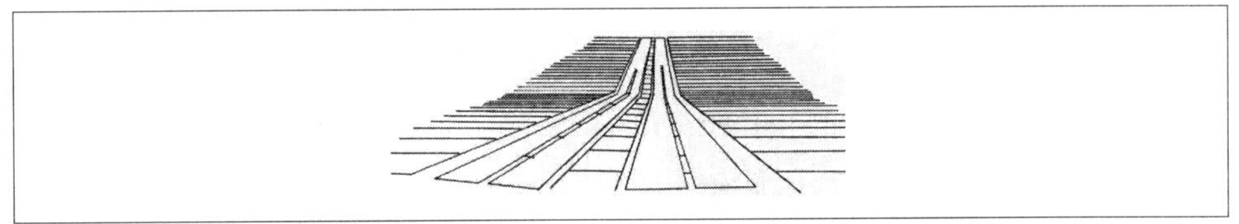

- 같은 방향으로 굴곡하는 오목(凹)부분의 두 종단곡선 사이에 짧은 직선 구간을 두는 것은 피하여야 한다. 이 경우 두 종단곡선을 포함하는 큰 종단곡선을 설치하여야 함

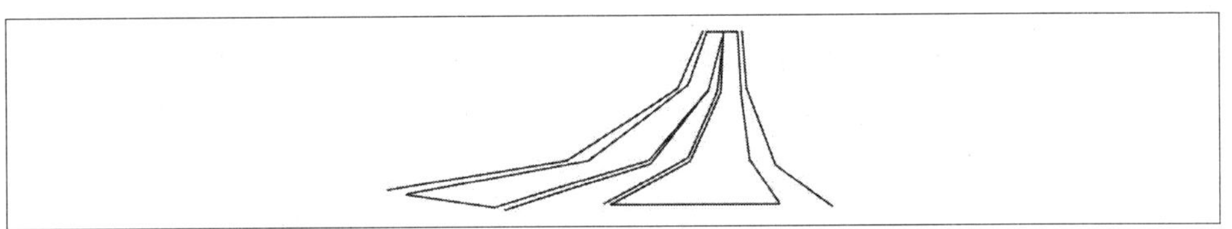

- 길이가 긴 상향 경사구간의 정상 부근에서는 경사를 비교적 완만하게 하는 편이 바람직

- 경사 변화가 작을 때의 종단곡선 변화 비율은 될 수 있는 대로 크게 하여야 함

- 최소 종단경사는 노면의 배수를 고려하여 0.3~0.5%로 하는 것이 바람직

- 절토부의 종단곡선 오목부는 배수에 문제가 발생할 수 있으므로 가급적 피하여야 함

3-4-2 Longitudinal Alignment Design

- Alignment shall be harmonious and appropriate in connection with the characteristics of the corresponding region.
- It is not desirable to have alignments with many up and down (凹凸) in short section.
- The alignment that visually appears disconnected due to deep groove in the middle of the road shall be avoided.

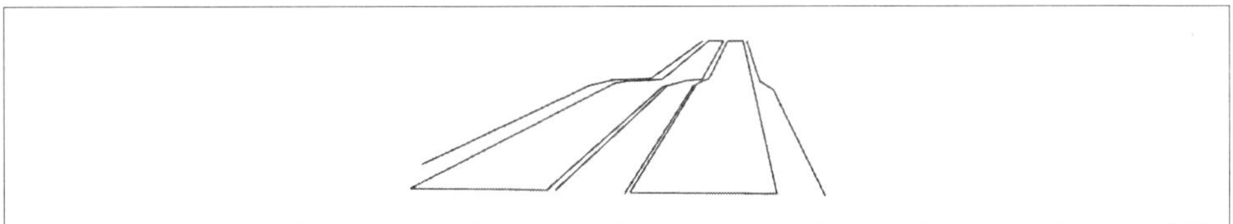

- The alignment with short length of longitudinal curve of concave section (凹) tends to appear sharply curved in visual point of view, therefore long longitudinal curve shall be installed to make a visually smooth alignment.

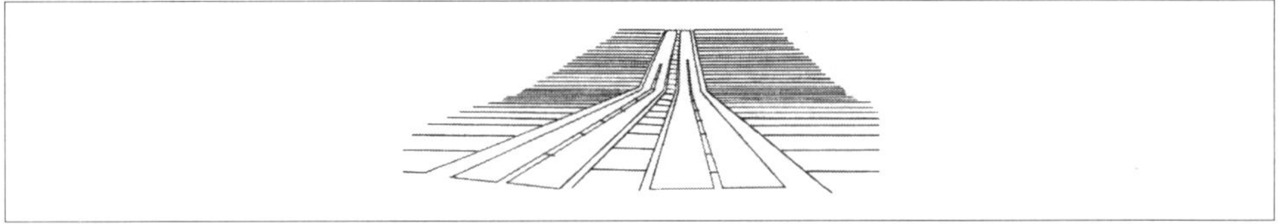

- Short linear section between two longitudinal curves of concave type(凹) heading for the same direction shall be avoided. In this case, big longitudinal curve including two longitudinal curves shall be installed.

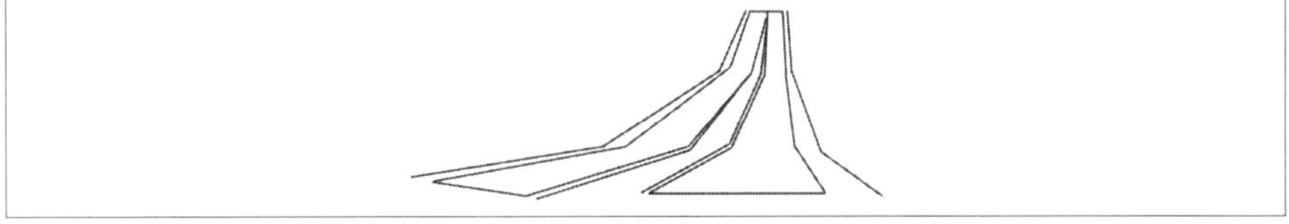

- It is desirable to have a gentle slope around the top of long upward incline section.
- The ratio of change of longitudinal curve in case of small changes of slope shall be as big as possible.
- Desirable value of minimum longitudinal incline is from 0.3 to 0.5% considering the drainage on road surface.
- Longitudinal curve of concave type in cutting area shall be avoided due to potential drainage problems.

3-4-3 평면선형과 종단선형의 조합

- 도로의 선형 설계는 노선 계획으로 시작해서 평면선형 설계, 종단선형 설계로 이어지고, 최종적으로 도로 환경과 조화를 이룬 평면선형과 종단선형의 결합(입체선형)으로 완료된다. 따라서, 평면선형과 종단선형의 조합은 실제로 도로를 주행하는 운전자의 시각인 3차원 투시도를 이용하여 검토

- 평면직선부의 종단선형 - 긴 연장의 일정한 경사구간에서 국부적인 작은 굴곡을 피하도록 함

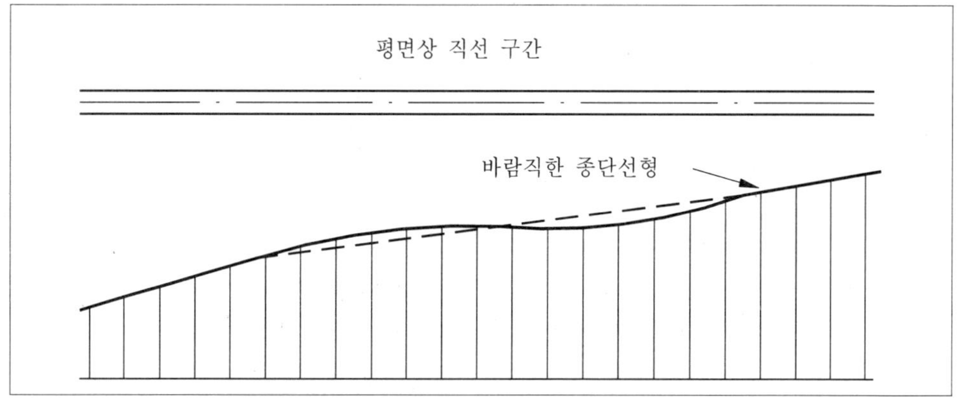

- 평면곡선부의 종단선형 - 짧은 구간의 둥근언덕 모양의 굴곡을 피하고 긴 구간에 걸쳐 종단경사를 일정하게 하여야 함

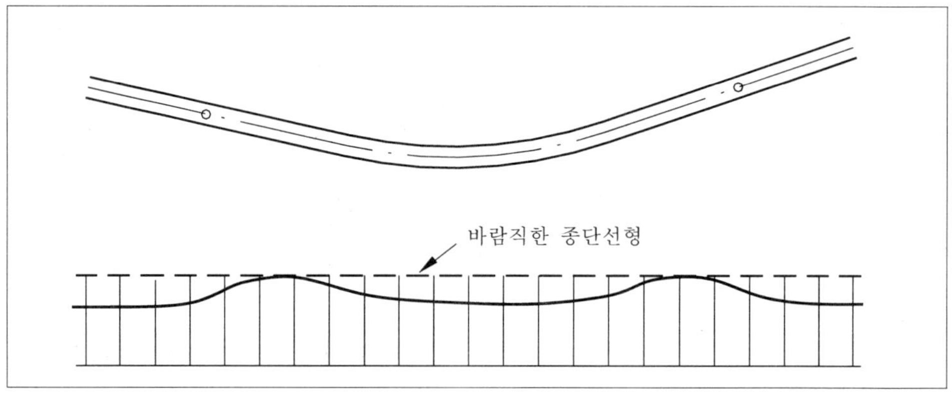

3-4-3 Combination of horizontal alignment and longitudinal alignment

- Road alignment design begins with route planning, through the horizontal and longitudinal alignment design. It is completed with combination of horizontal and longitudinal alignments (three dimensional alignment) that are in balance with road environment. Therefore, the combination of horizontal and longitudinal alignments shall be reviewed with the three dimensional projection diagram from the drivers' points of view.

- Longitudinal alignment of horizontal linear section - local and small irregulari ti es shall be avoided in long and constant incline section.

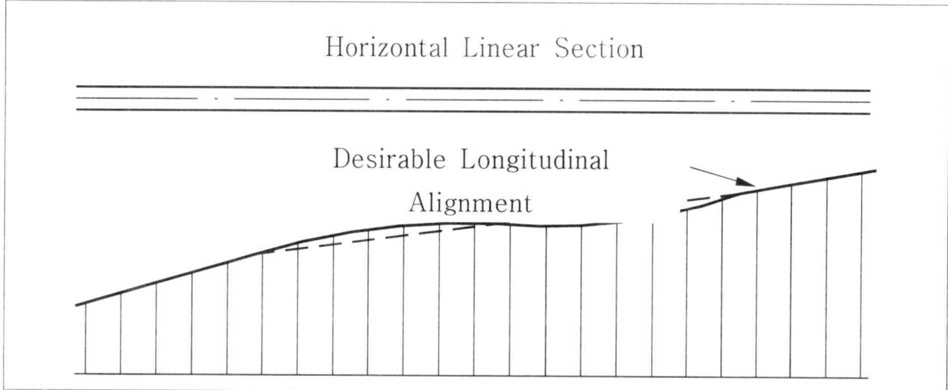

- Longitudinal alignment of horizontal curve section - Irregularity on short section of round-hill shape shall be avoided, and the longitudinal incline over the long section shall be maintained constant.

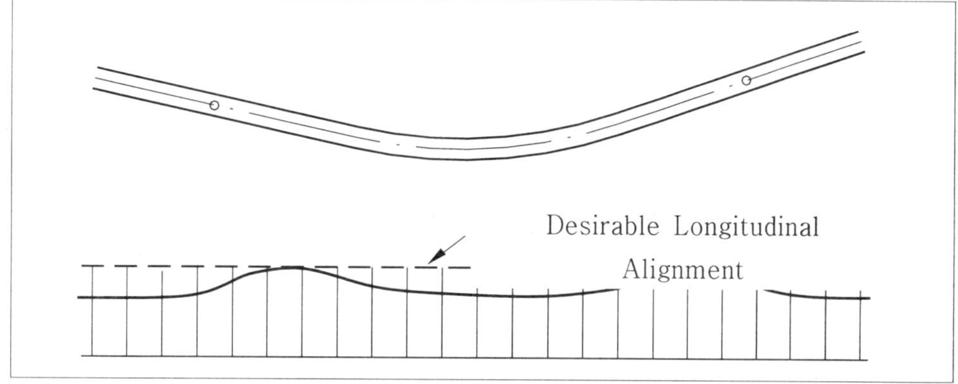

- 두 평면곡선 사이의 짧은 직선구간과 종단선형의 정점부에서 반대방향의 평면곡선 설치를 피할 것

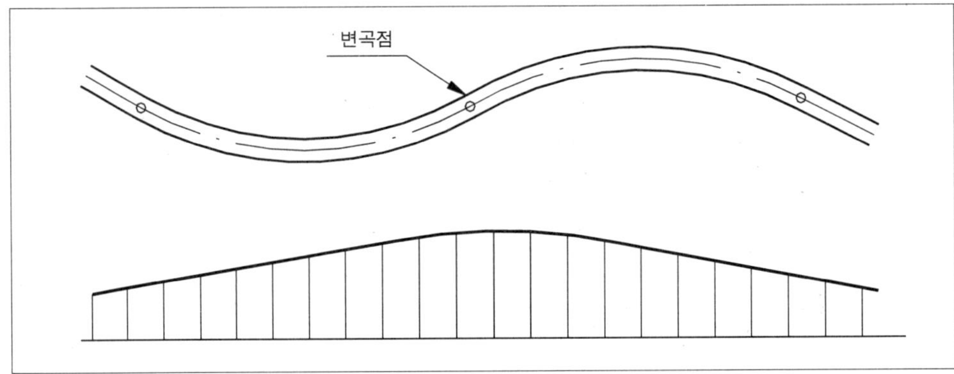

- 오목형 종단곡선상의 저점부에 평면곡선의 변곡점 설치를 피할 것 - 노면배수가 원활하지 못한 경우가 발생

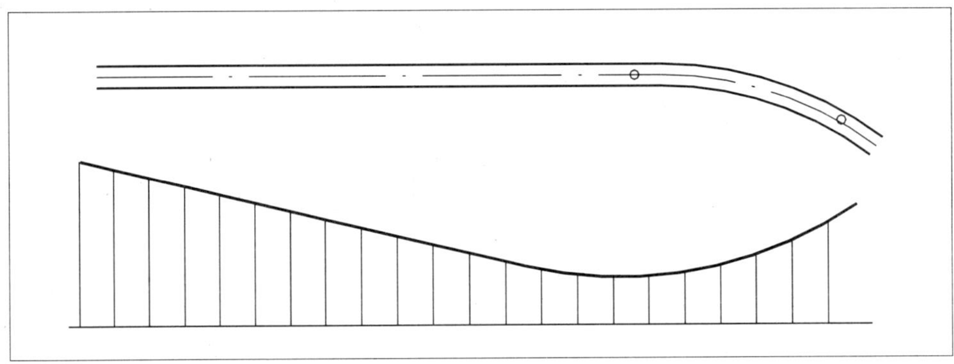

- 불연속 효과 - 언덕 등에 의하여 도로의 일부가 보이지 않아서 도로가 불연속된 것처럼 보임.

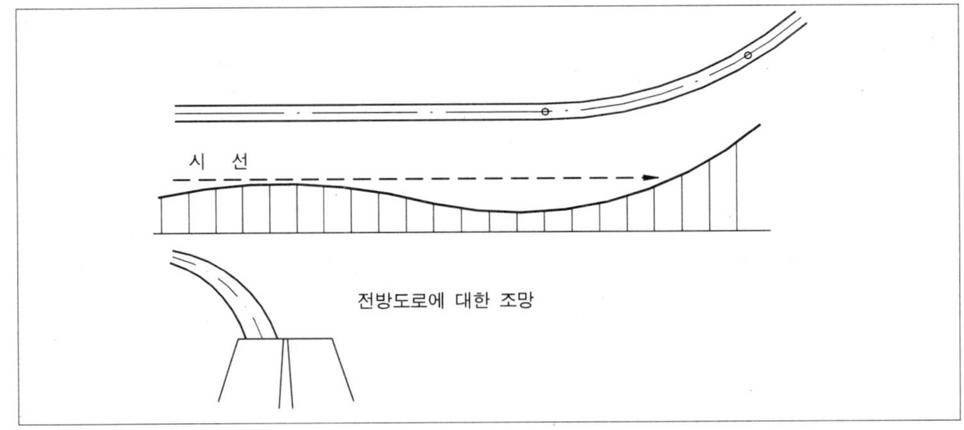

- The short linear section between two horizontal curves and horizontal curve in opposite direction at the top of longitudinal alignment shall be avoided.

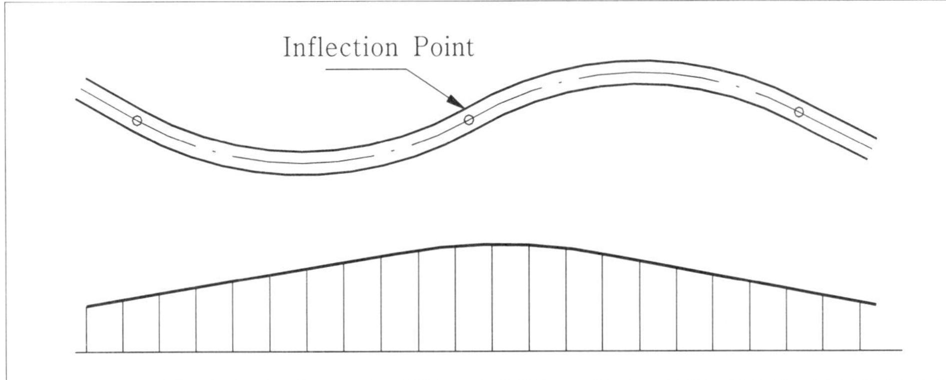

- Installation of inflection point to horizontal curve at the low end of concave type longitudinal curve shall be avoided - Drainage might not be smooth.

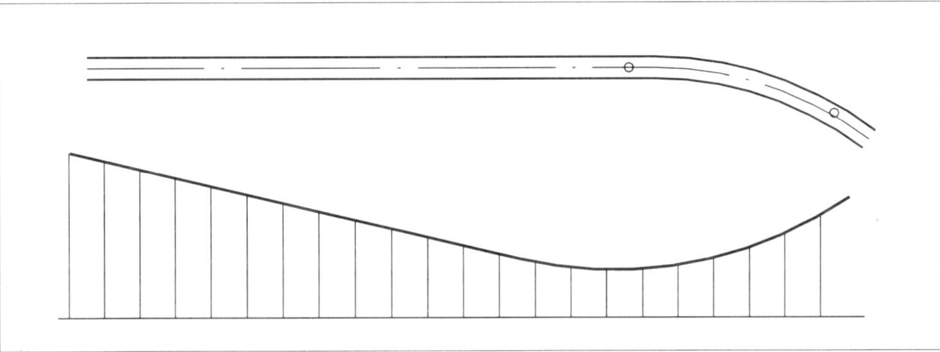

- Discontinuity effect - Parts of the road is not visible because of hills and other obstacles and consequently the road looks discontinuous.

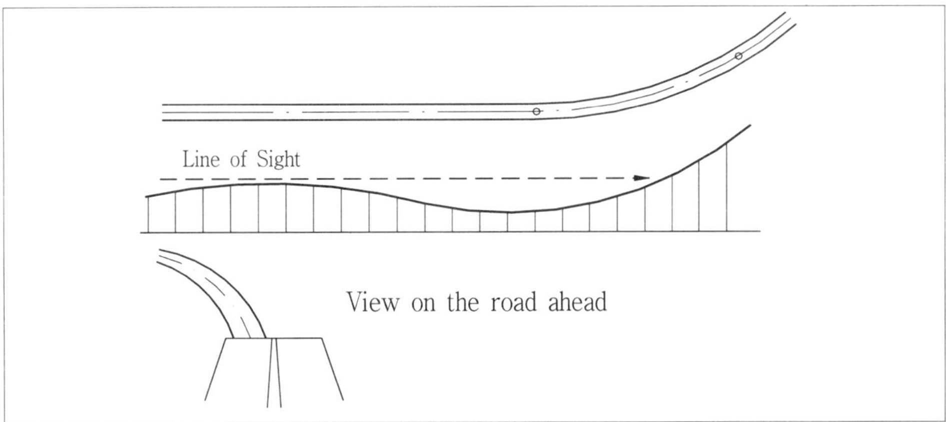

◉ 긴 평면 직선부 구간에서 종단 곡선의 반복된 굴곡은 피하는 것이 바람직

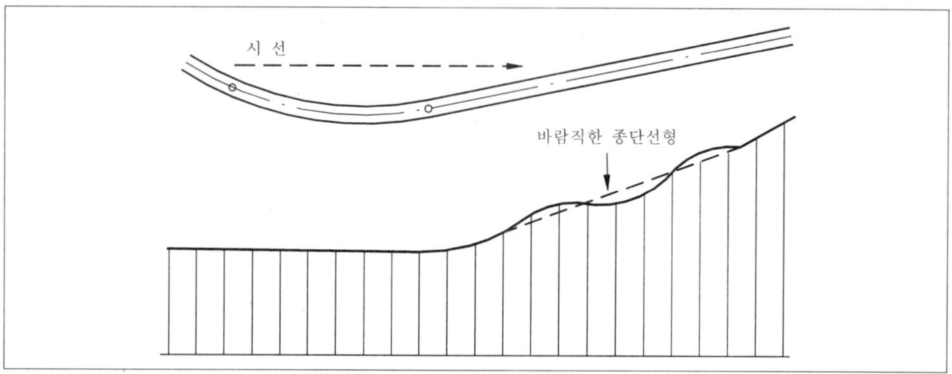

◉ 평면곡선과 종단곡선이 같은 방향 또는 다른 방향으로 대응하여 균형된 도로의 경우에 시각적 효과가 좋음

【평면선형과 종단선형의 조합 예】

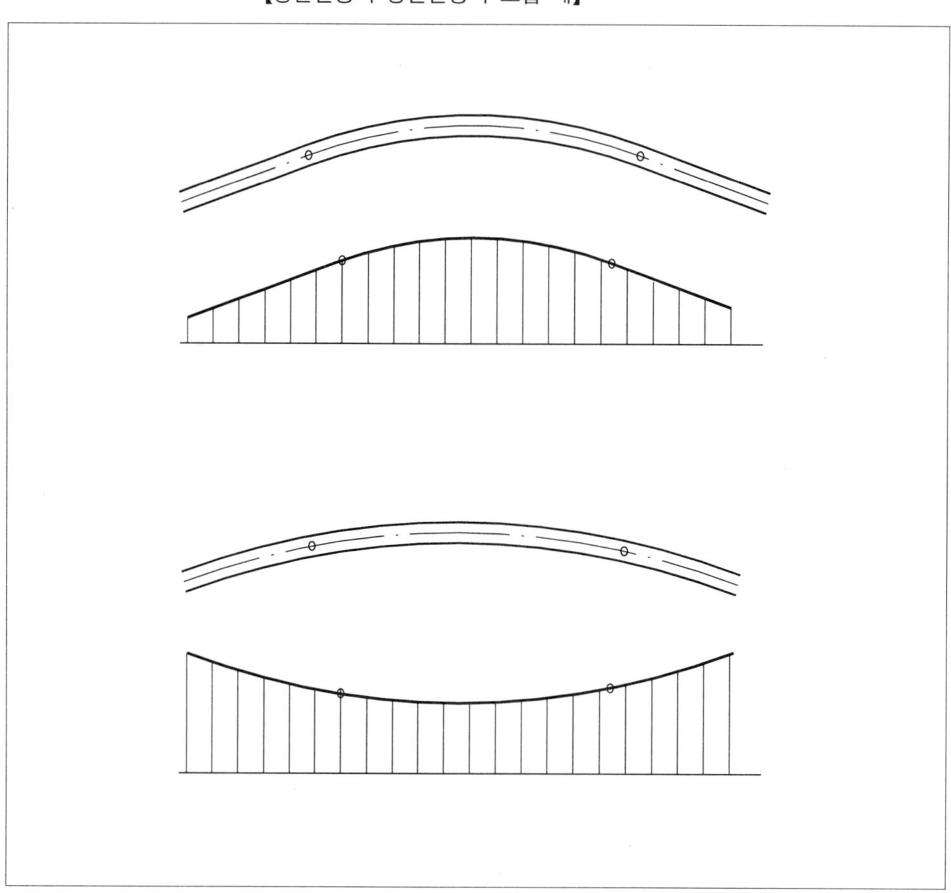

- It is desirable to avoid repetitive irregularities of longitudinal curve in long horizontal linear section.

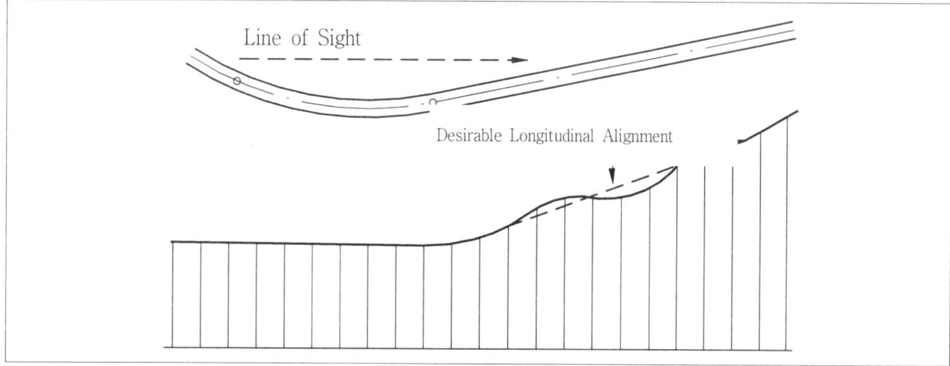

- Visual effects are good if the road is well balanced with horizontal curve and longitudinal curve aligned in both same and different directions.

【Example of the combination of horizontal and longitudinal alignment】

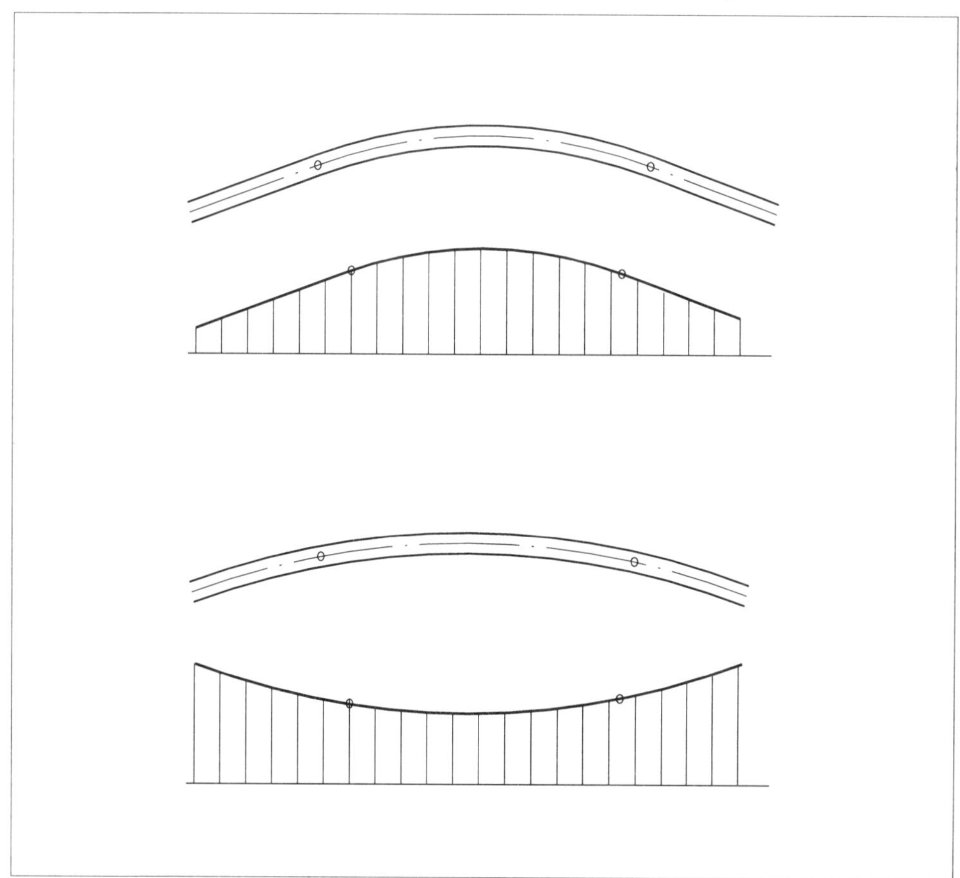

- 평면곡선반지름의 교각이 작을 때에는 작은 평면곡선반지름보다 큰 곡선반지름을 설치하면 시거가 양호해짐

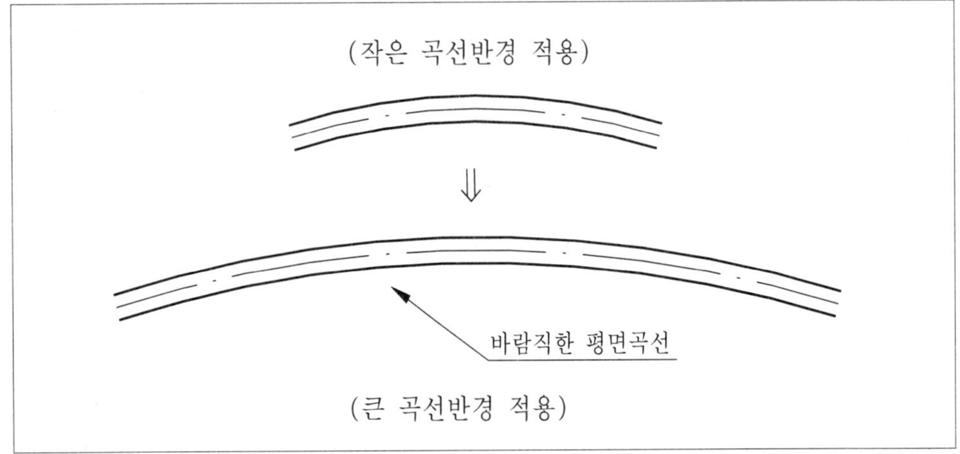

- 원활한 평면선형 - 긴 길이의 평면 직선부와 평면곡선반지름이 작은 것의 조합은 원활하지 못하므로 직선부와 곡선부 사이에 완화구간을 설치하고 큰 평면곡선반지름을 적용하여 원활한 평면선형으로 설계

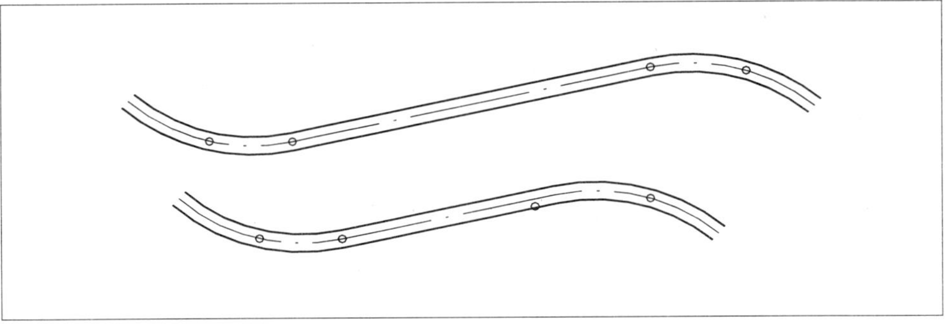

- 볼록형 종단곡선의 정점부 또는 오목형 종단곡선의 저점부에 급한 평면곡선반지름의 삽입은 피할 것

- 볼록형 종단곡선의 정점부 또는 오목형 종단곡선의 저점부에 배향곡선의 변곡점을 두는 것은 피할 것

- 하나의 평면곡선 내에서 종단선형이 볼록과 오목을 반복하는 것은 피할 것

- 같은 방향으로 굴곡하는 두 곡선 사이에 짧은 직선의 삽입은 피할 것

- When the angle of intersection with horizontal curve radius is small, installation of curvature radius bigger than the horizontal curvature radius will enhance the stopping sight distance.

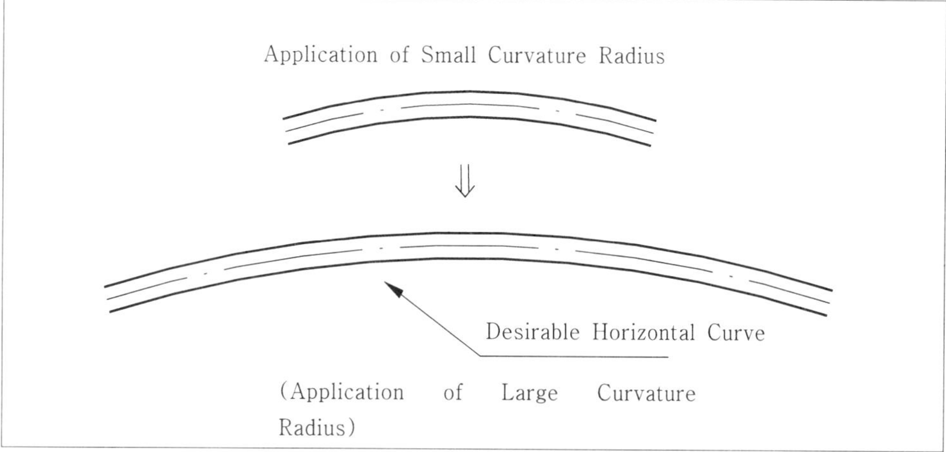

- Smooth horizontal alignment - As the combination of long length of horizontal linear section and small radius of horizontal curve is not smooth, transition curve shall be installed between the linear section and curve. Also large radius of horizontal curve shall be applied to design smooth horizontal alignment.

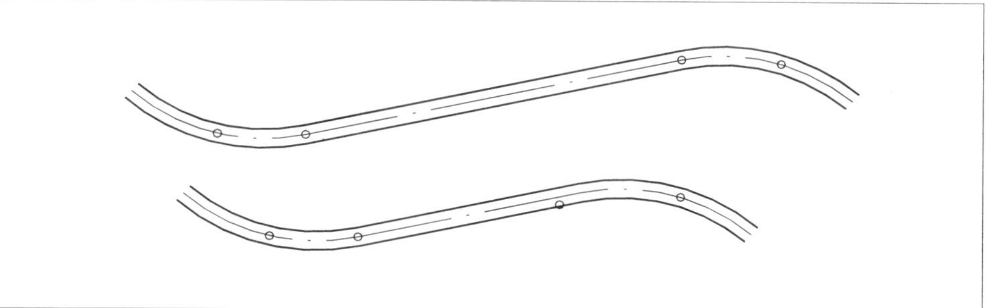

- Insertion of steep radius of horizontal curve around the top of convex type longitudinal curve or low end of concave type longitudinal curve shall be avoided.

- Installation of the inflection point of reverse curve around the top of convex type longitudinal curve or low end of concave type longitudinal curve shall be avoided.

- Repeating convex and convex longitudinal alignment within one horizontal curve shall be avoided.

- Insertion of short linear section between two curves in the same direction shall be avoided.

Chapter 7

Access Facilities

1. 개 요

- 출입시설이란 도로의 진입과 진출을 목적으로 설치되는 시설로 평면교차 시설과 입체교차 시설이 있으며, 구분방법에 따라 다음과 같이 구분된다.

구분방법	구 분 내 용
일반적 구분	평면교차, 입체교차
교차도로의 등급	분기점(JCT), 인터체인지(I.C)
동선처리방법	완전입체교차, 불완전입체교차, 엇갈림형식 직결형, 준직결형, 루프형

- 본 매뉴얼에서는 고속도로에 거의 적용되지 않는 평면교차에 대해서는 다루지 않도록 한다.

2. 설치 계획

2-1 절 차

구 분		내 용
기본계획	배치계획	•개략적인 설치 위치 결정
	위치선정	•노선 선정 관련 개략 계획 •설치계획에 대한 추가, 삭제 여부 검토 •개략적 출입교통량의 추정 •교통량 분배계획
	형식 결정	•접속도로 결정 •구체적 설치 위치의 확정 •이용교통량의 상세한 추정 •출입시설 형식 검토 •개략적인 공사비 산출
설계	기본설계	•기본 선형결정 •시설 배치계획 •개략공사비 산정 •접속도로의 정비계획
	실시설계	•토공, 배수, 구조물, 포장, 교통관리시설, 건축 등의 결정 •보상비의 산정 •용지 및 지장물 현황조사

1. Overview

- Access facility is installed for the purpose of entering and exiting the road, and there are 'at-grade' & 'grade-separated' intersection facilities which are classified as followings :

Classification Method	Classification Details
General classification	At-grade Crossing, Grade-Separated Crossing
Grade of crossings	Junction (JCT), Interchange (I.C)
Traffic line process method	Highway Grade Separation, Partial Grade Separation, Weaving Type, Directional Interchange, Semi Directional Interchange, Loop Type

- The present manual does not cover At-grade Crossing, which is rarely applied to expressway system.

2. Installation Plan

2-1 Procedure

Classification		Details
Basic Plan	Placement plan	• To determine the rough installation location
	Position selection	• Draft plan related to route selection • To review items to be added and/or deleted in installation plan • Rough estimate of access traffic volume • Traffic volume distribution plan
	Type decision	• To determine the connecting road • To determine the detailed installation location • Detailed estimate of traffic volume • To review the type of access facility • To estimate the rough construction cost
Design	Basic Design	• Basic alignment determination • Facility placement plan • Calculation of estimated construction cost • Maintenance of connection road
	Execution Design	• Determination of earthwork, drainage, structure, pavement, traffic control facilities, building, etc • Calculation of land acquisition cost • Survey on land and obstacles

2-2 배치기준

- 국도 등 주요 도로와 교차하거나 가까운 지점
- 인구가 3만명 이상인 도시 부근이나 출입시설의 세력권 인구가 5만~10만명 정도가 되도록 배치

【도시 인구에 따른 출입시설의 표준 설치 수】

도 시 인 구 (명)	1노선당 출입시설의 표준 설치 수
100,000 미만	1
100,000 ~ 300,000	1 ~ 2
300,000 ~ 500,000	2 ~ 3
500,000 이상	3

- 항만, 비행장, 유통시설, 또는 주요 관광지 등으로 통하는 주요도로와 교차하거나 가까운 지점
- 출입시설의 출입 교통량이 30,000대/일 이하가 되도록 배치
- 출입시설간의 간격이 최소 2km, 최대 30km가 되도록 배치

【출입시설간 표준 간격범위】

지 역	표 준 간 격(km)
대도시 도시고속도로	2 ~ 5
대도시 주변 주요 공업지대	5 ~ 10
소도시가 존재하고 있는 평야지대	15 ~ 25
지방 촌락, 산간지	20 ~ 30

- 본선과 출입시설에 관한 총 비용 대 편익비가 최대가 되도록 배치

2-2 Placement Standard

- Crossing point or near to crossing point with main roads such as National Road

- Access facilities shall be placed near a city with population over 30,000 habitants, or the population of the area affected by the access facilities needs to be around 5,000 ~ 10,000 habitants.

【Standard number of access facilities depending on the city population】

City population (habitants)	Standard number of access facility installation per route
Below 100,000	1
100,000 ~ 300,000	1 ~ 2
300,000 ~ 500,000	2 ~ 3
Over 500,000	3

- Access facility shall be placed at or near to the crossing with the major roads leading to ports, airports, distribution facilities, major tourist attractions, etc.

- Traffic volume per access facility shall be lower than 30,000 vehicles per day.

- The distance between access facilities shall be at least 2km, and at most 30 km.

【Standard distance between access facilities】

Region	Standard Distance (km)
City highway in big cities	2 ~ 5
Major industrial areas around big cities	5 ~ 10
Plain areas around small cities	15 ~ 25
Small villages at mountain areas	20 ~ 30

- The placement shall be done to maximize the merit for given total cost of construction of main lines and access facilities.

2-3 위치선정

- 출입시설의 위치 선정은 지역계획 및 광역적인 교통운용계획과의 관련을 바탕으로 입지조사, 접속도로의 조건, 타시설과의 간격, 본선 선형, 관리·운영과의 관계 등을 고려하여 계획

- 입지조사 : 인터체인지 위치를 결정함에 있어서는 교통상의 조건, 사회적 조건, 자연적 조건 등에 대하여 충분한 입지조사를 실시함

- 접속도로의 조건
  - 충분한 도로 교통 용량 확보
  - 주요 교통발생원(시가지, 공항, 항만, 관광지 등)과의 접근성
  - 출입 교통량의 기존 도로망 이용 및 배분의 적정성

- 타 시설과의 간격 : 인접 시설물과의 간격은 아래의 거리 이상이어야 하나, 부득이 표의 간격을 확보할 수 없는 경우에는 충분한 안전시설(표시판 등)을 설치

시설의 명칭	최소간격 (km)	기 준 점
인터체인지 상호간	2	I.C 중심간 간격
인터체인지와 휴게소	2	I.C 중심과 휴게소 중심간 간격
인터체인지와 주차장	1	I.C 중심과 주차장 중심간 간격
인터체인지와 버스정류장	1	I.C 중심과 버스정류장 중심간 간격

※ 터널 출구부에서 인터체인지 변이구간의 시점까지의 소요이격 거리 산정

$$L = l_1 + l_2 + L_3 = \frac{V \cdot t_1}{3.6} + \frac{V \cdot t_2}{3.6} + \frac{V \cdot t_3 \cdot (n-1)}{3.6}$$

여기서, L : 소요이격거리(m)　　t_1 : 조도순응시간(3sec)

l_1 : 조도순응거리　　t_2 : 인지반응시간(4sec)

l_2 : 인지반응거리　　t_3 : 차로변경시간(차로당 10sec)

l_3 : 차로변경거리　　n : 차로 수

V : 설계속도(km/h)

2-3 Location Selection

- The access facility location shall be planned while considering survey results of site conditions, connection road conditions, distance from other facilities, main road alignments, relation with maintenance and operation based on the local development plan and relation with broad traffic operation plan.

- Survey of site conditions : Sufficient site survey shall be done in order to find out the traffic, socialand natural conditions when determining the location of interchange.

- Conditions of connection roads
  - To secure enough traffic volume capacity of the roads
  - Proximity to the major traffic volume creating facilities (downtown, airport, port, tourist attractions, etc)
  - Suitability of usage and distribution of the existing road networks for the access traffic volumes.

- Distance from other facilities : The distance from nearby facilities shall be longer than the following values. If the specified distance cannot be achieved for any unavoidable reasons, the safety facilities (traffic signs) shall be installed.

Facilities	Minimum distance (km)	Reference point
Between interchanges	2	Distance between center point of I.C
Between interchange and service area	2	Distance between the center of I.C and center of service area
Between interchange and parking lot	1	Distance between the center of I.C and center of parking lot
Between interchange and bus station	1	Distance between the center of I.C and center of bus station

※ Calculation of the required distance from tunnel exit to the beginning of interchange transition section

$$L = l_1 + l_2 + L_3 = \frac{V \cdot t_1}{3.6} + \frac{V \cdot t_2}{3.6} + \frac{V \cdot t_3 \cdot (n-1)}{3.6}$$

여기서, L : Required distance(m)

t_1 : Illumination adaptation time(3sec)

l_1 : Illumination adaptation distance

t_2 : Recognition reaction time(4sec)

l_2 : Recognition reaction distance

t_3 : Lane change time (10sec per lane)

l_3 : Lane change distance

n : Number of lanes

V : Design velocity(km/h)

3. 인터체인지

3-1 인터체인지의 구성

- 인터체인지의 종류마다 형식을 규정하고 교통 운영상의 차이를 초래하는 기본적인 요소는 동선결합이며, 이 동선결합은 기본 동선 결합, 연결로결합, 접속단 결합으로 구성

- 기본 동선 결합(두 개 교통류의 상호 결합관계)

분 류(Diverging)	합 류(Merging)	엇갈림(Weaving)	교 차(Crossing)

3. Interchange

3-1 Configuration of Interchange

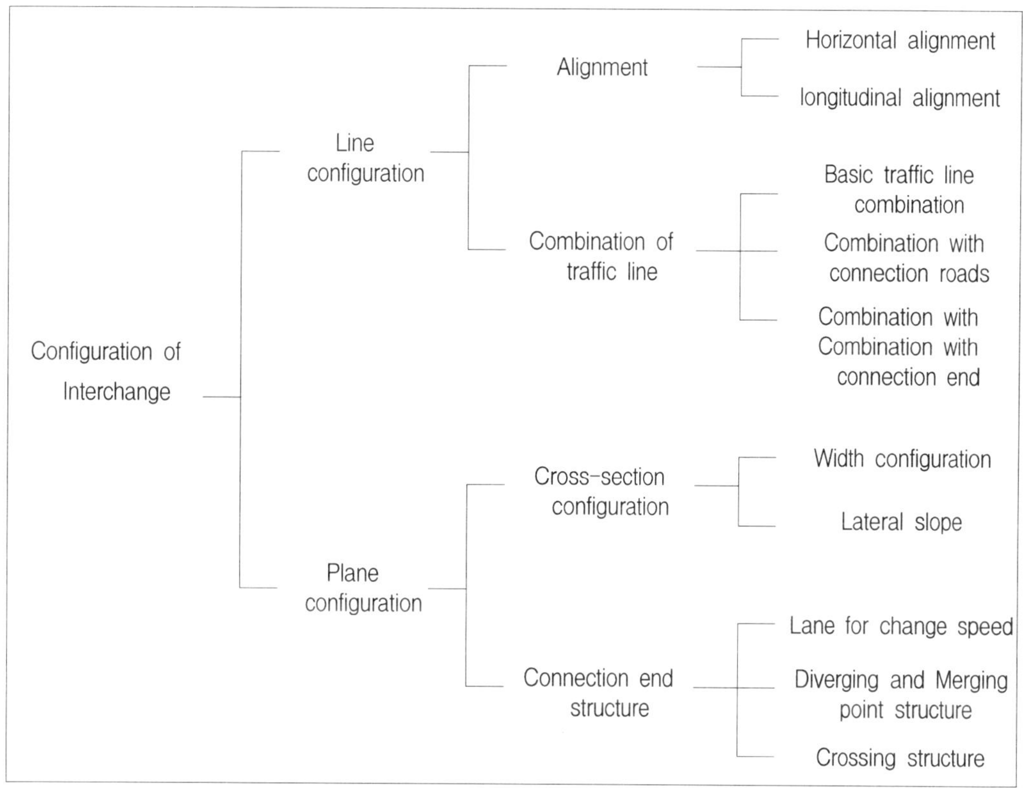

- The basic element determining the form for each type of interchange and causing the difference in traffic operation is traffic line combination. There are three types of traffic line combination ; basic traffic line combination, connection road combination, and type, connection end combination.

- Combination of basic traffic lines (Mutual combination relation of two traffic flows)

Diverging	Merging	Weaving	Crossing

◎ 연결로 결합(교차하는 두 개의 주동선 사이의 동선결합)

우회전 연결로	좌회전 연결로				
우직결 연결로 (Direct ramp)	준직결 연결로(Semi-direct)		좌직결 연결로		루프(LOOP)
• 본선 우측에서 분류 후 약90° 우회전하여 우측 합류	• 우측에서 분류한 후 완만하게 좌측으로 방향 전환하여 좌회전		• 좌측에서 직접 분류하여 좌회전		• 우측에서 분류한 후 약270° 우회전하여 우측 합류

◎ 접속단 결합(두 접속단의 상호관계)

연속분류(DD)	연속합류(MM)	합 분류(MD)	분 합류(DM)

※ 여기서, M은 합류 D는 분류

3-2 인터체인지의 형식구분

3-2-1 형식 평가요인

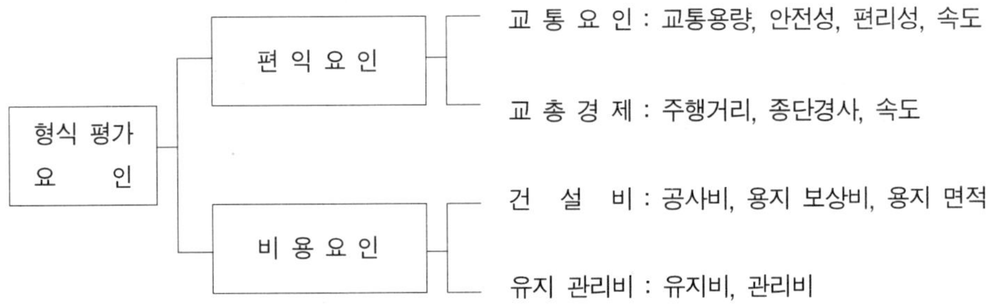

- Combination with connection roads (Combining traffic lines between two crossing major traffic lines)

Right-turn connection road	Left turn connection road				
Right direct ramp	Semi-direct ramp		Left direct ramp		LOOP
• Diverging from the right of the main line, turn around 90° to merge from the right	• Diverging from the right of the main line, and making slow left turn to the left		• Diverging from the left and turning left		• Diverging from the right, making 270° right turn, and merge from the right

- Combination of connection end (relationship between two connection ends)

Continuous Diverging (DD)	Continuous Merging(MM)	Merging Diverging(MD)	Diverging Merging(DM)

※ Where M is merging, and D is diverging

3-2 Classification of Interchange Type

3-2-1 Type Evaluation Factors

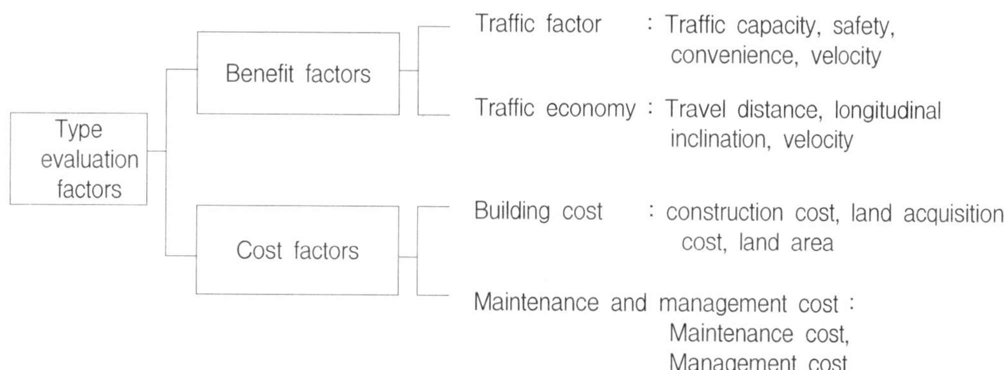

3-2-2 인터체인지의 형식적용

①불완전 입체교차

- 개 요 : 평면교차를 1개 이상 포함하는 형식
- 특 징 : 교통의 연속성 및 안전성 불리, 경제성 유리, 우회거리 감소

- 다이아몬드 형

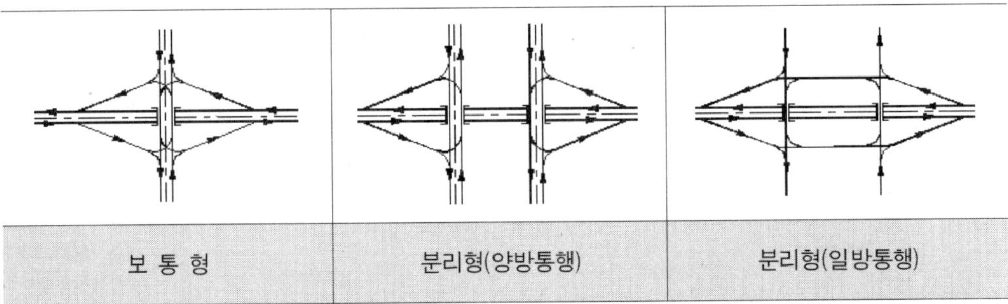

보 통 형	분리형(양방통행)	분리형(일방통행)

장 점	단 점
• 가장 단순한 형식으로 용지가 가장 적게 소요 • 횡단구조물 불필요로 건설비 저렴 • 우회거리 짧아 교통경제상 유리	• 평면교차부 교통용량 감소 • 영업소 설치 시 분리설치 필요 • 연결로 선형에 따라 사고 위험성 내포

- 불완전 클로버형

: 다이아몬드형보다 건설비는 높지만 교통용량 측면에서 유리

유출입부가 교차도로 전방위치	유출입부가 교차도로 후방위치	교차도로의 한 쪽에 연결로 위치

3-2-2 Interchange Type Application

① Partial Grade Separation

- Overview : This includes more than one at-grade crossing
- Characteristics : Disadvantage in the continuity and safety, lower construction cost, reduction of detour distance

- Diamond shape

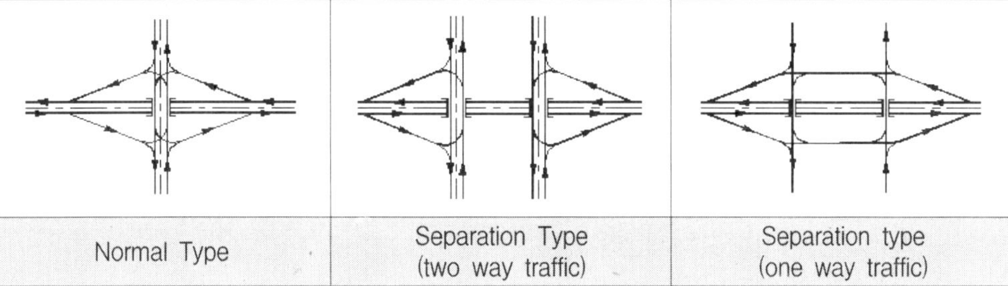

Advantages	Disadvantages
• This is the simplest type which requires the smallest area.	• Reduced traffic capacity at at-grade crossing areas
• Cost efficient since cross-sectional structures are not needed.	• Separate installation required when installing tollgates
• Short detour distance translating into superior traffic economy	• Potential danger depending on the alignment of connection roads

- Partial Cloverleaf Interchange

: higher construction cost than diamond type, but advantage on traffic capacity

Access road is ahead of crossing	Access road behind the crossing road	Connection road is at one end of crossing road
N, W, E, S	N, W, E, S	N, W, E, S

② 로타리형 입체교차

- 개　요 : 연결로를 전부 독립으로 하지 않고 엇갈림이 발생하는 형식
- 특　징 : 교통량이 적은 5지 교차 이상에 적용, 엇갈림 발생

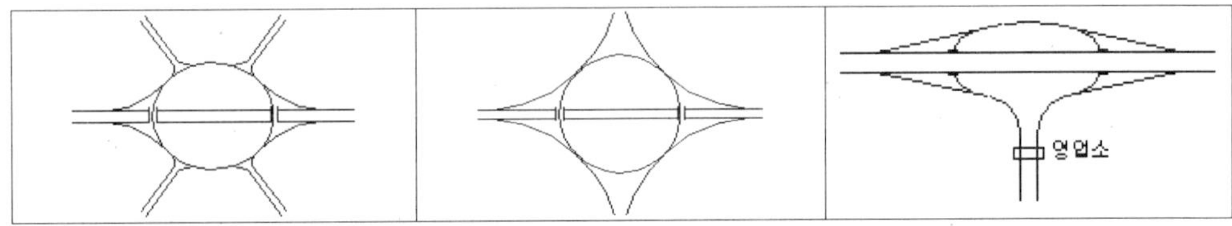

③ 로타리형 입체교차 완전 입체교차

- 개　요 : 각 연결로가 전부 독립되어 있는 형식
- 특　징 : 교통의 연속성 및 안전성 증대, 공사비 고가, 엇갈림 미 발생

- 직결형 완전 입체교차(세갈래 교차)

【직결 Y형 입체교차】

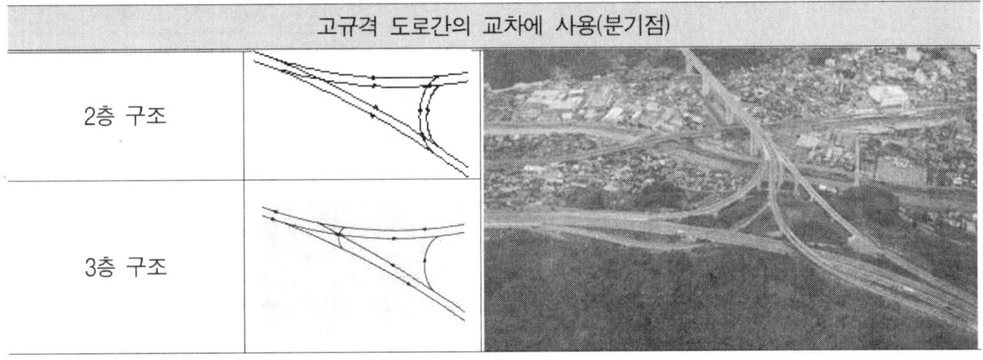

【준 직결 Y형 입체교차】

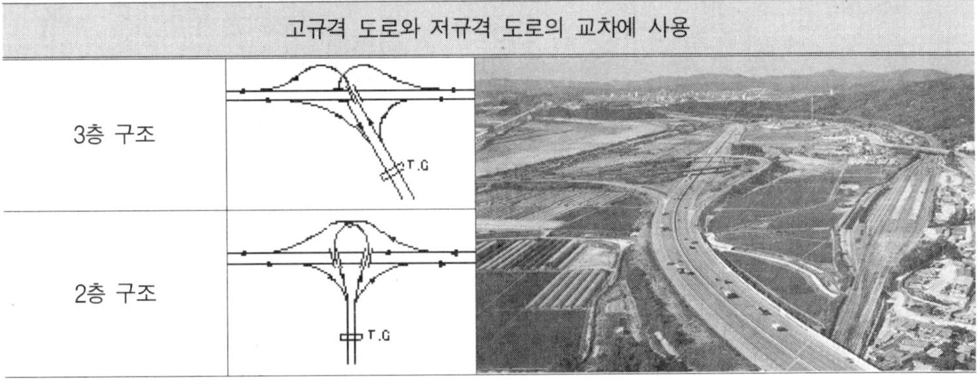

② Rotary-type Grade Separation

- Overview : connections are not all independent, and weaving type.
- Characteristics : applied to small traffic five or more branch crossing, weaving type.

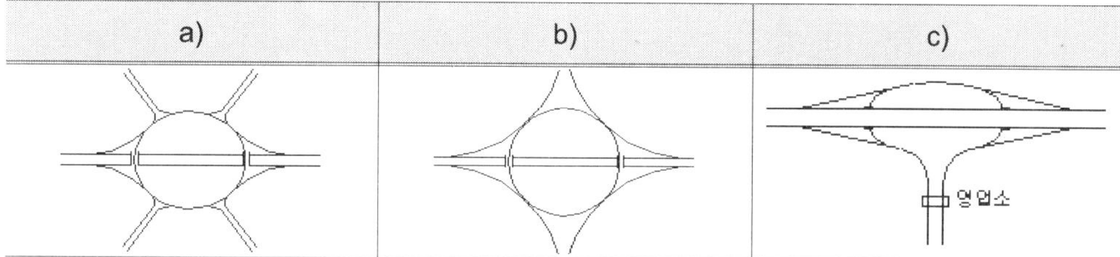

③ Rotary-type Grade Separation and Highway Grade Separation

- Overview : Each connection is completely independent
- Characteristics : Higher traffic continuity and safety, higher construction cost, no weaving

- Complete Directional Interchange (three-branch crossing)

【Direct Y Interchange】

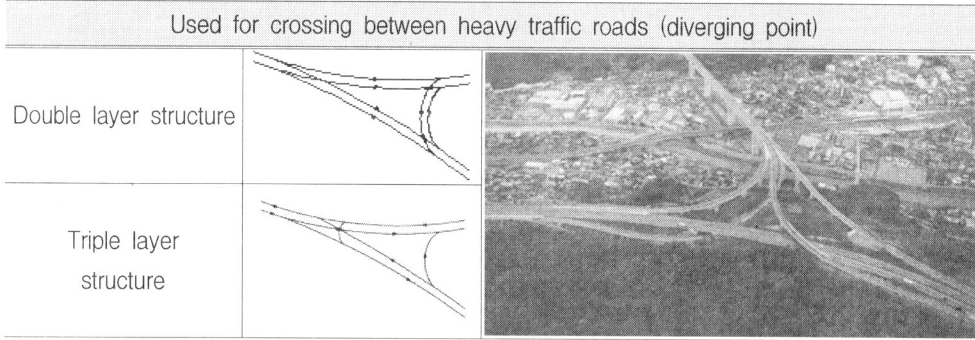

【Semi Direct Y Interchange】

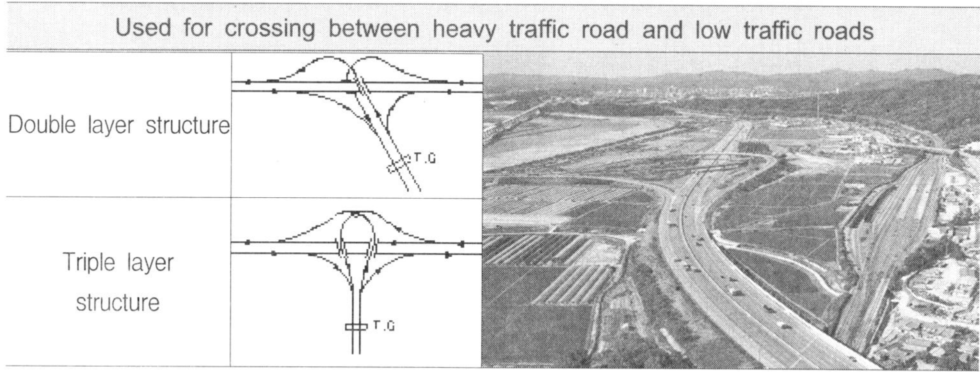

- 트럼펫 형(네 갈래 교차)
  - ▶ 폐쇄식 영업체계로 운영되는 유로도로의 대표적 형식
  - ▶ 영업소 한 곳에 집약 가능

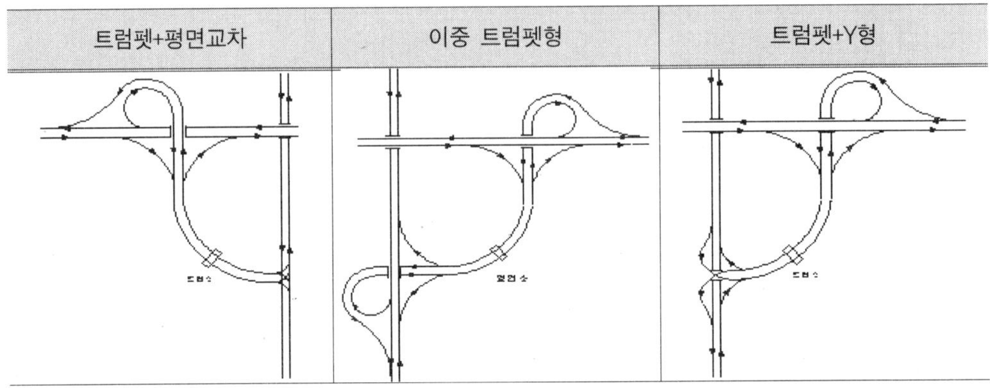

- 트럼펫 형 완전입체교차(세 갈래 교차)

【루프 연결로에 따른 형식 분류】

구　분	A 형	B 형
개　요	• 루프를 교차구조물의 전방에 설치하여 유입 연결로로 사용	• 루프를 유출 연결로로 사용
	ⓐ ⓑ ⓒ T.G	ⓐ ⓑ ⓒ T.G
적　용	• 본선에 고가차도(Overpass)로 운전자의 사전 인지 유도 • ⓐⓒ 교통량이 ⓑⓒ 교통량과 비슷하거나 많을 때 적용	• Loop 연결로 교통량이 작은 경우 사용 • 고속에서 루프구간 저속으로 속도 전환이 어려우며 시거 확보면에서 불리함

- Trumpet Interchange (four-branch crossing)
  - ▸ The most typical type of tolled road with closed operation system
  - ▸ Possible to concentrate tollgates in single location

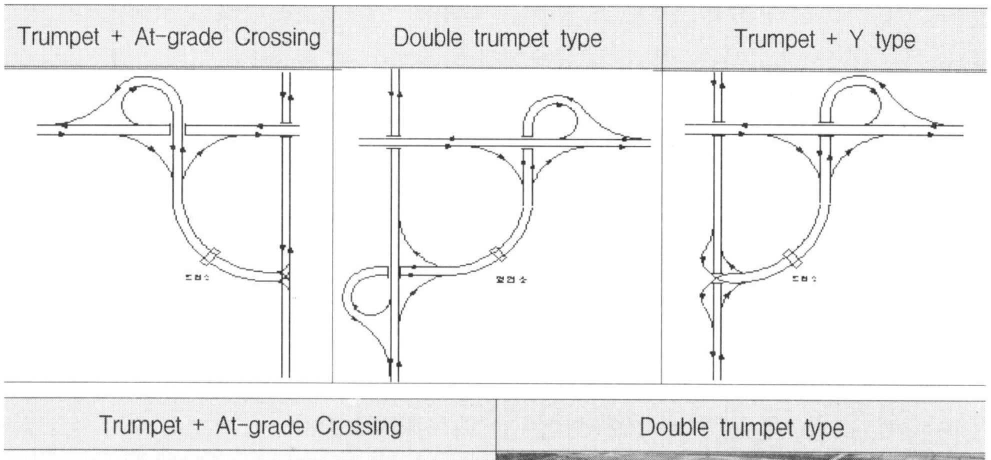

- Complete Trumpet Interchange (three-branch crossing)

【Type classification by loop connection road】

Classification	A type	B type
Overview	• Loop is installed in front of the crossing structure for connecting road for incoming traffic	• Loop is used for connection for outgoing traffic
	ⓒ T.G ⓐ ⓑ	ⓒ T.G ⓐ ⓑ
Application	• Overpass on the main line leads the pre-recognition of drivers • When traffics on ⓐⓒ are similar or higher than the traffics on ⓑⓒ.	• Loop connection is used when traffic volume is low • Not easy to change the speed from high speed to low speed in loop section, and disadvantage in securing sight distance

【교차각에 따른 분류】

구 분	직 각 교 차	사 각 교 차
개 요	• 교차도로와 직각으로 교차	• 교차도로와 경사져서 교차
	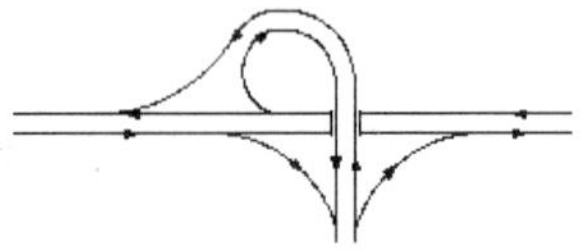	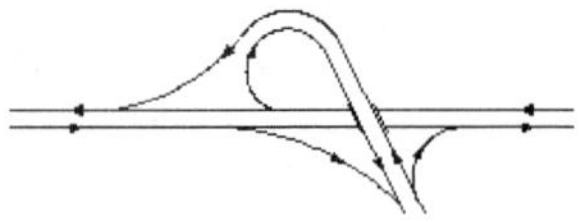
특 성	직각교차에 비하여 사각교차의 장점 • 좌회전 교통류에 완만한 곡선반경 제공 • 좌회전 교통류 회전각 작다. • 좌회전 교통류 주행거리 짧다.	

【루프연결로 선형형태에 따른 형식】

구 분	단 순 원 형	난 형
개 요 도	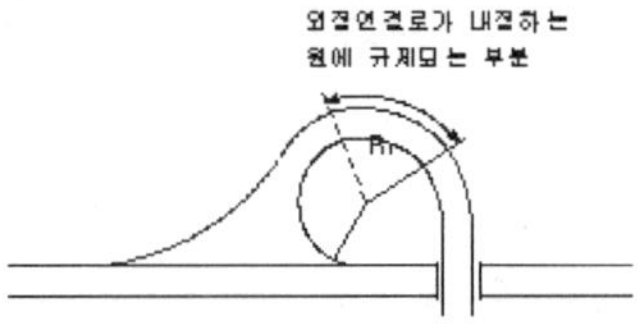 외접연결로가 내접하는 원에 규제되는 부분	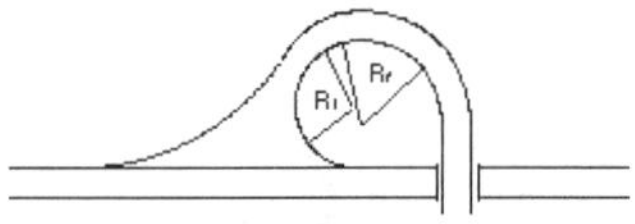
특 성	• 루프의 원이 갖는 교각이 180° • 외접 S 연결로의 루프의 최소곡선 반경에 따라 결정	• 루프가 2개원 이상으로 구성 • 외접 S 연결로 루프의 최소곡선 반경에 영향 안 받음

- 직결형 완전 입체교차(네 갈래 교차)
  ▸ 하나 이상의 직결 또는 준직결 연결로 존재
  ▸ 공사비 과다

보통형	터빈형	분리 터빈형
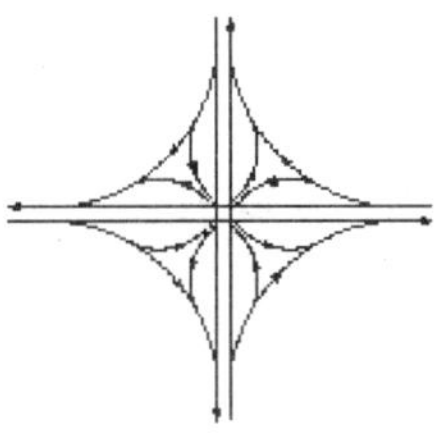	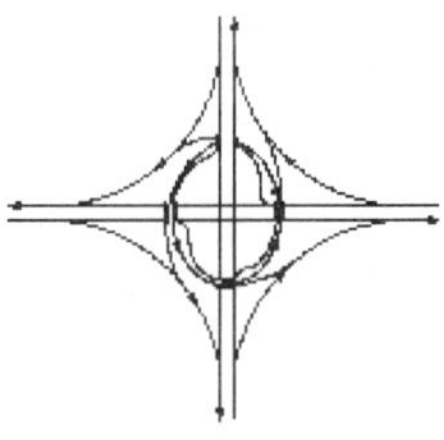	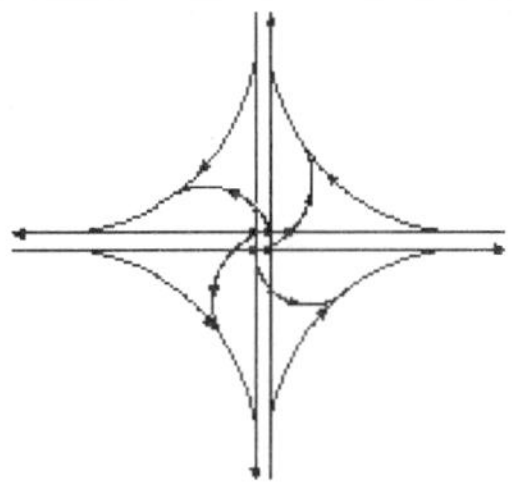

【Classification by angle of intersection】

Classification	Right angle crossing	Oblique angle crossing
Overview	• Crossing with crossing road at right angle	• Crossing with crossing road at oblique angle
	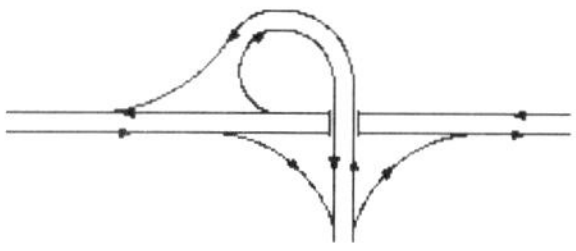	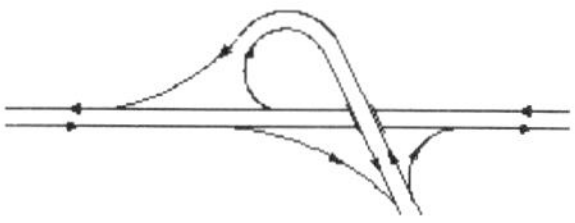
Characteristics	Benefits of oblique angle crossing compared to right angle crossing • Provision of gentle curvature for the left turning traffic flow • Smaller turning angle for left turning traffic flow • Shorter driving distance for left turning traffic flow	

【Type based on the path type of loop connection road】

Classification	Simple circle	Oval shape
Overview	Section where circumscribing connection road is restricted by the inscribing circle 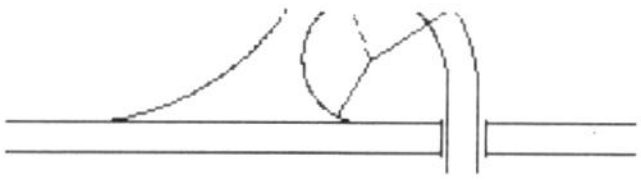	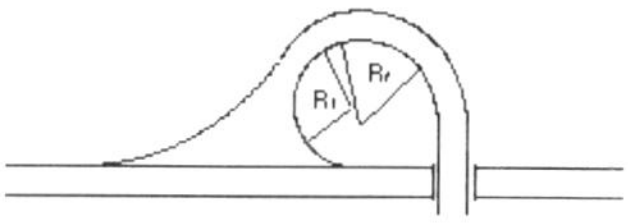
Characteristics	• Intersection angle of the loop circle is 180° • Determined by the minimum curvature of the loop of circumscribing S connection	• Loop consists of more than two circles • Not affected by the minimum curvature of the loop of circumscribing S connection

- Complete Directional Interchange (four-branches crossing)
  - ▸ Exist as more than one direct or semi-direct connection
  - ▸ Excessive construction cost

Normal Type	Turbine Type	Separated turbine type
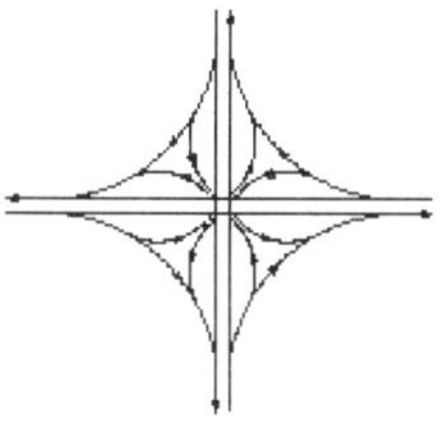	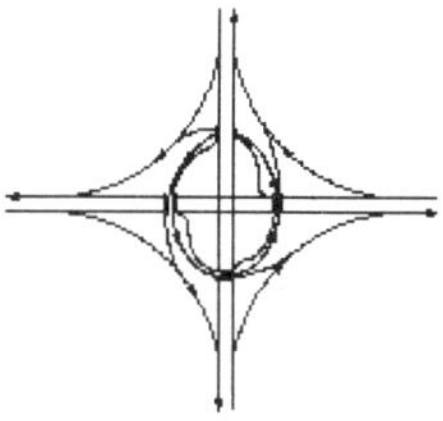	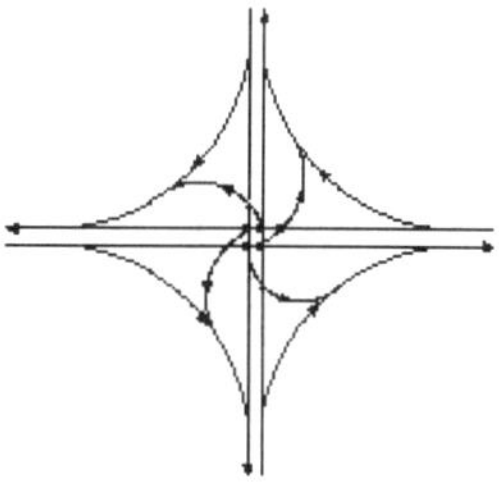

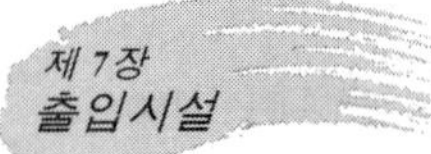

- 클로버형 완전입체교차
  - ▶ 입체교차구조물이 한 개 뿐인 입체교차
  - ▶ 루프 연결로 설치에 편입 용지과다
  - ▶ 인접한 루프구간 엇갈림 발생 → 집산로 설치

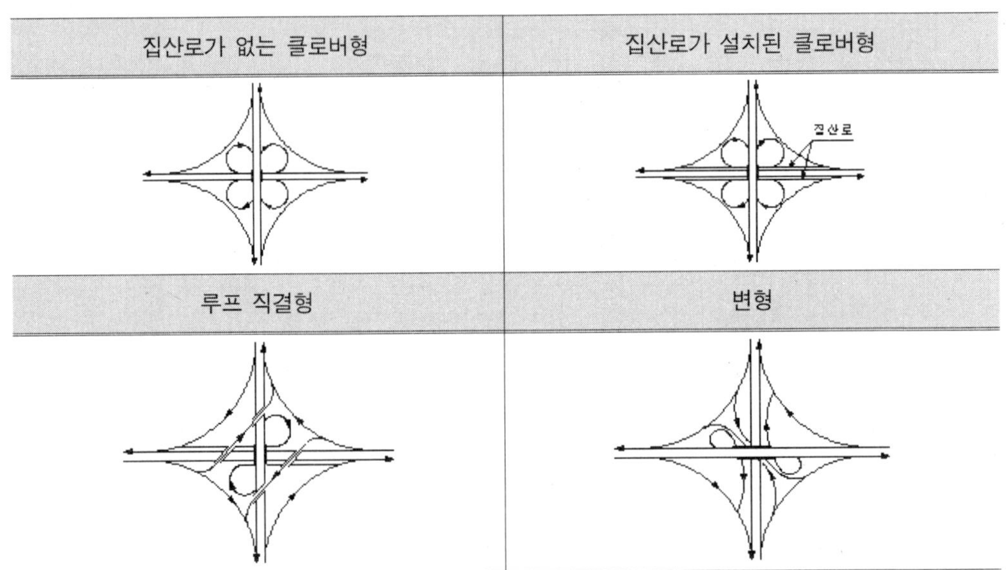

- Complete Cloverleaf Interchange
  - ▸ Grade separation with only one interchange structures
  - ▸ Excessive land requirement for installing loop connection roads
  - ▸ Weaving with adjacent loop section → Collector/distributer road installation

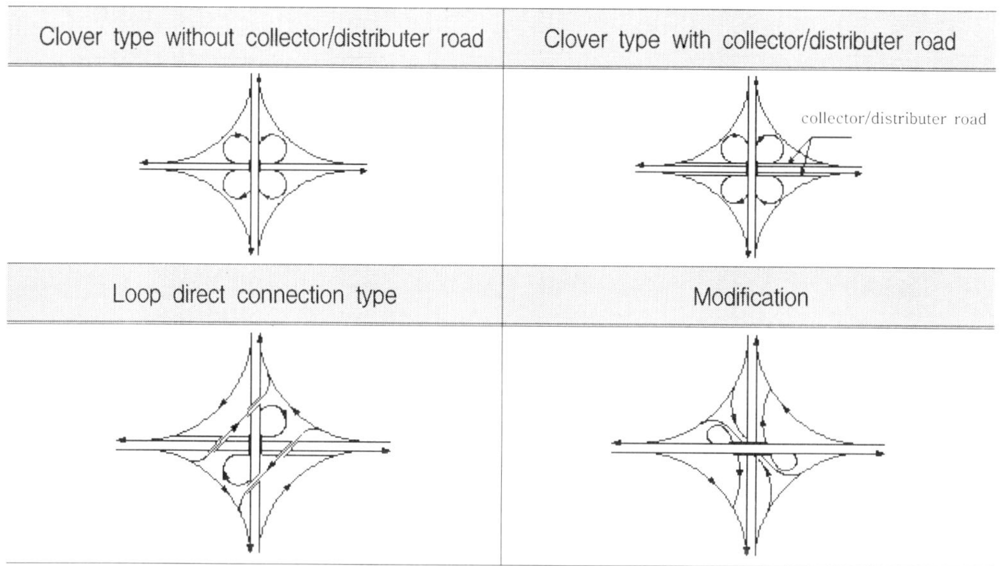

Hak-eu Junction (Loop direct connection type)

An-hyeon IC (modified clover type)

Ho-beob IC (modified clover type)

Japan (direct connection type)

Chapter 8
Earth Work

1. 일반사항

1-1 토공부의 구성

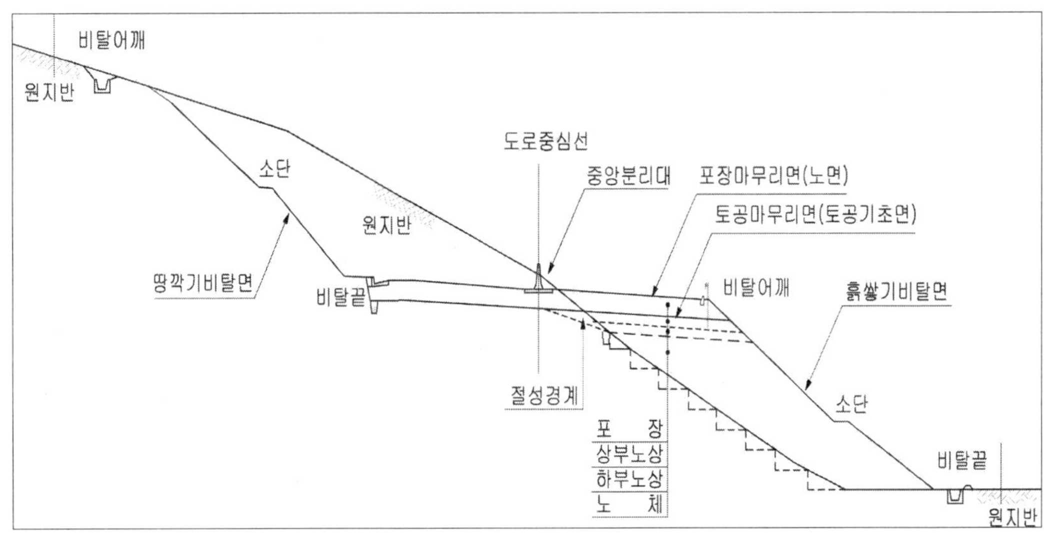

1-1-1 흙쌓기부

- 원지반부터 노상면까지 흙을 쌓아올린 부분

1-1-2 땅깎기부

- 원지반부터 노상면까지 원지반의 흙을 굴착한 부분

1-1-3 노 체

- 흙쌓기부에서 포장 및 노상 이외의 부분을 말하며 노상 및 포장층을 지지하는 역할

1-1-4 노 상

- 포장층 아래 두께 약 1m의 거의 균일한 토층, 포장층에 전달되는 교통하중을 지지하거나 보조기층에 대한 균등한 지지력을 확보하는 역할

1-1-5 포 장

- 노면으로부터 노상 윗면까지의 부분, 교통하중을 지지하고 하중을 분산시키는 역할

1-1-6 비탈면

- 흙쌓기 및 땅깎기에 의하여 형성되는 비탈면

1. General Considerations

1-1 Configuration of Earth Work

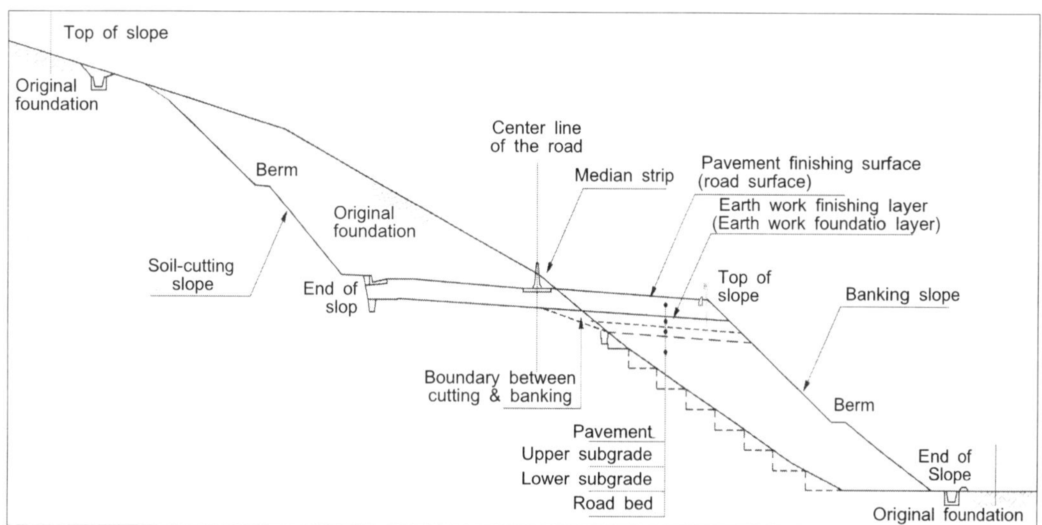

1-1-1 Banking Part

- From the orignal foundation up to the road surface, where filled with soil

1-1-2 Cutting Part

- From the orignal foundation down to road surface, where the slope soils are digged out.

1-1-3 Road Bed

- Soil banking part expect pavement and subgrade, and supporting subgrade and pavement layer.

1-1-4 Subgrade

- Uniform soil layer with the thickness of about one meter under the pavement layer. This layer supports the traffic load transferred on the pavement layer and secures the uniform support for subbase.

1-1-5 Pavement

- From the road surface to upper subgrade. This layer supports traffic loads, and distributes the load.

1-1-6 Slope Surface

- Slope surface formed by cutting or banking

1-2 토공 설계 시 주요 고려사항

1-2-1 현지조건(자연조건 및 사회조건 등)

① 지형, 토질 및 지질조사는 토공의 설계·시공에 미치는 영향이 크므로 조사결과를 상세하고 충분히 검토

1-2-2 경제성 및 시공성

① 건설비와 유지관리비를 포함한 종합적인 비교·검토를 통하여 효율적이고 경제적인 공법 선정

1-2-3 외적작용(교통하중, 강우 등)에 대한 안정성 확보

① 노상 및 노체 시공 시 부등침하 방지대책
② 연약지반 흙쌓기 시 압밀침하에 대한 합리적인 대책
③ 강우 등에 따른 비탈면의 붕괴방지대책 등

1-2-4 주변 환경과의 조화

① 흙쌓기부 및 땅깎기부가 주변 환경과 순응·조화될 수 있도록 설계
② 공사 시 수반되는 소음, 진동, 먼지, 지반변동, 수질오염, 지하수위 변화 등에 대한 사항에 대하여 충분히 검토

1-2-5 유지관리 효율성

① 도로 공용 후 노면의 부등침하나 비탈면 붕괴가 발생하지 않아야 하며 변화가 발생 시 간단히 복구할 수 있는 구조가 될 수 있도록 검토

2. 토공계획의 수립

2-1 타당성 및 기본설계 단계

- 경제성 분석 및 교통량 조사 등에 근거하여 가능노선을 다수 선정하고 사회적·경제적 타당성, 교통운용 적합성, 땅깎기·흙쌓기량의 균형 및 시공성 등을 비교·검토하여 최적의 노선을 검토한다.

1-2 Major Considerations During Earth Work

1-2-1 Site Conditions (natural conditions, social conditions, etc)

① Survey of topography, soil, and geological features have a great impact to design and construction, therefore the survey results shall be sufficiently reviewed.

1-2-2 Economic Efficiency and Ease of Construction

① To choose the efficient and economic construction method through the comprehensive comparative review including construction and maintenance cost.

1-2-3 Security of the Stability Against External Action (traffic load, rainfall, etc)

① Countermeasure to prevent differential settlement when building subgrade and road bed

② Reasonable countermeasure against consolidated settlement during banking soils on soft foundation

③ Countermeasure to prevent the collapse of slope by rainfall and etc.

1-2-4 Balance with Surrounding Environment

① Banking and cutting parts shall be designed to be balanced and accommodated by adjacent environment.

② Noise, vibration, dust, foundation fluctuation, water pollution, changes in underwater level, and etc that are accompanying with the construction shall be sufficiently reviewed.

1-2-5 Efficiency of Maintenance and Control

① There shall be no differential settlement or slope collapse after public usage of the road, and the structure shall be reviewed so that the changes can be recovered simply.

2. Establishment of Earth Work Plan

2-1 Feasibility and Basic Design Stage

- The optimum route shall be reviewed by comparing and review the social and economic feasibility, suitability of traffic operation, balance between cutting and banking, and ease of construction after selecting multiple possible routes based on the economic analysis and traffic volume survey.

2-2 실시설계 단계

2-2-1 흙쌓기부 설계 시 고려사항

① 지형 · 지질에 적합한 흙쌓기부의 설계
② 흙쌓기 재료적 특성 파악
③ 기상에 관한 정보(땅깎기부 설계 시에도 해당)
④ 흙쌓기부 안정검토

2-2-2 땅깎기부 설계 시 고려사항

① 지형 · 지질에 적합한 비탈면의 설계
② 비탈면 안정검토
③ 법령 지정구역의 규제 · 제약조건

2-2-3 시공계획에 따른 고려사항

① 시공법
② 토석 운반, 공사용 도로
③ 토취장 및 사토장의 선정 등

2-2-4 공사용 도로 계획 시 고려사항

① 현장 내 공사용 도로

- 최대한 본선 및 부체도로를 우선적으로 사용하는 것을 원칙
- 공사용 도로의 위치 및 계획고는 땅깎기 및 흙쌓기의 공정을 고려하여 설정

② 현장 외 공사용 도로

- 기존도로를 이용 또는 개량하는 경우
  - 장래계획의 유무, 공사방법, 용지취득 등에 대한 사전 검토
  - 도로 구성, 폭, 노면상황, 교통량, 연도의 이용상황, 교통안전시설 설치상황, 지하매설물 등에 대한 사전조사
  - 도로관리기관과의 협의

2-2 Detail Design Stage

2-2-1 Design Considerations During Designing Soil Banking Part

① Design of banking part suitable for topography and geological features
② Identifying the material characteristics of soil banking
③ Information on weather (also applied for soil cutting design)
④ Stability review on banking part

2-2-2 Design Considerations During Designing Soil Cutting Part

① Design of slop suitable for topography and geological features
② Slope stability review
③ Regulation and limited condition of the area designated by law

2-2-3 Considerations According to the Construction Plan

① Construction method
② Road for construction and moving sand & stones
③ Selection of borrow pit and spoil bank

2-2-4 Considerations While Planning Road For Construction

① On-site roads for construction

- In principle, main line and side road shall be used preferably as much as possible.

- The position and planed height of the road for construction shall be determined considering the soil banking and cutting process.

② Roads for construction outside the construction sites

- When using or improving the existing roads
  - Prior review about existence of future plan, construction method, acquisition of land, etc.

  - Prior research on road configuration, width, road surface status, traffic volume, usage by years, traffic safety facility installation status, understand laying materials, etc.

- Agreement with road maintenance organizations

- 도로를 신설하는 경우
  - 현지상황에 의하여 기존도로를 사용할 수 없는 경우 계획
  - 이용목적에 부합되도록 가능한 경제적인 노선 선택
  - 사용 후 철거하는 경우 지형의 원상복구가 불가능하므로 최종 존치형상을 고려하여 설계하고 관계기관과 협의

2-3 측량조사

- 측량법 및 기타 관계규정에 의거하여 중심선 및 종·횡단측량 실시

- 현지상황과의 일치성을 검토하여 보완측량을 실시하며, 구조물의 위치는 상세히 조사

2-4 재료원 조사 및 경제성 검토

- 공사에 소요되는 포장용, 구조물용, 동상방지, 보조기층용 세골재·조골재와 쌓기용 재료의 사용성을 검토하기 위하여 계획노선 인근의 재료원(세골재, 조골재, 레미콘, 아스콘 등)의 현황파악 및 현장 발생암에 대한 품질시험 실시

- 경제성 분석을 통하여 암유용에 대한 사용성 여부 등 재료의 공급계획 수립

3. 토공의 세부설계

3-1 지반조사

3-1-1 지반조사의 목적

① 노선선정, 토공, 구조물공, 배수공의 계획·설계·시공 및 유지관리를 합리적이고 경제적으로 진행하기 위하여 자료를 수집하는 단계로서, 대상지반의 지층분포와 토질·암석 및 암반 등 지반의 공학적 성질을 명확하게 파악

- When building new roads
  - Building new road shall be planned when the existing roads cannot be used by the site conditions.
  - Possible economic route shall be selected to meet the purpose.
  - When the roads are to be removed after finishing the construction, as the recovery of the topography to the original state is not possible, the design shall be done considering the final shape after the recovery and the agreement shall be made with the related organizations.

2-3 Measurement Investigation

- Execute the traversing and longitudinal measurement in accordance with Land Survey Act and other related regulations.

- Execute the complementary survey after reviewing the consentaneity with the site conditions, and positions of the structures shall be examined thoroughly in detail.

2-4 Material Source Investigation and Economic Efficiency Evaluation

- The status of adjacent material source (fine aggregate, coarse aggregate, ready mixed concrete, ascon) shall be identified in order to evaluate the usability of the materials for pavement, structures, fine and coarse aggregates for subbase, and soils for banking, and quality of the rocks from the site shall be tested.

- Establish the material supply plan including usability of the rock utilization through the economic efficiency analysis.

3. Detailed Design of Earth Work

3-1 Site Soil Research

3-1-1 Purpose of Site Soil Research

① This is a stage of collecting data for reasonable and economic execution of route selection, earth work, structure construction, plan of water drain work, design, construction, and maintenance and control. Engineering properties of the foundation including distribution of soil layers, characteristics of soil, rocks, and rock foundation shall be cleary investigated.

3-1-2 설계단계별 지반조사 흐름

<table>
<tr><th colspan="2">조사단계</th><th>시 기</th><th>목 적</th><th>내 용</th><th>범 위</th></tr>
<tr><td rowspan="3">예
비
조
사</td><td>자료조사</td><td rowspan="2">사업구상에서
구체적
으로 계획</td><td rowspan="2">도로노선 계획</td><td rowspan="2">• 각종 자료조사 분석
• 지표답사</td><td rowspan="2">대상구간의
광범위한 지역</td></tr>
<tr><td>현장답사</td></tr>
<tr><td>개략조사</td><td>비교노선
검토에서부터
노선 결정
까지</td><td>도로노선의 선정</td><td>• 지형 및 지질조건
에 대한 개략조사
• 환경 및 입지조건
에 대한 광역조사</td><td>계획노선 및 비교
노선을 포함한
광범위한 지역</td></tr>
<tr><td colspan="2">본 조 사</td><td>노선 결정
이후부터
공사 착공
까지</td><td>설계 및 시공계획 의
수립
• 설계입력 자료
평가
• 공법의 선정
• 단면해석
• 공사비 산출 등</td><td>• 상세한 지질 및
지반 조사
• 주변 환경 및 공사
에 필요한 제설비,
법규 등 조사</td><td>결정된 노선 및
주변지역</td></tr>
<tr><td colspan="2">시 공 중
보완조사</td><td>시 공 중</td><td>시공 관리, 공사 통제</td><td>• 현장계측
• 주변 환경의 영향
정도 조사
• 시공관리 조사</td><td>시공으로 영향을
받을 우려가 있는
주변지역</td></tr>
</table>

3-1-3 지반조사의 방법 및 내용

① 구조물의 종류, 규모 등에 따라 지반조사의 방법과 범위가 달라지므로 조사대상의 지형, 지질의 특성 등을 고려한 후 조사목적에 적합한 방법을 선정

② 현장조사와 실내조사로 구분

- 현장조사 : 현장답사, 지표지질조사, 지구물리탐사, 시추조사, 사운딩 등
- 실내조사 : 자료 및 지형분석, 실내시험 등

3-1-2 Site Soil Survey Flow for Each Stage of Design

Investigation stage		Period	Purpose	Details	Scope
Pre-investigation	Data investigation	From the project idea to solid planning	Planning road route	• Various data investigation and analysis • Site survey of the surface	Wide range of target section
	Site survey				
	Outline investigation	From the route comparison to the determination of the route	Route selection	• Outline investigation of topography and geological conditions • Wide area investigation on environment and conditions of location	Wide range of area including the planned route and comparison routes
Main investigation		From route selection to the start of the construction	Establishment of design and construction plan • Design input data evaluation • Selection of construction method • Section analysis • Construction cost calculation	• Detailed investigation on geological condition and foundation • Investigation on adjacent environment and various facilities required for the construction, and regulations	Determined route and adjacent area
Complementary investigation during construction		During construction	Construction control and management	• Site measurement • Investigation of the effects of the construction on adjacent environment • Investigation of construction control	Adjacent ares that could be affected by the construction

3-1-3 Details and Method of Site Soil Survey

① Since the method and range of soil survey change depending on the type of structure and scale, select the method appropriate for the survey purpose considering the topography and geological features of the survey target.

② Site investigation and indoor investigation

- Site investigation : Site survey, surface geological features investigation, geophysical exploration, drilling investigation, sounding, etc.
- Indoor investigation: Data and topography analysis, indoor test, etc

[지반조사 및 시험의 방법 및 내용]

조사·시험	조사·시험 내용	조사항목
지표지질조사	• 암반의 풍화상태, 절리의 방향성, 연속성, 간격 및 거칠기 등 • 충진물의 분포상태 • 단층선등의 구조대 발달 여부	지형·지질파악
시험굴	• 지질파악	지질파악
시추조사	• 지층 분포상태 • 절리면의 발달상태(절리간격 및 경사각) • 암석코아의 강도 및 RQD • 파쇄구간의 존재여부 • 시추공전단강도시험(풍화암층 C, ø 측정)	지질파악 원지반의 공학적·물리적 특성파악 지하수 상황파악
물리탐사	• 탄성파 탐사, 전기비저항, 토모그래피 : 탄성파 속도 및 전기비저항의 분포에 의하여 지하지질구조 및 파쇄대의 존재여부 파악 • 시추공영상처리기법(BIPS, 텔레뷰어 등) : 불연속면의 방향성 파악	지질파악 원지반의 공학적·물리적 특성파악 지하수상황파악
현장시험	• Schmidt Hammer Test • Point Load Test • Profile Gauge Measurement	원지반의 공학적·물리적 특성파악
실내시험	• 토성시험 : 함수비, 비중, 입도 및 액·소성 한계 등 • 암석시험 : 일축압축강도(풍화암 점하중시험으로 측정) 절리면전단강도(C, ø 측정)	원지반의 공학적·물리적 특성파악
성과분석	• 주절리군의 방향성 • 암반의 공학적 특성 : 강도 정수 산출(C, ø) • 암반비탈면의 경사 결정 : 예상파괴형태 및 구간별 안정성 검토 • 대책공법 선정	

[Site soil survey and test method & details]

Survey · Test	Details of survey and test	Survey items
Surface geological features investigation	• Efflorescence status of rock floor, direction of joint, continuity, gap and roughness, etc • Distribution status of filling material • Development of dislocation lines and other structures	Topography and geological features investigation
Test tunnel	• Geological features investigation	Geological features investigation
Drilling investigation	• Distribution status of soil layers • Development status of dislocation surface (dislocation gap and inclination) • Strength of rock core and RQD • Existence of spall section • Shear strength test of drill (efflorescence rock layer C, ø measurement)	Geological feature investigation/ Engineering&physical features investigation for original foundation/ Underground water status investigation
Physical exploration	• Seismic exploration, electric resistivity, tomography : To investigate the underground geological features and existence of spall section through the seismic wave velocity and electric resistivity distribution • Borehole image process technique (BIPS, televiewer, etc) : Investigate the direction of discontinuous face	Geological feature investigation/ Engineering&physical features investigation for original foundation
Site test	• Schmidt Hammer Test • Point Load Test • Profile Gauge Measurement	Engineering&physical features investigation for original foundation
Indoor test	• Soil property test: moisture content, specific gravity, grain size and liquid plasticity limit, etc • Rock test: Single axis compression strength (measure by efflorescence rock concentrated load test) dislocation surface shear strength test (C, ø measure)	Engineering&physical features investigation for original foundation
Results Analysis	• Directional property of the set of major joint • Engineering characteristics of rock floor : strength constant calculation (C, ø) • Determine the inclination of rock floor slope: Evaluation of expected demolition shape and stability for each section • Select countermeasure of construction method	

3-1-4 지반조사의 항목별 기준

조사위치		조사항목	최소 조사빈도	심 도	비 고
흙쌓기부	일반구간	핸드오거	300m	3 ~ 5m	
	연약지반	핸드오거	200m	3 ~ 5m	
		피에조콘관입시험	100m	필요깊이까지	
		시추조사	200m	지지층(풍화암)확인	• 콘관입시험 위치
		베인시험	400m	불교란시료 채취심도	• 시추조사 위치 • Borehole, Field vane식
		간극수압소산시험	200m(1회/CPTu 2회)	불교란시료 채취심도	
땅깎기부		시험굴조사	200m	1 ~ 2m	
		시추조사	개소 당 1개소 이상 (조사등급별 분류에 따라 조정)	계획고 하 2m	• SPT : 1m마다
		탄성파탐사 또는 전기비저항탐사			
		화상정보시험			
구조물부	교량부	시추조사	교대 및 교각마다 1개소	경암 1m, 연암 3m, 풍화암 7m 까지	• SPT : 1m마다 • 상 · 하행선 분리 시 모두 시행
		화상정보시험	터파기 시 비탈면 안정성 검토가 필요한 경우	암반구간	
		공내재하시험	• 풍화대가 깊어 직접 기초 심도 결정 시 침하량 산정이 필요한 경우 • 연약층이 깊어 말뚝기초의 수평방향변위 반력계수 산정이 필요한 경우	층별(토사,풍화암,암반) 각각 1회	
		탄성파탐사 또는 전기비저항 탐사	• 직접기초위치의 풍화암 심도가 급변하는 경우 • 현장여건상 민원 등으로 시추조사가 불가한 경우	• 직접기초 예상심도 이상깊이 까지	교축방향 2개 측선 또는 교축 및 교축 직각방향
		토모그래피 탐사	상부구조물에 큰 영향을 미칠 공동 발견 시	• 직접기초 : 푸팅 폭의 3배 깊이 까지, • 말뚝기초 : 기초폭의 1~3배 깊이 까지	
	터널부	시추조사	터널 입출구부 각각 2개소 및 300m 당 1개소	• 계획고 하 2m까지	• 시추조사하여 지층이 토사층 또는 연약층대 등이 존재시 풍화암층 또는 필요한 심도까지
		탄성파 탐사	터널 개소당 입·출구부 필요한 연장 (Cross Hole Test)		
		전기비저항탐사	터널 전 연장		
		암반수압시험	터널 내 용출수 산정에 필요한 시추조사 위치		
		화상정보시험	터널 개소당 입출구부 각각 2개소	암반구간	땅깎기부 기준 준용
재료원	석산	시추조사	3개소 이상	필요깊이까지	
	하천골재원	시추조사	필요 시	필요깊이까지	
	토취장	시추조사	3개소 이상	필요깊이까지	
		시험굴조사	5개소 이상	1 ~ 2m	

3-1-4 Standard of Ground Survey by Item

Investigation location		Investigation item	Minimum investigation frequency	Depth	Remarks
Banking part	General section	Hand auger	300m	3 ~ 5m	
	Soft foundation	Hand auger	200m	3 ~ 5m	
		Piezo-cone penetration test	100m	To required depth	
		Drilling test	200m	Check support layer (efflorescence rock)	• Cone penetration test position
		Vane test	400m	To the depth of undisturbed samples	• Drilling test position • Borehole, Field vane type
		Pore pressure dissipation test	200m(Once /CPTu Twice)	To the depth of undisturbed samples	
Cutting part		Test tunnel investigation	200m	1 ~ 2m	
		Drilling test	More than one point per site (Adjustable for the classification by Investigation grade)	Planned height 2m	• SPT : every 1m
		Seismic wave exploration or electric resistivity exploration			
		Image information test			
Structure parts	Bridge Parts	Drilling test	One per abutment and pier	Hard rock 1m, soft rock 3m, efflorescence rock up to 7m	• SPT : every 1m • Both up and down lanes if separated
		Image information test	When slope stability evaluation is required during pit excavation	Rock section	
		Pressure meter test	• When settlement needs to be calculated in determining depth due to deep efflorescence layers • When horizontal displacement reaction coefficients of need to be calculated in horizontal pile base due to deep soft layers	Once for each layer(Soil, efflorescence rock, rock floors)	
		Seismic wave exploration and electric resistivity exploration	• When depth of efflorescence rock changes quickly over the direct foundation. • When drilling test is not allowed due to site conditions like civil claims	• Deeper than expected depth of direct foundation	Two of side lines in bridge axis direction , bridge axis, or perpendicular to bridge axis
		Tomography exploration	When holes that could affect the upper structures are found	• Direct foundation: Depth of three times putting width, • Pile base: To the dept of 1~3 times of base width	
	Tunnel Parts	Drilling test	Two locations for each of entrance and exit of the tunnel, and one per 300m	• Planned height up to 2m	• Down to efflorescence rock layer of required depth if soil layers or soft layers are found from the drilling test
		Seismic wave exploration	Entrance and exit of tunnel Required length (Cross Hole Test)		
		Electric resistivity test	Tunnel entire length		
		Rock floor water pressure	Position of drilling test required for the calculation of the water in the tunnel		
		Image information test	Two sites for each of entrance and exit of each tunnel	Rock section	Cutting part standard is to be applied
Material source	Rock mountain	Drilling test	More than three locations	To the necessary depth	
	River aggregate source	Drilling test	when necessary	To the necessary depth	
	Borrow pit	Drilling test	More than three locations	To the necessary depth	
		Test tunnel test	More than five locations	1 ~ 2m	

3-1-5 지반조사 자료의 활용

① 지반조사를 토대로 암선을 추정하고, 비탈면 안정성 · 연약지반 · 구조물 안정성 검토 등을 통하여 최적공법을 선정

② 설계단계에서 선정된 공법은 국한된 지반조사 자료에 의한 검토 결과이므로 시공 전 전문가에 의한 재검토를 통하여 적합한 대체공법을 수립

구 분	내 용
비탈면 안정성 검토	• 기본계획의 적정성 검토 - 현장 지표지질조사 검토 - 물리탐사(탄성파 및 전기비저항탐사 등) 및 시추조사자료 등의 분석 - 비탈면의 안정성을 검토 • 비탈면 경사, 보강구간 및 보강공법, 비탈면 보호공법(식재공)을 선정 • 추가조사(보링)의 판단 및 대체공법 검토
연약지반 검토	• 지반활동에 대한 안전율 결정 • 한계성 토고의 결정 • 총 침하량의 산정 • 연약지반 처리공법의 결정
구조물 안정성 검토	• 지반조사 결과(지지층의 토성, 지지층의 심도, 중간층의 토성)를 토대로 지지력, 허용변위, 단면을 만족하는 기초형식을 선정

3-2 비탈면 설계

3-2-1 흙쌓기 비탈면

① 흙쌓기 비탈면 설계 시 주요고려사항

- 지형조건에 따른 흙쌓기 계획
- 비탈면 안정해석 및 경사와 소단의 결정
- 지하수 및 지표수의 배수계획
- 장기적인 비탈면 표면보호 방법
- 유지관리를 위한 점검시설
- 시공 중 관리방안

3-1-5 Utilization of Ground Survey Data

① Rock lines shall be estimated based on the result of foundation investigation, and optimum construction method shall be selected through the review of slope stability, soft foundation, and structure stability.

② Since the construction method determined during the design stage is the result of review based on the limited foundation investigation data, the appropriate alternative construction method shall be established through the re-evaluation by the specialists.

Classification	Contents
Slope stability evaluatiozn	• Evaluation of suitability of basic plan - Evaluation of site surface & geological properties - Physical exploration (Seismic wave and electric resistivity) and analysis of drilling test data - Slope stability evaluation • Selection of slope inclination, reinforcement section and reinforcement construction method, and slope protection construction method (plating method) • Determination on the need of additional investigation (boring) and evaluation of alternative construction method
Soft foundation evaluation	• Determination of the safety rate for the activities on the foundation • Determination of the maximum height of soil banking • Calculation of total settlement • Determination of the construction method on soft foundation
Structure stability evaluation	• Selection of basic type that meets the support load, allowed displacement, and cross-section based on the foundation investigation results (soil property of support layer, depth of support layer, soil property of the middle layer)

3-2 Slope Design

3-2-1 Soil Banking Slope

① Major considerations on the design of soiling banking slope

- Filling plan according to the topographic conditions
- Stability analysis of slope and determination of inclination and berm
- Drain plan for underground and surface water
- Long-term slop surface protection method
- Monitoring facility for the maintenance and control
- Control method while the construction

② 흙쌓기 비탈면 경사

- 표준경사와 다른 경우 또는 비탈면의 높이가 10m를 초과한 경우는 별도의 안정해석을 통하여 경사를 결정

【흙쌓기 비탈면의 표준경사】

흙쌓기 재료	비탈면 높이(m)	비탈면 상·하부에 고정시설물이 없는 경우(도로, 철도 등)	비탈면 상·하부에 고정시설물이 있는 경우(주택, 건물 등)
입도분포가 좋은 양질의 모래, 모래자갈 암괴, 암버럭	0 ~ 5	1 : 1.5	1 : 1.5
	5 ~ 10	1 : 1.8	1 : 1.8
	10 초과	별도 검토	별도 검토
입도분포가 나쁜 모래, 점토질, 사질토, 점성토	0 ~ 5	1 : 1.5	1 : 1.5
	5 ~ 10	1 : 1.8	1 : 1.8
	10 초과	별도 검토	별도 검토

③ 흙쌓기 비탈면의 안정검토 대상

- 비탈면 높이가 10m를 초과하는 경우
- 비탈면 경사가 표준경사보다 급한 경우
- 흙쌓기 재료의 함수특성이 높고 전단강도가 낮은 경우
- 붕괴 시 복구가 장시간 소요되거나 주변 인접시설물에 중대한 인명, 재산상 피해를 주는 경우
- 지형특성으로 인하여 흙쌓기 토체 내부로 지속적인 지하수의 유입이 발생하는 경우(경사지반, 계곡부 흙쌓기)
- 홍수 시 비탈면이 침수되거나 비탈끝이 침식되는 경우
- 흙쌓기 비탈면의 원지반이 양호하지 못한 경우(연약지반 등)
- 내진안정해석이 필요한 경우
- 위 조건 외에 설계자가 필요하다고 판단하는 경우

② Soil banking slope inclination

- When the inclination is different from the standard slope or when the height of the slope is over 10 meters, the angel shall be determined through the separate stability analysis

【Standard inclination of soil banking slope】

Filling material	Slope height(m)	When there are not fixed facilities over and below the slope (road, railroad, etc)	When there are fixed facilities over or below the slope (house, building, etc)
Good-quality sand, sand gravel, rock, and crushed-rock with good distribution of grain sizes	0 ~ 5	1 : 1.5	1 : 1.5
	5 ~ 10	1 : 1.8	1 : 1.8
	Over 10	Separate evaluation	Separate evaluation
Clay, sandy soil, and cohesive soil with poor distribution of gain sizes	0 ~ 5	1 : 1.5	1 : 1.5
	5 ~ 10	1 : 1.8	1 : 1.8
	Over 10	Separate evaluation	Separate evaluation

③ Stability of the banking slope has to be evaluated for the following cases

- When the height of the slop is over 10 meters
- When the slope inclination is steeper than standard inclination
- When moisture contents of the banking material is high and shear strength is low
- When the recovery from the collapse could take long time, or the collapse could result serious damage to lives and properties.
- When continuous inflow of underground water into the soil structured due to the characteristics in topography (Sloped foundation, banking on valley area)
- When slope surface of end of slope is submerged during flood
- When original foundation for banking slope is poor. (software foundation, etc)
- When earthquake-proof analysis is required
- When designers decide the evaluation is necessary besides the above cases

④ 흙쌓기 비탈면의 안정대책 수립

- 흙쌓기 높이가 높아 비탈면이 불안정한 것으로 분석되는 구간(단, 기초지반이 연약지반이 아닌 경우)은 투수성이 양호하고 강도정수가 큰 재료를 선별하여 사용하거나 설계 비탈면 경사를 완화 또는 소단 폭의 증가 등의 방법으로 안정대책을 수립

- 비탈면 하단에 하천이나 도로, 철도 등의 중요한 시설이 위치하여 설계 비탈면 경사를 완화하거나 소단 폭을 증가시킬 수 없는 경우에는 EPS 성토, 콘크리트 옹벽, 보강토 옹벽, 개비온(gabion) 옹벽, 돌쌓기 등의 방법으로 비탈면의 안정을 유지

- 비탈면 안정대책 공법이 결정되면 대책공법의 설계에 대한 안정검토를 수행

⑤ 흙쌓기부의 기초지반 대책

- 토질조사 등의 결과에 의하여 흙쌓기의 안정에 영향을 미치는 불안정한 지반(연약지반 및 산허리 붕괴지, 낭떠러지, 비탈면 붕괴지역 등의 불안정지반 등)은 그 규모에 따라 적절한 대책을 수립

- 기초지반이 급한 경사를 이루고 있는 구간은 층따기, 배수공 등을 반영

- 기초지반 내에 폐광 등의 공동이 있는 경우 충분한 사전조사 후 채움 등의 적절한 조치계획 수립

⑥ 흙쌓기부 배수대책

- 흙쌓기부의 붕괴는 침투수 및 용수가 원인이 되어 발생하는 경우가 많으므로 필터층 및 지하 배수공 등의 배수시설을 설치

- 한쪽깎기·한쪽쌓기 구간은 노체면 또는 땅깎기면에 지하 배수공을 설치하는 것을 원칙

- 용수구간이 많은 경우 흙쌓기와 원지반의 경계에 지하 배수용 필터를 설치하여 물의 침투를 방지하고, 필터 재료로 투수성이 좋은 흙과 모래를 혼합하여 사용하거나 용수량에 따라서는 유공관의 매설계획 수립

- 오목부에 흙쌓기 하는 경우에는 오목부를 매립하여 표면에 배수공을 설치하고 흙쌓기 저부를 통과하는 충분한 배수층을 설치

④ Establishment of the measure to stabilize the banking slope

- Stability countermeasure shall be established by using the materials with high permeability and high strength constant, reducing the slope inclination, or increasing the width of berm for the section where the slope is considered to be unstable because of the high banking. (Only when the basic foundation is not soft foundation)

- When it is not possible to reduce the slope inclination or increasing the width of berm because the important facilities like river, road, railroad are located at the end of the slope surface, then stability of the slope shall be maintained by EPS banking, cocreate retaining wall, reinforcement soil retaining wall, gabion retaining wall, stone banking, etc.

- Once the slope stability countermeasure construction method is determined, execute the stability evaluation for the design of countermeasure construction plan.

⑤ Countermeasure to the basic foundation of soil banking parts

- Depending on the result of the soil property investigation, the unstable foundation which affects the stability of the soil banking (soft foundation, mountainside collapse area, cliff, slope collapse area, and other unstable foundation) shall have the appropriate countermeasure depending on the scale.

- Bench cut or drip hole shall be reflected when base foundation is forming a steep slope.

- When there is a cavity like abandoned mine in the base foundation, proper measure and plan shall be established such as banking after sufficient pre-investigation.

⑥ Soil banking are drain plan

- As collapse of the banking part is largely due to the penetrating water or rain water, drain facilities like filtering layer and underground drip holes shall be installed.

- In cutting and embankment section, underground drain holes shall be installed on the road bed or soil cutting parts in principle.

- When there are much of rain water, underground drain filter shall be installed at the boundary of soil banking and original foundation to prevent the penetration of water, and the soil and sand with good permeability shall be used for filter materials. Depending on the amount of rain water, burial of perforated drain pipes might be considered and planned.

- 오목부에 흙쌓기하는 경우에는 오목부를 매립하여 표면에 배수공을 설치하고 흙쌓기 저부를 통과하는 충분한 배수층을 설치

【흙쌓기 비탈면의 배수시설】

지표수 배수시설	지하수 배수시설
• 비탈어깨배수구 • 종배수구 • 소단배수구 • 비탈끝배수구	• 지하배수구(맹암거) • 수평배수층(블랭킷) • 돌망태배수공

⑦ 높은 흙쌓기부(흙쌓기 높이 15m 이상)

- 높은 흙쌓기부 설계 시 흙쌓기에 따른 압축침하 및 안정대책에 대한 검토 필요

- 높은 흙쌓기부의 비탈면 경사는 완만할수록 유리하나 현지조건과 공사비와의 관계를 고려하여 적정한 경사를 검토

- 흙쌓기 재료는 여러 가지 성질을 갖는 현지 발생재를 각부의 적재적소에 배분하여 안정성이 높도록 하는 것이 바람직하며, 투수성이 좋은 재료(암편 등)는 최하부 필터층(배수층)에 사용하도록 배려

- 높은 흙쌓기부의 배수처리는 흙쌓기의 안정, 공용 후 침하, 붕괴방지 등에 영향을 미치므로 세심한 배수처리 방안 검토 필요

- When banking on concave area, concave area is filled and the drip holes shall be installed on the surface, and sufficient drain layer passing through the lower part of soil banking part shall be installed.

【Water drain facility of banking slope】

Surface water drain facility	Underground water drain facility
• Top of the slope drain holes • Vertical drain hole • Berm drain holes • End of slope drain holes	• Underground drain holes (stone filled dummy drain) • Horizontal drain layer(blanket) • Wire cylinder drain holes

⑦ Elevated banking part (soil banking with the height over 15 meters)

- When designing elevated banking part, evaluation on the compressed settlement and stability countermeasure for the soil banking is required.

- Lower inclination on the slop of elevated banking part is better but the appropriate inclination has to be evaluated considering the relation between site condition and construction cost.

- It is desirable to increase the stability by distributing local materials that have various properties over the proper places on each part, and high permeability materials (flat rock pieces) shall be considered to be used at the bottom filter layer (water drain layer).

- Since the drain facility of the elevated banking part greatly affects the stability of banking, settlement after completion, and preventing collapsing, water drain plan shall be carefully evaluated.

【흙쌓기 비탈면의 주요 배수시설】

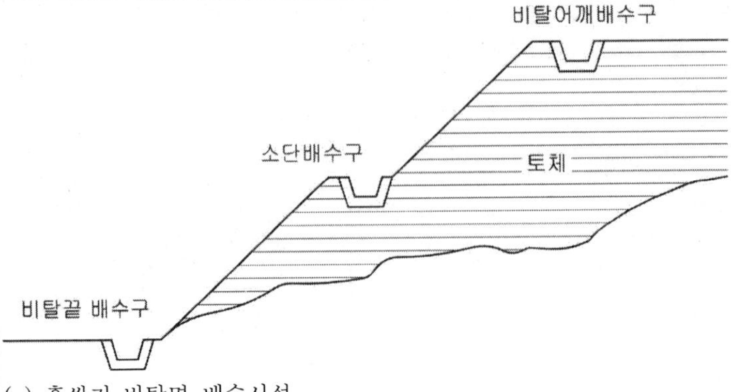

비탈어깨배수구
소단배수구
토체
비탈끝 배수구

(a) 흙쌓기 비탈면 배수시설

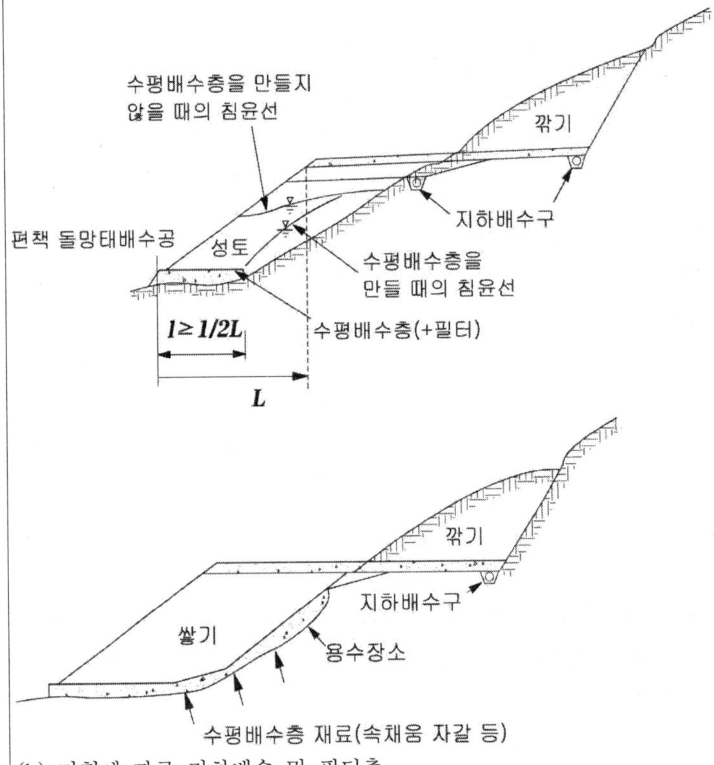

수평배수층을 만들지 않을 때의 침윤선
깎기
지하배수구
편책 돌망태배수공
성토
수평배수층을 만들 때의 침윤선
l≥1/2L
수평배수층(+필터)
L

깎기
지하배수구
쌓기
용수장소
수평배수층 재료(속채움 자갈 등)

(b) 지형에 따른 지하배수 및 필터층

⑧ 층따기

- 1:4 이상의 경사지반 위에 흙쌓기를 하는 경우 원지반 표면에 층따기를 반영하여 흙쌓기와 원지반과의 밀착을 도모하여 흙쌓기의 활동을 방지

- 지표수가 경사지반과 흙쌓기의 경계에 침입하여 활동을 발생시키는 경우가 있으므로 경사지반 상단에 배수구를 설치하고, 층따기면에 3~5%의 횡단경사 설치

【Major water drain facility of banking slope】

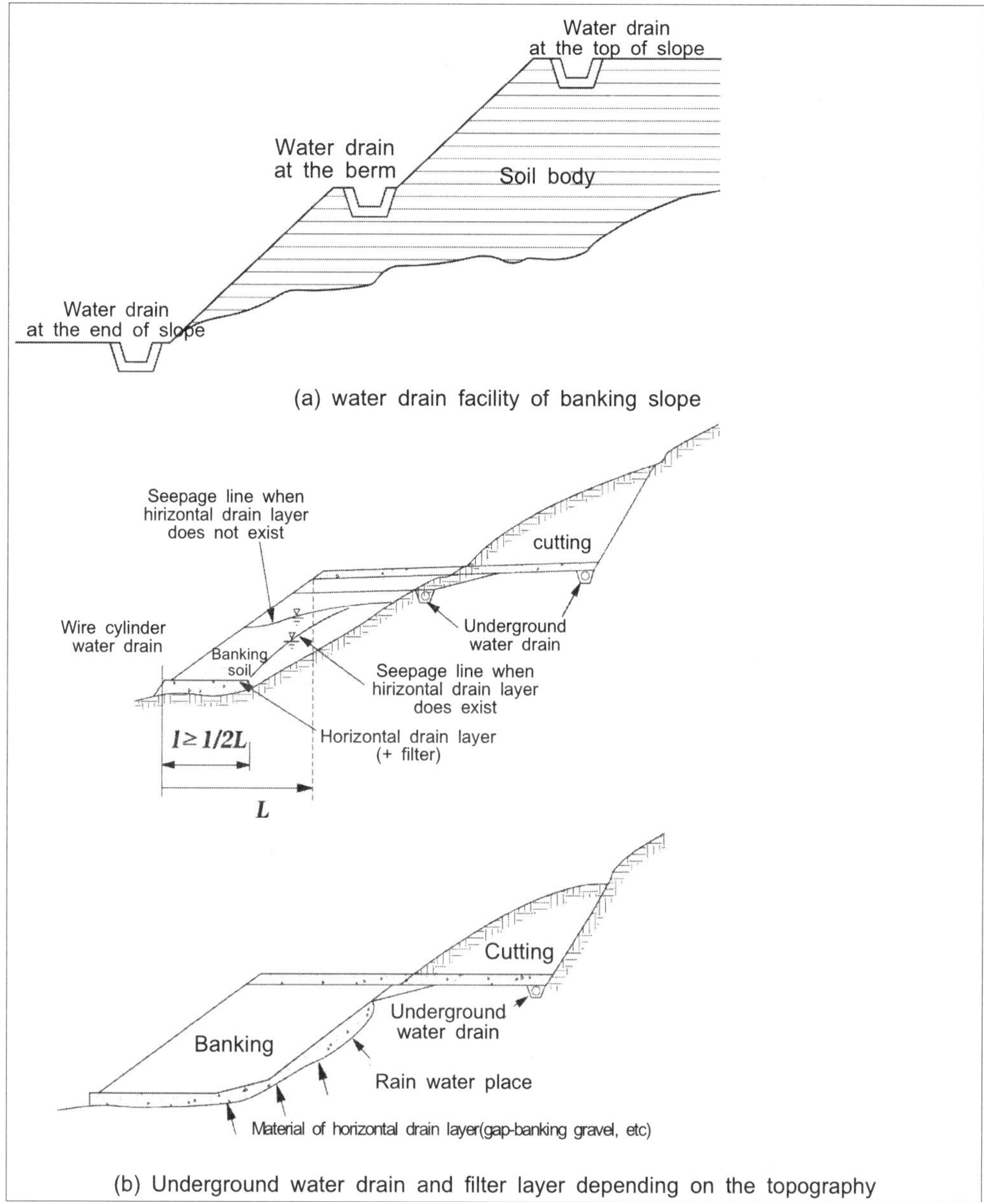

(a) water drain facility of banking slope

(b) Underground water drain and filter layer depending on the topography

⑧ Bench cutting

- In case of banking over sloped foundation over 1:4, bench cutting shall be done to enhance the attachment between the banking layer and original foundation to prevent the activity of banking layer.
- As surface water could penetrate to the border of the sloped foundation and banking layer to trigger an activity, water drains shall be installed above the sloped foundation and traversing slope of 3~5% shall be installed in the bench cutting surface.

【경사지반에서의 흙쌓기 단면】

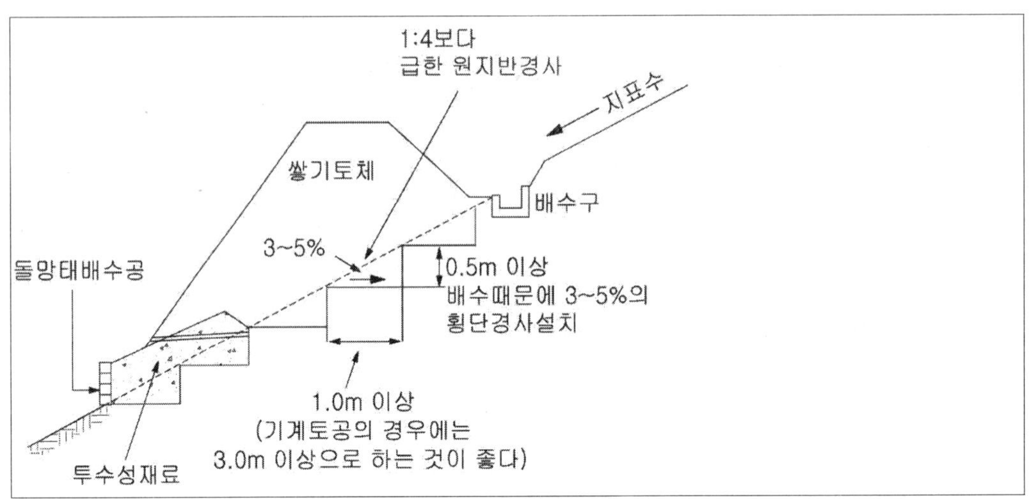

⑨ 한쪽깎기한쪽쌓기부

- 땅깎기 단부에서 흙쌓기부 노상 저면의 깊이까지 깎는 것을 원칙으로 하며, 1:4 정도의 경사로 땅깎기부 노상 저면에 접속

- 한쪽깎기한쪽쌓기부에서는 단차로 인하여 포장에 균열이 발생하기 쉬우므로 이를 방지하기 위하여 땅깎기부 노상저면(땅깎기면)과 흙쌓기부 노상 저면(상부 노체면)을 연결하는 접속구간을 설치

- 상부노체면 또는 땅깎기면에 지하배수구를 설치하고 배수 유출구로 유도 배수

【한쪽깎기한쪽쌓기부 층따기와 배수처리】

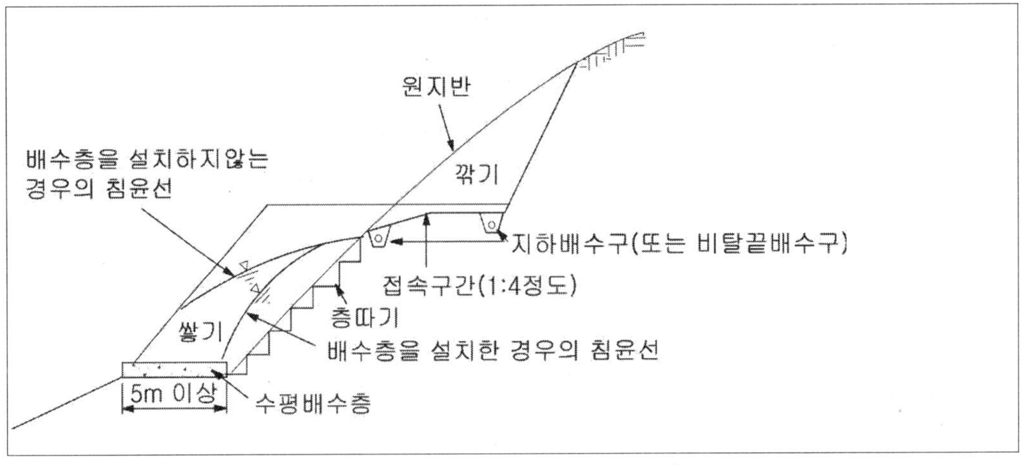

【Cross-section of the banking on sloped foundation】

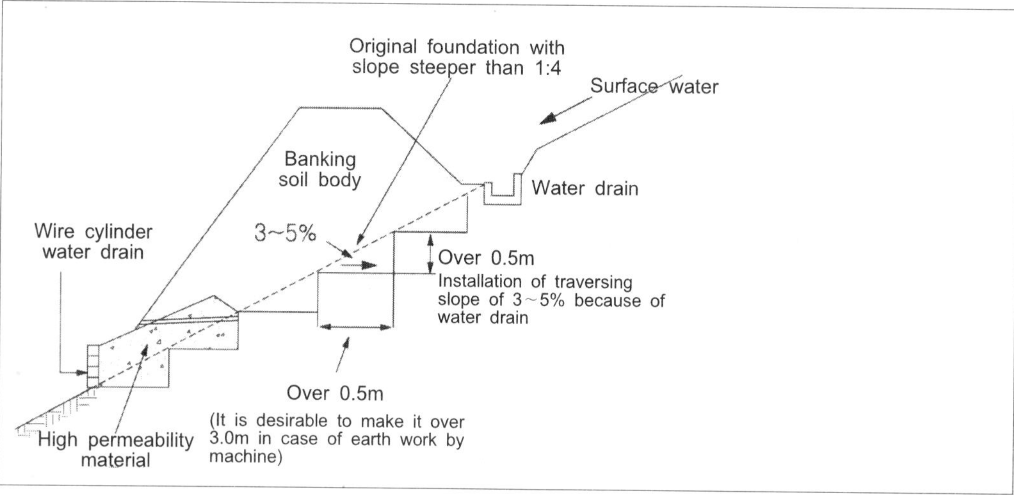

⑨ Cutting and embankment

• In principle, cutting shall be done from the end of soil cutting part to the depth of the lower road surface of soil banking area, and connected to the lower road surface of the cutting part with the slope of about 1:4.

• Pavement can be easily cracked because of the gap in this structure, and connection section shall be installed in order to prevent the crack between the lower road surface of cutting area (cutting surface) and lower road surface of banking area (upper road bed).

• Underground water drain shall be installed in the upper road bed or cutting surface, and drain water will be leaded to drain exit.

【Cutting and embankment part bench cutting and water drain facilities】

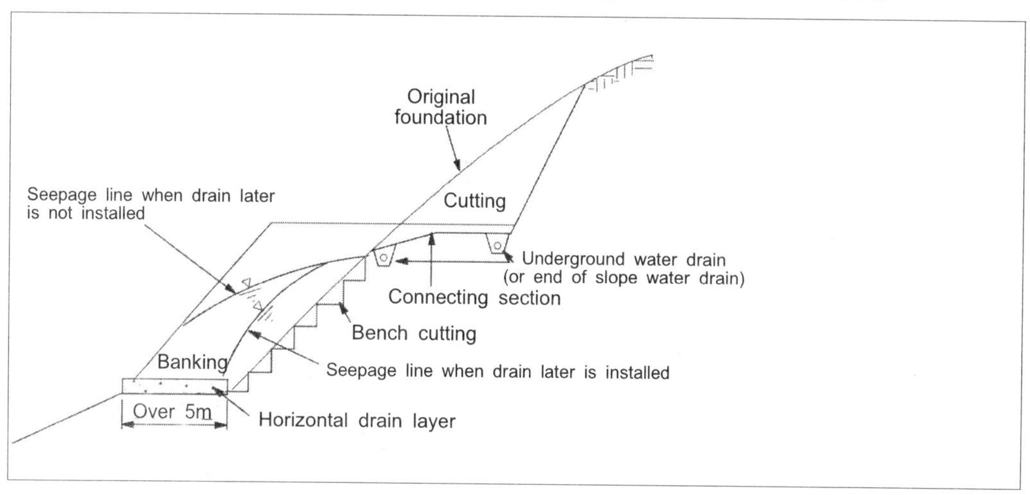

⑩ 땅깎기흙쌓기 경계부

- 흙쌓기부 노상 저면의 깊이까지 기존 지반을 굴착하는 것을 원칙으로 하고 굴착깊이를 서서히 감소시켜 땅깎기부 노상저면에 접속

【땅깎기흙쌓기 경계부 접속부 처리】

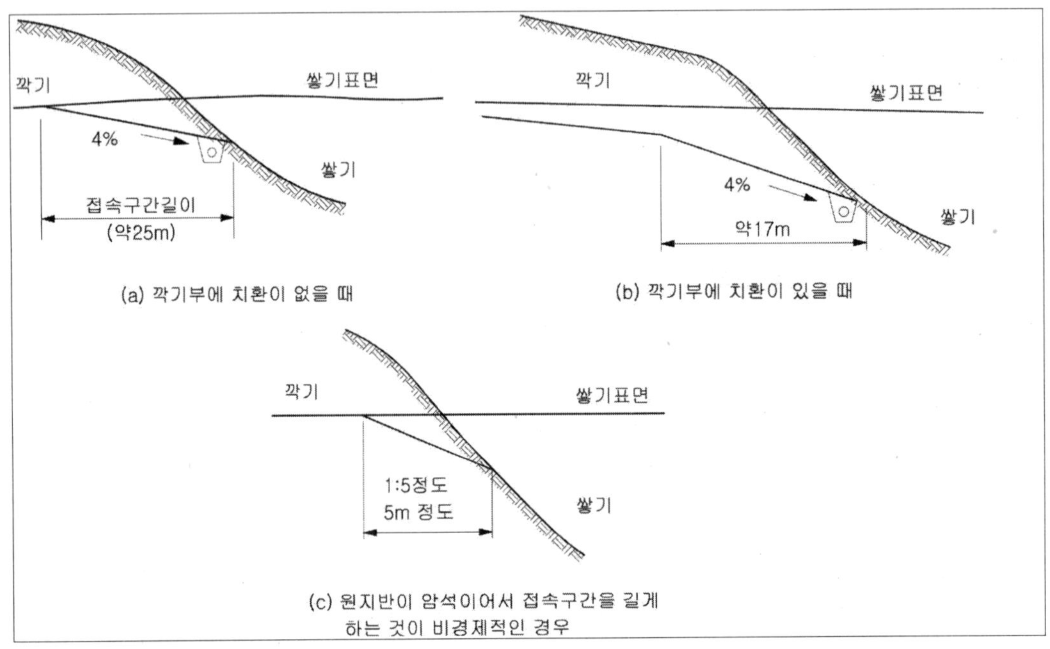

3-2-2 땅깎기 비탈면

① 땅깎기 비탈면 설계 시 주요고려사항

- 지형조건에 따른 땅깎기 계획
- 비탈면 안정해석 및 경사와 소단의 결정
- 지하수 및 지표수의 배수계획
- 장기적인 비탈면 표면보호 방법
- 유지관리를 위한 점검시설
- 시공 중 관리방안

② 땅깎기부 암발파 및 적용기준

- 발파의 영향으로 인한 소음, 진동, 비석 등의 환경공해를 저감시킬 수 있는 적정 발파공법의 적용기준을 설정
- 발파공사 시행 전 반드시 시험발파를 통하여 발파진동 추정식을 산출
- 시공성, 경제성 및 인근 보안물건의 안전성 등을 종합적으로 검토하여 적정 발파공법을 적용

⑩ Boundary of cu tting and banking

- In principle, cutting of original foundation shall be done to the depth of lower road surface of the banking area, and digging depth shall be slowly reduced to connect with the lower road surface of the cutting area

【Connection part at the boundary of banking and cutting area】

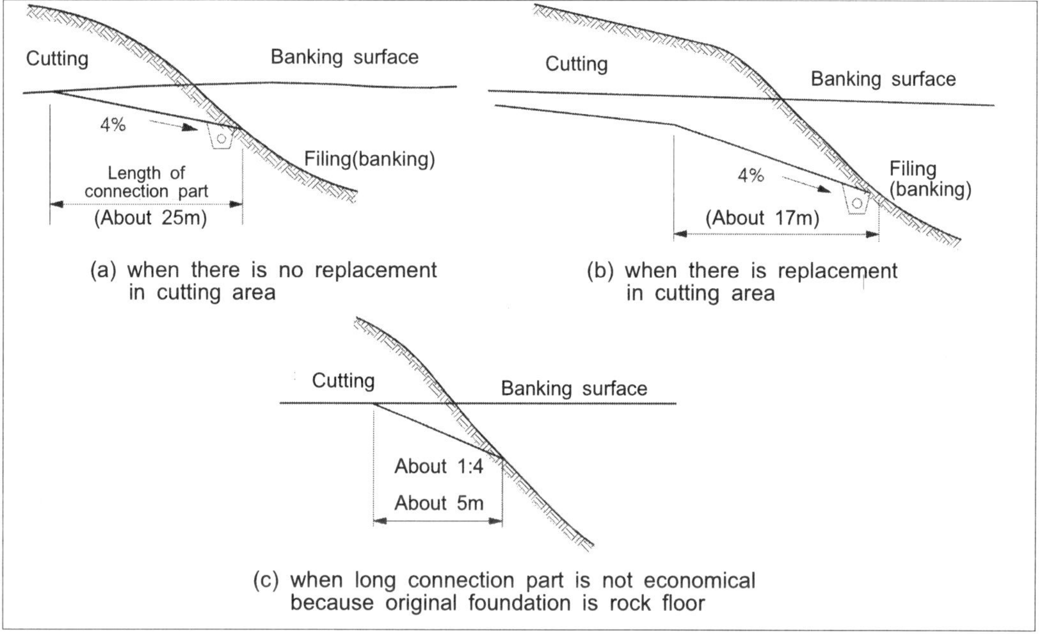

3-2-2 Soil Cutting Slope

① Major considerations in designing cutting slope

- Cutting plan depending on the topographic conditions
- Slope stability analysis and determination of inclination and berm
- Underground and surface water drain plan
- Long term slope surface protection method
- Monitoring facility for maintenance and control
- Control plan during construction

② Blasting at cutting area and application standard

- Application standard of appropriate blasting method to reduce the noise, vibration, stone debris, and other environmental pollution shall be set.
- Test blasting must be performed to draw the blasting vibration estimation equation before the main blasting.
- Appropriate blasting method shall be applied considering total factors including ease of construction, economic efficiency, safety of security building.

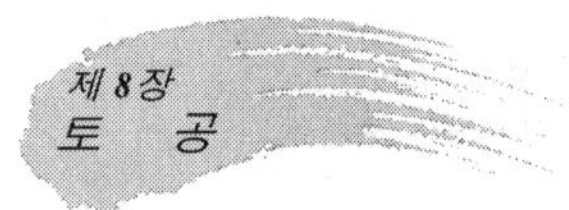

③ 땅깎기 비탈면 경사

- 원지반의 종류, 상태 및 비탈면 높이에 따라서 일반적인 지질의 경우 표준경사를 적용하며, 붕괴성 요인을 가진 지질부는 별도의 검토를 통하여 종합적 판단

- 암반 비탈면의 경우 지표지질조사 및 시추조사에 의하여 파악된 절리의 방향성 및 발달상태에 따라 안정해석을 실시하여 비탈면 경사를 우선적으로 결정

[토사 원지반 땅깎기 비탈면 표준경사]

원 지 반 의 토 질		비탈면 높이	경 사	분류기호 (통일분류)
모 래			1:1.5	SW, SP
사 질 토	밀실할 것	5m 이하	1:0.8~1:1.0	SM, SP
		5~10m	1:1.0~1:1.2	
	밀실하지 않고 입도분포가 나쁨	5m 이하	1:1.0~1:1.2	
		5~10m	1:1.2~1:1.5	
자갈 또는 암괴 섞인 사질토	밀실하고 입도 분포가 좋은 것	10m 이하	1:0.8~1:1.0	SM, SC
		10~15m	1:1.0~1:1.2	
	밀실하지 않고 입도분포가 나쁨	10m 이하	1:1.0~1:1.2	
		10~15m	1:1.2~1:1.5	
점 성 토		0~10m	1:0.8~1:1.2	ML, MH, CL, CH
암괴 또는 호박돌 섞인 점성토		5m 이하	1:1.0~1:1.2	GM, GC
		5~10m	1:1.2~1:1.5	

③ Cutting are slope inclination

- Standard slope shall be applied for the general soil properties depending on the type & state of original foundation and height of slope, and soil with potential collapse danger shall be evaluated comprehensively through the separate evaluation.

- In case of rock foundation slope, stability analysis shall be done depending on the directional property and development state of the dislocation identified by drilling test & surface geological feature investigation, and slope inclination shall be determined first.

【Standard inclination of cutting-slope in Original foundation】

<table>
<tr><th colspan="2">Soil property of original foundation</th><th>Height of slope</th><th>Inclination</th><th>Classification symbol (Unified classification)</th></tr>
<tr><td colspan="2">Sand</td><td></td><td>1:1.5</td><td>SW, SP</td></tr>
<tr><td rowspan="4">Sandy soil</td><td rowspan="2">Dense</td><td>Below 5m</td><td>1:0.8～1:1.0</td><td rowspan="4">SM, SP</td></tr>
<tr><td>5～10m</td><td>1:1.0～1:1.2</td></tr>
<tr><td rowspan="2">Not dense, poor grain size distribution</td><td>Below 5m</td><td>1:1.0～1:1.2</td></tr>
<tr><td>5～10m</td><td>1:1.2～1:1.5</td></tr>
<tr><td rowspan="4">Sandy soil mixed with gravel and rocks</td><td rowspan="2">Dense and good grain size distribution</td><td>Below 10m</td><td>1:0.8～1:1.0</td><td rowspan="4">SM, SC</td></tr>
<tr><td>10～15m</td><td>1:1.0～1:1.2</td></tr>
<tr><td rowspan="2">Not dense, poor grain size distribution</td><td>Below 10m</td><td>1:1.0～1:1.2</td></tr>
<tr><td>10～15m</td><td>1:1.2～1:1.5</td></tr>
<tr><td colspan="2">clay soil</td><td>0～10m</td><td>1:0.8～1:1.2</td><td>ML, MH, CL, CH</td></tr>
<tr><td colspan="2" rowspan="2">clay soil mixed with rocks and boulder</td><td>Below 5m</td><td>1:1.0～1:1.2</td><td rowspan="2">GM, GC</td></tr>
<tr><td>5～10m</td><td>1:1.2～1:1.5</td></tr>
</table>

[암반의 특성에 따른 비탈면 표준경사]

암반 구분 (굴착난이도)	암반 파쇄 상태 NX 시추 시(BX) TCR(%)	 RQD(%)	굴 착 난이도	경 사	소단 설치	비 고
풍화암 또는 연·경암으로 파쇄가 극심한 경우	20% 이하 (5% 이하)	10 %이하 (0%)	리핑암	1:1.0 ~ 1:1.2	H=5m마다 1m 폭	*최하단기준 매20m마다 3m 소단설치 *발파암과 리핑암 사이에는 소단을 설치하지 않음 *소단사이에 토사와 리핑 구분선의 발생 시 많은 쪽 비탈면 경사를 적용
강한 풍화암으로 파쇄가 거의 없는 경우와 대부분의 연·경암	20~40% (10~30%)	10~25% (0~10%)	발파암 (연암)	1:0.8 ~ 1:1.0	H=10m마다 1~2m 폭	
	40~60% (30~50%)	25~50% (10~40%)	발파암 (보통암)	1:0.7		
	60% 이상 (50%이상)	50% 이상 (40%이상)	발파암 (경암)	1:0.5	H=20m마다 3m 폭	

④ 땅깎기 비탈면의 안정검토 대상

- 비탈면 높이가 10m 이상인 경우
- 비탈면 경사가 표준경사보다 급한 경우
- 비탈면 지반이 붕적토로 이루어진 경우
- 암반의 풍화가 심하고 용수가 많은 경우
- 붕괴 시 복구가 장시간 소요되거나 주변 인접 시설물에 중대한 인명, 재산상 피해를 주는 경우
- 불연속면이 비탈면 방향으로 경사진 지질조건의 경우
- 내진안정해석이 필요한 경우
- 불안정 요인을 갖는 지형·지질 조건인 경우
- 위 조건 외에 설계자가 필요하다고 판단하는 경우

[Standard inclination of slope according to the characteristics of rock floor]

<table>
<tr><th rowspan="3">Rock floor classification (digging difficulty)</th><th colspan="2">Rock floor spall status</th><th rowspan="3">Digging difficulty</th><th rowspan="3">Inclination</th><th rowspan="3">Berm installation</th><th rowspan="3">Remarks</th></tr>
<tr><th colspan="2">When NX drilling (BX)</th></tr>
<tr><th>TCR(%)</th><th>RQD(%)</th></tr>
<tr><td>Efflorescence rock or hard · soft rock, where spallation is significant</td><td>Below 20%
(Below 5%)</td><td>Below 10 %
(0%)</td><td>Ripping rock</td><td>1:1.0
~
1:1.2</td><td>H=every 5m
1m width</td><td rowspan="4">* With the lowest part as reference, 3m berm shall be installed at every 20 meters.
* Berm shall not be installed between blasting rock and ripping rock
* When soil and ripping lines are formed between berms, apply the inclination of the slope with more lines.</td></tr>
<tr><td rowspan="3">Strong efflorescence with trivial level of spallation, and most of soft and hard rock</td><td>20~40%
(10~30%)</td><td>10~25%
(0~10%)</td><td>Blasting rock (soft rock)</td><td>1:0.8
~
1:1.0</td><td rowspan="2">H=every 10m
1~2m width</td></tr>
<tr><td>40~60%
(30~50%)</td><td>25~50%
(10~40%)</td><td>Blasting rock (soft rock)</td><td>1:0.7</td></tr>
<tr><td>Over 60%
(Over 50%)</td><td>Over 50%
(Over 40%)</td><td>Blasting rock (hard rock)</td><td>1:0.5</td><td>H=every 20m
3m width</td></tr>
</table>

④ The stability of soil cutting slope has to be evaluated for the following cases

- When the height of the slope is over 10 m
- When the inclination of the slope is higher than standard inclination
- When the foundation of slope surface is made of colluvial deposit
- Rock floor is heavily effervescent, and contains large amount of rain water
- When the recovery from the collapse could take long time, or the collapse could result serious damage to lives and properties.
- Geological conditions with the discontinued surface are sloped in the direction of slope surface
- When earthquake-proof analysis is required
- When topographical or geological features have unstable factors.
- When designers decide that the evaluation is necessary besides the above cases

⑤ 땅깎기 비탈면의 안정 검토 시 고려사항

- 비탈면 안정 해석 시 지하수 조건은 지반조사 결과를 종합적으로 판단하여 안정성에 가장 불리한 상태 발생조건에 대해서 수행

- 땅깎기 비탈면 안정해석은 비탈면 내의 지하수위와 및 시공속도에 따른 장·단기적인 배수 조건을 고려하여 유효응력해석 또는 전응력해석을 수행

- 실트 및 모래질로 구성된 땅깎기 비탈면은 강우를 고려한 침투해석을 실시한 후 얻어지는 간극수압분포를 이용하여 안정해석을 수행

- 불연속면에 기인한 파괴가 예상되는 경우 불연속면의 전단강도를 이용하여 안정해석을 수행

⑥ 땅깎기 비탈면의 붕괴유형

지반구성물질	붕괴유형
토사 비탈면	원호파괴, 표층파괴, 표층유실
혼합 비탈면	원호파괴, 표층파괴 표층유실, 낙석, 평면·쐐기파괴
암반 비탈면	낙석, 평면·쐐기파괴, 전도파괴, 표층파괴, 표층유실

붕괴 유형	붕괴 개요	모식도	붕괴 사례
평면파괴	비탈면 경사방향과 동일한 방향을 가지고 있는 불연속면을 따라 파괴		
쐐기파괴	두 불연속면이 경사면을 비스듬히 가로질러 교차하고 그 교차선을 따라 파괴		
원호파괴	지반이 매우 약하거나 절리의 발달이 아주 심한 파쇄암반에서 원호 혹은 원형의 경로를 따라 발생하는 활동파괴		
전도파괴	암반 블록들이 비탈면 경사방향으로 회전하여 파괴		

⑤ Considerations during the stability evaluation of the soil cutting slope

- During the stability analysis of the slope, the underground water condition shall be analyzed for the worst case scenarios in terms of stability by comprehensive identification based on the foundation investigation result.

- The stability analysis of the cutting slope shall be done for effective stress analysis and total stress analysis considering the short term and long term drain conditions depending on the construction speed and level of underground water inside the slope.

- For the cutting slope made of silt and sandy soil, penetration analysis considering rainfall and stability analysis shall be executed with using obtained pore water pressure distribution.

- When failure due to discontinuous surface is expected, stability analysis using the shear strength of the discontinuous surface shall be performed.

⑥ Type of collapse on soil cutting slope

Foundation components materials	Type of collapse
Soil slope surface	Arc failure, outer layer failure, sweep of outer layer
Mixed slope surface	Arc failure, outer layer failure, sweep of outer layer, rockfall, horizontal/wedge failure
Rock floor slope surface	Rockfall, horizontal/wedge failure, overturning failure, outer layer failure, sweep of outer layer

Type of collapse	Collapse overview	Schematic drawing	Collapse example
Horizontal failure	Failure along the discontinuity surface that has the same direction with slope inclination		
Wedge failure	Two discontinuous surfaces are crossing with an angle with slope, and the failures are along the crossline lines		
Arc failure	In soft foundation or serious spallation foundation because of well developed dislocations, this is activity failure with the failures along the arc or circular path		
Overturning failure	Failures resulted by rock floor blocks rotating along the direction of the slope		

⑦ 조사 및 시험

- 땅깎기 비탈면의 붕괴 위험정도에 따라 암종을 4등급으로 분류하여 안정성해석을 위한 기초자료로 적용

[조사등급의 분류]

등급	해 당 지 역
A급	경상도 일대 퇴적암 지역 및 사전 지표지질조사 결과 지역인근에서 산사태성 대규모 붕괴가 다수 발생한 지역으로 비탈면의 높이가 20m 이상인 경우
B급	경상도 일대 퇴적암 및 사전 지표지질조사 결과 지역인근에서 절리방향이 불량하여 표준 비탈면 경사로 시공 시 붕괴가 다수 발생되는 지역으로 비탈면의 높이가 20m 이상인 경우
C급	절리방향이 불량하여 인근에서 비탈면 붕괴가 종종 발생되는 지역, A·B등급에 해당하는 지역으로 비탈면의 높이가 20m 미만인 경우, 특별히 안정검토가 필요한 경우
D급	비탈면이 안정한 지역으로 소규모의 붕괴 또는 낙석발생이 우려되는 지역 및 단순 확장구간으로 기 노출된 비탈면으로 절리의 방향 및 거칠기 등을 추정할 수 있는 구간

⑦ Investigation and test

- Cutting slope are classified into four classes depending on the level of collapse danger, and they are applied as base data for stability analysis.

【Classification of the investigation class】

Class	Corresponding area
Class A	Sedimentary rock area around Gyeongsang-do and the area where large-scaled landslide-type collapse frequently occurred from the result of pre-investigation for the surface geological feature survey, and height of the slope is over 20 meters
Class B	Sedimentary rock area around Gyeongsang-do and frequent collapse areas during the construction of standard slope because of the poor dislocation direction from the result of pre-investigation for the surface geological feature survey , and height of the slop is over 20 meters.
Class C	Area where slope collapse happens often because of poor dislocation direction. Could be classified to A and B class except the case that the height of the slope is lower than 20 meters. When specially stability evaluation is required.
Class D	Area where slops are stable, and small scale collapse and rockfall could happen. Simple expansion section, where the direction of location and coarseness can be estimated from the existing exposed slope

[현장 조사항목 및 빈도]

구분	시추조사 및 현장시험
A급	• 지표지질조사 • 필요 시 전기비저항탐사 또는 탄성파탐사 • 시추조사 3공(1공에 대하여는 풍화암층을 3중관 샘플러 사용 불교란시료 채취, 시추공 전단강도 시험) • 시추공 화상정보 시험(BIPS)
B급	• 지표지질조사 • 필요 시 전기비저항탐사 또는 탄성파탐사 • 시추조사 2공(1공에 대하여는 풍화암층을 3중관 샘플러 사용 불교란시료 채취, 시추공 전단강도 시험) • 시추공 화상정보 시험(BIPS)
C급	• 지표지질조사 • 필요 시 전기비저항탐사 또는 탄성파탐사 • 시추조사 2공(1공에 대하여는 풍화암층이 5m 이상인 경우 3중관 샘플러 사용 풍화암층 불교란시료채취, 시추공 전단강도 시험) • 시추공 화상정보 시험(BIPS)
D급	• 지표지질조사 • 필요 시 전기비저항탐사 또는 탄성파탐사 • 시추조사 1공(풍화암층이 5m 이상인 경우 풍화암층을 3중관 시추기를 이용하여 불료란시료 채취) • 시추공 화상정보 시험(BIPS)

⑧ 땅깎기 비탈면의 표면수 및 용수 처리

- 표면수나 용수에 의하여 비탈면의 세굴 또는 붕괴 우려가 있을 경우 비탈어깨나 소단에 배수구를 설치

- 땅깎기 비탈면의 배수시설은 비탈면 안정에 있어서 매우 중요한 역할을 수행하므로 땅깎기 비탈면 내부로 유입되는 표면수를 억제시키고, 표면수와 지하수는 신속하게 배수시키는 시설을 설치하여 표면수 및 용수를 적절히 처리할 수 있도록 설계

[땅깎기 비탈면의 배수시설]

지표수 배수시설	지하수 배수시설
• 산마루배수구 • 소단배수구 • 종배수구 • 비탈끝배수구	• 수평배수공 • 수직배수공(배수 우물) • 배수터널

【Site survey items and frequency】

Classification	Drilling test and field test
Class A	• Surface soil property investigation • Electric resistivity and seismic wave exploration if necessary • Drilling test, 3 holes (To take undisturbed sample of efflorescence rock layer using three fold tube sampler for one hole, drilling hole shear strength test) • Drilling hole image information test (BIPS)
Class B	• Surface soil property investigation • Electric resistivity and seismic wave exploration if necessary • Drilling test, 2 holes (To take undisturbed sample of efflorescence rock layer using three fold tube sampler for one hole, drilling hole shear strength test) • Drilling hole image information test (BIPS
Class C	• Surface soil property investigation • Electric resistivity and seismic wave exploration if necessary • Drilling test, 2 holes (To take undisturbed sample of efflorescence rock layer using three fold tube sampler for one hole, drilling hole shear strength test) • Drilling hole image information test (BIPS
Class D	• Surface soil property investigation • Electric resistivity and seismic wave exploration if necessary • Drilling test, 1 one(If the efflorescence rock layer is over 5 meters, undisturbed sample from the efflorescence rock layer shall be taken using the three fold tube driller) • Drilling hole image information test (BIPS)

⑧ Treatment of surface water and rain water on the cutting slope

- When there is a concern on erosion or collapse of slope due to surface water or rain water, water drain shall be installed either at the top of slope or at berm.

- Since water drain at cutting slope is very important for the stability of the slope, water drain facilities shall be designed to suppress the inflow of surface water into the slope and drain the surface and underground water quickly out of the slope to process the surface and rain water appropriately.

【Water drain facility of cutting slope surface】

Surface water drain facilities	Underground water drain facilities
• Mountain ridge drain • Berm drain • Vertical drain • Slope bottom drain	• Horizontal drip hole • Vertical drip hole(drain well) • Drain tunnel

[땅깎기 비탈면의 배수시설]

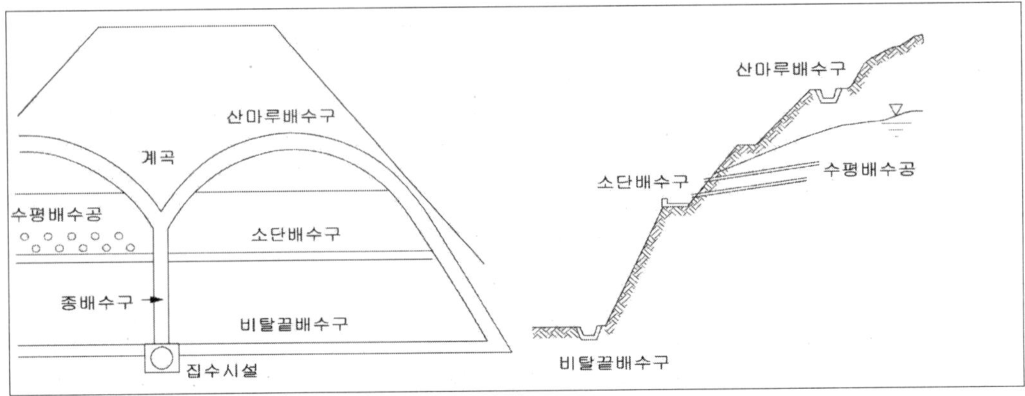

⑨ 대규모 땅깎기 비탈면(비탈면 높이 20m 이상)

- 장기적으로 불안정해질 가능성이 높으므로 정기적인 유지관리를 위하여 비탈면의 주요 부위에 접근이 가능하도록 점검로를 설치

- 비탈면 내에 점검로로 이용될 수 있는 약 3m 정도 폭의 소단을 설치하는 것이 바람직

3-2-3 비탈면 보호공

① 식생공을 원칙으로 적용

- 식생공이 적절하지 않은 경우 또는 식생공만으로는 비탈면의 안정을 확보할 수 없는 경우 여건을 고려하여 구조물에 의한 보호공 적용

- 구조물에 의한 보호공을 적용하는 경우 공사비가 증가하므로 지질, 비탈면의 안정성, 경제성, 미관, 유지보수, 기타 상황 등을 충분히 고려한 후 적절한 보호공을 결정

구 분	보 호 공	주 요 목 적
식생공	씨앗 뿜어붙이기공, 식생 매트공, 식생줄떼공, 줄떼공, 식생판공, 식생망태공, 부분 객토 식생공, 목본류 식생공	식생에 의한 비탈면 보호, 녹화, 구조물에 의한 비탈면 보호공과의 병용
구조물에 의한 보호공	콘크리트 블럭 격자공, 모르타르 뿜어 붙이기공, 블럭 붙임공, 돌붙임공	비탈표면의 풍화 침식 및 동상 등의 방지
	현장타설 콘크리트 격자공, 콘크리트 붙임공, 비탈면 앵커공	비탈 표면부의 붕락 방지, 약간의 토압을 받는 흙막이
	비탈면 돌망태공, 콘크리트 블록 정형공	용수가 많은 곳 부등침하가 예상되는 곳 또는 다소 튀어나올 염려가 있는 곳의 흙막이
	블록쌓기공, 돌쌓기공	흙막이

[Drain facilities of cutting slope]

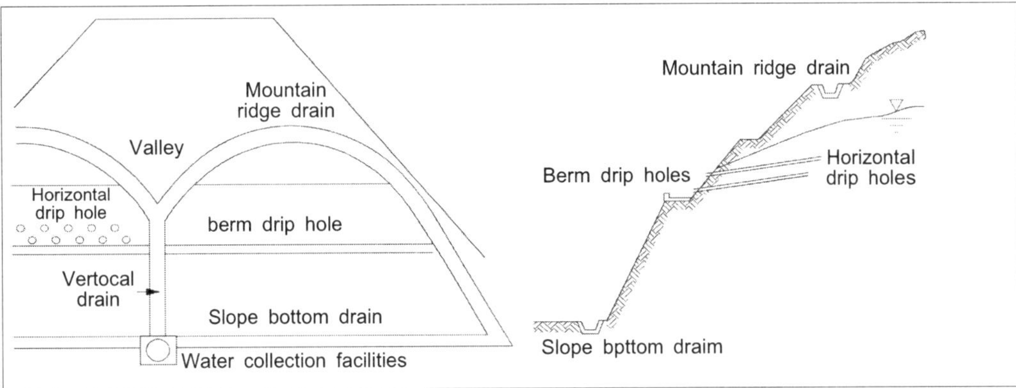

⑨ Giant cutting slope (slope height over 20 meters)

- As this structure has a possibility to be unstable in long term, service road to access the major parts of the slops shall be installed.

- It is desirable to install the berm with the width of 3 meters that can be used as a service road of the slope

3-2-3 Slope Protection Method

① Sodding planting shall be applied in principle

- When sodding planting is not appropriate or sodding planting does not provide enough stability of the slope, protection by structure shall be applied considering the conditions.

- Since the protection method by structure increases the construction cost, appropriate protection method shall be determined after sufficient considerations on soil property, stability of slope, economic efficiency, appearance, maintenance and repair, and other situations.

Classification	Slope Protection	Major Purpose
Sodding planting	Seed blasting method, planting method, planting line sodding method, line sodding method, planting plate method, planting basket method, partial soil dressing planting method, woody plant sodding planting method	Protection of slope by sodding planting, afforestation, combination with the protection of slope by structure
Slope protection by structures	Concrete block, grid method, mortar blasting method, block pitching, stone pitching	Efflorescence erosion and prevention of frost of slope surface
	Field concrete grid method, concrete pitching method, slope surface anchor method	Prevention of slope surface collapse, sheeting for certain level for soil pressure
	Slope surface wire cylinder method, concrete block fixed shape method	Sheeting where large amount of rain water exist, different settlement is expected, and there is concern of pop out
	Block stacking method, stone stacking method	Sheeting

3-3 산사태 대책 수립

3-3-1 산사태 발생우려 지역

① 산사태 발생우려가 있는 지역은 가능한 우회하는 것을 원칙

- 우회가 곤란한 경우 또는 공사 중 및 공용 중 산사태가 발생할 우려가 있는 경우 안정성, 경제성, 시공성을 고려하여 적절한 대책을 강구

3-3-2 산사태의 안정해석

① 산사태 안정해석은 비탈면의 안정 확보에 필요한 대책공사의 규모를 결정하기 위하여 실시

② 산사태 안정해석은 절편 분할법을 사용하는 것을 원칙

3-3-3 산사태의 대책공법

① 산사태 대책공법은 억제공을 기본으로 설계

② 산사태 대책공법은 억제공과 억지공으로 대별되나 각 공법의 특성, 산사태 발생우려 구간의 지형, 지질, 발생 메카니즘 등을 고려하여 적절한 공법을 조합

[주요 산사태 대책공법의 종류]

구분	억제공(안전율 유지법)	억지공(안전율 증가법)
개요	산사태 지형, 지하수 상태 등 자연조건을 변화시켜 안전율을 유지	구조물을 산사태 지역에 시공하여 비탈면의 안전율을 증가
종류	• 지하수 배제공 (수로공, 침투방지공) • 지하수 배제공 • 천층 지하수 배제공 (지하배수공 등) • 심층 지하수 배제공 (물빼기 보링공, 집수정공, 배수터널 등)	• 활동력 감소법 (배토공, 압성토공) • 저항력 증가법 (억지말뚝공, 앵커공)

3-3 Establishment of Plan for Landslide

3-3-1 Area in danger of landslide

① In principle, area in danger of landslide shall be avoided if possible.

- When detour is difficult and there is a danger of landslide during or after the construction, appropriate countermeasure shall be established considering the stability, economic efficiency, and ease of construction.

3-3-2 Stability Analysis of Landslide

① Landslide stability analysis shall be executed to determine the scale of countermeasure construction for securing the stability of the slope.

② Slice division method shall be used for landslide stability analysis in principle.

3-3-3 Landslide Countermeasure Construction Method

① Landslide countermeasure method shall be designed with control method as a basic requirement.

② Landslide countermeasure construction method are largely divided into control method and prevention method, and proper method shall be combined considering the characteristics of each method, topography of the area in danger of landslide, soil property, landslide mechanism, etc.

【Type of major landslide countermeasure method】

Classification	Control method (Keeping the safety rate)	Prevention method (Improving the safety rate)
Overview	To keep the safety rate by changing the natural conditions like topography, underground water conditions	To improve the safety rate by installing structures in the area of danger
Type	• Underground water removal method (Water drain method, penetration prevention method) • Underground water removal method • Superficial layer underground water removal method (Underground water drain method) • Deep underground water removal method (Water drain boring, water collection method, drain tunnel, etc)	• Activity reduction method (Earth removal work, loading berm method) • Resistance increase method (Suppression pile method, anchor method)

3-4 연약지반의 설계

3-4-1 연약지반의 정의

① 점토, 실트와 같은 미세한 입자의 흙이나 간극이 큰 유기질토, 느슨한 모래 등으로 이루어진 토층으로 구성

② 지하수위가 높고, 제체 및 구조물의 안정과 침하문제를 발생시키는 지반

3-4-2 연약지반의 판정기준

① 포화된 점토층, 느슨한 사질토층, 유기질 성분이 다량 함유된 지층이 일반적인 범주의 연약지반에 해당

[연약지반의 판정기준]

구 분	이탄질 지반 및 점토질 지반		사질토 지반
연약층 두께	10m 미만	10m 이상	-
N값	4 이하	6 이하	10 이하
일축압축강도 qu(kgf/㎠)	0.6 이하	1.0 이하	-
콘관입저항값 qc(kgf/㎠)	8 이하	12 이하	40 이하

3-4 Soft Foundation Design

3-4-1 Definition of soft foundation

① Soil layer formed by fine grains soil like clay or silt, organic soil with big void gaps, or loose sand.

② Type of foundation with high level of underground water, and causes troubles in stability and settlement of the entire road bed and structures.

3-4-2 Soft Foundation Judgement Criterion

① Saturated clay soil layer, loose sandy soil layer, and the layers of abundant organic components are generally considered as soft foundation

【Judgement criteria of soft foundation】

Classification	Peat soil and clay soil foundation		Sandy soil foundation
Thickness of soft layer	Below 10m	Over 10m	-
N value	Below 4	Below 6	Below 10
Unconfined compression qu(kgf/cm^2)	Below 0.6	Below 1.0	-
Cone penetration resistivity qc(kgf/cm^2)	Below 8	Below 12	Below 40

[토질특성에 따른 연약지반 판정기준]

지반구분	토층 및 토질구분			토질정수 wn (%)	토질정수 eo	토질정수 qu(kPa)	토질정수 N값
이탄지반	고유기질토	이탄토	섬유질 고압축토	300 이상	75 이상	40 이하	1 이하
		흑 니	분해가 진척된 고유기질토	300~200	7.5~5.0		
점토지반	세립토	유기질토	소성도 A선 이하의 유기질토	200~100	5.0~2.5	100 이하	4 이하
		화산회질 점토	소성도 A선 이상의 화산회질 2차 퇴적 점성토				
		실 트	소성도 A선 이하, Dilatancy가 큰 세립토	100~50	2.5~1.25		
		점 토	소성도 A선 이상 Dilatancy가 큰 세립토				
사질지반	사질토	SM, SC	0.074mm체 통과량 15~20%	50~30	1.25~0.8	≒ 0	10 이하
		SP-SC SW-SM	0.074mm체 통과량 15% 이하	30 이하	0.8 이하		

[구조물의 종류에 따른 개략적인 연약지반 판정기준]

구조물의 종류	지반상태 토 질	지반상태 N값	지반상태 qu (kPa)	지반상태 qc (kPa)	판 정
도 로	-	2 이하	25 이하	125 이하	초연약
		2~4	25~50	125~250	연 약
		4~8	50~100	250~500	보 통
고속도로	이탄층	4 이하	50 이하	-	연약지반
	점성토	4 이하	50 이하	-	
	사질토	10 이하	-	-	

여기서, qu : 일축압축강도, qc : 콘관입 지지력, qb : 장기 허용지내력

【Judgement criteria of soft foundation based on the soil property】

Foundation classification	Soil layer and soil property classification			Soil property constant			
				wn (%)	eo	qu(kPa)	N value
Peat foundation	High organic soil	Peat soil	Fiber high compression soil	Over 300	Over 75	Below 40	Below 1
		Much	High organic soil with the progression of decomposition	300~200	7.5~5.0		
Clay foundation	Fine grain soil	Organic soil	Organic soil with plasticity lower than A line	200~100	5.0~2.5	Below 100	Below 4
		Volcanic ash clay soil	Volcanic ash secondary deposit clay layer with the plasticity higher than A line				
		Silt	Fine grain soil with plasticity lower than A line and high dilatancy	100~50	2.5~1.25		
		Clay	Fine grain soil with plasticity higher than A line and high dilatancy				
Sandy soil foundation	Sandy soil	SM, SC	0.074mm grain pass rate 15~20%	50~30	1.25~0.8	≒ 0	Below 10
		SP-SC SW-SM	0.074mm grain pass rate below 15%	below 30	Below 0.8		

【High level soft foundation judgement criteria based on the type of structure】

Type of structures	Foundation status				Judgement
	Soil property	N value	qu (kPa)	qc (kPa)	
Road	-	Below 2	Below 25	Below 125	ultra soft
		2~4	25~50	125~250	soft
		4~8	50~100	250~500	normal
Expressway	peat soil	Below 4	Below 50	-	soft foundation
	clay soil	Below 4	Below 50	-	
	sandy soil	Below 10	-	-	

Where, qu : Unconfined compression strength, qc : Con penetration support force,
qb : long term allowable soil pressure

3-4-3 연약지반의 공학적 문제점

① 일반적인 특징 : 전단강도가 작고, 압축성이 큼

- 연약한 점성토층 - 흙 입자 사이의 간극이 크고, 자연함수비가 매우 높으며, 세립분의 함량이 많아 소성성(plasticity)이 큼

- 느슨한 사질토층 - 흙 입자 사이의 간극이 상대적으로 크고, 입자간 접촉 면적 작음

- 작은 충격에도 입자 구조가 쉽게 파괴되며, 큰 변형 발생

② 연약한 점토층 지반 위에 흙쌓기 하는 경우

- 흙쌓기 하중에 의한 장기간의 압밀 현상으로 많은 침하가 발생

- 하부 지반의 활동으로 흙쌓기체가 붕괴 가능

- 도로의 경우 구조물 주변 토공 구간에서 부등침하가 많이 발생하며, 기존 도로에 인접하여 확장되는 구간에서도 피해 발생

③ 말뚝기초 등으로 지지되는 구조물의 경우

- 점토층의 침하로 말뚝과 주변 지반의 상대변형이 발생하면서 지층이 말뚝을 아래로 끌어내리고자 하는 힘, 즉 부마찰력(negative skin friction)이 작용하여 하중이 추가

④ 흙의 강도 및 지지력이 작은 경우

- 흙쌓기 높이가 높아지면서 지중 응력이 증가하여 흙쌓기 하중에 의한 활동력이 지반의 전단 저항력을 초과하면 파괴가 발생

3-4-4 설계정수 결정

① 사운딩 등의 원위치시험, 보링주상도 및 토질시험 등으로 나누어 각 조사 지점마다 정리 한 후, 이들의 결과를 정리하여 토질단면도를 작성

② 토층 구분을 하여 검토 단면에 있어서 안정 및 침하계산을 위한 설계정수를 결정

3-4-3 Engineering Problem of Soft Foundation

① General characteristics: low shear strength, high compressibility

- Soft clay soil layer - void gap of soil grain is big, natural moisture content is very high, and high plasticity due to large contents of fine grained soil

- Loose sandy soil layer - gaps between soil grains are relatively big, and contact area between grains are small

- Grain structure can easily be destroyed by a small shock, and large displacement could occur

② In case of banking on the soft clay layer foundation

- Large settlement occur because of long term compression by the filing load.

- Danger of collapse of banking soil due to activity of the lower part of foundation

- In case of road, many differential settlements occur in the earth work section around the structures, and damage could occur to the extension road adjacent existing road.

③ In case of structures supported by pile bases

- Load is added because of the settlement of the clay layer, and negative skin friction - the force of soil layer to pull down the pile to the bottom while relative displacement of the foundation occurs

④ When the strength and support force of the soil is small

- When the height of the banking soil is high, inner soil stress increases and activity by the load of banking soil exceeds the shear resistance of the foundation, and this results failure.

3-4-4 Determination of Design Constant

① Test result from in-situs test like sounding, boring log, and soil test for each test points shall be summarized and the soil sectional diagram shall be prepared based on the results.

② Design constant for the calculation of stability and settlement with the evaluation section identifying each soil layer.

3-4-5 검토 단면도의 작성

① 토층의 구성, 층 두께, 기반의 상황(경사 등), 제체 높이 등에 따라서 대표적인 검토 지점을 선정

② 선정지점에 대하여 토성 종단도, 기반의 등고선도, 토성 일람도 등을 이용하여 검토 단면도 작성

3-4-6 연약지반의 안정검토

① 활동면은 단일 활동원호로 하고, 계산식은 전응력법에 따라 계산하는 것을 원칙

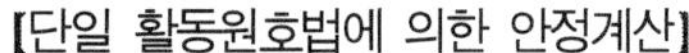
[단일 활동원호법에 의한 안정계산]

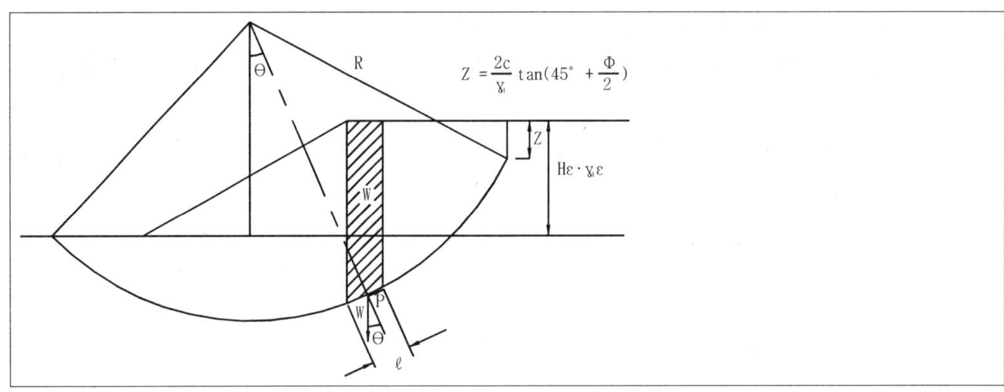

② 안정설계 시 활동파괴에 대한 최소 안전율은 공용초기의 1.3배 이상을 목표

③ 안정검토 계산 시 유의사항

- 연약층 하부의 기반이 경사져 있는 지반 또는 강도저하가 현저하고 강도회복이 늦은 지반은 안정상의 문제가 발생할 소지가 많으므로 세심한 검토 필요
- 제체재료의 전단강도 및 단위체적중량은 안정계산 결과에 미치는 영향이 크므로 실내 및 현장시험 결과 등을 검토하여 적정한 값을 사용

3-4-7 연약지반의 침하검토

① 연약지반상에 있는 제체의 침하를 검토하는 경우 시간 효과를 활용하여 운영 후의 도로에 미칠 침하를 예측하여 계획, 설계 및 시공하는 것을 기본

② 침하량 산정 시 고려사항

- 흙쌓기하중, 포장하중(노상+기층+표층), 침하토하중(침하보충토), 교통하중

3-4-5 Preparation of the Evaluation Sectional Diagram

① Select the representative evaluation point depending on the composition of soil layer, layer thickness, states (slope, etc) of the foundation, and height of the total objects.

② Evaluation sectional diagram shall be prepared for the selected point using soil profile, contour line diagram, and soil overview diagram.

3-4-6 Stability Evaluation of Soft Foundation

① Active surface shall be single active arc, and the calculation shall be based on the total stress method in principle.

【Calculation of stability based on the single active arc method】

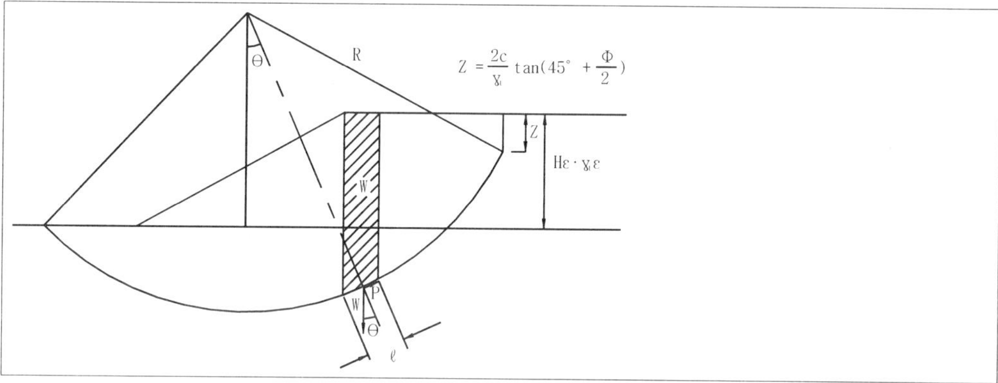

② During the stability design, the target minimum safety rate for the sliding failure shall be over 1.3 times the value of initial opening to public.

③ Caution during the calculation of stability evaluation

- Since the foundations with sloped bottom of the software foundation, noticeable strength degradation, or slow recovery of strength are susceptible to stability problems, thorough review is required.

- Since the shear strength and unit volume weight of the entire object have a big impact to the result of stability calculation, appropriate values shall be used through the evaluation of indoor and field test result.

3-4-7 Settlement Evaluation of Soft Foundation

① When evaluating the settlement of the objects over the soft foundation, planning, design, execution, and construction shall be done with the settlement effects predicted using the effects of time.

② Factors to be considered when calculating the amount of settlement

- Filling soil weight, pavement weight (subgrade + base course + surface), settlement soil weight (settlement banking soil), traffic weight

③ 침하 검토 항목

- 일반 제체부
  - 침하토량, 토공 여분량(여성토), 토공 폭원 여유

- 횡단 구조물
  - 여분량, 단면 여유

- 부속 구조물의 구조

- 포장 구조(잠정 포장)

④ 일시 흙쌓기 시 압밀 소요기간 검토

- 압밀소요기간 산정 방법

$$t = \frac{T_V \times H^2}{C_V}$$

여기서, t : 압밀소요시간

C_V : 압밀계수(㎠/s)

T_V : 시간계수(U=90%일 때 T_V=0.848)

0%<U<53% 일 경우 $T_V = \frac{\pi}{4} \times (\frac{U(\%)}{100})^2$

53%<U<100% 일 경우 $T_v = 1.781 - 0.933 \cdot \log(100 - U)$

- 압밀소요기간은 공사 기간에도 영향을 받으므로 시공기간을 고려하고 그에 맞는 대책공법 설계가 되도록 함

⑤ 일시 쌓기 시 침하량 검토

- 압밀도 90% 도달 시 잔류침하량이 허용침하량 기준 0.10m를 초과할 경우 압밀촉진을 위한 대책공법이 필요하며, 압밀기간 및 침하량 산정결과를 종합적으로 고려하여 압밀촉진을 위한 연약지반 대책공법을 선정하여야 함

③ Settlement evaluation items

- General objects
  - Amount of settlement soil, earth work spare amount(execissve banking soil), margin of earth work width

- Traversing structures
  - Spare amount, sectional margin

- Structure of auxiliary structures

- Structure of pavement (temporary pavement)

④ Evaluation of required time of consolidation during temporary banking

- Calculation method of required time for consolidation

$$t = \frac{T_V \times H^2}{C_V}$$

where, t : required time of consolidation

C_V : consolidation coefficient(㎠/s)

T_V : time coefficients (T_V=0.848 when U=90%)

when 0%<U<53%, $T_V = \frac{\pi}{4} \times (\frac{U(\%)}{100})^2$

when 53%<U<100%, $T_v = 1.781 - 0.933 \cdot \log(100 - U)$

- As the required time for consolidation is affected by the construction period, the construction method shall be designed accordingly depending on the construction period.

⑤ Evaluation of amount of settlement during temporary banking

- When the degree of consolidation reaches 90% or allowable settlement exceeds the standard of 0.10 m, the countermeasure construction method for the acceleration of consolidation, and the soft foundation countermeasure construction method for the acceleration of consolidation shall be selected comprehensively considering the calculation result of consolidation period and amount of settlement.

3-4-8 연약지반 개량공법 선정

① 연약지반 대책공법의 종류

<table>
<tr><th>개량원리</th><th colspan="2">공법의 명칭</th><th>개량목적</th><th>적용지반</th></tr>
<tr><td rowspan="5">다짐</td><td colspan="2">샌드 컴팩션 파일 공법</td><td rowspan="3">• 액상화 방지
• 침하감소
• 지반의 강도 증가</td><td rowspan="2">점성토, 사질토, 유기질토</td></tr>
<tr><td colspan="2">동다짐공법</td></tr>
<tr><td colspan="2">바이브로플로테이션 공법</td><td>사질토</td></tr>
<tr><td colspan="2">중추낙하다짐공법</td><td rowspan="2">• 침하감소
• 액상화 방지</td><td rowspan="2">사질토</td></tr>
<tr><td colspan="2">폭파다짐, 전기충격공법</td></tr>
<tr><td rowspan="5">고결열처리</td><td colspan="2">표층혼합처리공법</td><td>• 도로의 노상, 노반의 안정처리</td><td rowspan="5">점성토, 사질토, 유기질토</td></tr>
<tr><td colspan="2">심층혼합처리공법</td><td rowspan="4">• 활동파괴 방지
• 침하저지 및 감소
• 전단변형 방지
• 히빙 방지</td></tr>
<tr><td colspan="2">약액주입공법</td></tr>
<tr><td colspan="2">소결공법</td></tr>
<tr><td colspan="2">동결공법</td></tr>
<tr><td rowspan="2">보강</td><td colspan="2">복토공법</td><td>• 도로의 노상, 노반의 안정처리</td><td rowspan="2">점성토, 유기질토</td></tr>
<tr><td colspan="2">표층피복공법(시트, 매트, 필타)</td><td>• 국부파괴, 국부치하 방지</td></tr>
<tr><td>경량화</td><td colspan="2">경량 자재</td><td rowspan="6">• 지반의 지지력 향상
• 지반의 전단변형 억제
• 지반의 침하억제
• 활동 파괴의 방지
• 시공기계의 주행성 확보</td><td rowspan="6">점성토, 유기질토</td></tr>
<tr><td>하중균형</td><td colspan="2">압성토공법</td></tr>
<tr><td rowspan="4">하중분산</td><td colspan="2">침상공법</td></tr>
<tr><td colspan="2">시트넷공법</td></tr>
<tr><td colspan="2">샌드매트공법</td></tr>
<tr><td colspan="2">표층혼합처 공법</td></tr>
<tr><td rowspan="3">치환공법</td><td colspan="2">굴착치환공법</td><td rowspan="3">• 활동파괴방지
• 침하의 감소
• 지반 전단변형 억제</td><td rowspan="3">사질토, 점성토, 유기질토</td></tr>
<tr><td colspan="2">강제치환공법</td></tr>
<tr><td colspan="2">폭파치환공법</td></tr>
<tr><td rowspan="12">압밀배수</td><td colspan="2">프리로딩공법</td><td rowspan="6">• 잔류침하의 감소
• 지반의 강도 증가</td><td rowspan="4">점성토, 유기질토</td></tr>
<tr><td rowspan="3">연직배수 공법</td><td>샌드드레인공법</td></tr>
<tr><td>페이퍼드레인공법</td></tr>
<tr><td>팩드레인공법</td></tr>
<tr><td rowspan="2">지하수위 저하공법</td><td>웰포인트공법</td><td rowspan="2">사질토</td></tr>
<tr><td>깊은우물공법</td></tr>
<tr><td colspan="2">진공압밀공법</td><td rowspan="4">• 압밀촉진
• 잔류침하감소
• 지반의 강도 증가</td><td rowspan="4">점성토, 유기질토</td></tr>
<tr><td colspan="2">생석회말뚝공법</td></tr>
<tr><td colspan="2">전기침투공법</td></tr>
<tr><td colspan="2">반투막공법</td></tr>
<tr><td colspan="2">쇄석말뚝공법</td><td>• 액상화 방지</td><td>사질토</td></tr>
<tr><td colspan="2">표층배수공법</td><td>• 표층지반강도 증가</td><td>점성토, 유기질토</td></tr>
</table>

3-4-8 Selection of Improvement Construction Method of Soft Foundation

① Type of countermeasure construction method of soft foundation

Improvement principle	Name of method		Improvement purpose	Applied foundation
compacting	Sand compaction pile method		• Prevention of liquefaction • Settlement reduction • Increasing strength of foundation	Cohesive, sandy, organic soil
	Dynamic compaction method			
	Vibro-floatation method			Sandy soil
	Center filling compacting method		• Settlement reduction • Prevention of liquefaction	Sandy soil
	Blasting compacting, electric shock method			
Consolidation Heat Processing	Surface layer mixed process method		• Stability treatment of subgrade and road base	Cohesive, sandy, organic soil
	Deep layer mixed process method		• Prevention of sliding failure • Prevention and reduction of settlement • Prevention of shear deformation • Prevention of heaving	
	Chemical grout method			
	Sinter method			
	Freezing method			
Reinforcement	Molding method		• Stability treatment of subgrade and road base	cohesive soil, organic soil
	Surface layer cover method(sheet, mat, filter)		• Prevention of local failure, local sink	
Weight reduction	Light materials		• Improvement of support force of foundation • Suppression of shear deformation of foundation • Suppression of foundation settlement • Prevention of sliding failure • To secure the movability of the construction machine	cohesive soil, organic soil
Weight balance	loading berm method			
Weight distribution	mattress method			
	Sheetnet method			
	sand mat method			
	surface mixed treatment method			
Replacement method	excavation replacement method		• Sliding failure prevention • Settlement reduction • Suppression of shear deformation of foundation	sandy soil, cohesive soil, organic soil
	Forced replacement method			
	Blasting replacement method			
Consolidation water drain	Free loading method		• Remaining settlement reduction • Foundation strength Increase	cohesive soil, organic soil
	Vertical drain method	Sand drain method		
		Paper drain method		
		Pack drain method		
	Underground water level lowering method	Well drain method		sandy soil
		Deep well method		
	Vacuum consolidation method		• Consolidation Acceleration • Remaining settlement reduction • Foundation strength Increase	cohesive soil, organic soil
	Quick Lime Pile method			
	Electric penetration method			
	Semipermeable membrane method			
	Crushed stone pile method		• Liquefaction prevention	sandy soil
	Surface drain method		• Surface foundation strength Increase	cohesive soil, organic soil

① 대책공법 선정 시 유의사항

- 개량의 목적과 대상토 지반의 성질, 하중조건, 시공여건, 공사기간, 주변에 미치는 영향 등을 감안하여 시공성 및 경제성을 고려한 적합한 공법을 선정하여야 함

3-4-9 압밀촉진을 위한 연약지반 처리공법 : 주로 토공부에 적용

① Plastic Board Drain (P.B.D공법)

- 표준 단면

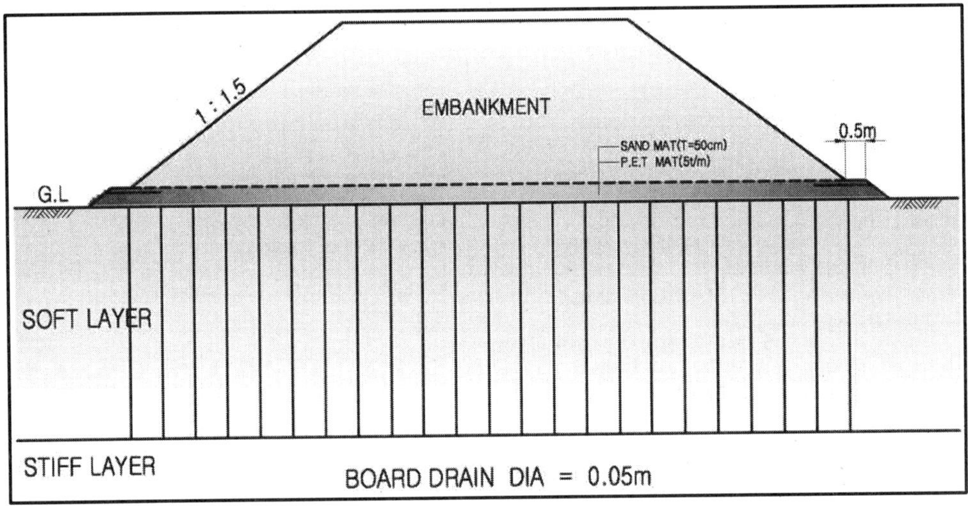

- 공법 개요
  - 연약층에 투수성이 양호한 Drain Board를 타입하여 간극수를 배출시키는 공법
  - 직경 : 50mm(환산직경)

- 관입 범위
  - 관입가능심도 : 최대 30m
  - N치 : 10 이하

- 장점
  - 타입 시 지반교란이 적음.
  - 재료의 구입이 용이하고, 중량이 가벼워 운반 및 취급용이
  - 품질이 균질하여 배수효과가 일정
  - 타 공법에 비하여 경제적인 공법임.
  - 시공실적이 많음.
  - 시공속도가 빠르고, 시공관리가 용이

- 단점
  - 불균질한 지층에서 시공성이 저하됨.
  - 재료강도에 따른 강도증가가 없음.
  - 재료강도 저하에 따른 피복재파열 가능성이 있음

① Cautions in the selection of countermeasure method

- Appropriate method shall be selected considering ease of construction and economic efficiency along with the purpose of the improvement, property of foundation, load conditions, construction environment, construction period, effect to surrounding environment.

3-4-9 Soft Foundation Treatment Method for Consolidation Acceleration: Applied Mainly for Earth Work Parts

① Plastic Board Drain (P.B.D method)

- Standard section

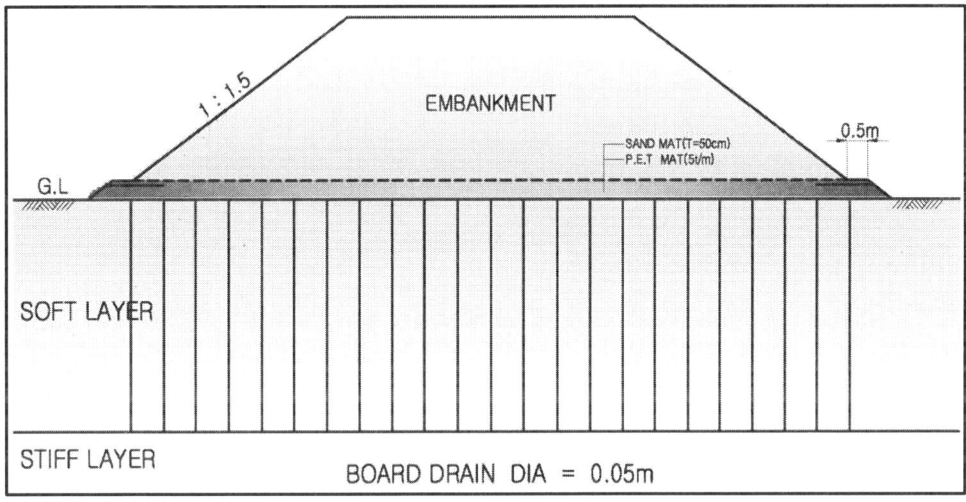

- Method overview
  - Drain board with superior permeability is driven into the soft layer to drain the pore water
  - Diameter : 50mm(converted diameter)
- Penetration range
  - Possible penetration depth: maximum 30m
  - N value : below 10
- Advantages
  - Low disturbance to foundation during driving
  - Wide availability of the material and ease of transportation and handling because light weight.
  - Constant effects of drain due to uniform quality
  - Cost effect compared to other method
  - Many construction histories
  - Shorter construction period, and easy construction control
- Disadvantage
  - Construction gets difficult if soil layer is not uniform
  - No increase in strength from the strength of material
  - Possible covering material failure due to the lowering of material strength

② Pack Drain

• 표준 단면

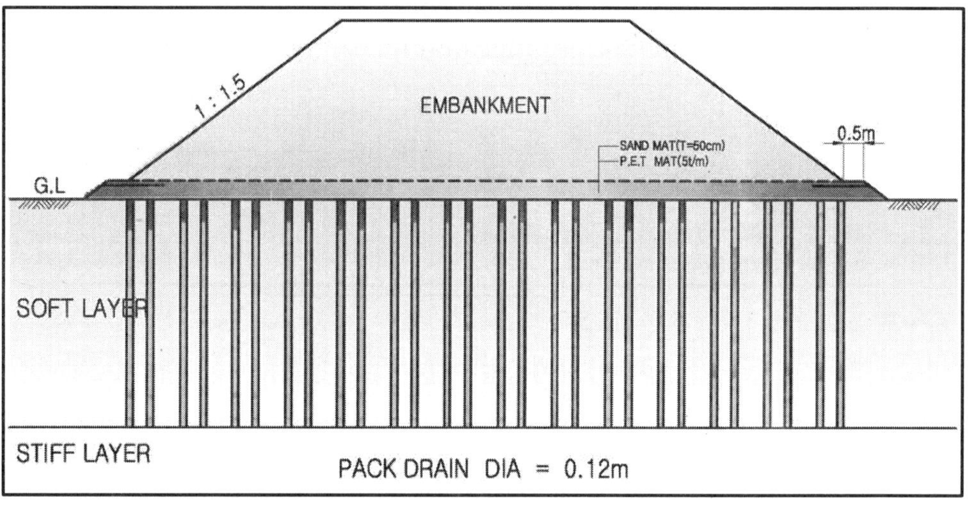

• 공법 개요

- Sand Drain을 개량한 공법으로서 먼저 합성섬유팩에 모래를 충진하여 압밀을 촉진시키는 공법.
- 직경 : 120mm

• 관입 범위

- 관입가능심도 : 최대 30m
- N치 : 20 이하

• 장점

- 배수효과의 신뢰성이 우수함.
- 타입 시 S.D에 비하여 지반교란이 적음.
- Drain재 절단가능성 적음.<투수성 포대 이용>
- S.D보다 모래의 양이 경감됨.
- 4본을 동시에 파설하므로 시공기간이 단축

• 단점

- 연약지반 심도가 불규칙한 지역에서 타설 심도 조절이 어려움.
- 연약층 심도가 불규칙한 경우 연약지반 개량효과가 떨어져, 잔류침하량이 과다하게 발생

② Pack Drain

• Standard section

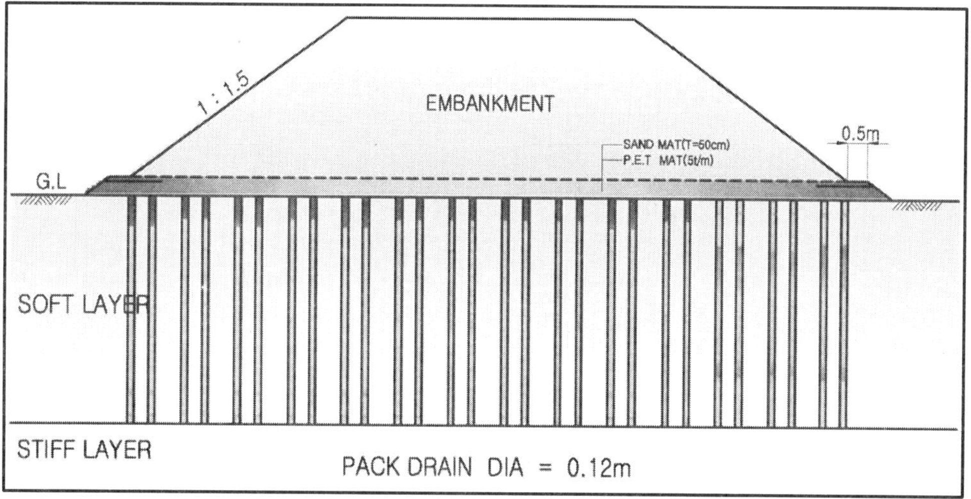

• Method overview

- This method is a modified method of sand drain, and sand is filled to synthetic fabric pack for consolidation acceleration
- Diameter : 120mm

• Penetration range

- Possible penetration depth: maximum 30m
- N value: below 20

• Advantage

- Excellent reliability of drain effects
- Lower disturbance in foundation during driving compared to S.D
- Lower possibility of cutting drain material <utilizing permeable sack>
- Requires less amount of sand than S.D
- Reduction in construction period because four sets are driven simultaneously.

• Disadvantage

- Difficult to control the depth when the depth of soft foundation is irregular.
- Improvement effects is reduced when the depth of soft foundation is irregular, and excessive amount of settlement is generated.

③ Sand Drain (S.D공법)

• 표준단면

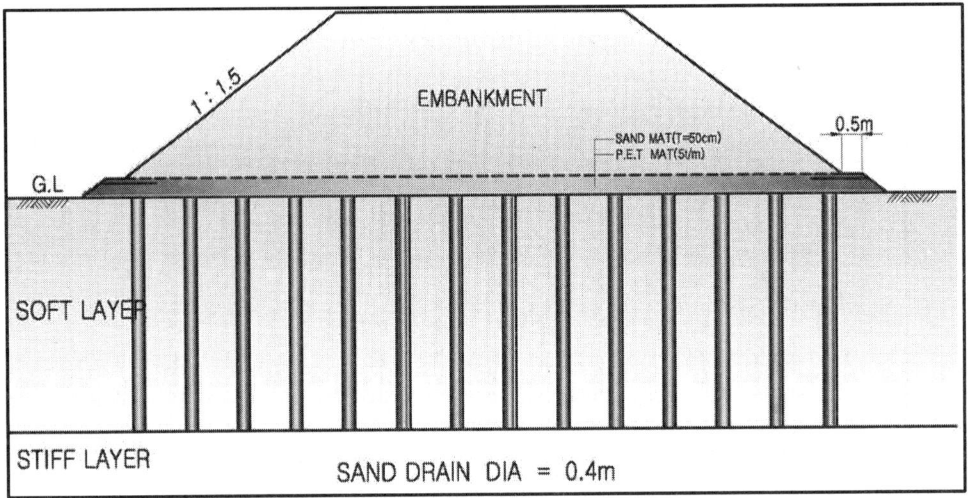

• 공법 개요
- 연약층에 Sand Pile을 타입하여 간극수를 배출시켜 압밀을 촉진시키는 공법
- 직경 : 400mm

• 관입 범위
- 관입가능심도 : 최대 30m
- N치 : 20~25 이하

• 장점
- 투수효과가 확실함.
- 불균질한 지층에서 작업성이 양호함.
- 시공실적이 많음.

• 단점
- 타 공법에 비하여 모래 사용량이 많아서 공사비가 고가임.
- Drain 절단가능성이 큼.
- 시공 시 Arching 효과로 모래기둥의 연속성에 신뢰도가 낮음.

③ Sand Drain (S.D method)

• Standard section

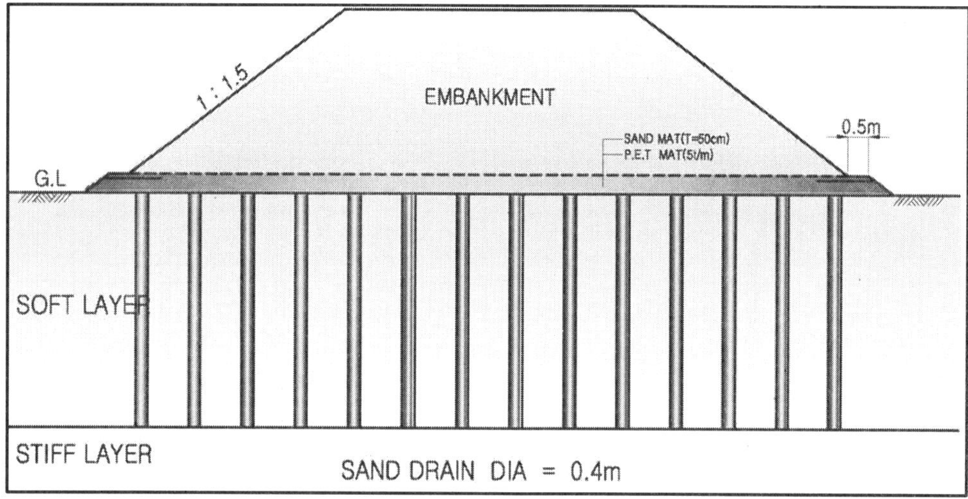

• Method overview
  - Sand piles are driven into the soft foundation to drain pore water for consolidation acceleration
  - Diameter : 400mm

• Penetration range
  - Possible penetration depth: maximum 30m
  - N value: below 20～25

• Advantage
  - Guaranteed effects of percolation
  - Ease of construction for irregular layers
  - Many proven construction histories

• Disadvantage
  - Cost higher than other method due the high usage of sand
  - High possibility of drain cut
  - Reliability of the continuity of sand pile is low because of arching effect during construction

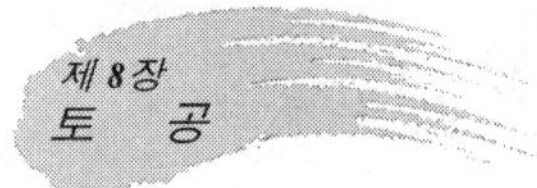

3-4-10 활동방지를 위한 연약지반 처리공법 : 주로 구조물 위치에 적용

① Vibrated Crushed-stone Compaction Pile (V.C.C.P 공법)

• 표준단면

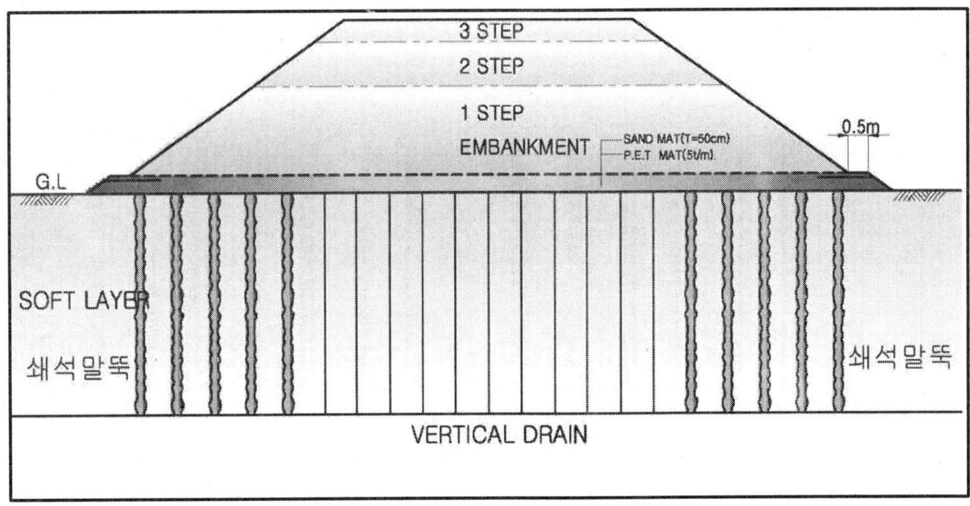

• 공법개요

- 직경 40mm인 케이싱을 지중에 압입시킨 후 진동·충격 하중을 이용하여 직경 70mm의 쇄석말뚝을 형성시켜 지반활동에 저항하는 공법

• 장점

- 모래부족으로 인한 재료수급의 문제 해결
- 과잉간극수압을 효과적으로 소산시키는 뛰어난 배수효과
- 쇄석기둥 주위의 응력집중으로 인한 침하저감 및 지지력 증대
- 쇄석기둥의 높은 마찰각과 하중집중효과로 인한 파괴면에서의 전단저항 증가로 인하여 비탈면 안정에 매우 효과적

• 단점

- 쇄석사용에 따른 시공장비의 마모가 큼
- 대형장비의 시공에 따른 충분한 작업공간 필요
- 시공실적이 적음

3-4-10 Soft Foundation Treatment Method for Preventing Sliding

: mainly applied to the position of structures

① Vibrated Crushed-stone Compaction Pile (V.C.C.P method)

• Standard section

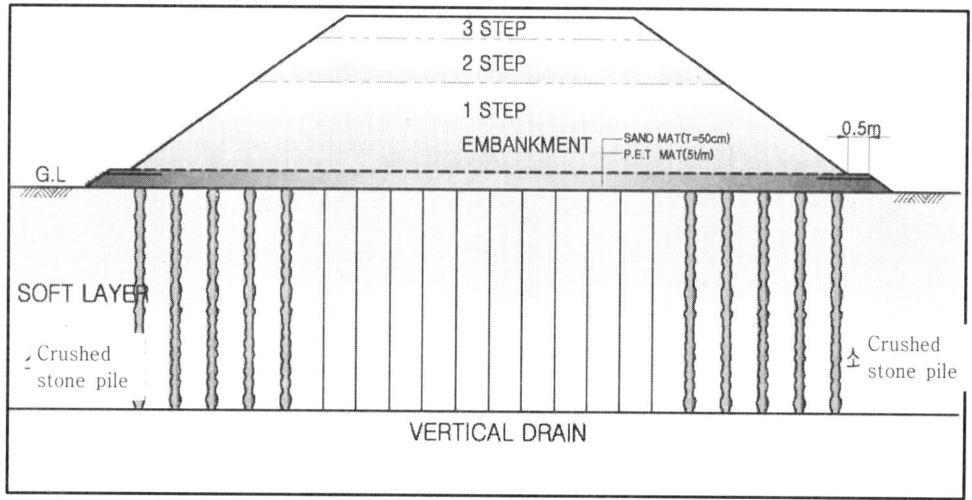

• Method overview

- Casings with the diameter of 40 mm are inserted under the ground and crushed stone piles with the diameters fo 70 mm are generated using the vibration, shock, and weight to resist against the sliding of the foundation

• Advantage

- Material supply problem (lack of sand) is resolved
- Excellent drain effect of decentralizing excessive pore pressure
- Decrease of settlement and increase of support force due to the concentration of the stress around the crushed stone piles
- Very efficient for the stability of the slope due to the increase of shear resistance in the failure surface by the load concentration effect and high friction angle of crushed stone piles.

• Disadvantage

- Wear level of construction equipment is high due to the usage of crushed stone
- Space for the construction with large scale equipment is required
- Not much construction histories

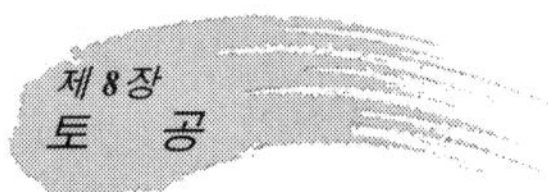

② Sand Compaction Pile (S.C.P 공법)

• 표준 단면

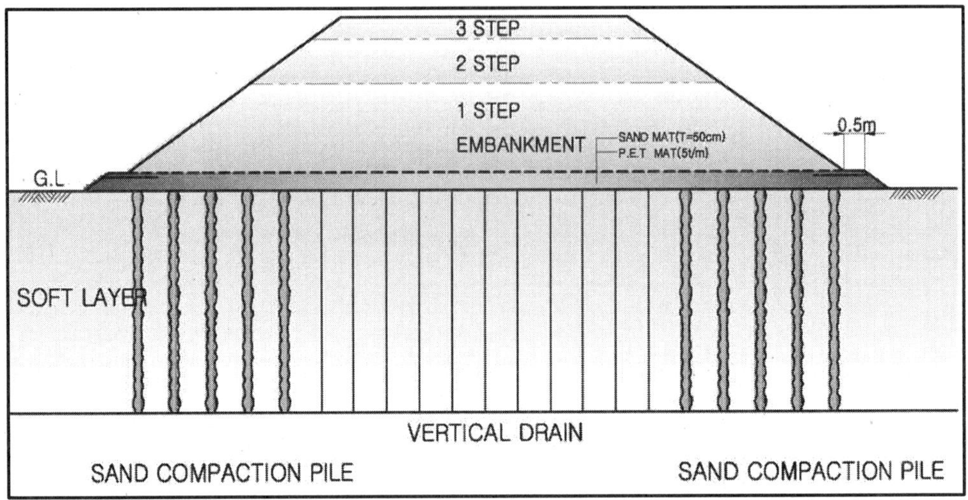

• 공법 개요

- 진동 및 충격하중을 이용하여 지반에 타설된 다짐모래말뚝에 의하여 지반활동에 저항하는 공법
- 직경 : 700mm

• 장점

- 압밀량 저감효과 및 잔류침하 조기완료
- 지반의 전단강도가 증가함.<복합지반>
- 지반의 측방유동방지 및 침하량이 경감됨.
- 지중에 작용하는 흙쌓기 하중이 모래말뚝과 연약층에 분리되어 적용함.

• 단점

- 시공관리가 어려움.
  (투입모래량, 다짐시간 측정 등)
- 타설장비가 대형이므로 진입에 문제가 있음.
- 타설 시 지반이 교란됨.

② Sand Compaction Pile (S.C.P method)

• Standard section

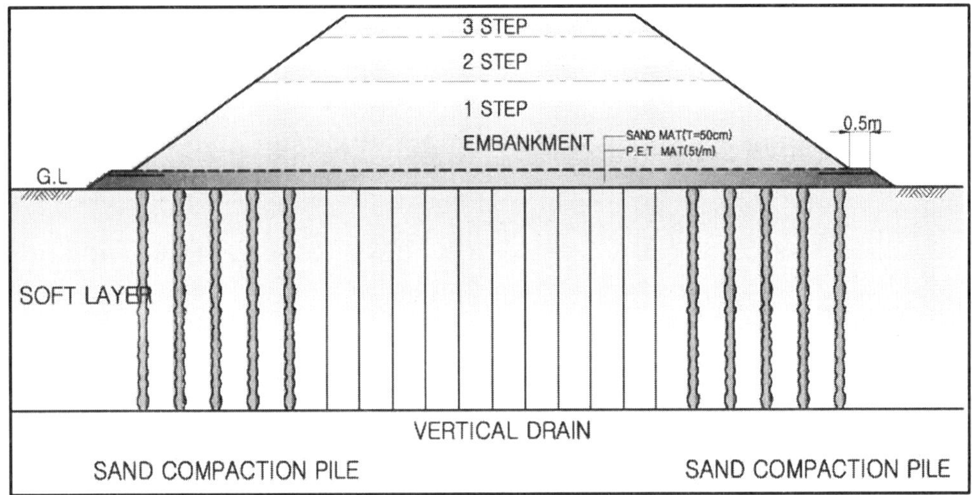

• Method overview

- This method is to resist the sliding of the foundation by the driven compacting sand piles into the foundation using vibration and shock weight.
- Diameter : 700mm

• Advantage

- Consolidation amount reduction effects and early progression of remaining settlement
- Shear strength of foundation is increased <compound foundation>
- Side fluctuation of the foundation is prevented and settlement amount is reduced.
- Weight of banking soil applied to the sane piles and soft ground separately.

• Disadvantage

- Difficult construction control .
  (Amount of sand usage, compacting time measurement, etc)
- Problem in moving driving equipment because of big size
- Foundation is disturbed during driving

3-4-11 연약지반의 안정 및 침하관리

① 연약지반의 안정관리

- 동태 관측 결과에 근거를 두고 안정한지 안정하지 않은지를 판단하여, 그에 따라 흙쌓기 속도를 관리
  - 침하량 측정에 의한 판단
  - 수평변위량의 측정에 의한 판단
  - 수직변위량의 측정에 의한 판단

- 일반적으로 흙쌓기 두께는 0.30m 이하를 원칙

- 방치 기간을 조정함에 따라 흙쌓기 속도를 관리

② 연약지반의 침하관리

- 동태 관측 결과에 근거한 실측 침하량으로부터 설계 시에 예측되는 침하량을 수정하여 장기 침하량의 추정 및 대책 수립 필요

- 지반 각 층의 침하량을 구하여 압밀의 진행 상황을 조사

- 장래 침하량 예측을 통하여 제체 형태의 수정, 토공 마무리 높이, 배수시설 등의 추가 흙쌓기 또는 포장구조를 결정

- 상재하중, 선행하중의 재하시기를 결정

- 장기 침하량을 추정하여, 교대와 제체와의 접속부의 단차 대책, 암거나 배수구조물의 침하 대책을 검토

4. 토량의 배분

4-1 일반사항

4-1-1 토량의 배분 시 유의사항

① 땅깎기 및 흙쌓기 설계에서는 교량, 터널 등의 구조물과의 관련을 고려한 후 가능한 한 땅깎기·흙쌓기량이 평행되도록 계획

② 땅깎기로 얻어지는 재료는 그 성상을 충분히 파악하여 흙쌓기 각부의 가장 적합한 장소로 배분

③ 토량의 균형에 주의하며 계획 시의 토량변화율이 타당한지 여부를 항상 점검하여 여분의 사토나 순성토의 발생을 방지

④ 땅깎기에서 발생하는 재료의 특성을 충분히 파악하여 노상 및 노체의 상부에는 다짐이 용이하고 압축성이 작은 재료를 배분하고, 입경이 큰 전석과 암괴, 소성지수가 높은 점성토 등은 흙쌓기의 하부에 배분되도록 계획

3-4-11 Stabilization of soft foundation and settlement control

① Soft foundation stabilization control

- From the degree of stability based on the movement observation, speed of soil banking is controlled.
  - Judgment based on the measurement of settlement
  - Judgment based on the measurement of horizontal displacement
  - Judgment based on the measurement of vertical displacement

- In general, banking (banking) height shall not exceed 0.30 m in principle.

- Filling speed is controlled by adjusting the period of placing them along

② Settlement control of soft foundation

- Prediction of long term settlement and establishment of countermeasure are required by modifying the settlement amount predicted from the design stage with the actual measurement of settlement based on the result of the movement observation.

- Progression status of consolidation shall be investigated by getting settlement of each layer.

- Extra banking or pavement structure such as modification of soil amount or shape, earth work finishing height, and drain shall be determined by predicting future settlement.

- To determine loading time of overburden loading and pre-loading

- By predicting future settlement, countermeasure for step between abutment and connection part, and for settlement of drain structures shall be evaluated.

4. Soil Distribution

4-1 General Considerations

4-1-1 Cautions in Distribution of Soil

① When designing banking and cutting, the amount of banking and cutting shall be planed to be parallel considering the relations with structures like bridge and tunnel.

② Materials from the cutting parts shall be fully evaluated and distributed to most appropriate place on banking parts.

③ Prevent the generation of spare waste soil or excessive banking by always checking the validity of the changes of planned amount soil while being careful on the balance of material.

④ The characteristics of the material from cutting shall be fully evaluated, and the material with easy compactness and low compressibility shall be distributed to the upper part of subgrade and road bed, and boulder with big grain, rocks, and cohesive soil with high plasticity shall be planned to be distributed to lower part of banking part.

4-2 토적도

4-2-1 토적도 작성기준

① 토량계산에 필요한 땅깎기의 단면적은 토사, 리핑암, 발파암으로 구분하여 산출

② 토량변화율은 토질에 따라 설계와 시공시가 크게 다른 경우가 있으므로 주변의 시공실적 등을 참고하여 결정

③ 토량배분은 흙의 운반거리가 최소화되도록 계획

④ 토적도는 종방향 토량 이동만을 표시하고 횡방향 이동은 반영되지 않으므로 횡방향 토량이 누락되지 않도록 주의

⑤ 횡단면도에 의거 토적표를 작성하며, 이때 땅깎기량은 다짐상태로 보정

[토량 배분의 예]

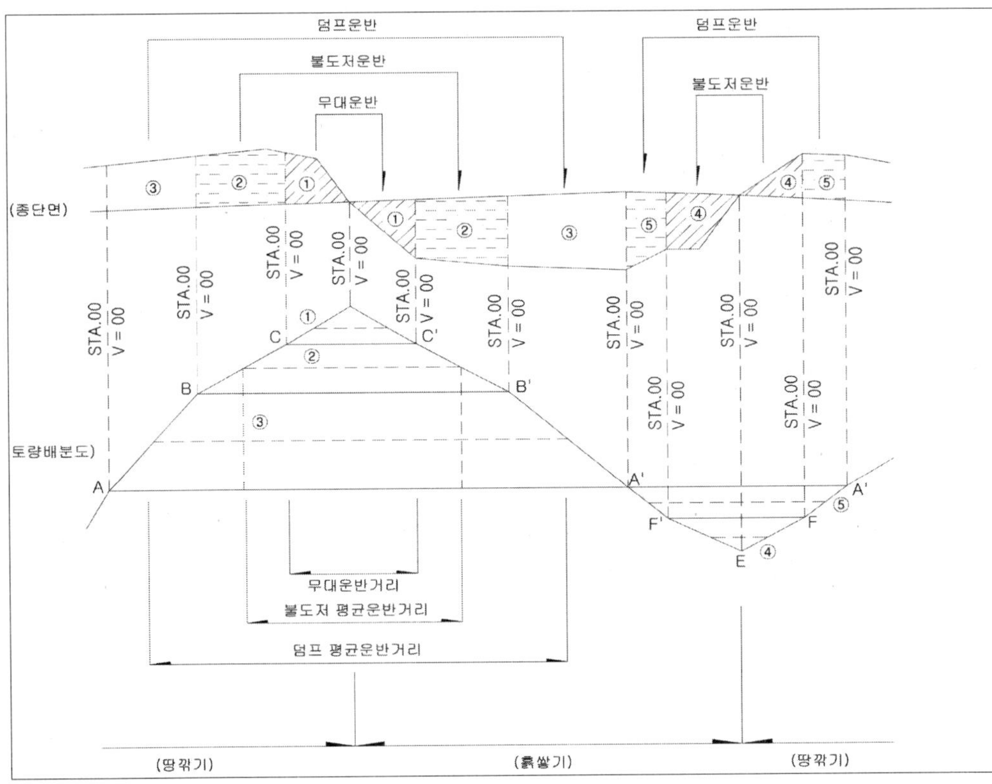

⑥ 흙쌓기 횡단면도 수량을 다짐상태수량으로 기입

⑦ 횡방향 토량란에는 땅깎기 및 흙쌓기 수량 중 작은 수량을 기입하여 유용무대 처리

4-2 Mass Curve

4-2-1 Mass Curve Preparation Standard

① The section area of cutting required for the calculation of soil shall be calculated separately for each of soil, ripping rock, and blasting rock.

② As there might be big difference in soil change rate from design stage to construction stage, it has to be determined by referring to construction histories around the site.

③ Soil distribution planning shall be done to minimize the oil moving distance.

④ Since mass curve displays soil movement in longitudinal direction only without showing traversing direction, be sure not to drop the traversing direction soil.

⑤ Mass curve shall be prepared based on the traversing section diagram , and cutting amount shall be corrected for compacting state

[Example of soil distribution]

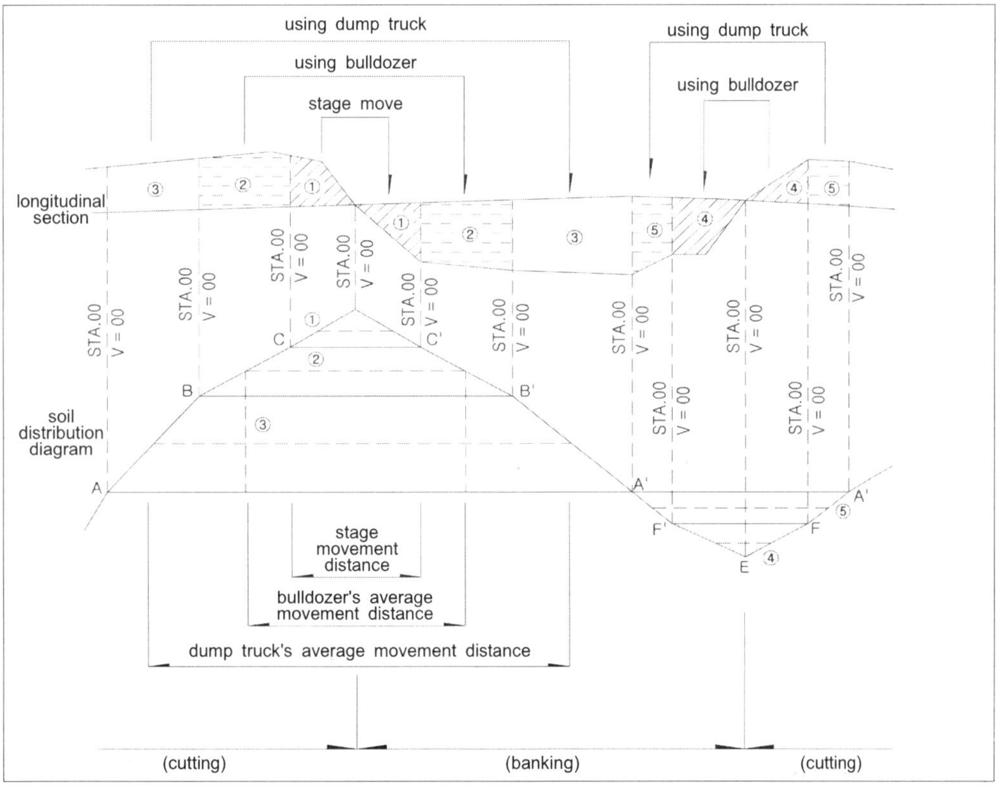

⑥ Specify the amount of banking traversing section as compacting state amount

⑦ In the field of traversing soil, write the smaller amount between cutting and banking, and it shall be treated as utility stage.

⑧ 차인 흙쌓기량란에는 땅깎기 수량과 흙쌓기 수량의 차이수량을 기입하며, 흙쌓기 수량이 땅깎기 수량보다 많은 경우에는 (-), 반대의 경우 (+) 기호를 붙임

⑨ 누가 토량란에는 첫 측점부터의 누계치를 기입하며 (+), (-) 부호에 유의

4-3 토공유동표

4-3-1 토공유동표 산정기준

① 운반거리 산정 시 모든 수량은 다짐상태로 환산하여 계산하되, 내역서 적용수량은 자연상태로 적용

② 토적도(Mass-Curve) 분석 시 반영하는 공종은 땅깎기(토사, 리핑암, 발파암), 측구 터파기, 흙쌓기(노상, 노체, 녹지대 흙쌓기)로 함

③ 표토제거는 무대에 삽입하고 다짐 환산계수 만큼의 차이는 덤프운반에 포함

④ 공제토는 덤프운반에서 제외

⑤ 토공 유동표에는 다음사항을 포함

- 표토제거, 공제토, 유용토, 교량암성토, 철거수량, 되메우기, 가도공, 본선암유용

⑥ 양측확장 구간은 좌·우측 별도 유동표를 작성하여 운반거리를 산출

⑦ 기타 제반사항

- 횡방향 및 종방향 무대처리토사는 토사→리핑암→발파암의 순서로 진행

- 녹지대 흙쌓기 구간은 토사만 사용하는 것을 원칙으로 하며, 사토현장으로서 현지 여건상 토사부족으로 흙쌓기와 높이 3m 이상일 때에는 암성토를 적용

- 흙쌓기용 재료항 중에 대하여, 순흙쌓기 구간은 노체, 노상, 녹지대 흙쌓기량 중 토사수량에 대하여 일정량을 토사운반수량(무대, 도저, 덤프)에 가산하여 토취장 땅깎기 및 운반에도 가산하며, 사토구간은 노체, 노상, 녹지대 흙쌓기량 중 토사수량에 대하여 일정량을 토사운반수량(무대, 도져, 덤프)에 가산하고 사토량에서 감함

⑧ In the field of differential amount of soil, write the difference between cutting and banking, and put (-) sign if the cutting soil is more than banking soil, and (+) in opposite case.

⑨ In the accumulated soil field, write the accumulated value from the first measurement point, and be careful on the (+) and (-) symbols.

4-3 Earth Work Flow Table

4-3-1 Calculation Standard of Earth Work Flow Table

① All amount for calculation of transportation shall be converted value of compacted state, and the applied amount in the breakdown detail tables shall be the value in natural state.

② Construction type reflected to the mass curve analysis shall be cutting (soil, ripping rock, blasting rock), side ditching, banking (subgrade, road bed, green section).

③ Removed surface soil is inserted to the stage, and the difference by the compacting conversion coefficient shall be included in dump truck movement

④ Deducted soil shall be excluded from the dump truck movement

⑤ The followings shall be included in the earth work flow table

- Surface soil removal, deducted soil, utility soil, bridge loading berm, removal amount, re-banking, utility soil, bridge loading berm, main road constriction, utilization of main road rocks

⑥ Separating left and right flow table shall be prepared for both side expansion section, and the movement distance shall be calculated.

⑦ Miscellaneous items

- The processing of the stage soil in traversing and longitudinal direction shall be in the order of soil→ripping rock→blasting rock

- Green section banking area shall use only soil in principle, and when there is not enough soil at spoil soil site and also when the height of banking is over 3m, rock banking shall be applied.

- Regarding extra materials for banking, certain fixed amount of soil of road bed, subgrade, and green section shall be added to the soil movement amount (stage, bulldozer, dump truck) and added to borrow pit cutting and movement as well for the net banking section, and certain fixed amount of soil of road bed, subgrade, and green section shall be added to the soil movement amount (stage, bulldozer, dump truck) and subtracted from spoil soil.

- 토량유동표는 다짐상태로 통일시켜야 하며, 아래의 원칙을 만족시켜야 함
  - 사토의 경우
    사토량 = 땅깎기량 - {(유대운반량 + 무대운반량) = 흙쌓기량}
  - 순흙쌓기의 경우
    순흙쌓기량 = {흙쌓기량 = (유대운반량 + 무대운반량)} - 땅깎기량

5. 토취장 및 사토장의 계획

5-1 토취장의 계획

5-1-1 일반사항

① 토취장을 선정할 때는 필요 토량에 맞도록 복수의 토취장을 후보지로 하여 각각에 대하여 지형, 채취 가능토량, 방재대책, 매립문화재, 법적규제, 흙 운반로, 보상관계, 주변환경, 토지 이용 등의 여러 제반조건에 대하여 조사하고 충분한 검토를 통하여 최적의 토취장을 선정

5-1-2 토취장 선정 시 고려사항

① 본선 인접지 부근에 후보지를 선정하여 운반거리를 최대한 최소화

② 다른 공공사업과 연계하여 효과적인 토취장 계획을 수립

③ 토량배분 계획 시 토량의 문제뿐만 아니라, 노상재·뒤채움재·비탈면 보호·운반로의 가설재·교통성 확보 등 필요한 재료를 얻을 수 있는 장소를 선정

④ 토취장 조사에서는 주변지역 및 운반로 등을 포함하여 광범위하게 실시하고 설계상의 문제점 등을 파악하도록 하며, 지형도·지질도·항공사진 등의 기존자료를 효과적으로 이용

⑤ 문화재 보호법 등의 법적규제를 받는 곳에서는 관계 공공단체와 협의하여 조속하게 해제절차 추진

⑥ 흙 운반로의 선정은 운반거리뿐만 아니라 연도상황, 교통량 및 보도 등을 고려하고 폭, 포장의 상황, 개량상황 등을 고려하여 종합적으로 판단

- Soil flow table shall be unified in compacted state, and it shall meet the following principles.
  - In case of waste soil
    Amount of waste = amount of cutting soil - {amount of paid movement + amount of free movement) = amount of banking}
  - In case of net banking
    Net banking amount = {banking amount = (amount of paid movement + amount of free movement)} - cutting amount

5. Borrow Pit and Spoil Bank Plan

5-1 Planning of Borrow Pit

5-1-1 General Considerations

① The optimum borrow pit shall be selected when selecting borrow pit from the multiple candidate borrow pits, and each site shall be fully evaluated for topography, available amount, collapse-proof countermeasure, buried cultural assets, regulation or law, soil transportation path, compensation, adjacent environment, land usage, and other various conditions.

5-1-2 Consideration in Selecting Borrow Pit

① Borrow pit shall be close to the main road to minimize the transportation distance.

② Establish the efficient borrow pit plan by connecting with other public projects

③ Select the place that materials for subgrade, post filling, slope protection, construction for movement road, and traffic access are also available in addition to the soil to move.

④ During the investigation of borrow pit, wide range of evaluation including the adjacent area and movement path shall be done, problems in the design shall be identified, and use the existing data such as topographical map, soil property map, and aerial photography efficiently.

⑤ For the region under legal control by cultural asset protection law, all procedure shall be executed promptly to agree with related public organizations.

⑥ Selection of soil transportation path shall be determined based on the comprehensive considerations including aging status, traffic volume, sidewalk, width, pavement statue, and improvement status in addition to the moving distance.

⑦ 땅깎기에 의하여 비탈면이 발생하는 경우 비탈면 경사와 비탈면 보호공 등을 검토하여 붕괴 등의 발생을 방지

5-2 사토장의 계획

5-2-1 일반사항

① 땅깎기 등에서 발생하는 잔토 및 물량토를 사토하는 경우는 가능 사토량, 토사유출 및 붕괴를 방지하기 위한 방재대책, 법적규제, 흙 운반로, 토지이용계획, 용지보상 등을 고려하여 후보지를 선정, 비교·검토한 후 가장 유리한 사토장을 선정

5-2-2 사토장 선정 시 고려사항

① 사토장의 위치는 운반작업을 공사현장 내에서 끝낼 수 있고, 잔토의 경우 미리 흙쌓기를 병행하여 작업을 진행시킬 수 있도록 가능한 도로의 부근에 선정

② 사토장을 계획하는 경우 장소에 따라서는 법적규제를 받을 수 있으므로 관계기관과 충분한 협의를 통하고 해제 절차를 추진

③ 사토장은 절벽, 경사지, 습지 등 일반적으로 지형·지질적으로 불량한 장소에 설치되는 경우가 많고 시공 중 및 시공 후 우수 등에 의하여 토사유출 및 붕괴가 발생할 우려가 있으므로 토류공 및 비탈면 보호공 등의 비탈면 안정대책과 배수대책 등이 필요

④ 흙 운반경로 선정 시 운반거리뿐만 아니라 연도상황, 교통량 및 보도 등을 고려하고 폭, 포장의 상황, 개량상황 등을 고려하여 종합적으로 판단

⑤ 사토장의 용량은 토량변화율, 토질·암질의 변화에 의한 땅깎기·흙쌓기량 및 사토량의 변화, 차량소통을 위하여 반입되는 모래 등의 토량 등도 함께 고려하여 여유량을 확보

⑦ When slope is generated by cutting, slope grade and slope protection construction shall be evaluated to prevent the collapse.

5-2 Spoil Bank Plan

5-2-1 General Considerations

① When wasting remaining soil and quantity soil from cutting and other process, select the candidate places considering the waste amount, countermeasure for preventing collapse or soil outflow, legal restriction, soil moving path, land utilization plan, land compensation, the choose the best place after comparison and evaluation.

5-2-2 Considerations in Selecting Spoil Bank

① The location of the spoil bank shall be near to the road so that the moving work can be completed with the construction site, and the remaining soil can be processed in parallel with banking in advance.

② As there might be a legal restriction depending on the location of the spoil bank, follow the release procedure through the sufficient coordination with the related institutes.

③ Usually spoil bank is installed in poor geologic area such as around cliff, slope, and wet area, and usually in danger of soil outflow or collapse by rain or other during or after the construction, countermeasure for slope stability and drain such as soil flow and slope protection construction shall be required.

④ Selection of soil transportation path shall be determined based on the comprehensive considerations including aging status, traffic volume, sidewalk, width, pavement statue, and improvement status in addition to the moving distance.

⑤ The capacity of the spoil bank shall be determined with the margin including the soil change rate, changes in cutting/banking and wastes by the changes of soil/rock property, and soil including sand brought in for vehicles clearance.

[토취장 및 사토장 계획의 흐름]

설계단계	목 적	조사내용	착 안 점
타당성 조사 및 기본설계 단계	• 최적노선에 맞추어 개략적인 토취장 및 사토장을 검토한다.	• 지형도, 지질도 등에 의한 도상 조사와 간단한 현지답사	• 법령(광업권, 보안림, 농진법 토사 채취 조례, 문화재)에 의한 개발 규제 유무의 파악 • 토량, 토성, 흙 운반로 등의 개략 검토 • 관련 공공사업의 파악
설계협의 단계	• 현지조건에 적합한 토취장 후보지 몇 곳을 선정하고 비교 검토해서 최적의 토취장 및 사토장을 선정한다.	• 가능 토량, 토성, 운반경로, 용지 등의 현지조사 • 유력한 후보지의 토질 조사 및 토질시험 • 관할 행정관청과 허가 유무를 사전협의	• 채취 가능 토량, 토성의 파악 • 사토 가능 토량 파악 • 관련 공공사업과의 연계 도모 • 흙 운반로의 사용 협의 • 토지 이용계획과의 조정 • 배수계획과 방재계획 검토
실시설계 단계	• 세밀한 조사·검토 및 설계협의에 따라 토취장 및 사토장을 최종 선정한다.	• 상세한 측량조사 • 용지보상을 위한 조사 • 토질 등의 보충조사 • 흙 운반로의 상세조사	• 보상 관계의 확인 • 흙 운반로의 개량 및 대책 • 배수, 방재대책 • 시공 시의 토량 변경에 유연하게 대처할 수 있도록 설계

[Flow of borrow pit and spoil bank plan]

Design stage	Purpose	Investigation contents	key points
Feasibility study and basic design stage	• Rough borrow pit and spoil bank shall be evaluated for the optimum route	• Investigation on papers using topographical map and geological map, and simple filed survey	• Understand the existence of the restrictions by Regulations (Mining right, secure forest, Agriculture Promotion Act, Soil sampling case, cultural assets) • High level evaluation of the soil amount, soil property, and moving path • Understanding of related public projects
Design agreement stage	• Select few candidate places and choose the best borrow pit and spoil bank places by comparative evaluation	• Field survey for available soil quantity, soil property, moving path, land • Soil test for strong candidate places • Pre agreement with administrative office whether the permit is required	• Understand available quantity to take, and soil property • Understand the avaliable space for waste soil • Plan the connection with other public projects • Agreement on the usage of soil moving path • Adjustment with land usage plan • Evaluation of drain plan and disaster-proof plan
Execution design stage	• Final selection of borrow pit and spoil bank based on the design agreement and detailed investigation and evaluation	• Detailed surveying • Survey for land acquisition • Additional survey on soil property • Detailed investigation on soil moving path	• Confirm compensation of land • Improvement and countermeasure of soil moving path • Countermeasure for drain and disaster-proof • Design to coupe with changes of soil quantity flexibly

Chapter 9
Drainage

1. 일반사항

1-1 목 적

- 도로시설물의 기능 보존
- 표면수 침투, 지하수 유입으로 인한 지반 지지력 저하 예방
- 노면배수 불량으로 인한 교통체증 및 사고예방
- 도로건설로 영향을 받을 수 있는 지표수의 흐름 파악, 보정하여 자연수로로 연결

1-2 배수시설의 구분

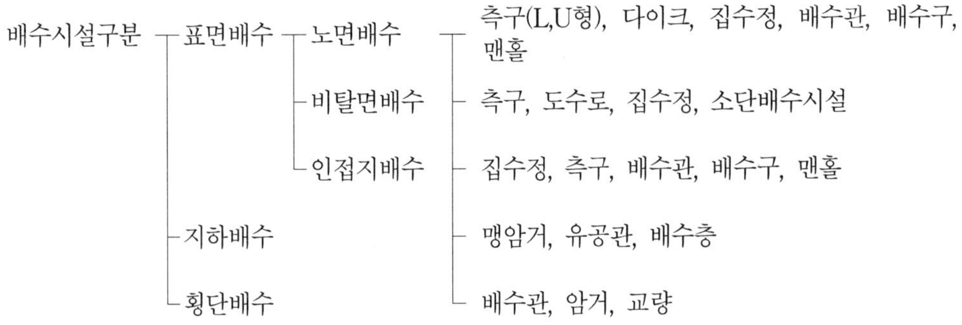

[도로 배수시설의 개요도]

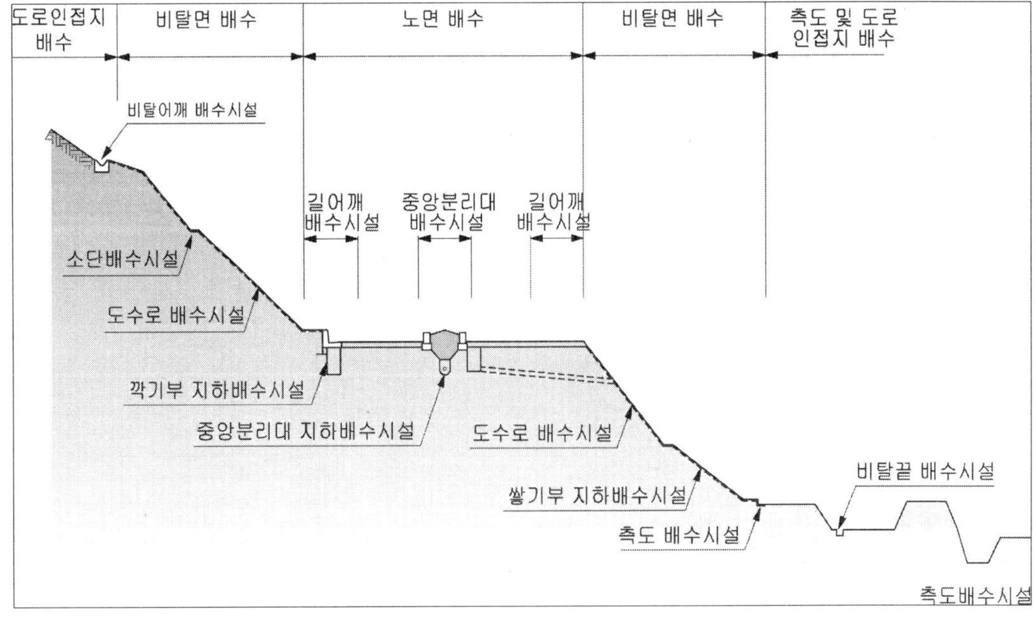

1. General Considerations

1-1 Purpose

- To preserve the function of road facilities
- To prevent the lowering of foundation support due to the inflow of underground water or penetration of surface water
- To prevent the traffic congestion and accident due to the poor road surface water drain
- To connect to natural water path with compensation after identifyining and revising the flow of surface water that be affected by the road construction

1-2 Classification of Drain Facilities

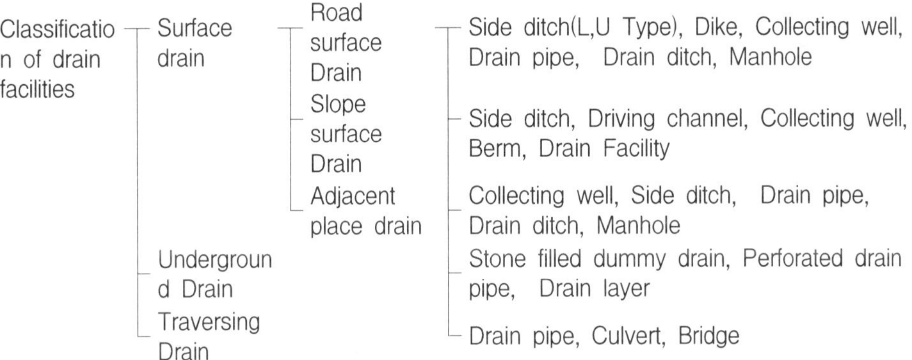

【Road Drain Facility Outline Drawing】

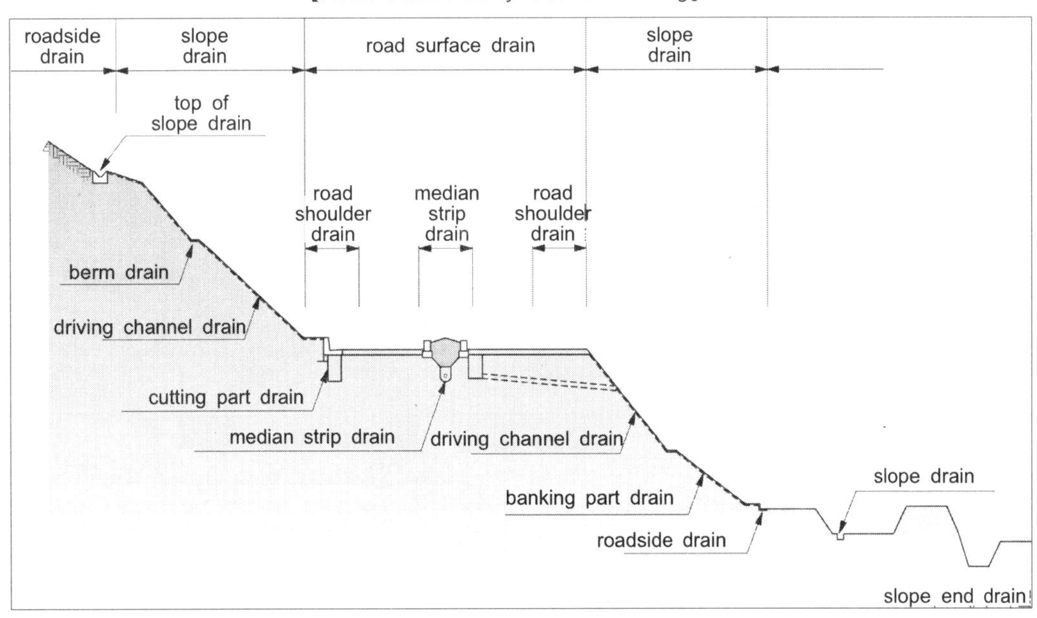

1-2-1 노면 배수시설

- 강우 시 교통안전을 도모하기 위하여 노면 또는 비탈면으로부터 유입되는 물을 원활히 배수하기 위하여 길어깨 또는 중앙분리대측에 설치하는 배수시설

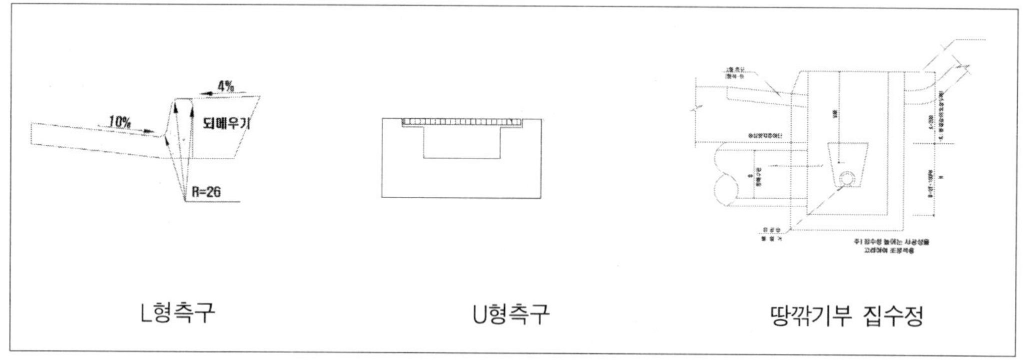

L형측구 U형측구 땅깎기부 집수정

1-2-2 비탈면 배수시설

- 도로 비탈면으로 유입되는 물의 배수를 위하여 흙쌓기 또는 땅깎기부 경사면에 설치하는 배수시설

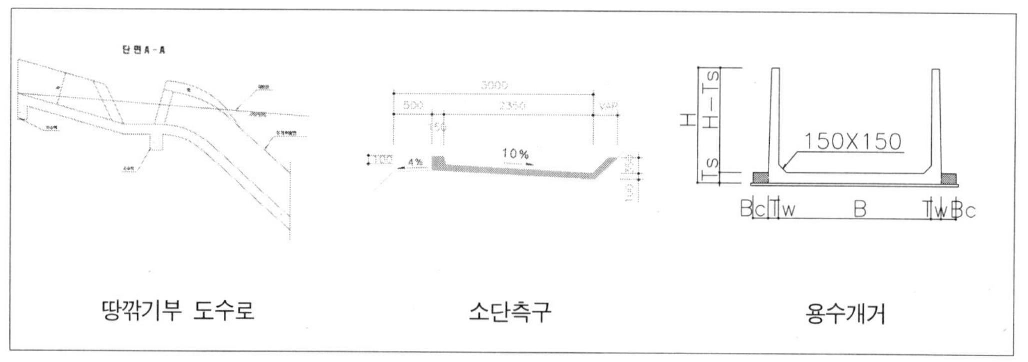

땅깎기부 도수로 소단측구 용수개거

1-2-3 인접지 배수시설

- 배수유역으로부터 도로 인접지로 유입되는 물을 배수하기 위하여 도로 끝단에 설치하는 표면배수시설

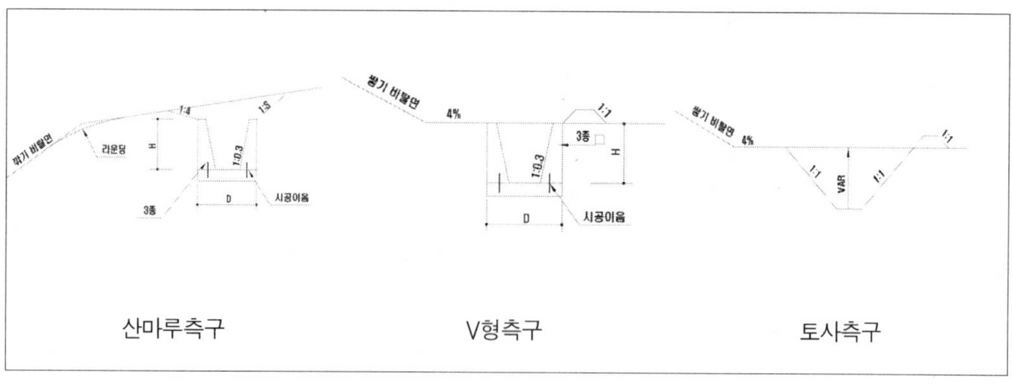

산마루측구 V형측구 토사측구

1-2-1 Road Surface Drain Facility

- Drain facilities installed on top of the slope or median strip to drain the water coming in from the slope surface or road surface for the traffic safety.

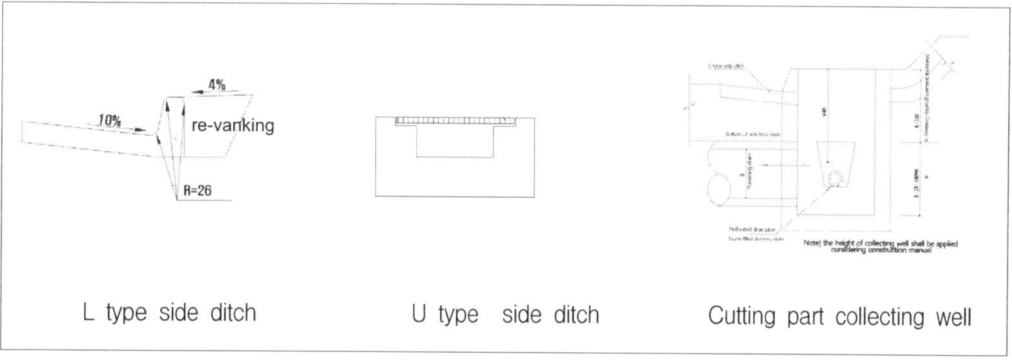

L type side ditch U type side ditch Cutting part collecting well

1-2-2 Slope Surface Drain Facility

- Drain facilities installed on the banking or cutting part to drain the water from the slope surface

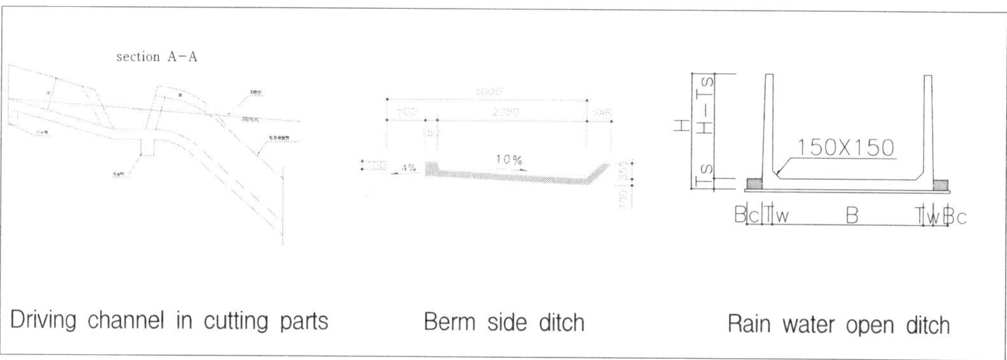

Driving channel in cutting parts Berm side ditch Rain water open ditch

1-2-3 Roadside Drain Facility

- Surface drain facilities installed at the end of the road to drain the water coming into roadside from a drainage basin.

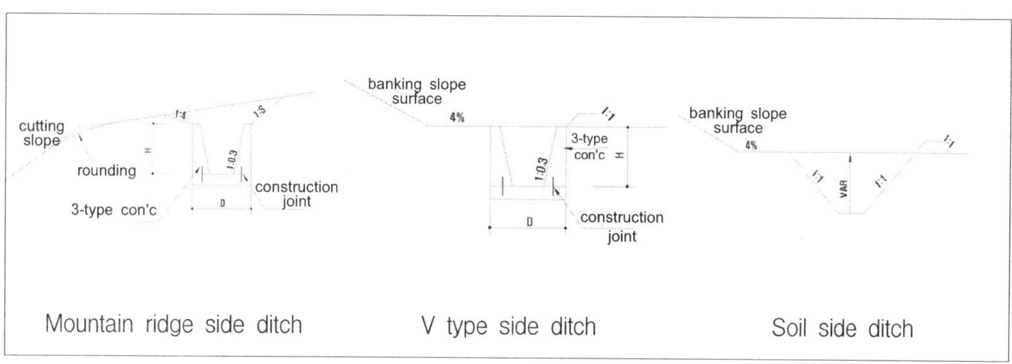

Mountain ridge side ditch V type side ditch Soil side ditch

1-2-4 지하 배수시설

- 지하수위가 높고 침투수가 많아 노체 안정에 위험이 있을 경우에 지하수위 저하 및 침투수 배제를 위하여 설치하는 지하배수 시설물

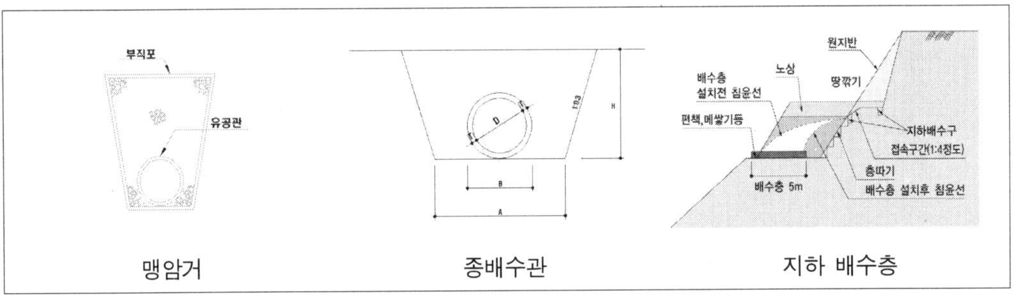

맹암거 종배수관 지하 배수층

1-2-5 횡단 배수시설

- 도로를 횡단하는 소하천 또는 수로를 위한 시설로서, 도로 본체의 보존과 도로 인접지의 호우에 대한 피해를 적절히 방지하기 위하여 설치하는 배수관, 배수암거 등의 배수시설물

횡배수관 배수암거

2. 배수계획의 수립

도로배수시설의 계획은 지형 및 조건, 도로의 종류, 주변 배수시설 등을 고려하여 경제적으로 수립한다.

각각의 배수시설을 계획할 때에는 배수목적, 배수시설의 입지조건, 계획유량을 초과한 경우에 예상되는 주변지역에 미치는 영향정도, 경제성, 지형 및 지질, 과거 홍수이력, 나뭇가지나 토석류 등 이물질에 의한 배수시설 막힘 등을 고려하여 규모를 결정하여야 한다.

1-2-4 Underground Drain Facility

- Underground drain facilities installed to remove the penetration water and to lower the underground water level when there is a danger to road bed safety due to a large amount of penetration water and high level of underground water.

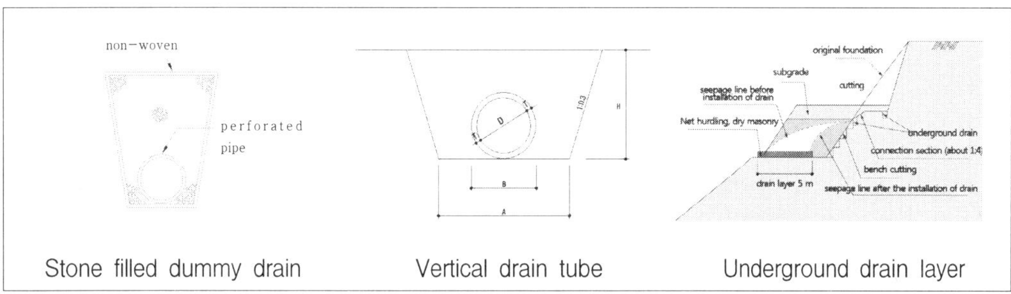

Stone filled dummy drain | Vertical drain tube | Underground drain layer

1-2-5 Traversing Drain Facility

- Facilities for small rivers or water paths traversing the road, and drain tubes, drain culvert, or drain facilities are installed to prevent the damages by roadside rain and to preserve the main road bed.

Traversing drain pipe | Drain culvert

2. Drain Plan Establishment

The drain facility plan shall be established ecomomically considering the topography, condition, type of road, surround drain facilities.

When planning each individual drain facility, the scale has to be determined based on the consideration of drain purpose, drain facility location condition, degree of expected effect when the planned flow is exceeded, economic efficiency, topography, geological features, past flood history, drain facility chocking with debris like branches or stones.

2-1 사전조사

2-1-1 조사항목

① 지형(집수면적) 및 지표면의 피복상황
② 기상, 강우
③ 토질, 지질, 투수성
④ 지하수면의 위치와 용수의 유무
⑤ 기존 도로의 배수설비
⑥ 수문조사, 과거 홍수이력 등
⑦ 지역주민 및 관계기관 탐문조사
⑧ 배수유역의 장래 토지사용 계획 등

2-1-2 조사 시 유의사항

① 표면수가 국부적으로 집중하여 흐르는 장소
② 비탈면으로부터의 용출수와 침투수가 많은 장소
③ 지하수 상황
④ 상습 침수지역
⑤ 나뭇가지, 토석류 등 유송잡물의 영향

2-2 설계 강우 발생빈도 산정

[설계 강우 발생빈도]

구 분		빈 도
평 지	암거 및 배수관(도심지, 집단가옥 등)	25년 (50년)
	노면 및 비탈면 배수시설	10년
	측도 및 도로 인접지 배수시설	10년
산 지	암거 및 배수관	100년
	노면 및 비탈면 배수시설	20년
	측도 및 도로 인접지 배수시설	20년
교량부	하천정비기본계획상의 계획빈도에 따름 단, 하천정비기본계획 미수립 시 관련기관과 협의하여 결정하거나 하천설계 기준에 따름	

2-1 Preliminary Investigation

2-1-1 Investigation Items

① Topography (water collecting area) and covering status of the surface
② Weather, rainfall
③ Soil property, Geological features, Permeability
④ Location of underground water surface and existence of rain water
⑤ Drain facilities of existing road
⑥ Flood gate investigation, past flood history
⑦ Visit survey with local residents and related institutions
⑧ Future land usage plan of the area

2-1-2 Cautions during Investigation

① Places where surface water flow is locally concentrated
② Places where lot of penetration water and effluent water from slope surface
③ Status of underground water
④ Frequent flood area
⑤ Effects of transpoted debris such as tree branched, sand and stones.

2-2 Calculation of Design Rainfall Generation Frequency

【Design rainfall generation frequency】

Classification		Frequency
Plain area	Culvert and drain pipe (city, dense residential area, etc)	25 year (50 year)
	Road surface and slope surface drain facility	10 year
	Sideroad and roadside drain facility	10 year
Mountain area	Culvert and drain pipe	100 year
	Road surface and slope surface drain facility	20 year
	Sideroad and roadside drain facility	20 year
Bride area	Follow the planned frequency in the river maintenance basic plan In case river maintenance plan is not established, follow the river design standard or determine by agreement with the related institutes	

- 산지의 적용범위(산악지 도로설계매뉴얼, 2007. 7, 국토해양부)
  - 산림청의 산사태 위험지도상 1, 2등급으로 분류되는 지역
  - 표고 400m 이상 산지를 접한 계곡 등 영향권 내의 지역
  - 산사태 및 토석류 등으로 피해가 발생한 지역
  - 산악지를 통과하는 지역으로 설계자문위원회에서 현행 설계기준보다 배수규격 확대 등 특별히 설계기준 강화가 필요하다고 판단되는 지역

- Application range of mountain area(Mountain area road design manual , 2007. 7, Ministry of Land, Transportation, and Maritime Affairs)

  - Area classified as class 1 or class 2 on avalanche danger map of the Office of Forestry

  - Area within the effected range such as valley adjacent to mountain of higher than 400 meters.

  - Area that had damages by avalanche or sand and stones

  - Area passing mountains where special design consideration is required such as expansion of drain standard over the existing design standard by design consultation committee.

2-3 설계강우강도 산정

설계강우강도는 도로 기능의 안전성과 도로 구조물의 안정성, 배수시설의 중요성 및 경제성에 의하여 결정되는 확률년의 확률별, 강우지속시간(강우도달시간)별 강우량의 크기를 의미하며 mm/hr의 단위로 표시한다.

2-3-1 강우도달시간의 산정

강우도달시간은 배수구역의 가장 멀리 떨어진 점에서부터 홍수량 산정지점까지 강우가 도달하는 시간을 의미한다, 강우도달시간은 유입시간과 유하시간의 합으로 표시된다. 유입시간은 배수구역의 가장 먼 지점에서 배수공 최상단류까지 강우가 유입되는 시간을 의미하며, 유하시간은 강우가 배수시설물이나 하천을 유하하는데 걸리는 시간을 의미한다.

① 유입시간의 산정

- 유입시간은 최소단위 배수구의 지표면 거리, 경사 및 조도계수 등에 의하여 변화

- 배수구역이 도심지 구역인 경우에는 Kerby 공식 사용을 사용하고, 배수구역이 일반적인 경우(하천)에는 Rizha 공식을 사용

② 유하시간의 산정

- 유하시간은 관거 구간마다의 거리와 계획유량에 대한 유속으로부터 구한 시간당 유하시간을 합계하여 구한다. 이를 위해서는 가상적인 관거의 배치와 크기가 필요

- 이 배치와 크기는 평균유속이 0.8~3.0m/초가 되도록 하며, 하류로 갈수록 경사는 완만하고 유속은 빠르며, 소류력(掃流力)을 크게 할 수 있도록 결정하되 반복계산을 통하여 계획 관거를 결정

- 배수구역이 도심지 구역인 경우에는 SCS 평균유속공식 사용하고, 배수구역이 일반적인 경우(하천)에는 3가지 방법(Kraven공식, Rizha 공식, 실측 수문자료를 분석)에 의하여 산정

2-3 Calculation of Design Rainfall Intensity

Design rainfall intensity is the amount of rainfall per probability of probability year, or per rainfall continuation period (rainfall reach time) determined by the importance and economic efficiency of drain facility, safety of road structures, and safety of road functions, and it is shown in the unit of mm/hr.

2-3-1 Calculation of the Rainfall Reaching Time

Rainfall reaching time means the time rainfall reaches from the farthest point in the drain area to the calculation point of floor amount. Rainfall reaching time is shown as the sum of inflow time and flowing time. Inflow time is the time taken for inflow of rainfall from the farthest point of drain area to uppermost flow of drain holes, and flowing time is the time the rainfall flows drain facilities or river.

① Calculation of inflow time

- Inflow time varies depending on the surface distance of minimum unit drain hole, slope and roughness coefficient

- Use Kerby equation if the drain area is in the city area, and Rizha equation if the drain is in general area (river).

② Calculation of flow time

- Flow time is calculated by adding the distance of each sewer area and flow time per unit time calculated from the flow velocity for planned flow rate. For this purpose, virtual sewer placement and size are necessary.

- This placement and size is to make the average flow velocity of 0.8~3.0m/second, and flow velocity gets faster as it reaches the lower flow with easier slope, and tractive force shall be determined to be maximum, and planned sewer shall be determined by repetitive calculation

- Use SCS average flow equation if the drain area is in the city area, and use three methods (Kraven equation, Rizha equation, actual measurement hydrological data analysis) if the drain is in general area (river) for the calculation.

2-3-2 확률강우강도의 산정

- 노면배수시설물 및 일반 구조물의 유출량 산출에 사용되는 강우강도표는 국토해양부에서 제시한 개정 I.D.F 곡선(intensity duration frequency, 강우강도-지속시간-생기빈도곡선)을 활용

- 측후소가 있는 지역은 지점별 강우강도표를 적용하고, 측후소가 없는 지역은 최인접 측후소의 강우강도표를 사용하되, 계획대상지점의 확률강우량도를 이용하여 강우강도표를 작성한 값과 최인접 측후소의 확률강우강도표와 비교 후 큰 값을 적용(단, 중요한 배수시설물은 관계 부서 및 감독관과 협의 후 설계강우강도를 정하여야 함)

2-4 배수유역도 작성

배수유역도의 작성은 국립지리원 발행의 1:5,000, 1:25,000, 1:50,000 축척의 지형도를 이용하며, 지도로부터 유역의 경계를 정확히 구분하기 어려울 경우에는 현장조사를 실시하여 보완하여야 한다. 조사를 통하여 토지이용도, 토질유형, 초목화 정도 등 유출과 관련된 항목을 도면에 표시하고 기록한다. 유역에 위치한 농업용 저수지, 소규모 저류시설은 대부분 홍수조절기능이 없어 유출에 영향을 주지 않으므로 무시하여도 가능하나, 유출량을 크게 변화시킬 수 있는 댐의 유무 등은 표시하여 두어야 한다. 또한, 도로 부근의 배수를 완벽하게 하기 위해서 강우강도에 따른 유출량과 유속, 방향 등을 표시한 축척 1/1,200 이나 1/5,000 또는 1/25,000(가급적 세밀한 축척을 사용)의 배수현황도를 작성한다.

2-5 횡단배수 위치 선정

도로계획의 선형설계가 완료되면 도로와 도로인접지역으로부터 유입되는 지표수가 모이는 위치를 선정하여 횡단배수시설을 설치하여야 하며, 수문분석에서 결정된 계획 홍수량으로부터 횡단배수시설 상류부의 수위를 과다하게 상승시키지 않은 상태에서 하류로 배제할 수 있는 경제적인 단면과 경사를 갖도록 계획한다.

2-3-2 Calculation of Probability Rainfall Intensity

- Revised IDF curve (intensity duration frequency, rainfall intensity - duration - frequency curve) provided by MTLM shall be used for the rainfall intensity table used for the calculation of run-off rate of the road surface drain facility and other general structure.

- Rainfall intensity table for each point shall be applied for the area with weather station, and use the nearby weather station's rainfall intensity table in case there is no weather station, and apply the bigger value among the probability rainfall table of the planned site and probability rainfall intensity table of the nearby weather station. (In case of important drain facilities, design rainfall intensity shall be determined after the agreement with related department and supervising officer)

2-4 Preparation Drainage Drawing

Preparation of drainage drawing shall use the topographical map with the scale of 1:5,000, 1:25,000, and 1:50,000 published by National Geographical Information Institute, and field survey shall be done to complement the design if the boundary of drainage ares on the map is not clear. Through the investigation, items related to run-off such as usage rate, soil property type, degree of making green shall be marked and recorded on the drawing. Most of reservoir for agriculture and small size low flow facilities located in the drainage area don't have flood control functions, and they can be ignored because they don't affect the run-off. However, existence of the dam that could make big changes of run-off shall be marked. And also drain status with run-off rate, flow velocity, and direction depending on the rainfall intensity marked on the map with the scale of 1/1,200, 1/5,000, or 1/25,000 (detailed scale if possible) shall be prepared in order to achieve perfect drain around the road.

2-5 Selection of Traversing Drain Location

When the alignment design of road plan is completed, traversing drain facility shall be installed by selecting the location where the surface water coming in from the road side area and the road is collected, and economical section and slope shall be planned, which can divert to the lower flow without increasing the level of subcritical flow of traversing drain facility excessively from the planned flood amount determined by the hydrological analysis.

2-6 최종 배수유출점 파악

도로 계획에 의한 배수시설로 인하여 기존 유역이 변경됨에 따라 도로시설 하류에 위치한 기존 배수구조물을 파악하여 원활한 배수가 이루어지는지 검토하여야 한다.

2-7 설계홍수량 산정

- 설계홍수량(설계유량) : 홍수특성, 홍수빈도, 홍수피해 가능성 및 사회적·경제적 요인을 함께 고려하여 최종적으로 설계기준으로 채택되는 첨두유량, 첨두수위 또는 설계강우량

- 설계홍수량 산정은 유역의 크기 및 유역 특성에 따라 산정

[설계홍수량 산정 방법]

유역분류	유역범위	설계홍수량 산정
소규모	4㎢ 이하	합리식 적용
중규모	4~250㎢ 이하	강우량, 유출자료 다수 존재 시 : 단위유량도법 자료가 미미한 미계측 유역 : 합성단위유량도법
대규모	250㎢ 이상	중규모 유역에 대한 유출계산 + 하도홍수 추적 및 합성

- 합리식 적용

$$Q_d = 0.2778 \cdot C \cdot I \cdot A$$

여기서, Q_d : 유역면적의 첨두유량(㎥/s)
C : 유출계수
I : 지속기간이 t인 강우강도(mm/hr)
A : 유역면적(km^2)

[노면상태별 유출계수]

구 분	C	구 분	C
포장면	0.9	도시지역	0.7
가파른 산지 및 비탈면	0.8	잡지	0.6
가파른 계곡 경작지	0.8	경작하는 평작지	0.5
논	0.8	경작하는 평계곡	0.6
완만한 산지	0.7	수림	0.3
완만한 경작지	0.7	밀림수풀과 덤풀숲	0.2

2-6 Identifying Final Drain Run-off Point

Smooth drain operation shall be evaluated by grasping the existing drain structures located in the road facility lower flow in accordance with the change of existing drainage due to the drain facility by road planning.

2-7 Calculation of Design Flood Rate

- Design flood flow(design flow) : peak flow, peak level, or design rainfall level adopted by final design standard considering the various factors such as floor characteristics, floor frequency, possibility of flood damage, and social and economic factors.
- Design flood flow shall be calculated based on the characteristics and size of the drainage.

[Calculation method of design flood rate]

Drainage classification	Drainage range	Calculation of design flood rate
Small scale	Below 4㎢	Apply rational formula
Mid scale	Below 4~250㎢	When a large amount of rainfall and run-off data are available: unit hydrograph method Unmeasured drainage with virtually no data: synthetic unit hydrograph method
large scale	Over 250㎢	Aalculation of run-off for the mid scale drainage + hydraulic flood tracking and synthesis

- Application of rational formula

$$Q_d = 0.2778 \cdot C \cdot I \cdot A$$

where, Q_d : peak flow in drainage area(㎥/s)

C : run-off coefficient

I : rainfall intensity with the duration t (mm/hr)

A : drainage area(km^2)

[Run-off coefficients for each state of road surface]

Classification	C	Classification	C
Paved surface	0.9	City area	0.7
Steep mountain and Slope surface	0.8	Miscellaneous area	0.6
Steep valley farming area	0.8	Plain farming area	0.5
Rice field	0.8	Plain farming valley	0.6
Gentle slope mountain	0.7	Forest	0.3
Gentle slope farming land	0.7	Dense forest woods and bush forest	0.2

3. 노면배수

3-1 노면배수시설의 설계흐름

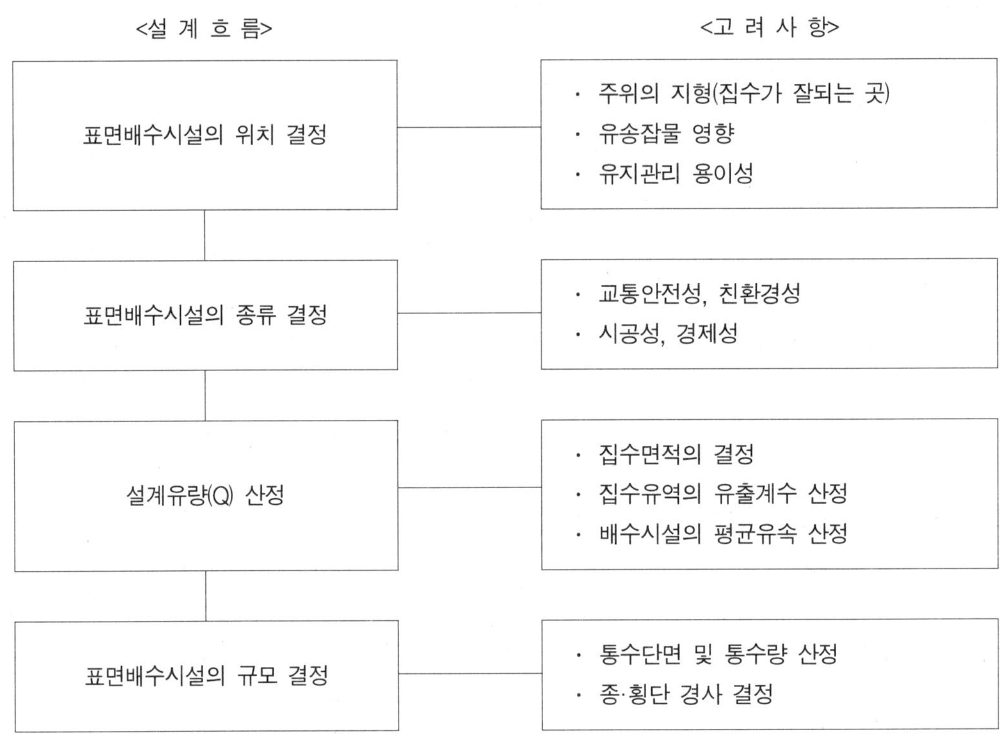

3-2 흙쌓기 구간 노면배수

3-2-1 다이크

◎ 노면에 내린 우수가 흙쌓기 비탈면으로 유하하여 비탈면이 유실되는 것을 방지

	형식1	형식2	형식3
	4%	2%	형식-3 4%

3. Road Surface Drain

3-1 Design Flow of Road Surface Drain Facility

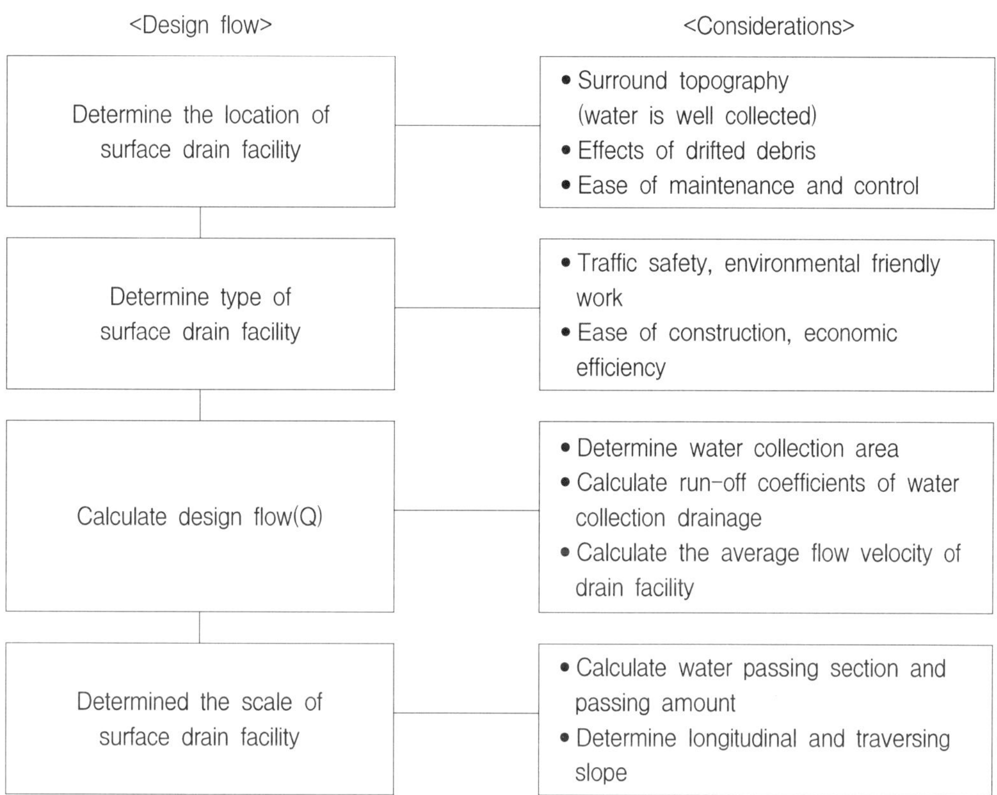

3-2 Road Surface Drain of Soil Banking Section

3-2-1 Dike

- To prevent the sweep of slope surface by the flow or rainfall on the road surface through slope surface.

	Type 1	Type 2	Type 3
	Road shoulder pavement 4%	2%	type 3 Road shoulder pavement 4%

3-2-2 집수거

- 도로 노면에서 집수되어 흐르는 물을 노면 외측으로 배수하기 위하여 도수로와 연결

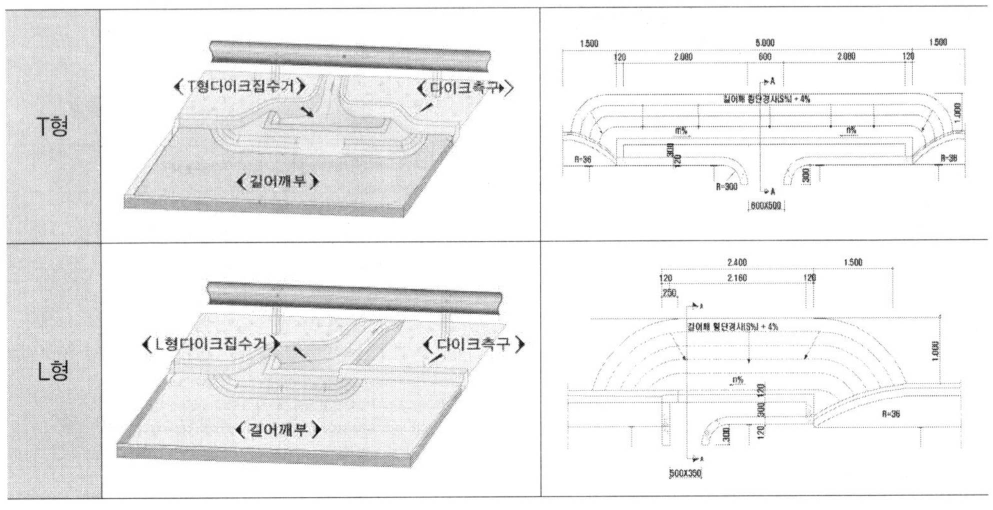

3-2-3 도수로

- 길어깨 또는 측구에 흐르는 물을 도로 인접지로 배수

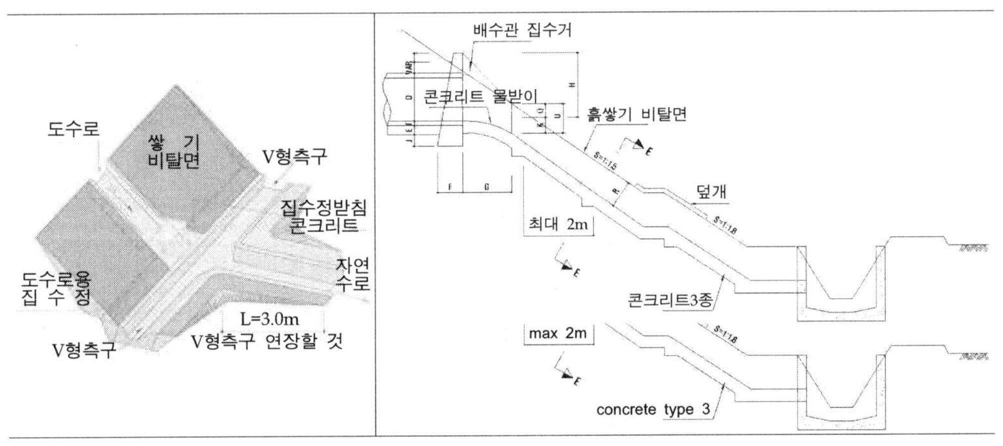

- 흙쌓기부 도수로 간격결정
  - 도수로 간격은 30m~150m의 범위
  - 길어깨의 빗물만의 배수시킬 경우 : 최대 300m
  - 산지부의 경우 흙쌓기부 도수로 최대 설치간격 : 70m

3-2-2 Collection Gutter

- Collection gutter is connected with driving channel to drain the collected and flowing water on the road surface to the outer side of road surface.

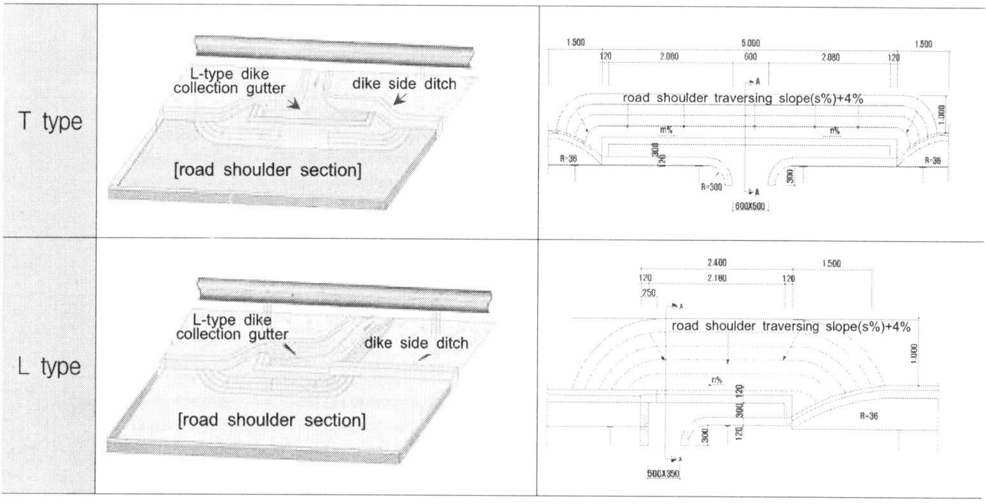

3-2-3 Driving channel

- This is to drain the water flowing on the road shoulder or side ditch into road side

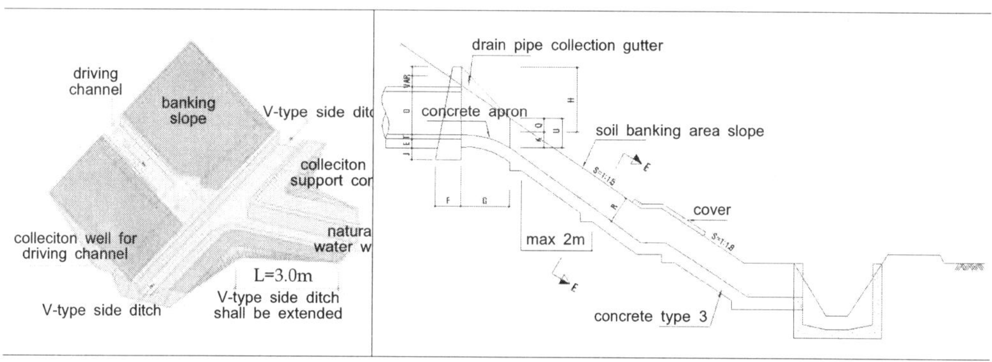

- Determination of the interval between driving channels of banking part
  - Interval of driving channel shall be in the range of 30m ~ 150m.
  - When draining only rain water on road shoulder : maximum 300m
  - In case of mountain area, maximum installation distance between driving channels on soil banking part : 70m

$$S = \frac{3.6 \times 10^6 \times Q}{C \times I \times W}$$

여기서 S : 흙쌓기부 도수로 간격(m)

Q : 길어깨 또는 측구의 허용통수량(m^3/sec)

C : 유출계수

I : 설계강우강도 (mm/hr)

W : 집수폭(m)

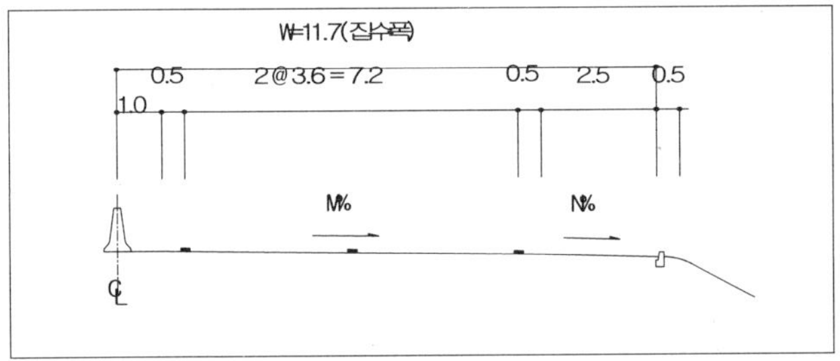

3-2-4 토사측구

- 용지확보가 용이하고 세굴의 영향이 적은 지역의 흙쌓기 비탈면 끝단에 설치

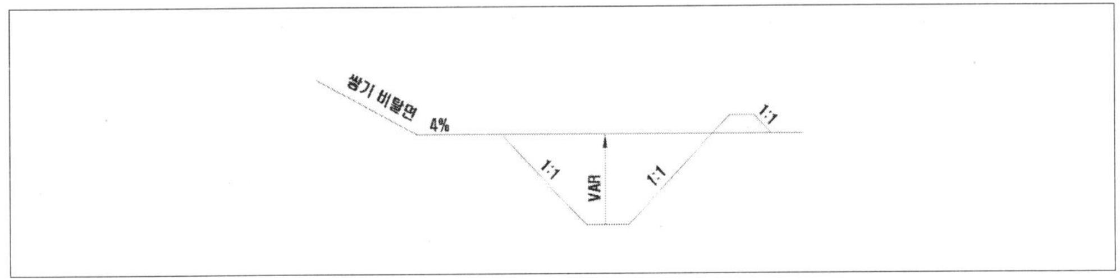

3-2-5 V형측구

- L형측구와 토사측구 연결부(땅깎기흙쌓기 경계부) 또는 침식에 의한 세굴이 예상되는 지역의 흙쌓기 비탈면 끝단에 설치

$$S = \frac{3.6 \times 10^6 \times Q}{C \times I \times W}$$

Where S : Distance between driving channels on soil banking parts(m)

Q : Allowable water passing rate on road shoulder or side ditch (m^3/sec)

C : Run-off coefficients

I : Design rainfall intensity (mm/hr)

W: Water collection width(m)

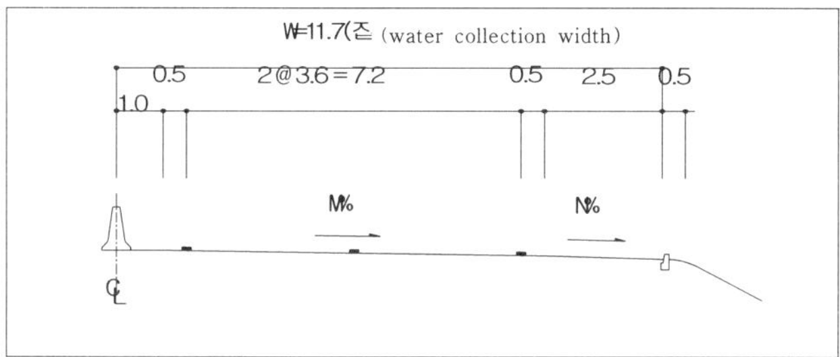

3-2-4 Soil Side Ditch

- This is installed at the end of banking slope surface where land acquisition is easy and effects of erosion are small

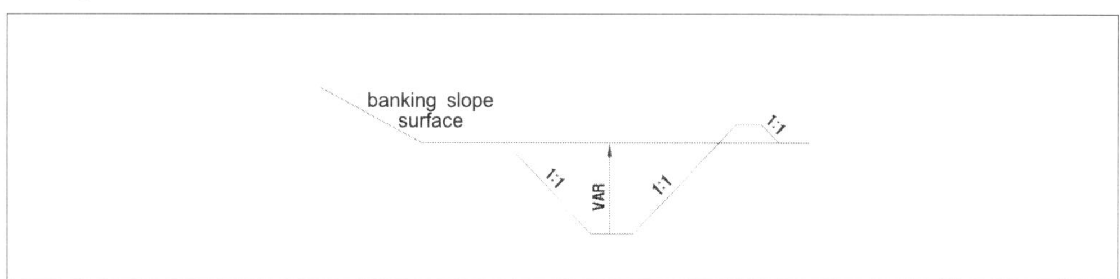

3-2-5 V-type Side Ditch

- This is installed at the end of banking slope surface where erosion is expected or at the connection part (banking/cutting boundary) of L type side ditch and soil side ditch.

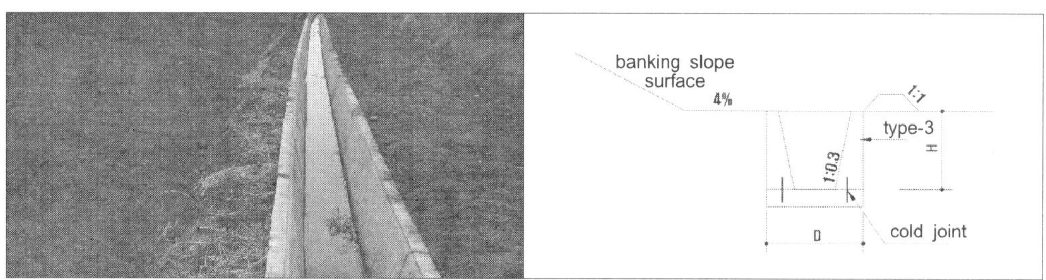

3-2-6 U형측구

- I.C, JCT 또는 분리차로의 녹지대, 부체도로, 휴게소, 도시계획도로 일부에 설치

3-3 땅깎기 구간 노면배수

3-3-1 L형측구

- 노면과 땅깎기 비탈면의 배수 및 도로시설의 보호

구 분	형 태	적용구간
형식-1 (H=0.5m)	4%, 10%, 되메우기, R=26	땅깎기면 토사 및 리핑암 전 구간 발파암 H=10.0m 미만인 경우 적용
형식-2 (H=1.2m)	P.V.C 관, R=26, 부직포, R=260, R=240, 9.35%, 10%, 1:0.42916, 뒷채움 (깬잡석)	땅깎기면 리핑+발파암 H=10.0 m 이상인 구간 또는 발파암 H=10.0m 이상 30.0m 미만인 구간에 설치
형식-3 (H=2.0m)	1:0.3, 되메우기, 1:0.304347, 부직포, P.V.C 관, 6.0%, 10%, 뒷채움 (깬잡석)	땅깎기면 발파암 H=30.0m 이상이 되는 구간 중에서 연장이 20m 이상되는 구간에 설치
형식-4 (H=0.35m)	R=30, 4%, 10%, 되메우기	부체도로에 설치

3-2-6 U-type Side Ditch

- This is installed at the IC, JCT, green area of median strip, side roads service area, and part of city express way.

3-3 Road Surface Drain in Soil Cutting Section

3-3-1 L-type Side Ditch

- This is for the protection of drain and road facility of road surface and soil cutting surface

Classification	Type	Applied section
Type-1 (H=0.5m)	4%, 10%, re-banking, R=26	Applied for soil, entire section of ripping rock, and blasting rock with H < 10.0m for cutting surface
Type -2 (H=1.2m)	P.V.C pipe, R=26, non-woven, R=260, R=240, 10%, 9.35%, 1:0.42916, backfilling (crushed stone)	Installed in ripping rock + blasting rock with H>10.0m, or blasting rock area with 10.0m < H < 30.m in soil cutting surface
Type -3 (H=2.0m)	1:0.3, re-banking, 1:0.304347, non-woven, P.V.C pipe, 6.0%, 10%, backfilling (crushed stone)	Among the section of blasting rock with H > 30.0 m section in cutting surface, this is installed in the section with the length over 20 meters.
Type-4 (H=0.35m)	R=30, 4%, 10%, re-banking	Installed in side road

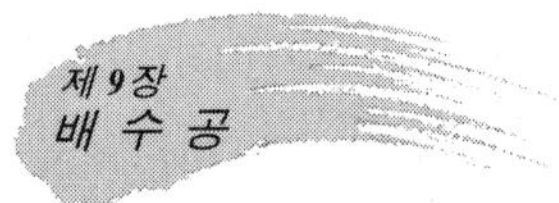

3-3-2 집수정

◎ 길어깨에 흐르는 우수를 배제

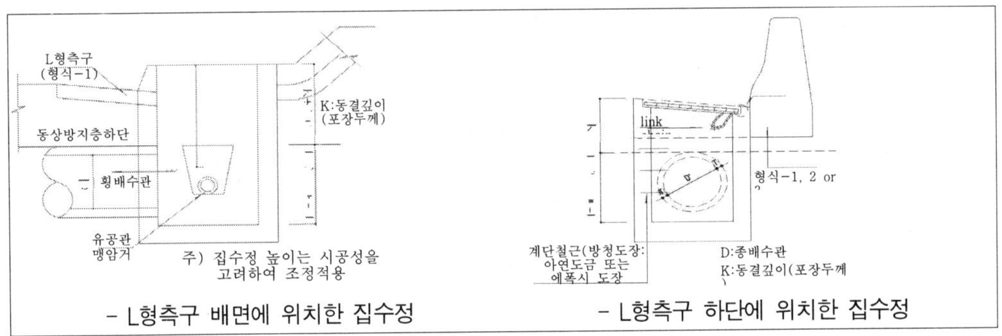

- L형측구 배면에 위치한 집수정　　　　- L형측구 하단에 위치한 집수정

◎ 땅깎기부 집수정 설치간격 결정

- 설치위치
  - 유량이 길어깨의 허용통수량과 같아지는 곳
  - 땅깎기부 비탈면 도수로 또는 종배수관이 연결되는 곳
  - 지하배수관 또는 종배수관 규격이 변화되는 곳
  - 종단선형 오목부의 최하단 지점
- 집수정 간격결정

$$S = \frac{3.6 \times 10^6 \times Q}{I(C_1 W_1 + C_2 W_2)}$$

여기서　S : 집수정 간격
　　　　I : 설계강우강도(mm/hr)
　　　　C_1 : 유출계수(포장부 0.9)
　　　　C_2 : 유출계수
　　　　W_1 : 집수폭(포장부)
　　　　W_2 : 집수폭(절개부)
　　　　Q : 허용통수량(m^3/sec)

- 청소와 유지관리를 고려한 최대 집수정 간격 : 50m
- 시공성을 고려한 최소 집수정 간격 : 5m

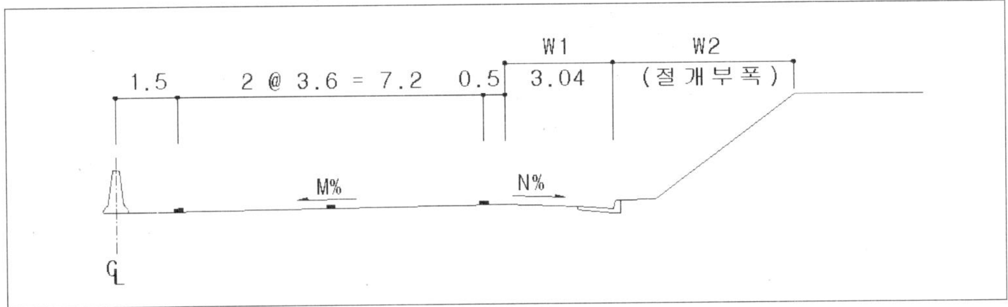

3-3-2 Collecting Well

- To drain the rain water flowing on road shoulder

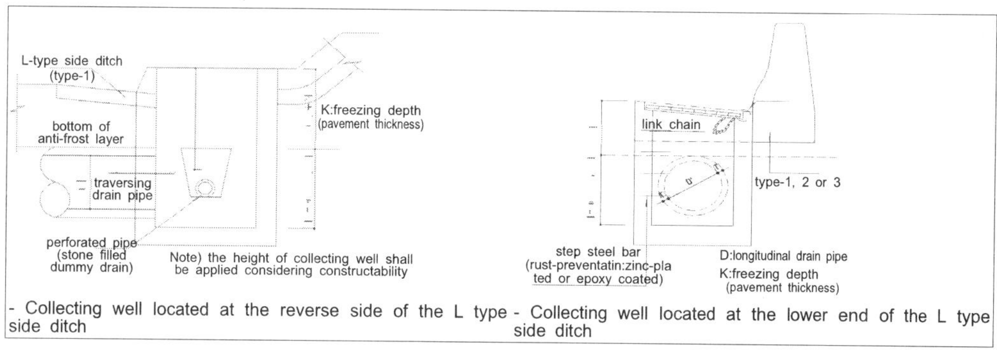

- Determination of installation interval of collecting well in cutting part
  - Installation location
    - Where flow becomes same as the allowable passing flow of road shoulder
    - Where cutting slope surface driving channel on cutting part or longitudinal drain pipe are connected
    - Where underground drain pipe or longitudinal pipe specification is changed
    - The lowest point of longitudinal path concave part
  - Collecting well distance determination

$$S = \frac{3.6 \times 10^6 \times Q}{I(C_1 W_1 + C_2 W_2)}$$

Where, S : Collecting well interval

I : Design rainfall intensity(mm/hr)

C_1 : run-off coefficient (pavement part 0.9)

C_2 : run-off coefficient

W_1 : Water collection width(Pavement part)

W_2 : Water collection width(Incision part)

Q : Allowable water passing rate(m^3/sec)

- Maximum connection well interval considering clean, maintenance, and control : 50m
- Minimum collecting well interval considering ease of construction : 5m

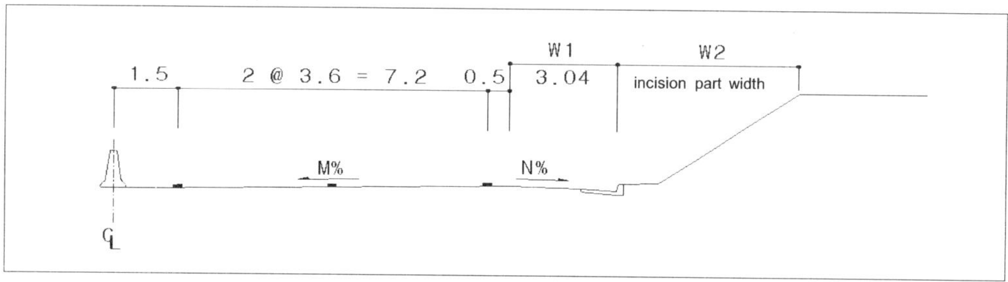

3-3-3 산마루측구

- 상부 비탈면에 내린 우수와 용출수를 비탈면에 유입시키지 않도록 비탈어깨에 설치

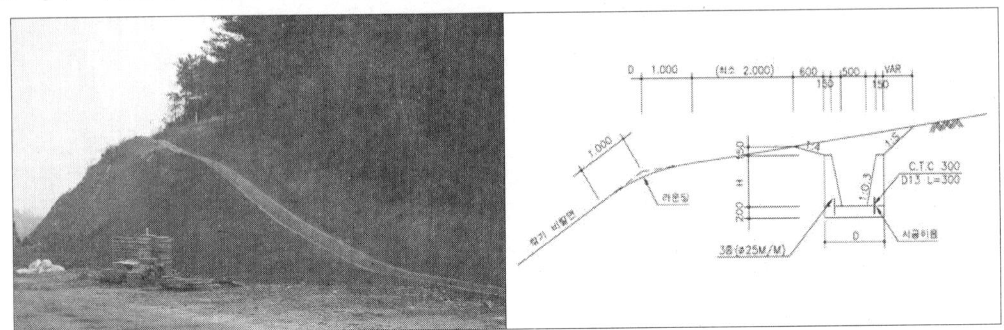

3-3-4 소단측구

- 비탈면에 흐르는 빗물이나 용출수에 의한 비탈면 침식 방지

4. 지하배수

4-1 지하배수시설의 기능

지하수를 집수하여 외부로 배출

4-2 지하배수시설의 분류

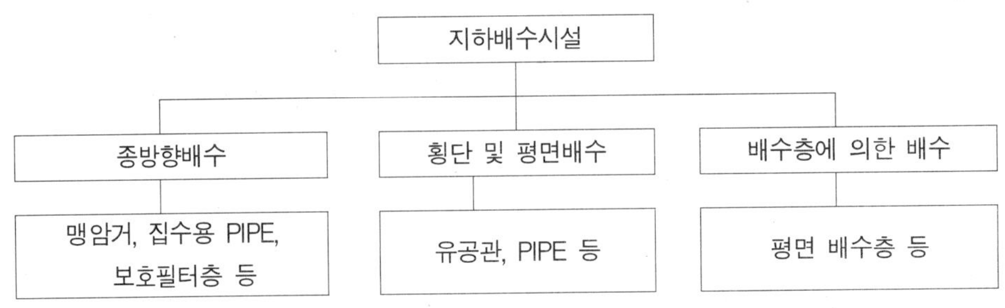

3-3-3 Mountain Ridge Side Ditch

- This is installed at the tope of the slope in order not to inflow the effluent water and rain water on the road surface to slope surface

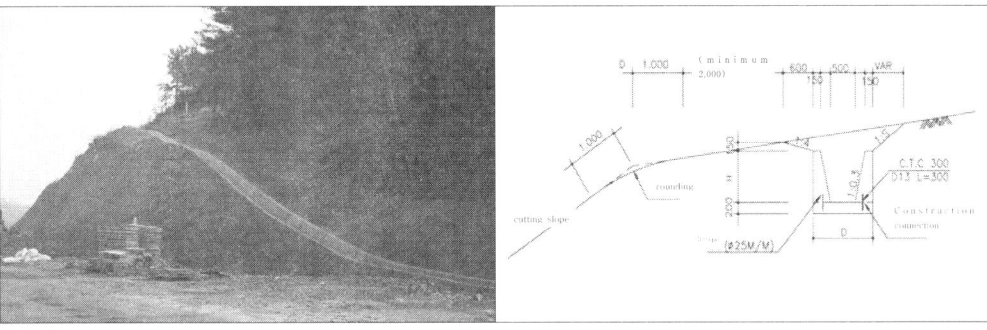

3-3-4 Berm Side Ditch

- This is installed to prevent the slope surface erosion by effluent water or rain water flowing over slope surface

4. Underground Drain

4-1 Function of Underground Drain Facility

To collection underground water and drain the water to outside

4-2 Classification of Underground Drain Facility

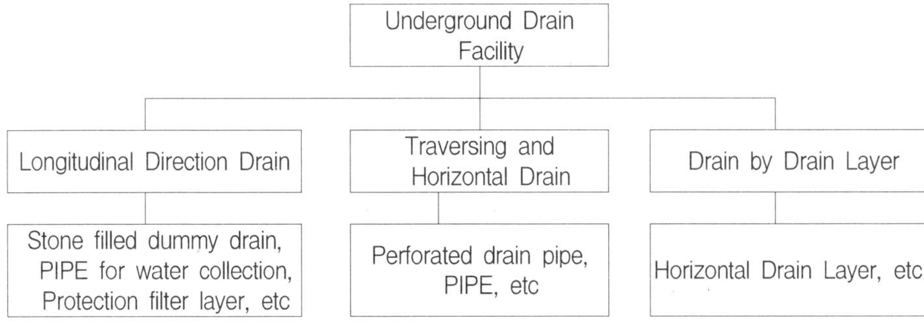

4-3 지하배수시설의 깊이

$$H = \frac{(S_w - S)B}{200} + h + D_w + h_p$$

여기서 H : 배수관의 최소 매설깊이(m)

S_w : 지하수면의 최소경사도

S : 노면의 경사(%)

h : 배수관 부근의 지하수위

B : 배수관의 간격(m)

D_w : 노상에서 지하수위까지의 깊이(0.60m 이상)

h_p : 노면과 노상과의 높이 차이(m)

4-4 지하배수시설의 소요통수량 산정

- 지하배수구는 투수계수가 높은 필터재로 구성된 구조를 원칙으로 하며, 필요에 따라 유공배수관을 설치

- 단면은 유공관 크기에 의하여 결정

$$A = \frac{Q}{V} = \frac{Fs \cdot q \cdot L}{V}$$

여기서 F_s : 안전율(보통 3으로 함)

q : 배수관 1m 당의 배수량 m^3/sec

L : 유공배수관의 길이(m)

V : 유공관내의 평균유속(m/sec)

4-3 Depth of Underground Drain Facility

$$H = \frac{(S_w - S)B}{200} + h + D_w + h_p$$

Where, H : Minimum burial depth of drain pipe(m)

S_W : Minimum slope of underground water surface

S : Road surface slope (%)

h : Underground water level around drain pipe

B : Distance between drain pipes (m)

D_W : Depth from sub grade to underground water level(over 0.60m)

h_p : Difference in height between road surface and sub grade (m)

4-4 Calculation of Required Pass Water Rate of Underground Drain Facilityl

- Underground drain ditch shall be of the structure with filter materials of high permeability coefficient in principle, and perforated drain pipe shall be installed if necessary

- Cross section shall be determined by the size of perforated drain pipe

$$A = \frac{Q}{V} = \frac{Fs \cdot q \cdot L}{V}$$

Where, F_s : Safety rate (normally set to 3)

q : Drain flow per 1 meter of drain pipe (m^3/sec)

L : Length of perforated drain pipe (m)

V : Average flow velocity inside the perforated drain pipe (m/sec)

4-5 맹암거

구분		적용위치	유공관 사용여부	부직포 사용여부
형식-1	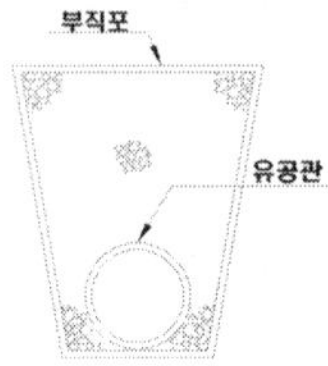	땅깎기부 L형측구 아래에 설치 (토사구간)	○	○
형식-2	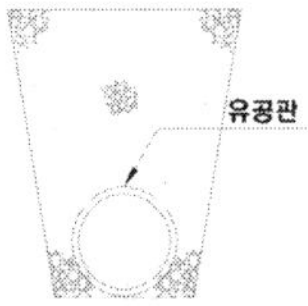	땅깎기부 L형측구 아래에 설치 (리핑암, 발파암 구간)	○	×
형식-3		한쪽깎기한쪽쌓기 구간 및 땅깎기·흙쌓기 경계구간에 설치(토사구간), 땅깎기부 비탈면 통수부에 설치(토사구간)	×	○
형식-4		형식-3과 동일 (리핑암, 발파암 구간)	×	×
형식-5		지하수 유출 및 용수 다발지역에 설치	다발 판넬 사용	×

4-5 Stone Filled Dummy Drain

Classification		Application location	Whether to use perforated pipe	Whether to use non-woven
Type-1	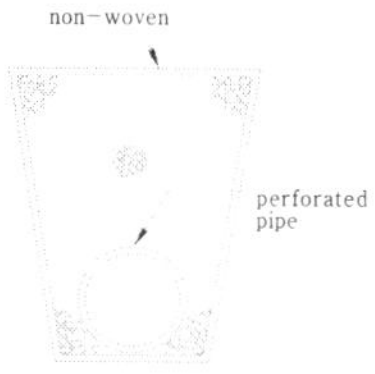 non-woven perforated pipe	Installed under the cutting area L type side ditch(soil area)	○	○
Type-2	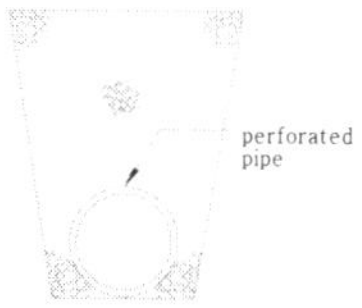 perforated pipe	Installed under the cutting area L type side ditch (ripping rock, blasting rock)	○	×
Type-3	non-woven	Installed at the cutting and embankment section, or at the boundary between cutting and banking part(Soil section), or cutting section slope surface water passing part (soil section)	×	○
Type-4		Same as Type-3 (Ripping rock, blasting rock section)	×	×
Type-5	 bundle bin	Installed at frequent underground water run-off and rainfall water	Use bundle panel	×

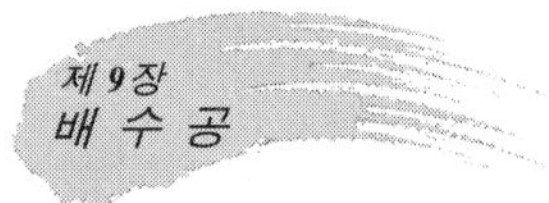

5. 횡단배수

5-1 횡단배수시설의 종류

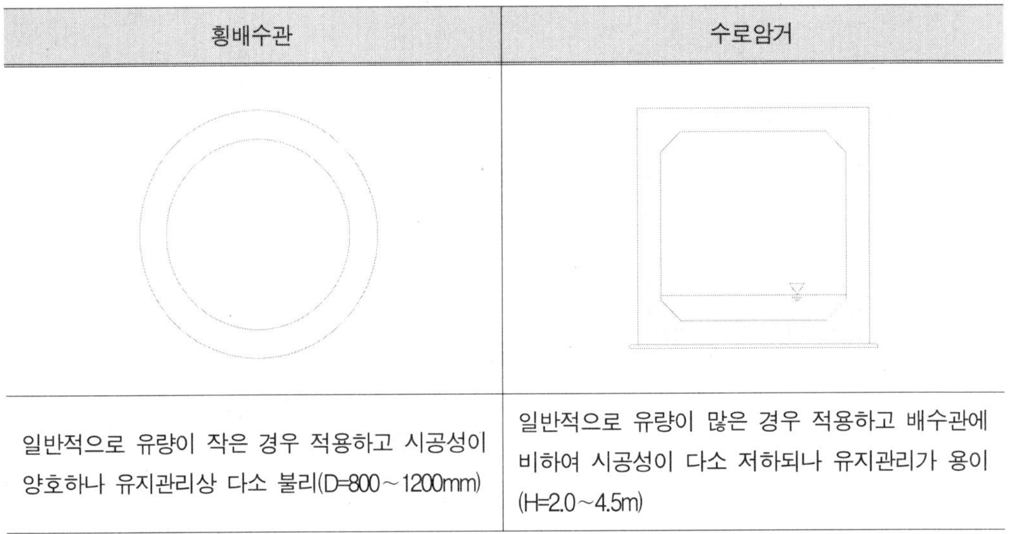

횡배수관	수로암거
일반적으로 유량이 작은 경우 적용하고 시공성이 양호하나 유지관리상 다소 불리(D=800～1200mm)	일반적으로 유량이 많은 경우 적용하고 배수관에 비하여 시공성이 다소 저하되나 유지관리가 용이 (H=2.0～4.5m)

5-2 횡단배수시설 설계흐름

[횡단배수시설 설계 흐름도]

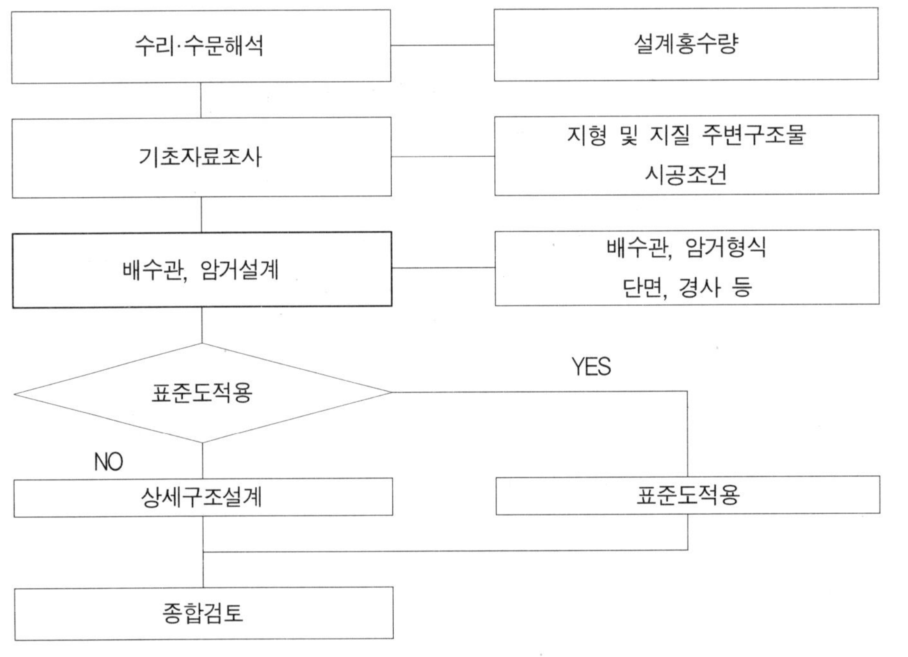

5. Traversing Drain

5-1 Type of Traversing Drain Facility

Traversing Drain Pipe	Water Way Culvert
Generally applied when the flow is small. Construction is easy, but little disadvantage in maintenance and control (D=800～1200mm)	Generally applied when the flow is high. Construction is more difficult than drain pipe, but maintenance and control is easier(H=2.0～4.5m)

5-2 Traversing Drain Facility Design Flow

[Traversing Drain Facility Design Flow]

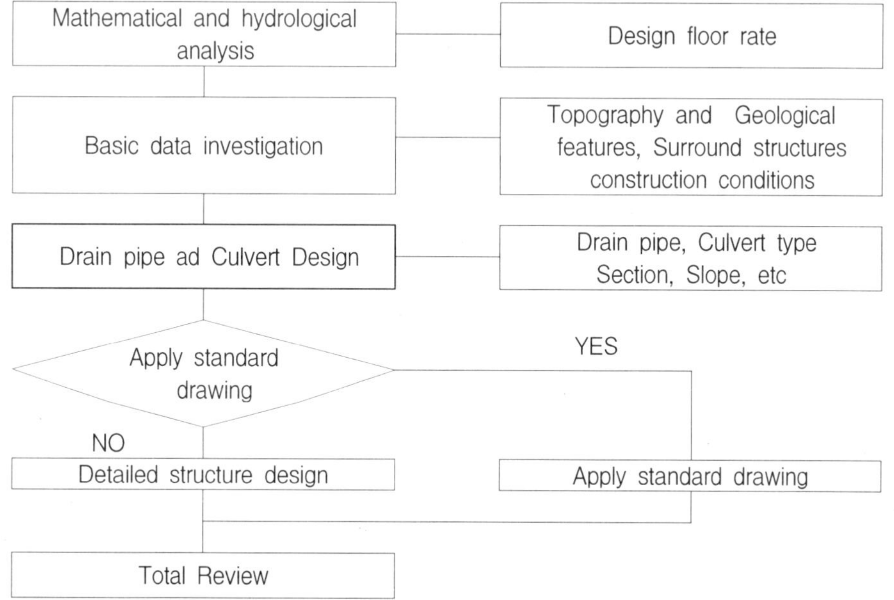

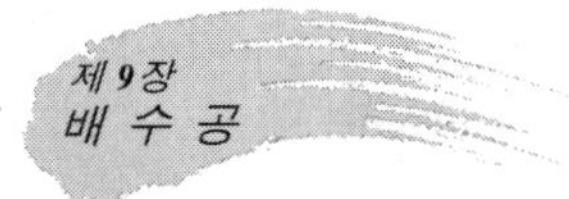

5-3 횡단배수시설의 단계별 설계절차

1단계	기초 수문자료 획득

① 유역면적 산정(A)
② 계획배수량(Q_d) 결정
③ 배수시설의 길이(L), 경사(S_o), 허용상류수심(AHW) 산정

[배수횡단 개요도]

단 면 도

EL= 134.5 m
AHW= 1.6 m HW
TW= 0.5 m
EL= 132.9 m
So= 0.0060 m/m
L= 30.0 m
EL= 132.8 m

2단계	암거의 초기단면 치수산정 (가정치)

① 일반적으로 A=Q/3을 기준으로 하거나 경험적으로 결정

3단계	유입부 조절을 가정하여 기 산정한 단면적에 대한 상류수심(HW)을 결정

① 유입부 조절: 배수유량이 유입부 단면적에 의하여 유입부에서 통제되는 것을 의미
② 유출부에서 $HW=H+h_o-S_o \cdot L$ 을 산정
여기서 h_o : $(d_c+D)/2$, TW, 0.75D(Pipe), 0.8D(Box) 중 최대값 적용
③ 유입부 HW와 유출부 HW 중 큰 값이 기준
④ HW>AHW일 경우 단면을 재산정하여 시산하고
HW<AHW일 경우의 산정된 제원으로 배수횡단시설 설계
- 여기서 AHW : 유입부에서의 포장하단부터 지반까지의 높이

5-3 Design Procedure for Each Step of Traversing Drain Facility

Step 1	Basic hydrological data collection

① Drainage area calculation (A)

② Determine designed drain amount (Q_d)

③ Calculation of length (L), slope (S_o), and allowable subcritical flow water depth (AHW) of drain facility

【 Drain Traversing High Level Drawing】

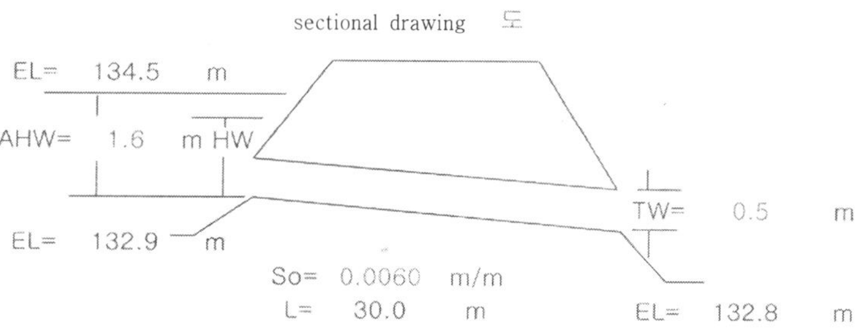

Step 2	Dimension calculation of initial section of culvert (assume value)

① In general, A=Q/3 shall be a reference or it is empirically determined.

Step 3	Determine the subcritical flow water depth (HW) for the pre-calculated section assuming inflow rate control

① Inflow control: This means that the drain flow is controlled at the inflow part by the section area of inflow part

② Calculate $HW=H+h_o-S_o \cdot L$ at run-off part

Where, h_o is the maximum value among $(d_c+D)/2$, TW, 0.75D(Pipe), and 0.8D(Box)

③ Bigger value among inflow part HW and run-off part HW becomes the reference

④ When HW>AHW: trial calculation by re-calculation of section area

When HW<AHW: Design drain traversing facility based on the calculated specification

- where, AHW: Height from the lower end of pavement to the foundation at inflow area

4단계	산정(가정)한 단면에 대한 규격 조정 후 시산 시행

5단계	유출부 유속을 검토하여 필요 시 수로 보호공 (콘크리트 라이닝, 감쇄공, 침전조 등)을 계획
【수로 보호공 설치 사례】	

6단계	설계 내용 요약 및 건의사항 (단면 치수·형상, 유속, 감쇄시설 필요 여부 등) 수록

Step 4	Execute trial calculation after adjustment of the specification for computed (assumption) section

Step 5	Review the flow velocity of run-off part, and plan water path protection method (concrete lining, end sill , settling tank, etc) if necessary.

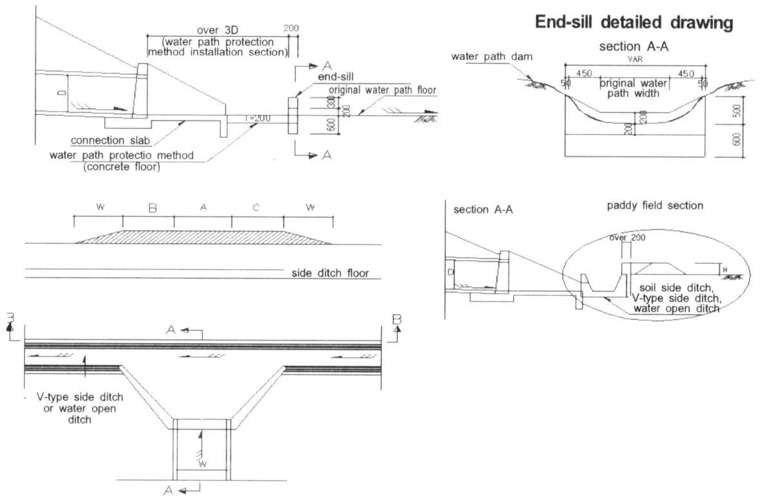

End-sill detailed drawing

Step 6	Summary of design details and suggestions (section dimension, shape, flow velocity, attenuation facility requirement, etc)

Chapter 10
Structure Construction

1. 교량의 계획

1-1 일반사항

교량의 계획단계에서는 교량의 선형, 가설상의 제약조건, 구조설계기준 만족, 내구성과 안전성 확보, 경제적이며 심미적인 교량형식 선정, 교장 및 경간분할, 주형 및 바닥틀의 배치계획, 하부구조의 형식 및 형상, 기초 형식 및 공법 결정 등이 이루어지게 된다. 계획단계에서 치밀한 조사를 행한 체계적인 계획 수립은 향후 교량의 설계 시 또는 공용수명 간 교량 성능에 큰 영향을 미치게 되므로, 교량계획은 아래의 사항 등을 원칙으로 신중하게 고려하여야 한다.

1-1-1 노선선형 및 교량위치의 적정성

- 토공 : 공사비 저렴 · 시공 및 유지관리 용이
- 교량 : 선형 및 위치 조건에 따른 형식 및 공법적용의 제한 → 경제성 · 시공성의 주요인자

1-1-2 교량계획의 외부적 제 요건 수용

- 지형 · 지질조건 - 기초형식 및 공법, 교대 · 교각 위치의 적합성 판단
- 통과(교차)조건 - 도로 · 철도 · 하천 등의 조건(시설한계 · 장래계획 · 하천시설 기준 등)수용

1-1-3 구조적 안정성

- 설계기준의 제 요건 만족, 구조계획의 합리성 · 명확성

1-1-4 경제성

- 외부적 제 요건의 수용 및 시공성과의 관계 속에서 투자가치 평가, 기존실적의 활용

1-1-5 시공성

- 외부적 제 요건의 수용, 안전성과 품질관리의 확실성

1-1-6 유지관리

- 초기투자비와 유지관리를 고려한 총비용(LCC)의 평가

1-1-7 주행의 안전성 및 쾌적성

- 상로교 및 연속교 원칙

1-1-8 미관 · 경관 설계

- 구조물 자체의 미관(구조 형상 · 재질 · 부속물 등), 주변 경관과의 조화

1-1-9 환경친화적 설계

- 개발과 보전의 균형, 개발의 세부적 실천

1. Plan of Bridge

1-1 General Items

In the bridge planning stage, alignment of bridge, limiting conditions on erection, meeting structure design standard, securing durability and stability, selection of economic and aesthetic bridge type, division of entire bridge length and span, cast and bottom frame placement plan, substructure type and shape, determining foundation type and construction method shall be considered. Bridge planning shall consider the following items carefully as a principle, since the establishment of the systematic plan with thorough investigation during the planning stage exercise great influence on the performance of bridge during the bridge design and after the completion.

1-1-1 Suitability of Route Alignment and Bridge Location

- Earthworks : low cost and easy construction, and easy maintenance and control
- Bridge : Limitation of the construction method application and type according to the alignment and location condition → Major factors of economic and construction

1-1-2 Accommodation of Various External Conditions for Bridge Plan

- Topographic and geological condition - judging suitability of foundation type and construction method, and location of abutment and pier.
- Pass(cross) condition : Accomodate the conditions(clearance, future plan, river facility standard etc) of road, railroad, river, and etc.

1-1-3 Structural Stability

- Meeting the various conditions of design standard, and rationality and clarify of structure plan.

1-1-4 Economic Efficiency

- Accommodation of various external conditions and evaluation of the investment value in relationship with construction efficiency, and utilization of exiting achievement history.

1-1-5 Construction Efficiency

- Accommodation of various external conditions, and reliability of stability and quality control

1-1-6 Maintenance Management

- Evaluation of total cost (LCC) considering initial investment and maintenance management

1-1-7 Stable and Pleasant Driving

- In principle, deck bridge and continuous bridge shall be installed

1-1-8 Aesthetic and Landscape Design

- Aesthetic structure itself (structure shape, material, accessories, etc), and balance with surrounding landscape

1-1-9 Environment Friendly Design

- Balance between development and preservation, detailed execution of development

1-2 교량 설계 시 조사항목

조사는 합리적이고 경제적인 계획, 설계, 시공을 하기 위하여 필요한 조건을 명확하게 하는 것으로 계획 예정 지점의 상황, 구조물의 규모 및 중요도를 충분히 고려하여 필요한 정밀도와 내용을 가져야 하며, 관련된 조사를 살펴보면 다음과 같다.

조사종류	조사내용	조사목적
지형조사	• 지형도 작성	• 교량 가설위치, 교량길이, 경간분할 결정
지질조사	• 지질역사, 지질자료 수집 • 지표지질조사, 지질도 작성 • 물리탐사	• 교량지지층의 지질조건 판단 • 하부공 위치선정과 구조계획
교차 도로 및 철도 조사	• 교차 도로, 철도 등의 폭원, 교량높이, 시설한계, 횡단구조, 종단경사 등의 상황 및 장래계획 조사 • 지하매설물의 조사 및 교통량 조사	• 교량길이, 경간분할, 다리밑 공간, 시공방법의 선정
하천조사	• 하천횡단형상 • 유량, 유속, 고수위, 저수위, 하천경사 등의 형상 및 장래계획 조사 • 세굴, 조류, 표층모래, 부유물 등의 변화에 대한 조사 • 항해선박의 조사	• 경간분할, 다리밑 공간, 교각형상, 기초, 근입깊이의 선정 • 시공시기, 시공방법의 선정 • 월류방지 소요단면의 결정 • 경간분할, 교각형상, 시공방법의 결정, 충돌하중의 결정 • 세굴방지공의 계획
바다, 호수, 저수지의 조사	• 조수위, 파고, 조류조사 • 항해선박 및 장래계획의 조사	• 계획고의 선정 • 수압, 시공시기, 시공방법의 선정 • 경간분할, 교각형상, 충돌하중의 결정 • 선박충돌 방지공 계획
토질조사	• 시추 • 표준관입시험 • 공내물리검층 • 공내재하시험 • 평판재하시험 • 물성시험 • 일축압축시험 • 삼축압축시험 • 압밀시험	• 하부공의 설계 • 지지층의 판단 • 허용지지력 및 침하량 산정 • 흙의 단위중량결정, 내부마찰각, 점착력의 결정 • 기초형식 선정 • 시공방법의 선정 • 말뚝기초의 경우 부마찰력 검토
지진조사	• 지진기록, 지진피해 기록의 조사 • 지반상시 미진동측정 • 지진기반의 확인	• 설계진도의 결정 • 내진, 구조형식 선정

1-2 Investigation Items during Bridge Design

Investigation is to clarify the rational and economic plan, design, and required condition for construction, and it shall contain the necessary precision and contents with sufficient consideration of status of the planed site and scale and importance of the structure. The followings show the related investigation.

Investigation type	Investigation details	Investigation purpose
Topographical investigation	• Preparation of topographical map	• Determine the erection location of bridge, length of bridge, span division
Geological survey	• Collection of geological history and geological data • Surface geological investigation, and preparing geological map • Physical exploration	• Judging geological condition of the bridge support layer • select the location of substructure and structure plan
intersection and railroad investigation	• Investigation of status and future plan such as the width of intersection, railroad, height of bridge, building limit, and cross structure, vertical slope • Investigation of underground buried facilities and traffic volume	• Determine bridge length, span division, space under the bridge, construction method
River investigation	• River cross shape • current level, current speed, high water level, low water level, river slope, and other shape and future plan investigation • Investigation on the changes of scour, current, surface layer sand, suspended solid • Investigation of a service ships	• Determine span division, space under the bridge, pier shape, foundation, and penetration depth • Determine construction period and method • Determine required section to prevent overflow • Determine span division, pier shape, construction method, and collision load • Plan for scour protection
Sea, lake, and reservoir investigation	• Tide level, wave height, and current investigation • Passing ships and future plan investigation	• Calculation of plan height • Determine water pressure, construction period and construction method • Determine span division, pier shape, collision load • Ship collision prevention construction plan
Soil investigation	• Drilling • Standard penetration test • Borehole physical layer test • Borehole pressuremeter test • Flat plate pressuremeter test • Physical property test • Unconfined compression test • Triaxial compression test • Consolidation test	• Substructure work design • Judgment of bearing layer • Allowable support force and settlement calculation • Determine unit weight of soil, internal friction angle, and adhesion force • Determine foundation type • Determine construction method • Negative friction evaluation in case of pile foundation
Earthquake investigation	• Earthquake recording, and earthquake damage recoding investigation • Measurement of regular micro vibration of ground • Confirm earthquake foundation	• Determine design seismic intensity • Selection of earthquake-proof, and structural type

조사종류	조사내용	조사목적
기상조사	• 기상관측 기록의 조사 • 풍속, 온도, 강설, 천후 등	• 온도변화의 결정 • 풍하중, 설하중의 결정 • 피복두께의 선정, 재료의 선정 • 시공시기 및 방법의 선정
부식조사	• 기존구조물의 부식상황 조사(염해 등) • 유기물 조사, pH 조사 • 도료에 대한 시험	• 사용재료의 선정 • 방청방법의 결정
교량 부착물 조사	• 교량위치의 기설 지하매설물조사 • 상수도, 하수도, 전력, 통신선 등의 신설 계획 조사	• 부착물의 제원, 중량의 결정
재료조사	• 콘크리트용 골재 및 물의 품질과 물량조사	• 재료의 선정 및 적산
시공조사	• 재료 및 부재 운반로의 현황조사 • 제작장, 조립장의 현황조사 • 주변건물의 조사 • 양식장 축산농가의 조사 • 사적, 문화재 등의 유무	• 교량형식의 결정 • 운반될 최대부재의 결정 • 공사용 도로계획, 타 공사와의 관련 • 교차물(도로, 철도, 하천)에 의한 제약조건 • 제작장, 조립장 위치 선정 • 시공 중 소음, 진동 등의 주변 영향 파악 • 환경보존 대책수립
주변 환경조사	• 교량의 조망권 조사 • 생태환경 보호구역 유무조사 · 교량주변의 지형조사 • 교량주변의 문화적, 역사적 유적지 유무조사 • 주변 개발계획 조사	• 교량의 경관설계 • 교량형식 및 시공방법의 선정
지역특성 조사	• 비행 고도한계(항공장애) 유무 조사 • 군작전지역 유무 조사	• 교량형식 및 시공방법의 선정 · 교량계획고의 결정 • 거부시설의 유무 결정

Investigation type	Investigation details	Investigation purpose
Weather investigation	• investigate the meteorological observation record • Wind speed, temperature, snow fall, weather, etc	• Determine temperature change • Determine wind load and snow load • Selection of cover thickness and material • Selection of construction period and method
Corrosion investigation	• Investigation of the corrosion of existing structure (salt damage, etc) • Organic investigation, pH investigation • Test on coating material	• Determine materials to use • Determine corrosion-proof method
Bridge attached facilities investigation	• Investigation of existing underground buried facilities in the location for the bridge • Investigation of new construction plan such as water supply, sewer, power line, and communication lines	• Determine specification and weight of attached facilities
Material investigation	• Investigation of aggregate for concrete, quality and quantity of water	• Selection and addition of materials
Construction investigation	• Iinvestigation of status of material and supplementary material moving path • Investigation of status of assembly place and manufacturing place • Adjacent building investigation • Farm and stock breeders investigation • Existence of historic site and cultural assets	• Determine bridge type • Determine biggest materials to move • Plan for construction road, relation with other constructions • Limiting conditions by crossing structure (road, railroad, river) • Selection of location of manufacturing and assembly site • Grasp the effect to surroundings during the construction construcnoise and vibration • Set up a plan to preserve the environment
Surrounding environment investigation	• Investigation of view right of bridge • Investigation of the existence of ecology environment preserve area • Investigation of topology around the bridge • Investigation of cultural and historic site around the bridge • Investigation of development plan of surrounding area	• Bridge view design • Bridge type and construction method selection
Regional characteristics investigation	• Investigation of the existence of flight altitude control (flight obstacles) • Investigation of the existence of military operation area	• Bridge type and construction method selection Determine bridge planned height • Determine the existence of rejection facility

1-3 교량가설위치 및 경간장 결정

1-3-1 교량가설위치 결정 시 검토사항

구 분	내 용
관련 계획에 의한 지역	계획도로와 교차하는 도로, 철도, 택지, 단지조성계획에 따른 신설도로 등의 각종 장래계획에 대하여 검토 후 교량가설 여부를 결정한다.
도로의 흙쌓기 높이가 높은 지역	도로 설계 시의 흙쌓기 높이가 높은 경우, 도로 토공 공사비 및 용지 보상비와 교량건설비를 비교 검토하여 교량가설 여부를 결정하여야 한다.
땅깎기 및 흙쌓기가 곤란한 산간지역	땅깎기 및 흙쌓기 시 비탈면의 높이가 높은 조건에서는 자연파괴와 비탈면 안전성 등에 대하여 종합적으로 판단하여 교량가설 여부 및 가설 위치를 결정하여야 한다.
산간 지역 깊은 계곡부	계곡을 메워야 하거나 대규모 흙쌓기로 인하여 자연파괴가 많은 지형조건에는 교량으로 계획하는 것이 바람직하다.
농경지 횡단 시	절대 농지 지역에 도로를 개설하는 경우에 식량의 자원화 측면에서 가급적이면 농경지를 최소로 점용하는 것이 좋다.
저수지 횡단	토공으로 도로를 개설할 경우에는 담수능력 저하에 따른 수자원 이용자의 손실이 크므로 이런 지역은 교량으로 계획하여야 한다.
연약지반 심도가 깊은 지역	연약지반 처리에 따른 제반 영향을 충분히 비교 검토하여 교량으로 건설하는 것이 타당성이 있는 것으로 판단될 경우에만 교량으로 계획한다.
지하시설물이 있는 지역	이설 불가능한 지하시설물이 있는 경우에는 사전에 관련기관에 협조 요청하여 검토한 뒤 교량계획 여부를 결정하여야 한다.

1-3-2 경간장 및 형하고 결정

교량은 그 기능상 도로 횡단 육교, 산악부 통과교량, 하천 횡단교량 등 3종으로 계획하며, 본선 연계의 통과목적상 교량을 가설토록 계획된 구간에 있어 그 구조물의 설치연장과 계획고 결정은 시공성과 기능성 및 경제성을 종합하여 다음과 같이 기준을 설정한다.

① 도로 횡단육교

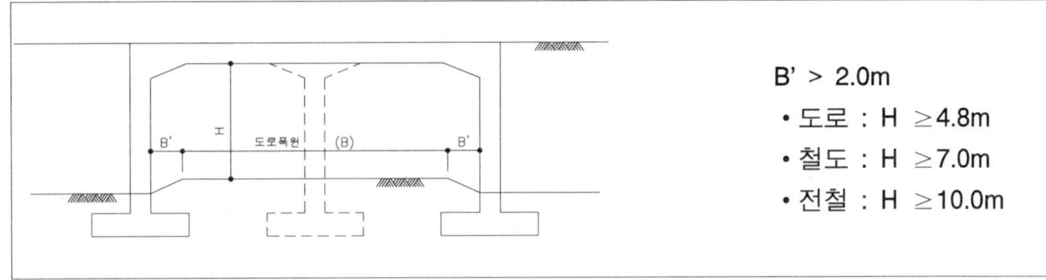

1-3 Bridge Erection Position and Span Length Determination

1-3-1 Items to Evaluate when Determining Bridge Erection Position

Classification	Contents
Region by related plan	Determine the erection of the bridge after reviewing various future plan including crossing road and railroad, house building lot, new road for the lot setup plan
Road with high banking height	When the height of banking soil is high during the road design, determine the erection of the bridge by comparing and reviewing the road earth work construction cost, land acquisition cost and bridge construction cost
Mountain area where cutting and banking are difficult	For the conditions of tall slope during cutting and banking, natural destruction and stability shall be comprehensively reviewed to make the decision whether to erect the bridge and where to erect.
Deep mountain valley	For the topological conditions that require filing valley area or many natural destructions due to large banking, planning bridge is desirable.
Crossing agriculture area	When installing the road at absolute agriculture area, efforts shall be made to use minimum agriculture land from the view point of strategic importance of food resources.
Crossing reservoir	Since installing road using earth work in reservoir area result huge loss in the water retaining capacity, bridge shall be planned in this case.
Where depth of soft foundation is deep	Plan the bridge only when it is suitable from the sufficient comparative review of various effects by soft foundation treatments
Where underground facility exists	When there are underground facilities that cannot be moved, bridge planning shall be decided after making the request to the relevant institution in advance and reviewing.

1-3-2 Determination of Span Length and Clearance

Three functional types of bridge are considered - road crossing viaduct, mountain area passing bridge, and river passing bridge. For the section where the bridge erection is planned for the purpose of main road passing and connection, the following standard combining ease of construction, functionality, and economic efficiency shall be applied in determining installation expansion of the structure and planned height.

① Road crossing viaduct

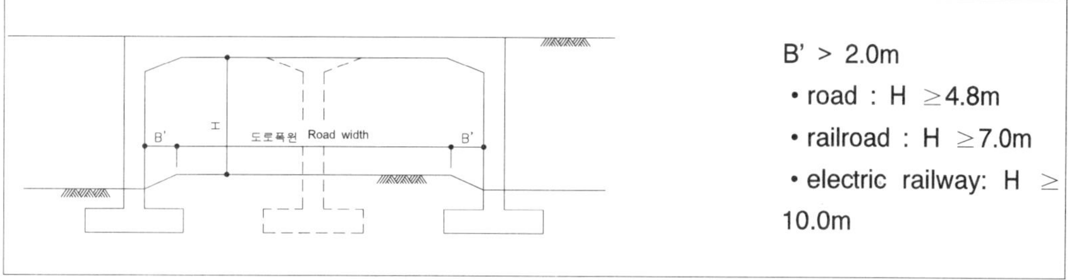

② 산악부 통과교량

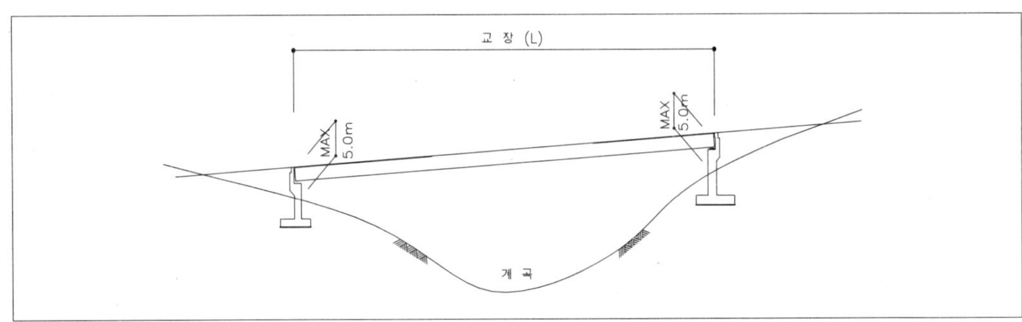

③ 하천 횡단교량

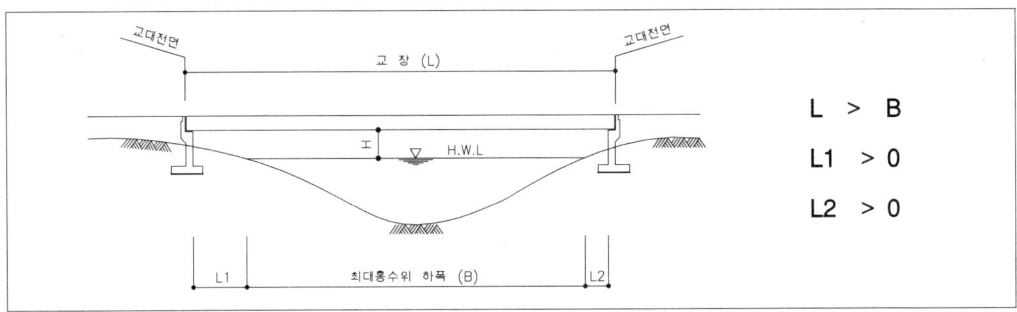

④ 계획홍수량과 다리밑 공간 및 경간장 간의 관계 (( )는 하천시설기준의 값)

계획홍수량(㎥/sec)	다리밑 공간 (m)	경 간 장 (m)
200 이하	0.6 이상	
200 ~ 500	0.8 이상	15(15) 이상
500 ~ 2,000	1.0 이상	20(20) 이상
2,000 ~ 5,000	1.2 이상	45(45) 이상
5,000~ 10,000	1.5 이상	70(70)
10,000 이상	2.0 이상	70(70)

⑤ 경간장은 하천설계기준(2005)에 따라 다음과 같이 결정

- 교량의 길이는 하천폭 이상
- 경간장은 산간 개착부, 그 외 하천의 상황, 지형의 상황 등에 따라 치수상 지장이 없다고 인정되는 경우를 제외하고는 다음 식으로 얻어지는 값 이상으로 결정
  (단, 그 값이 70m를 넘는 경우에는 70m로 적용)
  $L = 20 + 0.005Q$ (여기서, L : 경간장(m), Q : 계획홍수량(㎥/sec))

② Mountain area passing bridge

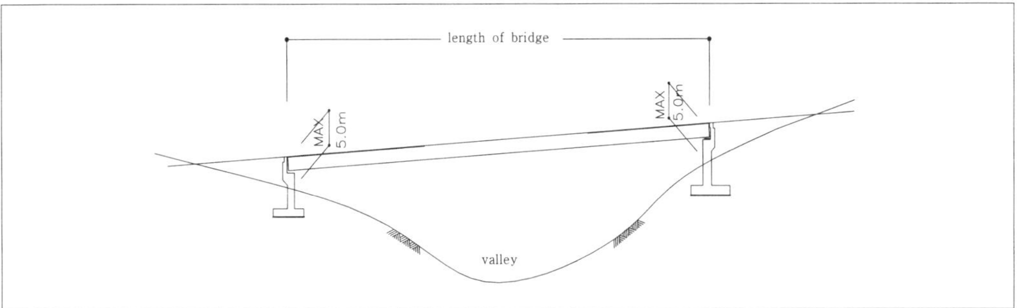

③ River passing bridge

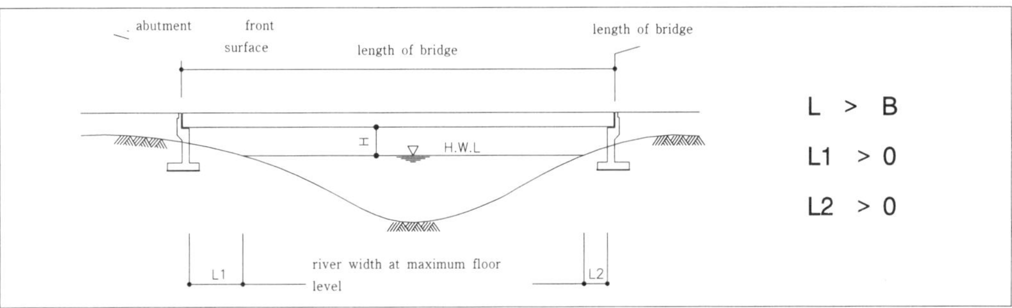

L > B

L1 > 0

L2 > 0

④ Relationship between planned flood level, space under structure, and span length (Values inside ( ) are for river facilities)

Plan flood rate (m³/sec)	Space under bridge (m)	Span length (m)
Below 200	Above 0.6	
200 ~ 500	Above 0.8	Above 15(15)
500 ~ 2,000	Above 1.0	Above 20(20)
2,000 ~ 5,000	Above 1.2	Above 45(45)
5,000~ 10,000	Above 1.5	70(70)
Above 10,000	Above 2.0	70(70)

⑤ Span length shall be determined in accordance with the River Design Standard (2005)

- Bridge length shall be longer than river width
- Span length shall be bigger than the value taken from the following equation except the cases the numbers are accepted without problem depending on the situations such as mountain new development area or other river conditions. (70m will be applied if the value exceeds 70m)

$L = 20 + 0.005Q$ (Where , L: span length (m), Q : Plan flood rate (m³/sec))

2. 교량형식의 선정

2-1 상부구조 형식 선정

2-1-1 형식선정 흐름도

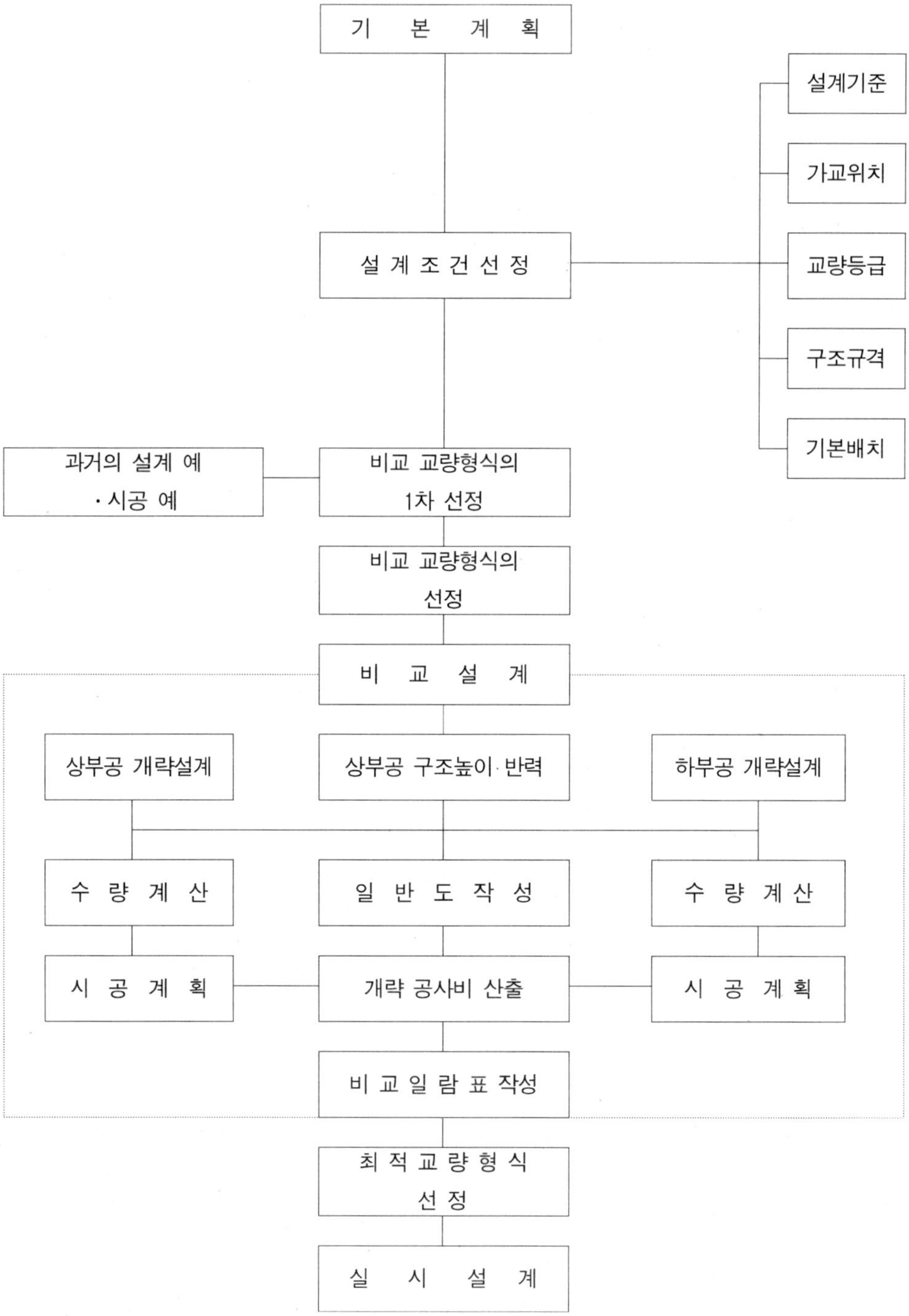

기 본 계 획

설 계 조 건 선 정 — 설계기준 / 가교위치 / 교량등급 / 구조규격 / 기본배치

비교 교량형식의 1차 선정 — 과거의 설계 예 · 시공 예

비교 교량형식의 선정

비 교 설 계

상부공 개략설계 | 상부공 구조높이·반력 | 하부공 개략설계

수 량 계 산 | 일 반 도 작 성 | 수 량 계 산

시 공 계 획 | 개략 공사비 산출 | 시 공 계 획

비 교 일 람 표 작성

최 적 교 량 형 식 선 정

실 시 설 계

2. Bridge Type Selection

2-1 Superstructure Type Selection

2-1-1 Type Selection Flow Chart

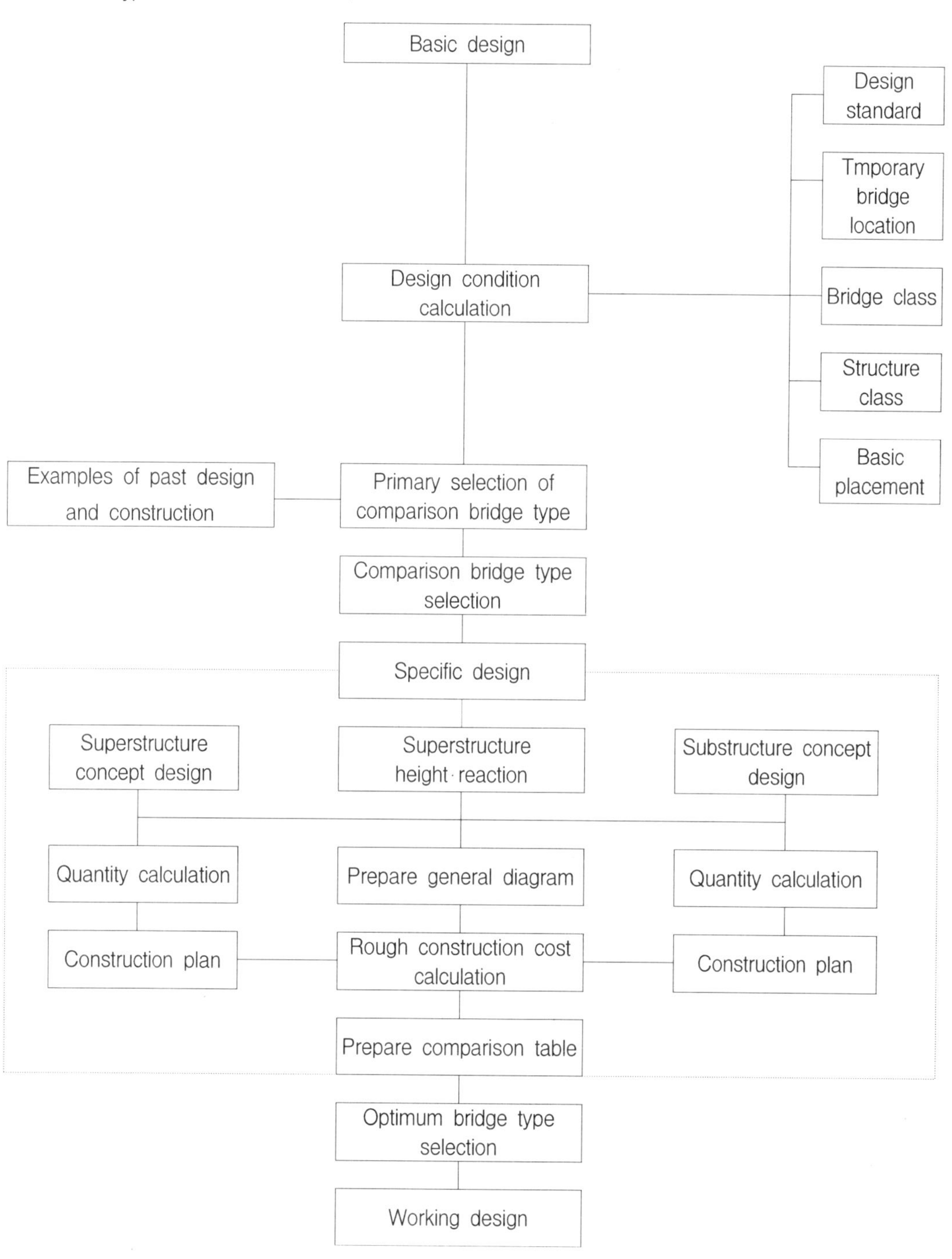

2-1-2 상부구조 형식별 선정기준

상부교량형식	적정 지간장(m)	가설위치 및 적용조건
1. RC 라멘 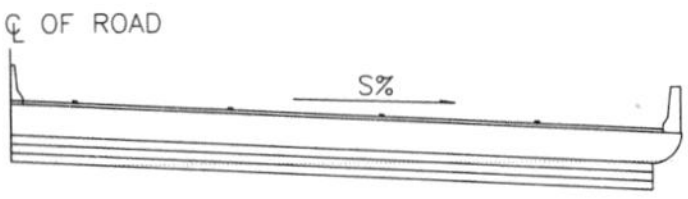	10 ~ 15m	• 소규모 도로 및 소하천 횡단 • 형하고 H=5~7m 내외 • 형하고 H=7.0m 이상의 경우 지중라멘 적용
2. 파형강판 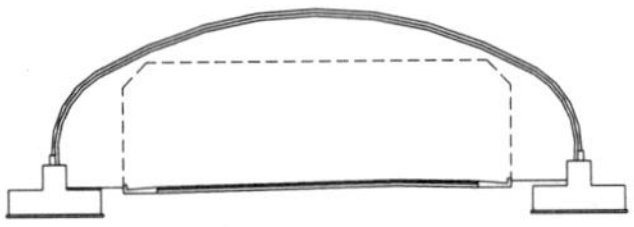	10 ~ 15m	• 소규모 도로 및 소하천 횡단 • 급속 시공이 필요한 지역
3. RC 슬래브 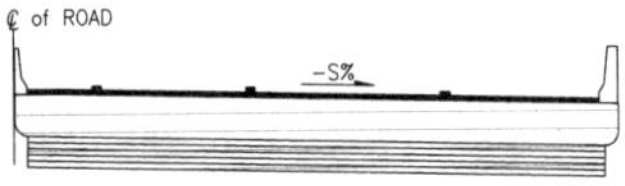	10 ~ 15m	• RC 라멘 적용이 곤란한 경우 • 형하고 H=5~12m 내외
4. PSC BEAM 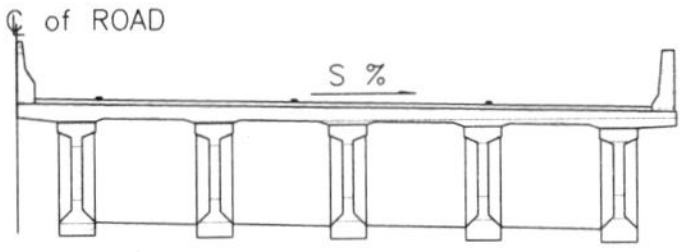	25 ~ 35m	• 미관이 중시되지 않는 중소 지간 • 선형상의 제약요인 고려 필요
5. π형 라멘 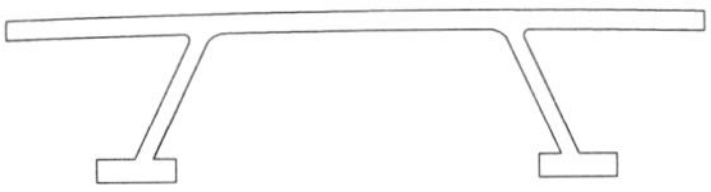	25 ~ 35m	• 미관이 중시되는 중소 지간 • 동바리 설치 가능구간
6.합성형 라멘 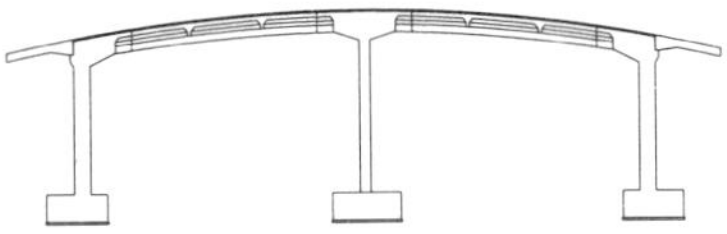	20 ~ 40m	• 종단계획상 형하고 확보가 어려운 구간 • 동바리 불필요
7. PSC e-BEAM 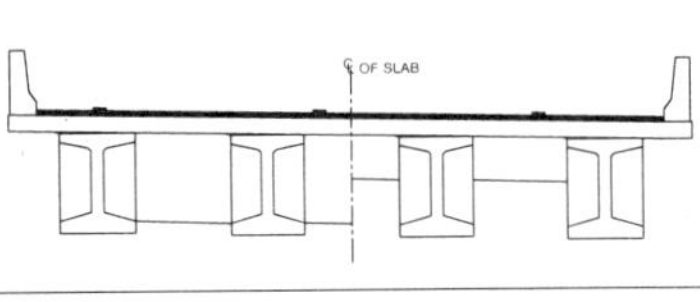	25 ~ 50m	• 종단계획상 형하고 확보가 어려운 구간 • 선형상의 제약요인 고려 필요

2-1-2 Selection Standard by Superstructure Type

Superstructure type	Appropriate span length(m)	Erection location and application condition
RC rahmen 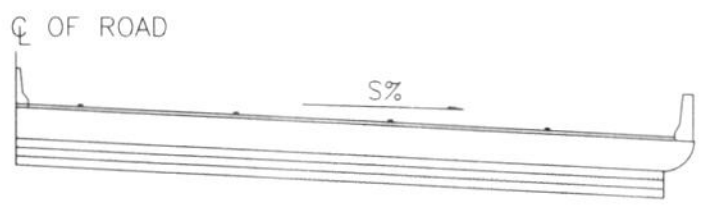	10 ~ 15m	• Small scale road and small river crossing • clearance is around H=5~7m • Apply underground rahmen when clearance is over H=7.0m
Corrugated steel plate 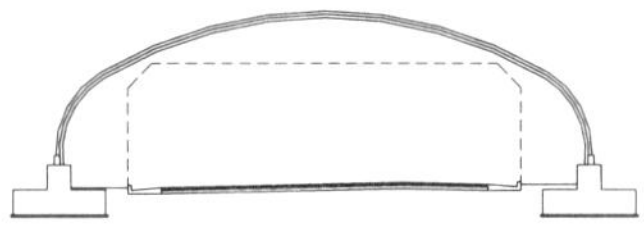	10 ~ 15m	• Small scale road and small river crossing • Where rapid construction is required
RC slab 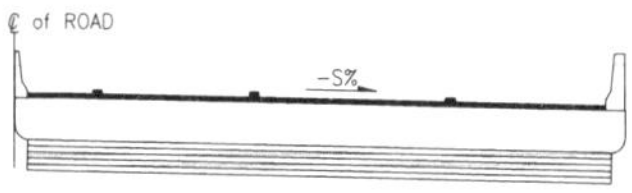	10 ~ 15m	• Where application of RC rahmen is difficult • clearance is around H=5~12m
PSC BEAM 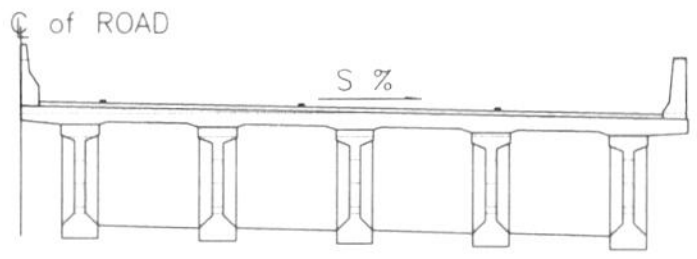	25 ~ 35m	• Mid-small span where appearance is not important • Consideration of the limit factor in alignment is required
π type rahmen 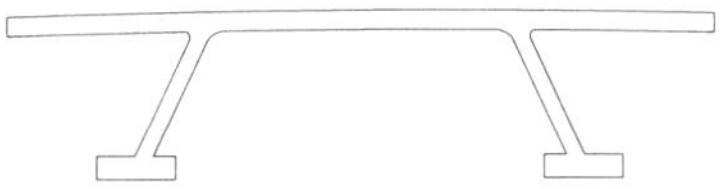	25 ~ 35m	• Mid-small span where appearance is important • Where timbering can be installed
Synthetic type rahmen 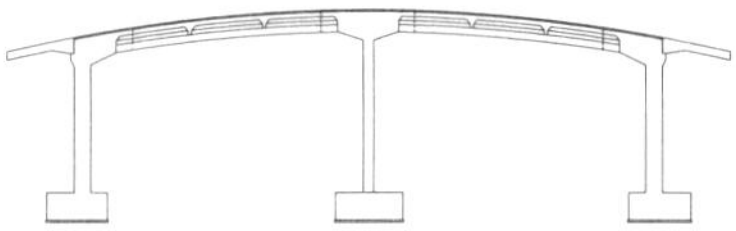	20 ~ 40m	• Zone where clearance is not secured in vertical plan • Timbering is not required
PSC e-BEAM 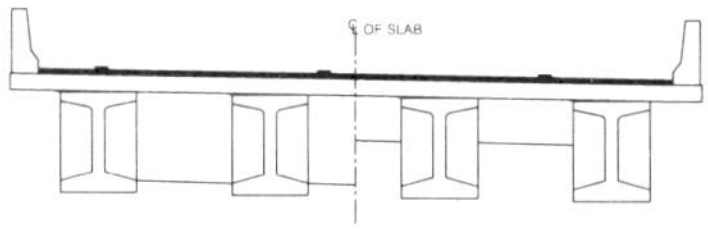	25 ~ 50m	• Zone where clearance is not secured in vertical plan • Consideration of the limit factor in alignment is required

상부교량형식	적정 지간장(m)	가설위치 및 적용조건
PF BEAM 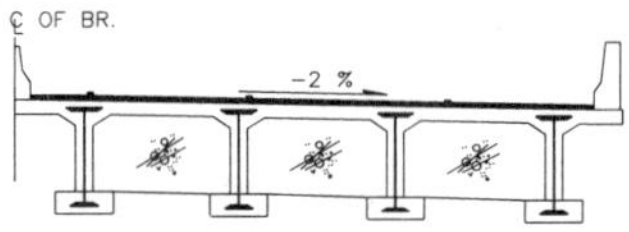	35 ~ 45m	● 종단계획상 형하고 확보가 어려운 구간 ● 선형상의 제약요인 고려 필요
IPC GIRDER 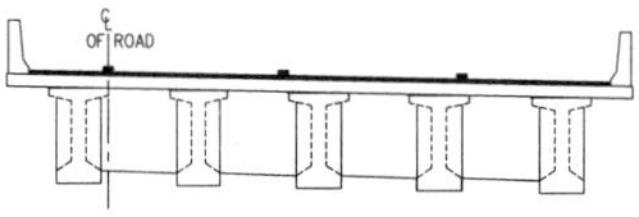	35 ~ 45m	● 미관이 중시되지 않는 중소 지간 ● 선형상의 제약요인 고려 필요
WPC GIRDER 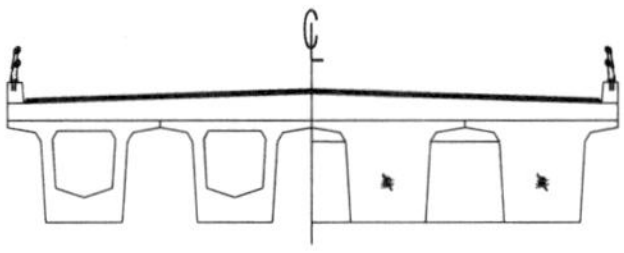	30 ~ 50m	● 미관이 중시되지 않는 중소 지간 ● 선형상의 제약요인 고려 필요
MSP 합성거더 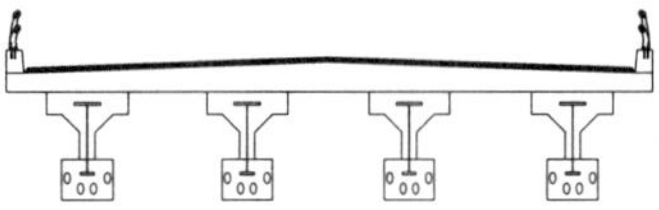	20 ~ 50m	● 종단계획상 형하고 확보가 어려운 구간 ● 선형상의 제약요인 고려 필요
RPF BEAM 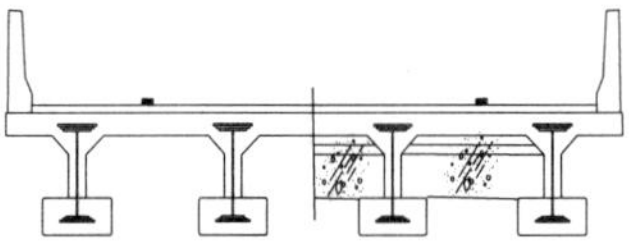	20~ 50	● 종단계획상 형하고 확보가 어려운 구간 ● 선형상의 제약요인 고려 필요
PSC BOX(FSM) 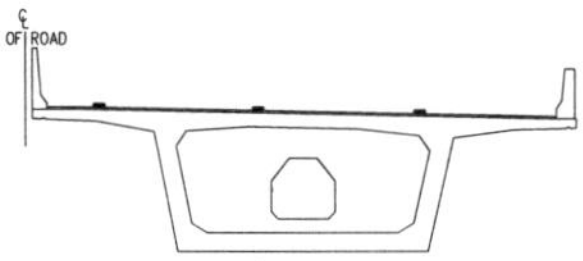	50m 내외	● 동바리 설치가 가능한 평탄한 지형 ● 형하고 H=20m 이하 구간에 적정
PSC BOX(ILM) 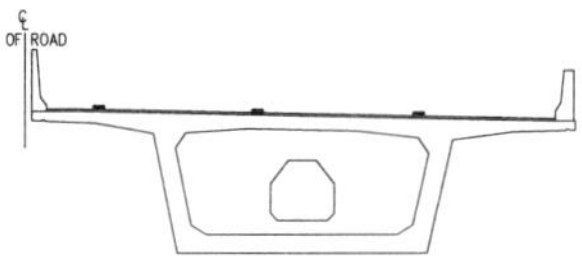	50m 내외	● 동바리설치가 곤란한 지역 ● 선형제약요인 고려 필요 ● 제작장 확보 가능 지역

Superstructure type	Appropriate span length(m)	Erection location and application condition
PF BEAM 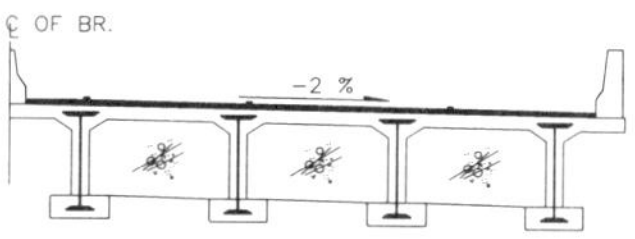	35 ~ 45m	• Zone where clearance is not secured in vertical plan • Consideration of the limit factor in alignment is required
IPC GIRDER 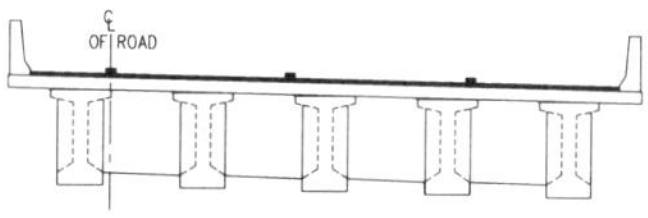	35 ~ 45m	• Mid-small span where appearance is not important • Consideration of the limit factor in alignment is required
WPC GIRDER 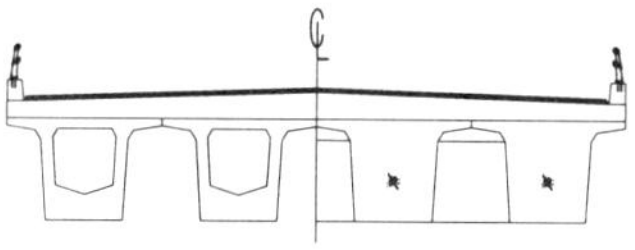	30 ~ 50m	• Mid-small span where appearance is not important • Consideration of the limit factor in alignment is required
MSP composite beam girder 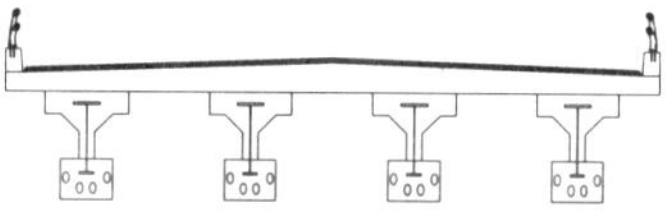	20 ~ 50m	• Zone where clearance is not secured in vertical plan • Consideration of the limit factor in alignment is required
RPF BEAM 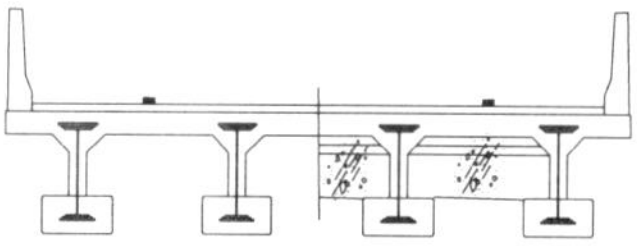	20~ 50	• Zone where overhead clearance is not secured in vertical plan • Consideration of the limit factor in alignment is required
PSC BOX(FSM) 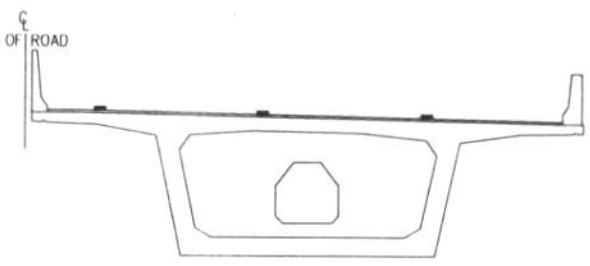	around 50m	• Flat area where timbering can be installed • Suitable for the section with clearance is below H=20m
PSC BOX(ILM) 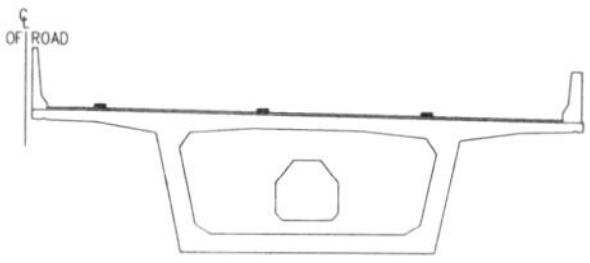	around 50m	• Where timbering installation is difficult • Consideration of the restriction factor in alignment is required • Where manufacturing place can be secured

상부교량형식	적정 지간장(m)	가설위치 및 적용조건
15.PSC BOX(MSS) 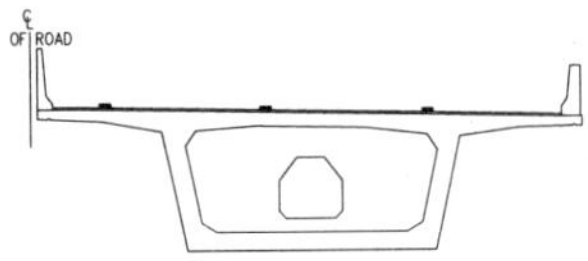	50m 내외	• 선형조건상 PSC BOX(ILM) 적용이 곤란한 경우
16. PSC BOX(FCM) 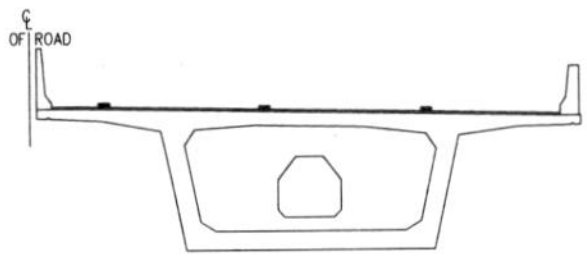	120m 내외	• 미관성 고려 구간 • 형하고 제약 없음 • 선형제약요인 고려 필요
17.PSC BOX(PSM) 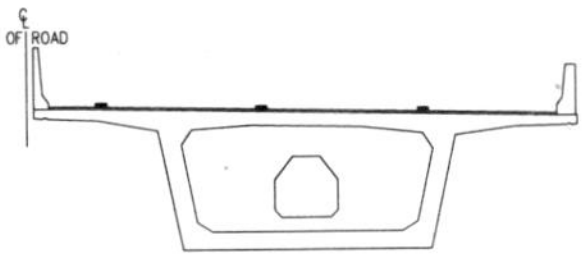	100m 내외	• 해상교량 등 대규모 공사에 유리 • 별도의 SEGMENT 제작장 및 야적장 필수
18. P.C.T 거더교 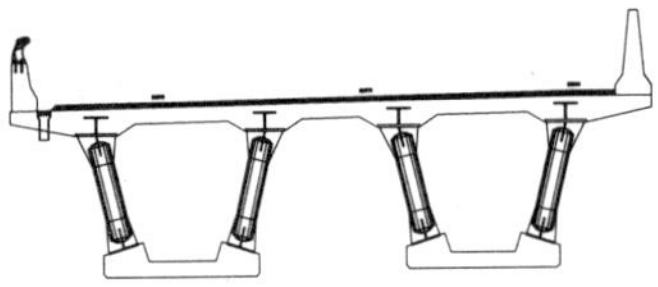	40 ~ 150m	• PSC BOX 또는 STEEL BOX 형식 적용이 곤란한 장지간 • 미관성 고려 구간
19. ST. BOX 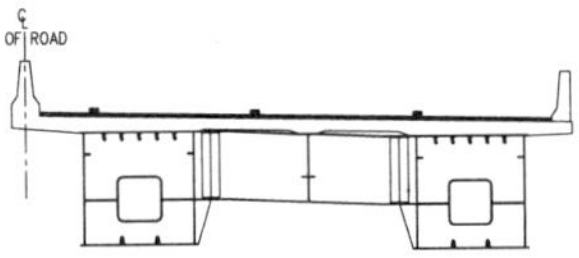	40 ~ 70m	• 미관성 고려 구간 • 선형상의 제약요인 거의 없음
20.소수거더교 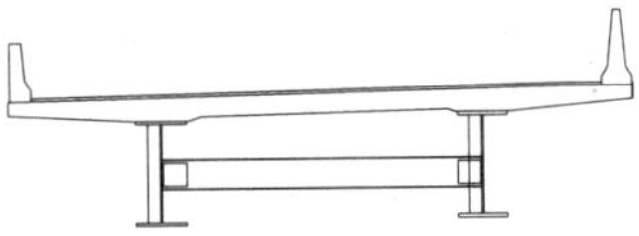	45 ~ 60m	• 미관성 고려 구간 • 선형제약요인 고려 필요
21. S.C.P 합성거더 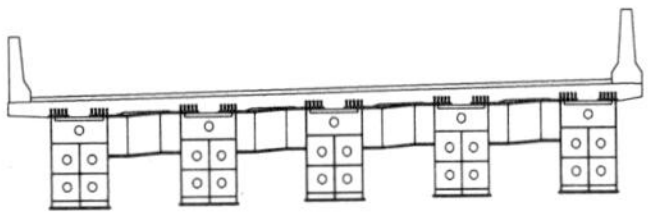	45~70m내외	• 종단계획상 형하고 확보가 어려운 구간 • 선형제약요인 고려 필요

Superstructure type	Appropriate span length(m)	Erection location and application condition
PSC BOX(MSS) 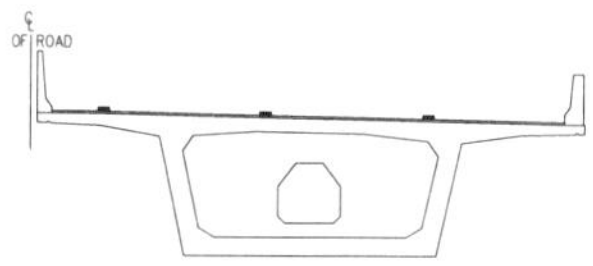	around 50m	• Where application of PSC BOX (ILM) is difficult because of alignment condition
PSC BOX(FCM) 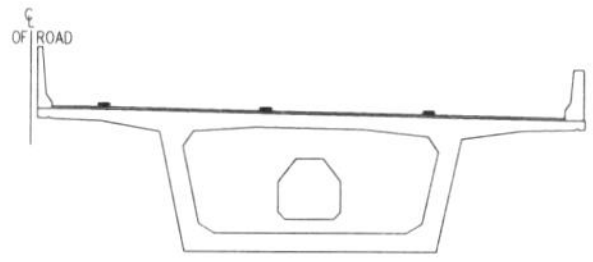	around 120m	• Where appearance is important • No clearance restriction • Consideration of the alignment restriction factor is required
PSC BOX(PSM) 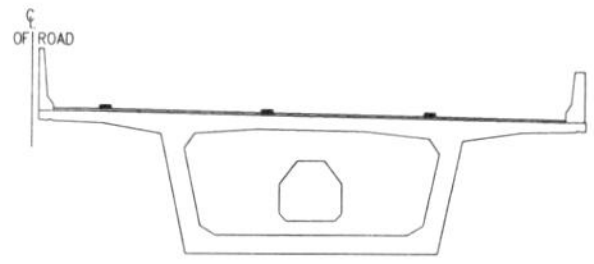	around 100m	• Advantage in large scale construction such as marine bridge • Separate segment manufacturing place and storage yard is essential
P.C.T girder bridge 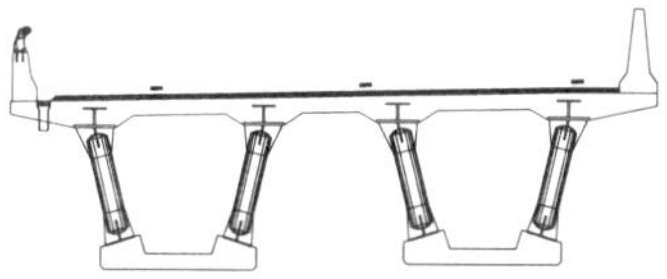	40 ~ 150m	• Where PSC BOX or Steel box cannot be applied because of long span • Where appearance is important
ST. BOX 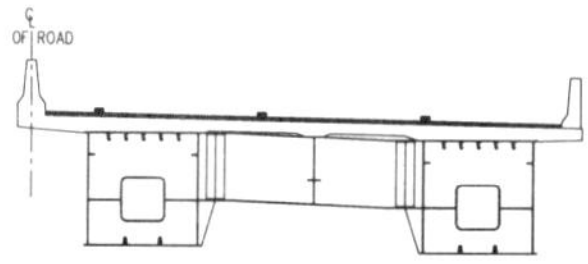	40 ~ 70m	• Where appearance is important • No restriction factor in alignment
Small girder number bridge 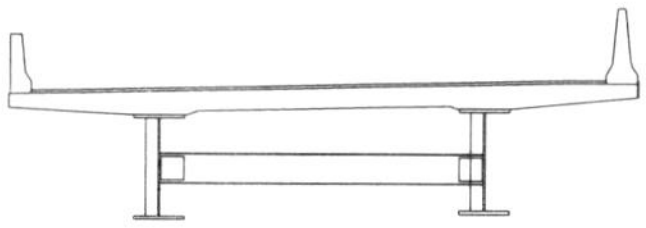	45 ~ 60m	• Where appearance is important • Consideration of the limit factor in alignment is required
S.C.P composite beam girder 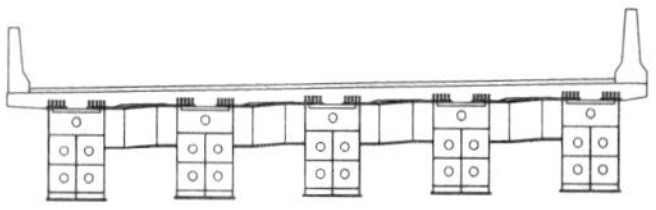	around 45~70m	• Zone where clearance is not secured in vertical plan • Consideration of the restriction factor in alignment is required

2-1-3 상부구조(특수교량) 형식별 선정기준

① 경제성을 고려한 특수교량 형식별 적용지간

교량형식				100	200	300	400	500	600	800	1000	2000
1	아치교		━	━	━							
2	E/D교			━								
3	사장교	콘크리트			━	━	━					
		강			━	━	━	━	━			
4	현수교	자정식			━							
		타정식						━	━	━		

② 특수교량의 형식별 특징 및 개요

상부교량형식	적용지간	가설위치 및 적용조건	특 징
아치교	50~300m	• 대규모 하천이나 해상 중소형 선박의 항로 횡단 조건 • 도회지 등 미관성 고려지역	• 미관성 우수 • 다양한 경관성 도출 가능 • 내풍안정성 상대적 우수 • 각별한 내진대책 필요
Extradosed교	100~200m	• 대규모 하천이나 해상 중소형 선박의 항로 횡단 조건 • 미관성 고려지역	• 미관성 우수 • Land Mark적 요소 미흡 • 사장교에 비하여 상대적 경제성 우수
사장교	200~800m	• 대규모 하천이나 대형선박의 항로 통과 구간 • Landmark적 상징성 고려지역 • 대규모 해상 연육교의 주교량	• 공사비 고가 • 다양한 가설 공법 적용 가능 • 각별한 내진 및 내풍대책 필요
현수교	200~2000m	• 대규모 하천이나 대형선박의 항로 통과 구간 • 수심이 깊어 하부공사비가 과다한 지역 • 대규모 해상 연육교의 주교량	• 공사비 매우 고가 • 선형상의 제약요인 고려 필요 • 각별한 내진 및 내풍대책 필요

2-1-3 Selection Standard by Superstructure (special bridge) Type

① Application span by type of special bridge considering economic efficiency

Bridge Type			100	200	300	400	500	600	800	1000	2000
1	Arch bridge		▬	▬	▬						
2	E/D Bridge		▬	▬							
3	Cable strayed bridge	Concrete		▬	▬	▬	▬				
		Steel		▬	▬	▬	▬	▬	▬		
4	Suspension bridge	Self-anchored		▬	▬						
		External-anchored					▬	▬	▬	▬	

② Features and overview by special bridge type

Superstructure Type	Applied span	Erection location and application condition	Features
Arch bridge 	50～300m	• Large scale river or mid-small scale marine ship passing condition • City or other area where appearance is to be considered	• Excellent appearance • Various scenary can be presented • Relatively good stability against wind • Special seismic design is required
Extradosed Bridge 	100～200m	• Large scale river or mid-small scale marine ship passing condition • City or other area where appearance is to be considered	• Excellent appearance • No compelling factors as landmark structure • Relatively good economic efficiency compared to cable strayed bridge
Cable strayed bridge 	200～800m	• Large scale river or large scale marine ship passing section • Area where symbolic landmark is considered • Large scale marine bridge connecting with main land	• High construction cost • Various erection method can be applied • Special earthquake-proof and wind-proof design is required
Suspension bridge 	200～2000m	• Large scale river or large scale marine ship passing section • Where construction cost of lower structure is high because of deep water level • Large scale marine main bridge connecting with main land	• Very high construction cost • Restriction factor in alignment is to be considered • Special earthquake-proof and wind-proof design are required

2-2 하부구조 형식 선정

하부구조 형식 선정은 상부구조 기본 조건에 적합하고 가교지점 지형의 특성과 흙쌓기 높이를 고려하면서 상부구조 단면과 조화하는 형식을 얻기 위해서 아래와 같은 체계화된 System을 도입하여 계획하여야 한다.

2-2-1 형식선정 흐름도

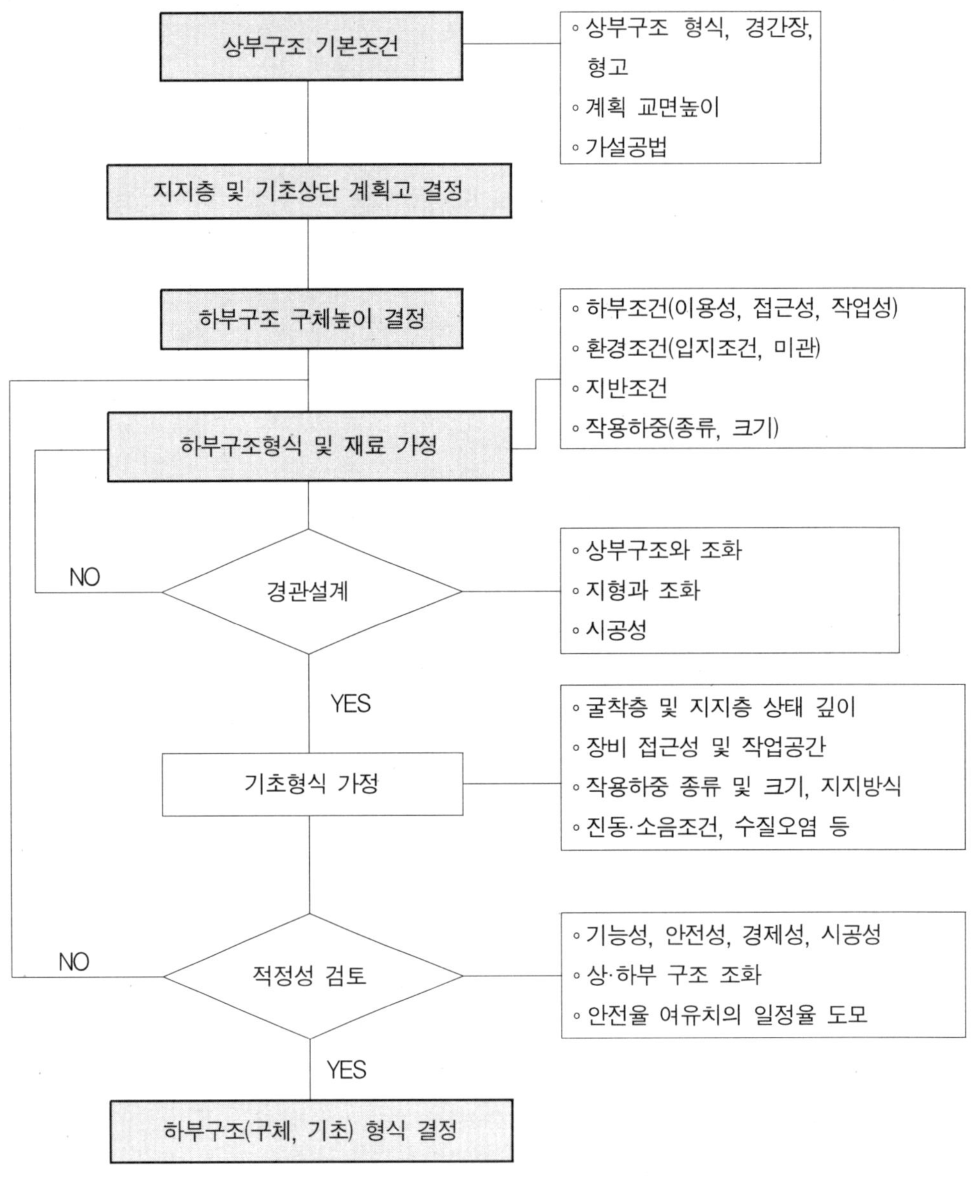

2-2 Substructure Type Selection

Substructure type selection shall meet the basic requirement of the superstructure, and the systematic plan shall be introduced as shown below to achieve the type that balances with the superstructure section while considering the characteristics of the bridge erection area and banking height.

2-2-1 Type Selection Flow Chart

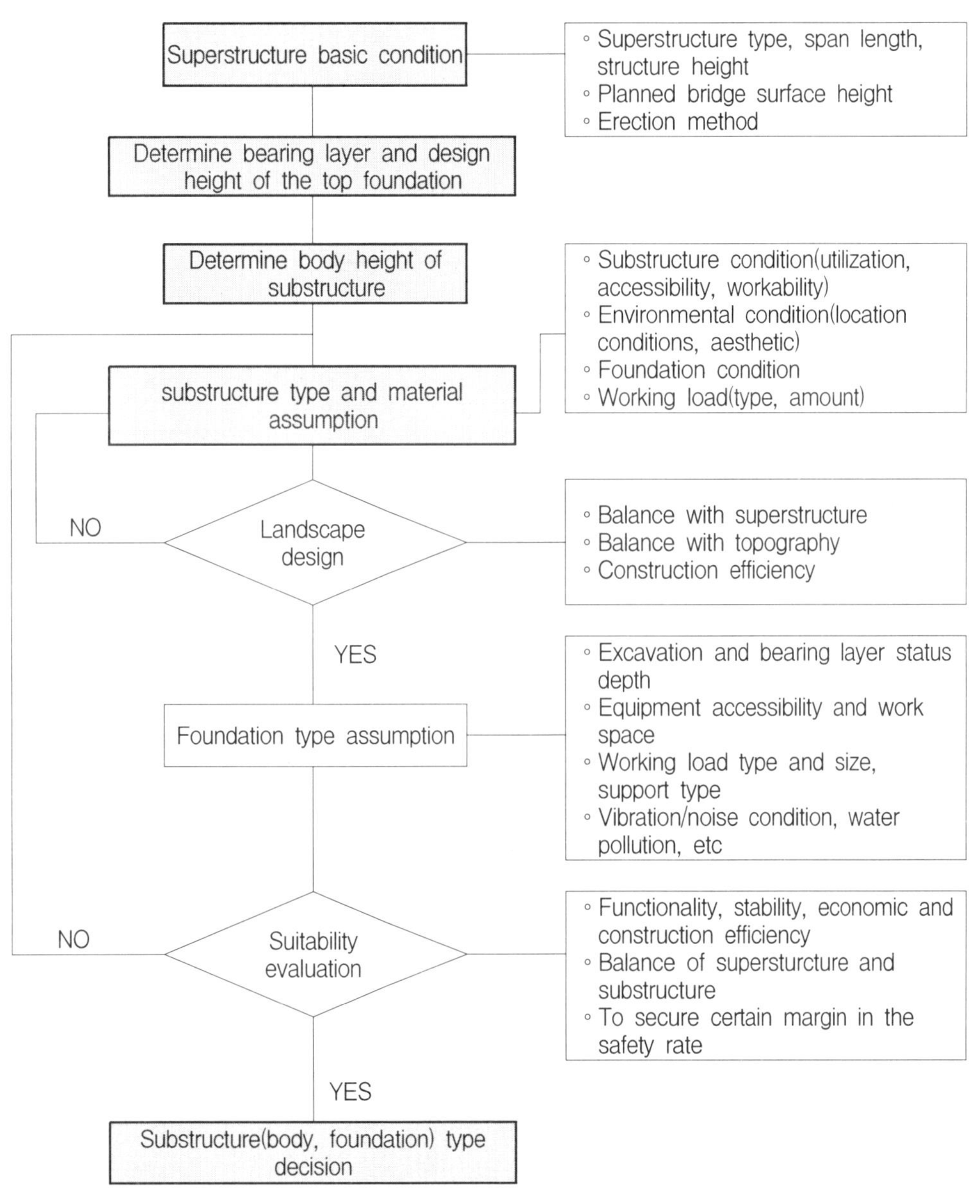

2-2-2 교대형식 선정

① 교대높이별 형식 선정기준

교대형식	높이 (m)			
		10	20	30
중력식	4			
반중력식	6			
역T형	6	13		
뒷부벽식		10 20		
라멘식		10 15		
박스형		12		
중간이음식		15		

② 주요 교대형식 개요

구분	역T형	뒷부벽식	반중력식
단면도	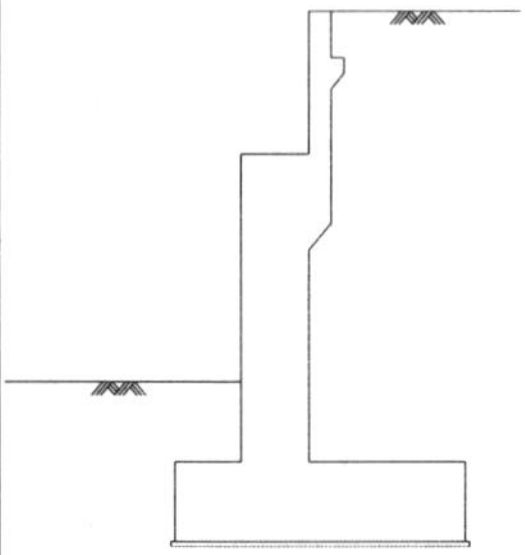	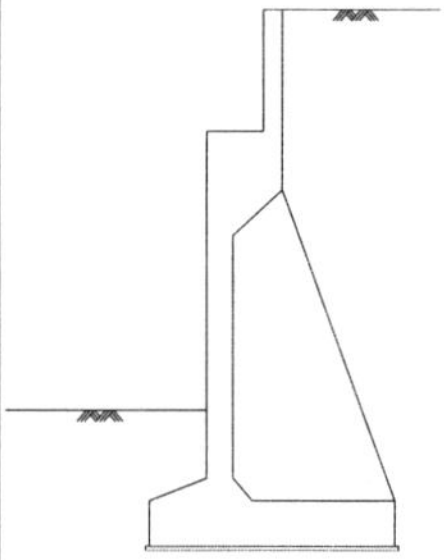	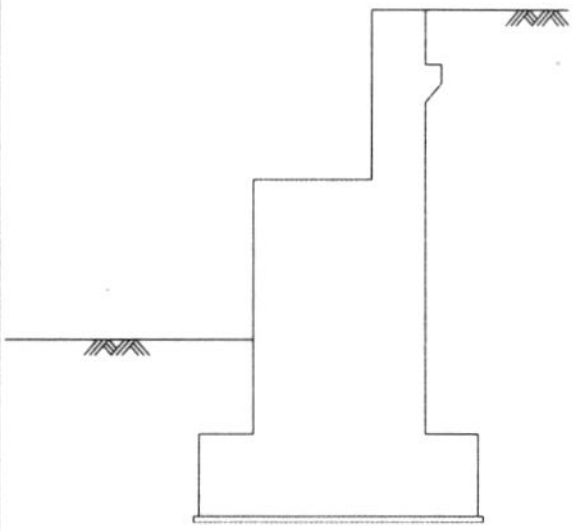
특성	• 높이 6m~13m 경제적 • 시공성 양호 • 뒤채움 다짐조건 양호 • 산지부 구릉지 적용 유리	• 높이 10m~20m 경제적 • 시공성 불량 • 뒤채움 다짐조건 불량 • 평지부, 도로 및 철도 횡단부 적용	• 높이 6m 미만 경제적 • 시공성 양호 • 뒤채움 다짐조건 양호 • 소하천 횡단교량 유리
적정상부형식 (경간규모)	SLAB / BEAM / BOX (중 · 대 규모)	BOX (중 · 대 규모)	SLAB / BEAM / BOX (소 · 중 규모)

2-2-2 Abutment Type Selection

① Type selection standard by the height of abutment

Abutment type	Height (m)		
	10	20	30
Gravity type	4		
Semi-gravity type	6		
Reverse T type	6 ~ 13		
Rear buttress type		10 ~ 20	
Rigid frame type		10 ~ 15	
Box type		12 ~ 20	
Middle connection type		15 ~ 20	

② Major abutment type overview

Classification	Revert T type	Rear buttress type	Semi-gravity type
Section	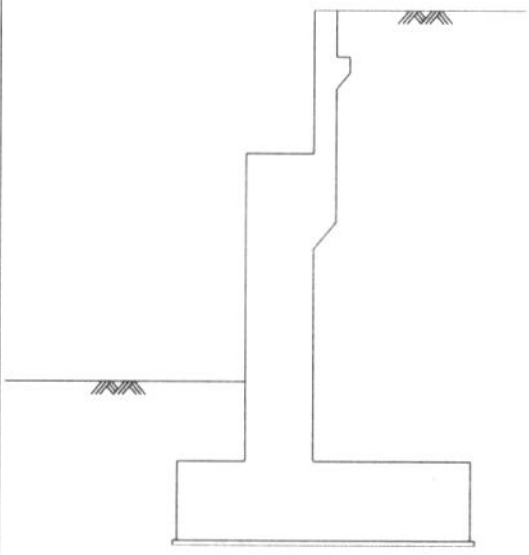	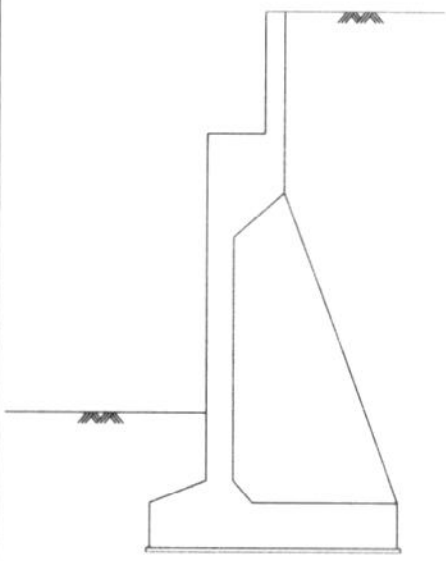	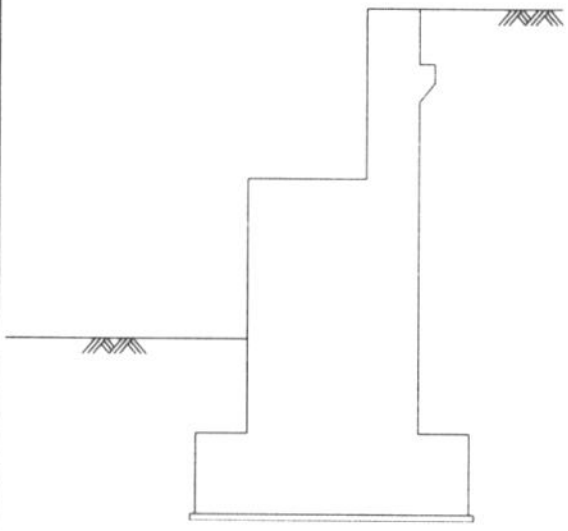
Feature	• Economic at the height of 6m～13m • Ease of construction • Good post banking and compacting condition • Advantage in application to mountain or hill area	• Economic at the height of 10m～20m • Difficult construction • Poor post banking and compacting condition • Applied to flat area, road and railroad crossing	• Economic at the height lower than 6m • Ease of construction • Good post banking and compacting condition • Advantage in small river crossing bridge
Appropriate superstructure type(span scale)	SLAB / BEAM / BOX (mid · large scale)	BOX (mid · large scale)	SLAB / BEAM / BOX (small · middle scale)

2-2-3 교각형식 선정

① 교각 높이별 형식선정 기준

교 대 형 식	높 이 (m) (눈금: 10, 20, 30)
라 멘 식 (1층)	5 ──── 15
π 형 교 각	5 ──────── 35
T 형 교 각	5 ─────── 30
라 멘 식 (2층)	15 ────── 35
벽 식 · Y 형	──────────

② 주요 교각 형식별 개요

라 멘 식 (1층)	π 형 교 각	T 형 교 각
▶ 적용 : STEEL BOX GIRDER (1SHOE/주형 사용의 경우) H=15m 이하	▶ 적용 : STEEL BOX GIRDER PSC BEAM, PREFLEX BEAM, RC 및 PSC SLAB H=5~35m 이하	▶ 적용 : STEEL BOX GIRDER PSC BEAM, PREFLEX BEAM, RC 및 PSC SLAB H=5~30m 이하
라 멘 식 (2층)	**중공식, 구주식, 벽식**	**Y 형 교 각**
▶ 적용 : STEEL BOX GIRDER (1SHOE/주형 사용의 경우) H=15~35m	▶ 적용 : PSC BOX GIRDER H=0~40m 이상	▶ 적용 : PSC BOX GIRDER H=10~40m 이상도 가능

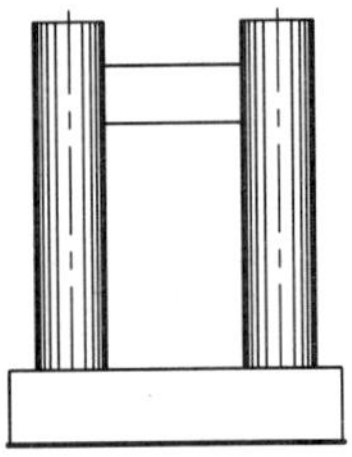

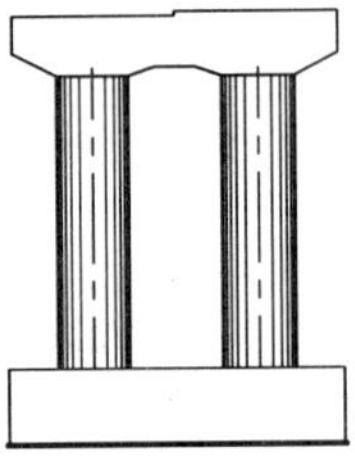

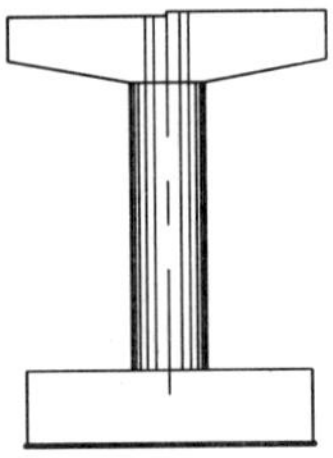

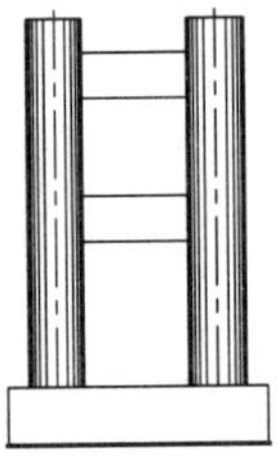

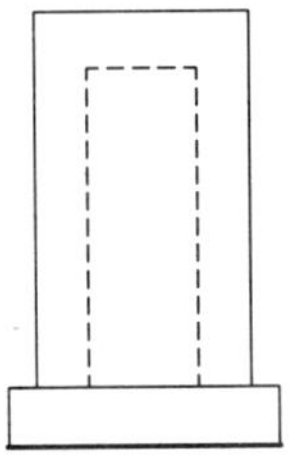

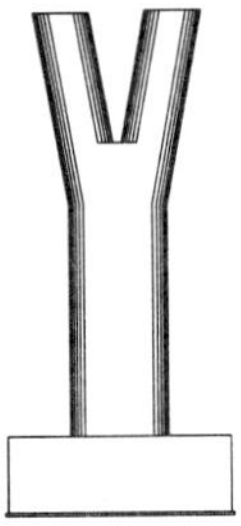

2-2-3 Pier Type Selection

① Type selection standard by the height of pier

Abutment type	Height(m)		
	10	20	30
Rigid frame type (one story)	5 — 15		
π type pier	5 — 35		
T type pier	5 — 30		
Rigid frame type (2 stories)	15 — 35		
Wall type · Y type	————		

② Major pier type overview

Rigid frame type (one story)	π type pier	T type pier
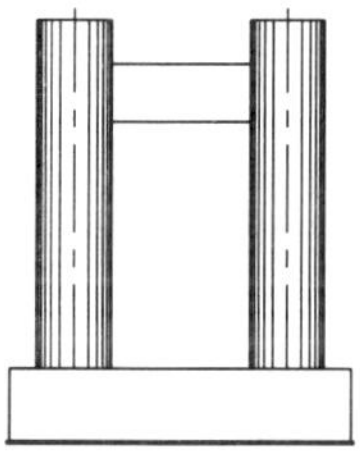	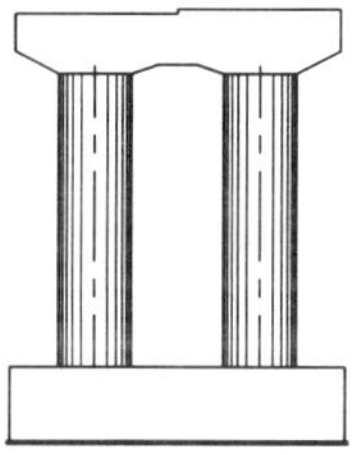	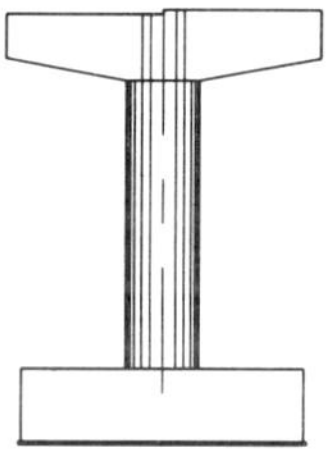
▶Application : STEEL BOX GIRDER (when using 1SHOE/main girder) H= below 15m	▶Application : STEEL BOX GIRDER PSC BEAM, PREFLEX BEAM, RC and PSC SLAB H=below 5~35m	▶Application : STEEL BOX GIRDER PSC BEAM, PREFLEX BEAM, RC and PSC SLAB H= below 5~30m
Rigid frame type (two stories)	**Cavity type, Pillar type, Wall type**	**Y type pier**
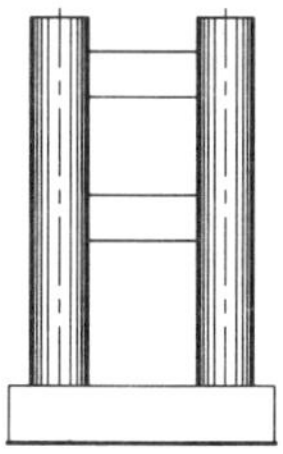	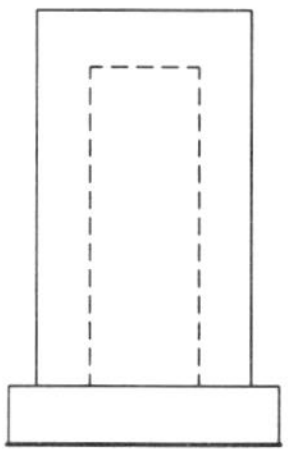	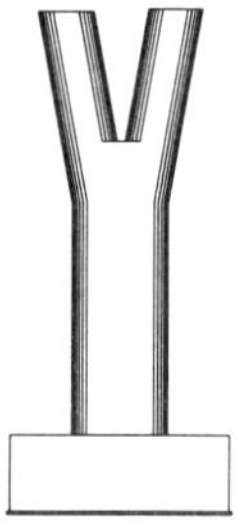
▶Application : STEEL BOX GIRDER (When using 1SHOE/main girder) H=15~35m	▶Application : PSC BOX GIRDER H= over 0~40m	▶Application : PSC BOX GIRDER H= over 10~40m is possible

2-3 기초형식 선정

교량의 기초 지지층 토질조건은 토질조사 결과에 의하여 지지층 심도 조건에 따른 선정 System과 Flow, 선정 시 고려사항을 감안하고, 전달되는 상부구조 하중 및 토질 조건에 적합한 기초 형식을 선정함을 원칙으로 한다.

2-3-1 기초형식 선정 시 고려사항

① 현장 기초 설치지점의 지형 및 지질조건 : 지지심도 2~5m
② 자연 및 환경 피해 최소화, 민원예방
③ 상부구조의 특성 (연직하중, 수평하중, 허용침하량)
④ 설계변경 최소화 : 지형 현황과 토질조건의 검토
⑤ 주변 환경을 고려한 시공성
⑥ 지진하중

2-3-2 기초형식의 적용성

구 분		장 · 단 점	적 용 성
직 접 기 초		• 지지지반 확인 용이 • 시공 간단 • 연직하중 및 수평하중에 대한 지지력 큼	• Df ≤ 5.0m 양질의 지지층 • 부등침하 및 세굴의 우려 없는 곳
말 뚝 기 초	기 성 말 뚝 기 초	• 시공이 비교적 간단 • 품질관리가 용이 • 연직하중 및 수평하중에 대한 지지력 작음	• 5.0m < Df • 세굴 우려 없고 지지층까지 항타 가능 • 연약층, 이토, 흙쌓기층 ┌ 5~15m : RC, PSC ├ 15~30m : steel, 현장타설 └ 30m 이상 : steel • 마찰저항 및 선단지지력 • 말뚝 항타 곤란한 층 천공 후 항타
	현 장 타 설 말 뚝 기 초	• 시공성 다소 복잡 • 품질관리 다소 복잡 • 연직하중 및 수평하중에 대한 지지력 보통	
케이슨 기 초	OPEN 케이슨 기 초	• 시공이 다소 복잡 • 품질관리 다소 복잡 • 연직하중 및 수평하중에 대한 지지력 보통	• 말뚝시공 곤란 지역(호박돌, 전석층, 풍화된 암석층) • 유속이 빨라 하상 세굴 우려 지역 • 수심이 깊은 지역(가도 및 축도 이용) • 작용하중이 큰 장대교량
	공기 케이슨 기 초	• 지지지반 확인 용이 • 시공이 매우 복잡 • 연직하중 및 수평하중에 대한 지지력 큼	
특 수 기 초		• 일반적인 기초형식으로 적용하기 곤란한 경우 특수 기초형식 적용검토	

2-3 Foundation Type Selection

Foundation bearing layer soil condition of the bridge shall be considered with selection system depending on the bearing layer depth condition based on the result of the soil investigation and considerations during the selection. And Foundation type suitable for the transferred load of superstructure and soil conditions shall be selected.

2-3-1 Considerations during Foundation Type Selection

① Topographic features and geological condition of the field foundation erection location: bearing depth 2~5m
② To minimize the damages of nature and environment destruction, prevent a civil appeals
③ Characteristics of superstructure (vertical load, horizontal load, allowable settlement)
④ To minimize the design changes: review the topographical status and soil conditions
⑤ Ease of construction considering surrounding environment
⑥ Earthquake load

2-3-2 Application of Foundation Type

Classification		Advantage · Disadvantage	Application
Spread foundation		• Easy to confirm bearing ground • Easy construction • High bearing capacity for vertical and horizontal load	• Df ≤ 5.0m bearing lyaer of high quality • Where there is no danger of differential settlement or scour
Pile foundation	Precast pile foundation	• Relatively simple construction • Easy quality control • Small bearing capacity to vertical and horizontal load	• 5.0m < Df • Where there is no danger of scour, and casting down to bearing layer is always possible • soft ground, blow-off, banking layer ┌ 5~15m : RC, PSC ├ 15~30m : steel, cast-in-place └ Over 30m : steel • Friction resistance and point support capacity • Driving pile after drilling if driving is difficult for the layer
	Cast-in-place pile foundation	• A little complicated construction • A little complicated quality control • Bearing capacity to vertical and horizontal load is average	
Caisson foundation	Open caisson foundation	• A little complicated construction • A little complicated quality control • Bearing capacity to vertical and horizontal load is average	• Where pile construction is difficult (boulder, boulder layer, weathered rock) • Where there is danger of river scour due to fast current • Deep depth of water area (sand island) • Large scale bridge under large working load
	Air caisson foundation	• Easy to confirm bearing ground • Complicated construction • High bearing capacity for vertical and horizontal load	
Special foundation		• Special foundation type shall be evaluated when general foundation type is difficult to be applied	

2-3-3 기초형식 선정 흐름도

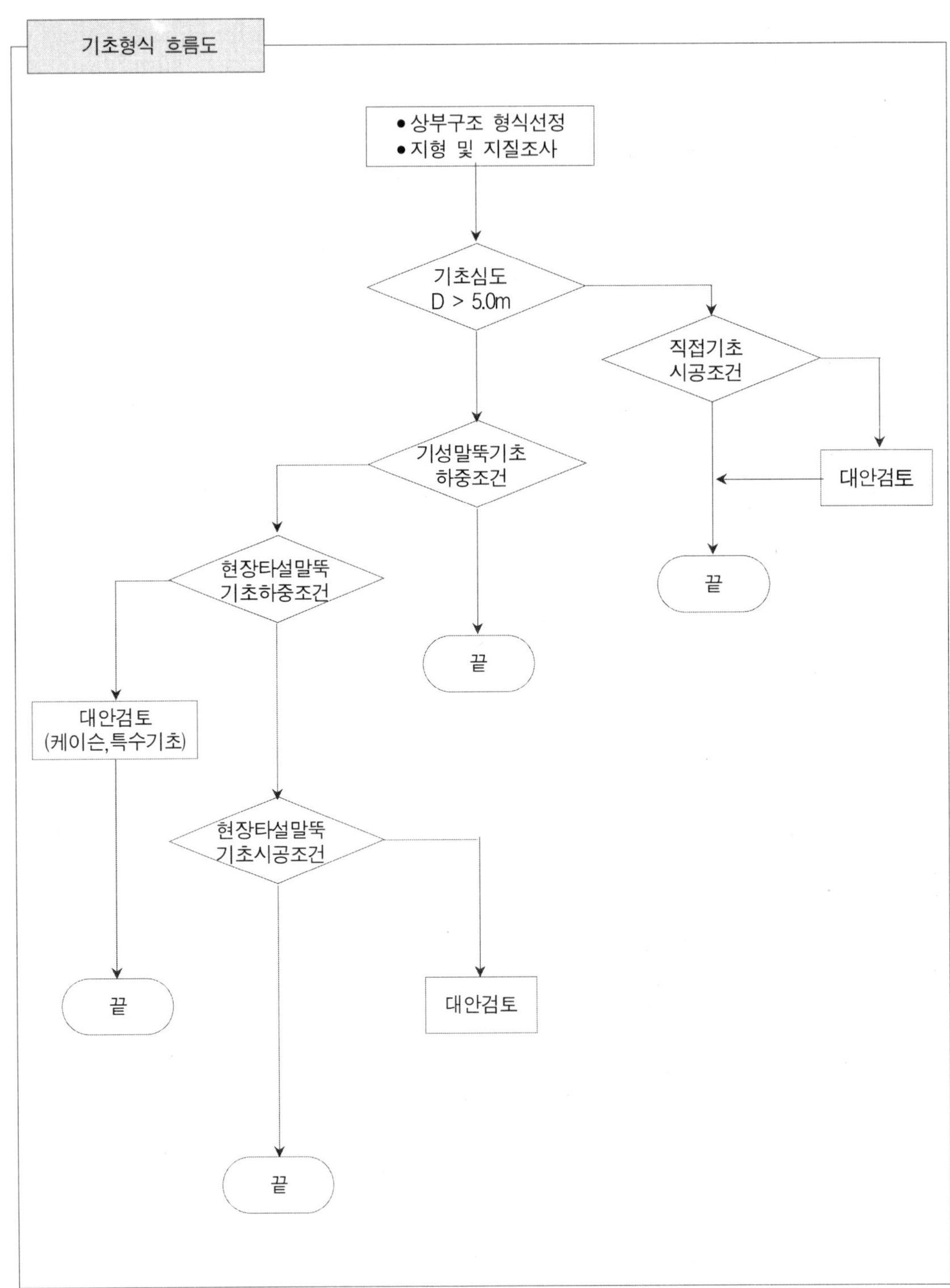

2-3-3 Foundation Type Selection Flow Chart

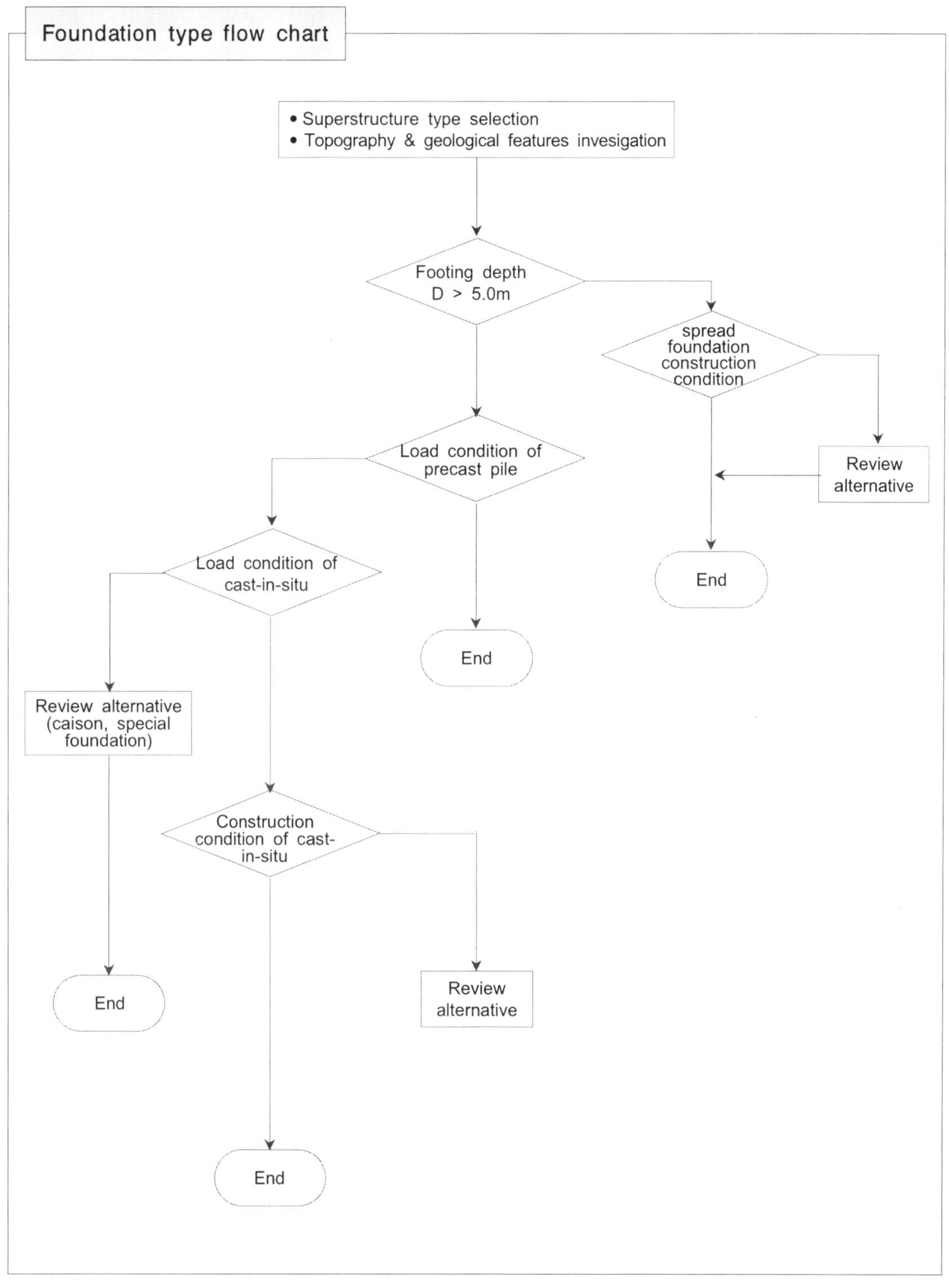

2-4 교량 시공법 선정

2-4-1 개 요

① 계획시설물의 규모, 위치 및 지형, 지질, 동원가능 장비 성능, 수효, 가설 제작장 설치 가능 여부, 공사소요기간, 가설공사비 등을 검토하여 계획

② 주변시설, 인근 주민에 미치는 소음, 진동, 교통혼잡을 최소화할 수 있는 공법 결정

③ 하천 또는 해상공사 중 토사유출 등에 의한 주변 오염 여부

2-4-2 교량가설 공법별 분류

구 분	강 교	콘크리트교(R.C, P.S.C)

적 용
공 법
	• 가Bent 공법
	  - 자주식 Crane, F/C 또는 Barge선
	• 케이블 가설 공법
	  - 수직형, 경사형
	• 압출공법(Incremental Launching)
	• 캔틸레버 공법
	• 대블럭 공법
		• 동바리 공법(Full Staging)
		• 프리캐스트 거더 공법
		  ┌ Crane 가설 공법
		  └ Erection Girder 공법
		• 캔틸레버 공법(F.C.M)
		• 압출공법(I.L.M)
		• 이동식비계 공법(M.S.S)

2-4-3 강교 및 Beam교 가설공법

가설공법 / 적용조건	BENT				가 설 BEAM (TRUSS) 공 법	밀어내기식 공법		편지지식 공법			일괄가설공법	
	자주식 CRANE	CABLE CRANE	트레벨러 CRANE	문형 CRANE		ILM 공법	가 설 BEAM (TRUSS) 공법	자주식 CRANE	CABLE CRANE	트레벨러 CRANE	자주식 CRANE	권상기에 의한
단순 빔 (STEEL PLATE BOX)	A	B	C	B	B	A					A	
연속 빔 (STEEL PLATE BOX)	A	B	B	B	B	A	B	B	B	B	C	C
곡선 빔	A	B	B		A	C	B	B			C	

주) A : 자주 사용하는 공법, B : 종종 사용하는 공법, C : 채용을 검토할 수 있는 공법

2-4 Bridge Construction Method Selection

2-4-1 Overview

① Set up plan with reviewing scale of planned facilities, location, topography, geological features, performance of available equipment, a number, erection possibility of erection manufacturing place, construction period, installation construction cost, etc

② Determine the method of construction to minimize the noise, vibration, and traffic congestions for residents and facilities

③ Whether adjacent area is contaminated due to the soil run-off during river or marine construction

2-4-2 Classification by Bridge Erection Method

Classification	Steel bridge	Concrete bridge(R.C, P.S.C)
Applied method	• Temporary bent method - Self moving crane, F/C, or Barge ship • Cable erection method - Vertical type, slope type • Incremental Launching • Cantilever method • Lift up barge method	• Timbering method(Full Staging) • Precast girder method ┌ Crane erection method └ Erection girder method • Free cantilever method(F.C.M) • Incremental launching method (I.L.M) • Movable scaffolding system(M.S.S)

2-4-3 Steel Bridge and Beam Bridge Installation Method

Installation method / Applied condition	BENT				Erection BEAM (TRUSS) method	Push out method		Cantilever method			Bundle installation method	
	Self moving CRANE	CABLE CRANE	Traveller CRANE	portal CRANE		ILM method	Installation BEAM (TRUSS) method	Self moving CRANE	CABLE CRANE	Traveller CRANE	Self moving CRANE	By coiling machine
Simple beam (STEEL PLATE BOX)	A	B	C	B	B	A					A	
Continuous beam (STEEL PLATE BOX)	A	B	B	B	B	A	B	B	B	B	C	C
Curve beam	A	B	B		A	C	B	B			C	

Note) A : Frequently used method, B : Method used often, C : Method that can be applied

3. 교량 상세설계

3-1 개 요

- 상세설계란 기본설계를 구체화하여 구조물을 정확하게 시공하는데 필요한 모든 설계로서, 시공을 위한 설계이다. 따라서 시공에 필요한 주의사항을 시공계획서에 기술하여 시공담당자에게 설계자의 의도를 충분히 전달할 수 있어야 한다.

- 상세설계단계는 설계기준 검토, 구조해석, 설계도면 작성, 공사시방서 작성, 건설사업비 산출 등으로 구성된다. 따라서 실시설계는 타당성조사 보고서·기본설계 검토결과·유사 규모의 시설물 자료 등 관련 설계자료를 이용하여 상세설계단계에서 구조계산서·실시설계도면·공사시방서·설계내역서 등을 작성하며, 설계심의단계에서 지적사항이 있을 경우에는 이를 반영하게 된다.

3-2 교량 설계기준

3-2-1 교량 설계 시 고려사항

① 소요차로, 폭원구성, 구조물 연장
② 구조물 설계방법과 내진설계 등급
③ 계획홍수량 및 계획홍수위
④ 설계하중의 종류, 크기 및 부가하중
⑤ 사용재료의 종류 및 특성(재질, 강도)
⑥ 조수간만의 차, 설계파고, 조류 속도
⑦ 설계기준, 시방서, 지침, 편람 등

- 주요 설계기준 : 도로교설계기준, 콘크리트구조설계기준, 구조물기초설계기준, 하천설계기준 등
- 주요 설계지침 : 강도로교 상세부 설계지침, 케이블 강교량 설계지침, 고속도로교량의 내진설계지침 등
- 주요 설계편람 : 도로교설계편람, 강구조설계편람 등
- 주요 설계요령 : 도로설계요령, 국도건설공사 설계실무요령, 강도로교 용접 및 도장 요령 등

3. Bridge Detail Design

3-1 Overview

- Detailed design contains all designs required for the accurate construction through embodiment of preliminary design, and this is design for construction. Therefore, caution items requires for the construction shall be described in the construction plan drawing and the intention of the designer shall be sufficiently transferred to construction engineers.

- Detailed design stage comprises design standard review, structure analysis, design drawing preparation, construction specification preparation, and construction project cost calculation. Therefore, structure calculation document, execution design drawing, construction specification, and design details are prepared with using related design data such as feasibility study report, basic design evaluation result, and similar scale facility data. If there is any correction required during the design deliberation, it shall be reflected.

3-2 Bridge Design Standard

3-2-1 Considerations on Bridge Design

① Required number of lanes, width components, structure expansion
② Structure design method and seismic design class
③ Plan flood rate and plan flood level
④ Type, size, of design load and additional load
⑤ Type and characteristics of used material (quality of material, strength)
⑥ Difference between ebb and flow, design wave height, current speed
⑦ Design standard, specification, instructions, handbook, etc

- Major design standard : Highway bridge design standard, Concrete structure design standard, structure foundation design standard, river design standard
- Major design instruction : steel highway bridge detailed design instruction, cable steel bridge design instruction, seismic design instructions on expressway bridge, etc.
- Major design handbook : Highway bridge design handbook, steel structure design handbook
- Major design technique : road design technique, Nation road construction design practical work technique, steel highway bridge welding and coating technique

3-2-2 설계하중

- 강도설계법에서는 계수하중을 적용하고, 허용응력설계법에서는 사용하중을 적용

- 구조계산 시의 하중조합은 고정하중, 활하중, 크리프, 건조수축, 온도변화, 지진력 그리고 지점의 부등침하 등의 하중 조합 중 가장 불리한 조합으로 설계

① 고정하중

- 고정하중 산출 시 단위중량은 다음 값을 기준으로 하되 실하중이 명백한 것은 그 값을 적용

재 료	단위중량 (kgf/m^3)	재 료	단위중량 (kgf/m^3)	재 료	단위중량 (kgf/m^3)
강재, 주강, 단강	7,850	무근 콘크리트	2,350	흙 (지하수위 이상)	1,800
알루미늄	2,800	시멘트 모르타르	2,150	흙 (지하수위 이하)	1,000
철근콘크리트	2,500	아스팔트 포장	2,300	뒤채움 (ø = 35°)	2,000
P.S 콘크리트	2,500	역청재(방수용)	1,100	뒤채움 (ø = 30°)	1,900
목 재	800	Latex Modified Concrete 포장	2,350		

② 활하중

<table>
<tr><th colspan="6">자 동 차 하 중</th></tr>
<tr><td colspan="3">DB-24 (표준트럭하중)</td><td colspan="3">DL-24 (표준차로하중)</td></tr>
<tr><td>구 분</td><td>전륜하중(Pf)</td><td>후륜하중(Pr)</td><td>구 분</td><td>모멘트계산(Pm)</td><td>전단력계산(Ps)</td></tr>
<tr><td>하중(kgf)</td><td>2,400</td><td>9,600</td><td>집중하중(P)</td><td>10,800kgf</td><td>15,600kgf</td></tr>
<tr><td colspan="3">종방향으로는 차로 당 1대
횡방향으로는 재하 가능한 대수</td><td colspan="3">지간에 따라 설계부재에 최대 응력이 발생하도록 단속적인 재하</td></tr>
<tr><td>노면 활하중</td><td colspan="5">교대 및 옹벽 배면에 작용하는 노면 활하중 : 1.0tonf/m²</td></tr>
<tr><td colspan="6">➔ DB-24, DL-24 하중에 대한 해석을 실시한 후 가장 불리한 단면력으로 설계</td></tr>
</table>

③ 충격하중

- 차량하중

$$i = \frac{15}{40+L} \leq 0.3, \quad \mathrm{L(m)} : \text{지간}$$

3-2-2 Design Load

- Strength design method applies coefficient load, and used load is applied in allowable stress design method.

- While structural calculation, the worst combination shall be used among fixed load, live load, creep, drying shrinkage, temperature change earthquake capacity, and differential settlement of location.

① Dead load

- When calculating dead load, unit weight shall be referring to the following values, and actual weight will be applied if it is obvious

Material	Unit weight (kgf/m^3)	Material	Unit weight (kgf/m^3)	Material	Unit weight (kgf/m^3)
Steel, cast steel, forged steel	7,850	coreless Concrete	2,350	Soil (over underground water level)	1,800
Aluminum	2,800	Cement Mortar	2,150	Soil (below underground water level)	1,000
Reinforced Concrete	2,500	Asphalt pavement	2,300	Post banking (ø = 35°)	2,000
P.S Concrete	2,500	Anti-corrosion material (Water-proof))	1,100	Post banking (ø = 30°)	1,900
Wood	800	Latex Modified Concrete pavement	2,350		

② Live load

Vehicle load					
DB-24 (standard truck load)			DL-24 (Standard vehicle load)		
Classification	Front wheel load(Pf)	Rear wheel load(Pr)	Classification	Moment calculation(Pm)	Shear force calculation(Ps)
Load(kgf)	2,400	9,600	Concentrated load(P)	10,800kgf	15,600kgf
One vehicle per lane in longitudinal direction Maximum possible loading in traversing direction			Intermittent loading to trigger maximum stree to the design material depending on span		
Road surface live load	Road live load applied to abutment and retaining wall reverse wall : $1.0tonf/m^2$				
→ After executing the analysis for DB-24, DL-24 load, and design with the worst case sectional force					

③ Impact load

- Vehicle load

$$i = \frac{15}{40+L} \leq 0.3, \quad L(m) : \text{Span}$$

④ 풍하중

- 교량의 상부구조에 작용하는 풍하중은 교축에 직각으로 작용하는 수평하중으로 하고, 고려하고 있는 부재에 가장 불리한 응력을 발생시키도록 재하
- 방음벽이 설치되어 있는 경우에는 바람의 특성 및 방음벽의 구조에 따라 풍하중을 저감시킬 수 있으나, 구조물의 안정성 향상을 위하여 저감하지 않도록 함

⑤ 토 압

구 분	안 정 검 토	단 면 설 계
적용방법	Rankine 토압	Coulomb 토압
교대의 뒤채움재는 Φ=35°→ γt=2.0 tonf/m³적용		옹벽의 뒤채움재는 Φ=30°→ γt=1.9 tonf/m³적용

※ 지진 시 Mononobe-Okabe의 식 적용

※ 교대 날개벽 : 정지토압계수 적용

3-2-3 재료의 강도

① 철 근

종 류	관련규격	인장시험(kgf/cm²)		허용응력 (kgf/cm²)	적 용 구 조 물
		항복점	인장강도		
SD40	KS D 3504 철근콘크리트용 봉강	4,000	5,700 이상	1,800	• 일반적인 부재, PSC 구조물
SD30		3,000	4,500 이상	1,500	• 정착부 덮개철근

② 강 재

강종	관련규격	인장시험(kgf/cm²)		적 용 구 조 물
		항복점	인장강도	
SWPC7B	KS D 7002 PS 강선 및 스트랜드	fpy = 16,000 이상 fpu = 19,000 이상		• 주형 및 바닥판의 강선 • 사재의 강선
SBPR785/1030	KS D 3505 PC 강봉	fpy = 8,000 fpu = 10,500		• 강봉
SM490	KS D 3515 용접구조용 압연강재	1,900	1,100	• 주부재
SS400	KS D 3503 일반구조용 압연강재	1,400	800	• 부부재 및 가설교량

④ Wind load

- Wind load applied to the superstructure of the bridge shall be horizontal load perpendicular to the bridge axis, and loading shall be done to generate the worst case stress for the structural material.
- When sound proof wall is installed, it reduced the wind load depending on the structure of the sound proof wall and wind characteristics, and it shall not be reduced for the enhancement of the stability of the structure

⑤ Soil pressure

Classification	Stability evaluation	Section design
Application method	Rankine soil pressure	Coulomb soil pressure
Post banking material of abutment is applied as Φ=35°→ ɣt=2.0 tonf/m³		Post banking materials of retaining wall is applied as Φ=30°→ ɣt=1.9 tonf/m³

※ Apply Mononobe-Okabe's formula during earthquake

※ Wing wall of abutment: Apply static soil pressure coefficients

3-2-3 Strength of Material

① Reinforcement

Type	Related specification	Elongation test(kgf/cm²)		Allowable stress (kgf/cm²)	Applied structure
		yield point	tensile strength		
SD40	KS D 3504 rod steel for reinforcement concrete	4,000	Over 5,700	1,800	• General material, PSC structure
SD30		3,000	Over 4,500	1,500	• covering reinforcement of anchorage part

② Steel

Steel type	Related standard	Tension test(kgf/cm²)		Applied structures
		Yielding point	Tensile strength	
SWPC7B	KS D 7002 PS steel wire and strands	fpy = over 16,000 fpu = over 19,000		• Cast type and steel wire of floor slab • Steel wire of diagonal member
SBPR785/1030	KS D 3505 PC steel rod	fpy = 8,000 fpu = 10,500		• Steel rod
SM490	KS D 3515 Rolling steel material for welding structure	1,900	1,100	• Main member
SS400	KS D 3503 Rolling steel material for general structure	1,400	800	• Sub member and bridge erection

③ 콘크리트

설계기준강도 (kgf/cm²)	골재최대지수 (mm)	슬럼프기준 (cm)	적 용 구 조 물
450	19	15	프리캐스트 콘크리트
400	19	15	PSC 현장타설 콘크리트
350	25	15	현장타설말뚝
270	25	15	바닥판 콘크리트, 접속슬래브, 교각, 교대
240	25	15	철근 콘크리트 옹벽, 통로 BOX 구조물
210	25	15	기초보강 콘크리트, 수로암거, 중력식옹벽
180	25	15	MASS 콘크리트, 부대시설기초
160	25	15	레벨링 콘크리트, 속채움 콘크리트

3-2-4 콘크리트 및 재료의 물리상수

① 콘크리트 탄성계수

압축강도(kgf/cm²)	단위중량 (tonf/m³)	탄성계수(kgf/cm²)
300 이하	1.45~2.5 2.3	$E_c = 4,270 \cdot W_c^{1.5} \cdot \sqrt{f_{ck}}$, $E_c = 15,000 \cdot \sqrt{f_{ck}}$
300 초과	1.45~2.5 2.3	$E_c = 3,000 \cdot W_c^{1.5} \cdot \sqrt{f_{ck}} + 70,000$, $E_c = 10,500 \cdot \sqrt{f_{ck}} + 70,000$

- 온도팽창계수 : $\alpha = 1.0 \times 10^{-5}$
- 건조수축 계수 : $\beta = 20 \times 10^{-5}$
- 크리프 계수 : ø=2.8 (조강시멘트), 3.0 (보통시멘트)

② 철근의 탄성계수

- Es = 2.0×106 (kgf/cm²)
- 온도팽창계수 : $\alpha = 1.2 \times 10^{-5}$

③ 강재의 탄성계수

- 탄성계수 : Es = 2.1×106 (k • 탄성계수 : Es = 2.1×106 (kgf/cm²)
- 온도팽창계수 : $\alpha = 1.2 \times 10^{-5}$

③ Concrete

Design standard strength (kgf/cm^2)	Aggregate maximum index (mm)	Slump standard (cm)	Applied structures
450	19	15	Precast concrete
400	19	15	PSC cast-in-place concrete
350	25	15	cast-in-place pile
270	25	15	Floor slab concrete, approach slab, pier, abutment
240	25	15	reinforced concrete retaining wall, corrodier BOX structure
210	25	15	Foundation reinforcement concrete, waterway culvert, gravity type retaining wall
180	25	15	MASS concrete, aux facility base
160	25	15	Levelling concrete, inside-filling concrete

3-2-4 Physical Constant of Concrete and Material

① Concrete elastic coefficient

Compression strength (kgf/cm^2)	Unit weight ($tonf/m^3$)	Modulus of elasticity (kgf/cm^2)
Below 300	1.45~2.5 2.3	$E_c = 4,270 \cdot W_c^{1.5} \cdot \sqrt{f_{ck}}$, $E_c = 15,000 \cdot \sqrt{f_{ck}}$
Over 300	1.45~2.5 2.3	$E_c = 3,000 \cdot W_c^{1.5} \cdot \sqrt{f_{ck}} + 70,000$, $E_c = 10,500 \cdot \sqrt{f_{ck}} + 70,000$

- Thermal expansion coefficients : $\alpha = 1.0 \times 10^{-5}$
- Drying shrinkage coefficients : $\beta = 20 \times 10^{-5}$
- Creep coefficients : ø=2.8 (high-early cement), 3.0 (normal cement)

② Modulus of elasticity of reinforcement

- Es = 2.0×106 (kgf/cm^2)
- Thermal expansion coefficients : $\alpha = 1.2 \times 10^{-5}$

③ Modulus of elasticity of steel

- Elastic coefficient: Es = 2.1×106 (k • Elastic coefficient : Es = 2.1×106 (kgf/cm^2)
- Thermal expansion coefficients : $\alpha = 1.2 \times 10^{-5}$

④ PS 강재의 탄성계수

- 탄성계수 : Eps = 2.0×106 (kgf/cm^2)

- 릴렉세이션율

PS 강재의 종류	겉보기 릴렉세이션율(%)
PS 강선, PS 강연선	5
PS 강봉	3

- 마찰계수

긴장재의 형태	덕트의 형태	파상마찰계수 (k/m)	곡률마찰계수 (μ/rad)
PS 강선 또는 도금되지 않은 PS 강연선	• 금속 쉬스	0.0066	0.30
	• 아연도금 금속 쉬스	0.0050	0.25
	• 아스팔트 또는 그리스로 코팅되고 피복된 것	0.0066	0.30
	• 아연도금된 강성 덕트	0.0007	0.25
고강도 강봉	• 금속 쉬스	0.0010	0.20
	• 아연도금된 금속 쉬스	0.0007	0.15

④ Modulus of elasticity of PS steel

- Modulus of elasticity: Eps = 2.0×106 (kgf/cm^2)

- Relaxation rate

Type of PS steel material	Apparent relaxation rate(%)
PS steel wire, PS steel strand	5
PS steel rod	3

- Friction coefficient

Type of tendon	Type of duct	Wobble friction coefficient (k/m)	Curvature friction coefficient (μ/rad)
PS steel wire or ungalvanized PS steel strand	• Metal sheath • Zinc-plated metal sheath • Coated and covered with asphalt or grease • Zinc-plated steel duct	0.0066 0.0050 0.0066 0.0007	0.30 0.25 0.30 0.25
High strength steel rod	• Metal sheath • Zinc-plated metal sheath	0.0010 0.0007	0.20 0.15

3-2-5 주요 사용재료의 허용응력

① 콘크리트의 허용응력

구 분	내 용		적 용
철근 콘크리트	허용 휨응력	허용 휨 압축응력	$f_{ca}=0.4f_{ck}$
		허용 휨 인장응력 (무근 확대기초, 벽체)	$f_{ta}=0.42\sqrt{f_{ck}}$
		휨강도(파괴계수) : 보통콘크리트	$f_r=2.0\sqrt{f_{ck}}$
	허용 압축응력(무근 확대기초, 벽체)		$f_{ca}=0.25f_{ck}$
	허용전단 응력	보 및 일방향 전단(일방향 슬래브, 확대기초)	
		콘크리트 허용 전단 응력	$v_{ca}=0.25\sqrt{f_{ck}}$
		전단보강이 있는 부재	$v_{ca}=1.15\sqrt{f_{ck}}$
	허용지압 응력		$f_{ba}=0.25f_{ck}\sqrt{\frac{A_c}{A_b}}\le 0.5f_{ck}$ A_c : 지지하는 콘크리트 전 면적 A_b : 지압을 받는 재하 면적
프리스트레스트 콘크리트	프리스트레스 도입 시 압축 강도		$f_{ci}'=0.8f_{ck}$
	크리프와 건조수축에 의한 손실이 일어나기 전		
	압축응력(포스트텐션)		$0.55f_{ci}'$
	인장응력	부착된 철근이 없는 인장 구역	14 kgf/㎠ 또는 $0.80\sqrt{f_{ci}'}$
		부착된 철근이 있는 인장 구역	$1.60\sqrt{f_{ci}'}$
	모든 손실이 일어난 후 사용하중 상태에서의 응력		
	압축응력		$0.4f_{ck}$
	인장응력	부착된 철근을 갖는 부재	$1.60\sqrt{f_{ck}}$
	균열응력(보통 콘크리트)		$2.0\sqrt{f_{ck}}$
	정착부의 지압응력	긴장재 정착 직후	$0.70f_{ci}'\sqrt{\frac{A_b'}{A_b}-0.2}\le 1.10f_{ci}'$
		프리스트레스 손실 직후	$0.50f_{ck}\sqrt{\frac{A_b'}{A_b}-0.2}\le 0.90f_{ck}$

② 강재의 허용응력

- 허용 축방향 인장응력 및 허용 휨 인장응력(kgf/cm^2)

강종 강재 / 판 두께(mm)	SS400 SM400 SMA41	SM490	SM490Y SM520 SMA50	SM570 SMA58
40 이하	1,400	1,900	2,100	2,600
40 초과 75 이하	1,300	1,750	2,000	2,500
75 초과 100 이하			1,950	2,450

3-2-5 Allowable Stress of Main Material

① Allowable stress of concrete

<table>
<tr><th>Classification</th><th colspan="2">Contents</th><th>Application</th></tr>
<tr><td rowspan="8">steel reinforcement Concrete</td><td rowspan="3">Allowable bending stress</td><td>Allowable bending compression stress</td><td>$f_{ca}=0.4f_{ck}$</td></tr>
<tr><td>Allowable bending tensile stress (No core expansion based, wall body)</td><td>$f_{ta}=0.42\sqrt{f_{ck}}$</td></tr>
<tr><td>Bending strength (modulus of rupture) : general concrete</td><td>$f_r=2.0\sqrt{f_{ck}}$</td></tr>
<tr><td colspan="2">Allowable compression stress(plain expansion foundation, wall)</td><td>$f_{ca}=0.25f_{ck}$</td></tr>
<tr><td rowspan="3">Allowable shear stress</td><td colspan="2">Beam and one directional shear(one direction slab, expansion base)</td></tr>
<tr><td>Concrete allowable shear stress</td><td>$v_{ca}=0.25\sqrt{f_{ck}}$</td></tr>
<tr><td>Shear reinforced member</td><td>$v_{ca}=1.15\sqrt{f_{ck}}$</td></tr>
<tr><td colspan="2">Allowable bearing stress</td><td>$f_{ba}=0.25f_{ck}\sqrt{\frac{A_c}{A_b}}\le 0.5f_{ck}$
A_c : total concrete area
A_b : loading area under bearing stress</td></tr>
<tr><td rowspan="11">prestress concrete</td><td colspan="2">compression strength when introducing prestress</td><td>$f_{ci}'=0.8f_{ck}$</td></tr>
<tr><td colspan="3">Before occurring the loss by creep and drying shrinkage</td></tr>
<tr><td colspan="2">Compression stress (post-tension)</td><td>$0.55f_{ci}'$</td></tr>
<tr><td rowspan="2">Tensile stress</td><td>Tensile area without bonded reinforcement</td><td>14 kgf/㎠ or $0.80\sqrt{f_{ci}'}$</td></tr>
<tr><td>Tensile area with bonded reinforcement</td><td>$1.60\sqrt{f_{ci}'}$</td></tr>
<tr><td colspan="3">Stress in the service load state after all loss occurs</td></tr>
<tr><td colspan="2">Compression stress</td><td>$0.4f_{ck}$</td></tr>
<tr><td>Tensile stress</td><td>member with bonded reinforcement</td><td>$1.60\sqrt{f_{ck}}$</td></tr>
<tr><td colspan="2">Crack stress (normal concrete)</td><td>$2.0\sqrt{f_{ck}}$</td></tr>
<tr><td rowspan="2">Bearing stress of anchorage part</td><td>Right after tendon anchor</td><td>$0.70f_{ci}'\sqrt{\frac{A_b'}{A_b}-0.2}\le 1.10f_{ci}'$</td></tr>
<tr><td>Right after prestress loss</td><td>$0.50f_{ck}\sqrt{\frac{A_b'}{A_b}-0.2}\le 0.90f_{ck}$</td></tr>
</table>

② Allowable stress of steel

- Allowable axial direction tensile stress and allowable bending tensile stress(kgf/cm^2)

<table>
<tr><th>Steel type
Steel plate
thickness(mm)</th><th>SS400
SM400
SMA41</th><th>SM490</th><th>SM490Y
SM520
SMA50</th><th>SM570
SMA58</th></tr>
<tr><td>below 40</td><td>1,400</td><td>1,900</td><td>2,100</td><td>2,600</td></tr>
<tr><td>40 ~ 75</td><td rowspan="2">1,300</td><td rowspan="2">1,750</td><td>2,000</td><td>2,500</td></tr>
<tr><td>75 ~ 100</td><td>1,950</td><td>2,450</td></tr>
</table>

• 허용 축방향 압축응력

$$f_{ca} = f_{cag} \cdot f_{cal} / f_{cao}$$

여기서, f_{ca} : 허용 축방향 압축응력(kgf/cm^2)

f_{cag} : 국부좌굴을 고려하지 않은 허용 축방향 압축응력(kgf/cm^2)

f_{cal} : 국부좌굴에 대한 허용응력(kgf/cm^2)

f_{cao} : 국부좌굴을 고려하지 않은 허용 축방향 압축응력의 상한값(kgf/cm^2)

• 국부좌굴을 고려하지 않은 허용 축방향 압축응력(kgf/cm^2)

<table>
<tr><th>강종
강재 판 두께(mm)</th><th>SS400
SM400
SMA41</th><th>SM490</th><th>SM490Y
SM520
SMA50</th><th>SM570
SMA58</th></tr>
<tr><td>40 이하</td><td>1,400 : $\frac{l}{r} \le 20$
1,400-8.4($\frac{l}{r}$-20) :
$20 < \frac{l}{r} \le 93$
$\frac{12,000,000}{6,700 + (\frac{l}{r})^2}$:
$93 < \frac{l}{r}$</td><td>1,900 : $\frac{l}{r} \le 15$
1,900-13($\frac{l}{r}$-15) :
$15 < \frac{l}{r} \le 80$
$\frac{12,000,000}{5,000 + (\frac{l}{r})^2}$:
$80 < \frac{l}{r}$</td><td>2,100 : $\frac{l}{r} \le 14$
2,100-15($\frac{l}{r}$-14) :
$14 < \frac{l}{r} \le 76$
$\frac{12,000,000}{4,500 + (\frac{l}{r})^2}$:
$76 < \frac{l}{r}$</td><td>2,600 : $\frac{l}{r} \le 18$
2,600-22($\frac{l}{r}$-18) :
$18 < \frac{l}{r} \le 67$
$\frac{12,000,000}{3,500 + (\frac{l}{r})^2}$:
$67 < \frac{l}{r}$</td></tr>
<tr><td>40 초과
75 이하</td><td rowspan="2">1,300 : $\frac{l}{r} \le 20$
1,300-7.5($\frac{l}{r}$-20) :
$20 < \frac{l}{r} \le 97$
$\frac{12,000,000}{7,300 + (\frac{l}{r})^2}$:
$97 < \frac{l}{r}$</td><td rowspan="2">1,750 : $\frac{l}{r} \le 15$
1,750-11($\frac{l}{r}$-15) :
$15 < \frac{l}{r} \le 83$
$\frac{12,000,000}{5,300 + (\frac{l}{r})^2}$:
$83 < \frac{l}{r}$</td><td>2,000 : $\frac{l}{r} \le 14$
2,000-14($\frac{l}{r}$-14) :
$14 < \frac{l}{r} \le 78$
$\frac{12,000,000}{4,700 + (\frac{l}{r})^2}$:
$78 < \frac{l}{r}$</td><td>2,500 : $\frac{l}{r} \le 18$
2,500-21($\frac{l}{r}$-18) :
$18 < \frac{l}{r} \le 69$
$\frac{12,000,000}{3,600 + (\frac{l}{r})^2}$:
$69 < \frac{l}{r}$</td></tr>
<tr><td>75 초과
100 이하</td><td>1,950 : $\frac{l}{r} \le 14$
1,950-13($\frac{l}{r}$-14) :
$14 < \frac{l}{r} \le 79$
$\frac{12,000,000}{4,900 + (\frac{l}{r})^2}$:
$79 < \frac{l}{r}$</td><td>2,450 : $\frac{l}{r} \le 18$
2,450-20($\frac{l}{r}$-18) :
$18 < \frac{l}{r} \le 69$
$\frac{12,000,000}{3,700 + (\frac{l}{r})^2}$:
$69 < \frac{l}{r}$</td></tr>
<tr><td>비 고</td><td colspan="4">l : 부재의 유효좌굴길이(cm)
r : 부재 총 단면의 단면회전반경(cm)</td></tr>
</table>

- Allowable axis direction compression stress

$$f_{ca} = f_{cag} \cdot f_{cal}/f_{cao}$$

Where, f_{ca} : Allowable axis direction compression stress(kgf/cm^2)

f_{cag} : Allowable axis direction compression stress without considering local buckling (kgf/cm^2)

f_{cal} : Allowable stress for local buckling (kgf/cm^2)

f_{cao} : Upper bound of allowable axis direction compression stress without considering local buckling(kgf/cm^2)

- Allowable axis direction compression stress without considering local buckling (kgf/cm^2)

Steel type / steel plate thickness (mm)	SS400 SM400 SMA41	SM490	SM490Y SM520 SMA50	SM570 SMA58
below 40	1,400 : $\frac{l}{r} \leq 20$ 1,400-8.4($\frac{l}{r}$-20) : $20 < \frac{l}{r} \leq 93$ $\frac{12,000,000}{6,700+(\frac{l}{r})^2}$: $93 < \frac{l}{r}$	1,900 : $\frac{l}{r} \leq 15$ 1,900-13($\frac{l}{r}$-15) : $15 < \frac{l}{r} \leq 80$ $\frac{12,000,000}{5,000+(\frac{l}{r})^2}$: $80 < \frac{l}{r}$	2,100 : $\frac{l}{r} \leq 14$ 2,100-15($\frac{l}{r}$-14) : $14 < \frac{l}{r} \leq 76$ $\frac{12,000,000}{4,500+(\frac{l}{r})^2}$: $76 < \frac{l}{r}$	2,600 : $\frac{l}{r} \leq 18$ 2,600-22($\frac{l}{r}$-18) : $18 < \frac{l}{r} \leq 67$ $\frac{12,000,000}{3,500+(\frac{l}{r})^2}$: $67 < \frac{l}{r}$
40 ~ 75	1,300 : $\frac{l}{r} \leq 20$ 1,300-7.5($\frac{l}{r}$-20) :	1,750 : $\frac{l}{r} \leq 15$ 1,750-11($\frac{l}{r}$-15) :	2,000 : $\frac{l}{r} \leq 14$ 2,000-14($\frac{l}{r}$-14) : $14 < \frac{l}{r} \leq 78$ $\frac{12,000,000}{4,700+(\frac{l}{r})^2}$: $78 < \frac{l}{r}$	2,500 : $\frac{l}{r} \leq 18$ 2,500-21($\frac{l}{r}$-18) : $18 < \frac{l}{r} \leq 69$ $\frac{12,000,000}{3,600+(\frac{l}{r})^2}$: $69 < \frac{l}{r}$
75 ~ 100	$20 < \frac{l}{r} \leq 97$ $\frac{12,000,000}{7,300+(\frac{l}{r})^2}$: $97 < \frac{l}{r}$	$15 < \frac{l}{r} \leq 83$ $\frac{12,000,000}{5,300+(\frac{l}{r})^2}$: $83 < \frac{l}{r}$	1,950 : $\frac{l}{r} \leq 14$ 1,950-13($\frac{l}{r}$-14) : $14 < \frac{l}{r} \leq 79$ $\frac{12,000,000}{4,900+(\frac{l}{r})^2}$: $79 < \frac{l}{r}$	2,450 : $\frac{l}{r} \leq 18$ 2,450-20($\frac{l}{r}$-18) : $18 < \frac{l}{r} \leq 69$ $\frac{12,000,000}{3,700+(\frac{l}{r})^2}$: $69 < \frac{l}{r}$
Remark	l : Effective buckling length of member(cm) r : Section radius of rotation of member total section(cm)			

- 허용 휨 압축응력(kgf/cm^2)
  - 압축플랜지가 콘크리트 상판 등에 직접 고정되어 있는 경우 및 상자형 단면, Ⅱ형 단면과 같이 휨 좌굴이 일어나기 어려운 경우

강종 강재 / 판 두께(mm)	SS400 SM400 SMA41	SM490	SM490Y SM520 SMA50	SM570 SMA58
40 이하	1,400	1,900	2,100	2,600
40 초과 75 이하	1,300	1,750	2,000	2,500
75 초과 100 이하			1,950	2,450

- 상기에서 규정한 이외의 경우의 허용 휨 압축응력(kgf/cm^2)

강종 강재 / 판 두께(mm)	SS400 SM400 SMA41	SM490	SM490Y SM520 SMA50	SM570 SMA58
$\frac{A_\omega}{A_c} \le 2$	40 이하	1,400 : $\frac{l}{b} \le 4.5$ 1,400-24($\frac{l}{b}$-4.5) : $4.5 < \frac{l}{b} \le 30$	1,900 : $\frac{l}{b} \le 4.0$ 1,900-38($\frac{l}{b}$-4.0) : $4.0 < \frac{l}{b} \le 30$	2,100 : $\frac{l}{b} \le 3.5$ 2,100-44($\frac{l}{b}$-3.5) : $3.5 < \frac{l}{b} \le 27$
45 초과 75 이하	1,300 : $\frac{l}{b} \le 5.0$ 1,300-22($\frac{l}{b}$-5.0) : $5.0 < \frac{l}{b} \le 30$	1,750 : $\frac{l}{b} \le 4.0$ 1,750-35($\frac{l}{b}$-4.0) : $4.0 < \frac{l}{b} \le 30$	2,000 : $\frac{l}{b} \le 4.0$ 2,000-42($\frac{l}{b}$-4.0) : $4.0 < \frac{l}{b} \le 27$	2,500 : $\frac{l}{b} \le 5.0$ 2,500-62($\frac{l}{b}$-5.0) : $5.0 < \frac{l}{b} \le 25$
75 초과 100 이하			1,950 : $\frac{l}{b} \le 4.0$ 1,950-40($\frac{l}{b}$-4.0) : $4.0 < \frac{l}{b} \le 27$	2,450 : $\frac{l}{b} \le 5.0$ 2,450-60($\frac{l}{b}$-5.0) : $5.0 < \frac{l}{b} \le 25$
$\frac{A_\omega}{A_c} > 2$	40 이하	1,400 : $\frac{l}{b} \le \frac{9}{K}$ 1,400-12($K\frac{l}{b}$-9) : $\frac{9}{K} < \frac{l}{b} \le 30$	1,900 : $\frac{l}{b} \le \frac{8}{K}$ 1,900-19($K\frac{l}{b}$-8) : $\frac{8}{K} < \frac{l}{b} \le 30$	2,100 : $\frac{l}{b} \le \frac{7}{K}$ 2,100-21($K\frac{l}{b}$-7) : $\frac{7}{K} < \frac{l}{b} \le 27$
40 초과 75 이하	1,300 : $\frac{l}{b} \le \frac{10}{K}$ 1,300-11($K\frac{l}{b}$-10) : $\frac{10}{K} < \frac{l}{b} \le 30$	1,750 : $\frac{l}{b} \le \frac{8}{K}$ 1,750-17.5($K\frac{l}{b}$-8) : $\frac{8}{K} < \frac{l}{b} \le 30$	2,000 : $\frac{l}{b} \le \frac{8}{K}$ 2,000-21($K\frac{l}{b}$-8) : $\frac{8}{K} < \frac{l}{b} \le 27$	2,500 : $\frac{l}{b} \le \frac{10}{K}$ 2,500-31($K\frac{l}{b}$-10) : $\frac{10}{K} < \frac{l}{b} \le 25$
75 초과 100 이하			1,950 : $\frac{l}{b} \le \frac{8}{K}$ 1,950-20($K\frac{l}{b}$-8) : $\frac{8}{K} < \frac{l}{b} \le 27$	2,450 : $\frac{l}{b} \le \frac{10}{K}$ 2,450-30($K\frac{l}{b}$-10) : $\frac{10}{K} < \frac{l}{b} \le 25$
비 고	A_ω : 복부판의 총 단면적(cm^2) A_c : 압축플랜지 총 단면적(cm^2) l : 압축플랜지의 고정점 간 거리(cm) b : 압축플랜지 폭(cm) $K=\sqrt{3+\frac{A_\omega}{2A_c}}$			

- Allowable bending compression stress (kgf/cm^2)

  ⇒ When compression flange is directly fixed to concrete upper plate, and low probability of flexural buckling due to box type section or Π type section.

Steel plate thickness(mm) \ Steel type	SS400 SM400 SMA41	SM490	SM490Y SM520 SMA50	SM570 SMA58
Below 40	1,400	1,900	2,100	2,600
40 ~ 75	1,300	1,750	2,000	2,500
75 ~ 100			1,950	2,450

- Allowable bending compression stress(kgf/cm^2) for the cases except of specified above

steel plate thickness (mm) \ steel type	SS400 SM400 SMA41	SM490	SM490Y SM520 SMA50	SM570 SMA58
$\frac{A_\omega}{A_c} \leq 2$	below 40	1,400 : $\frac{l}{b} \leq 4.5$ 1,400-24($\frac{l}{b}$-4.5) : $4.5 < \frac{l}{b} \leq 30$	1,900 : $\frac{l}{b} \leq 4.0$ 1,900-38($\frac{l}{b}$-4.0) : $4.0 < \frac{l}{b} \leq 30$	2,100 : $\frac{l}{b} \leq 3.5$ 2,100-44($\frac{l}{b}$-3.5) : $3.5 < \frac{l}{b} \leq 27$
45 ~ 75	1,300 : $\frac{l}{b} \leq 5.0$ 1,300-22($\frac{l}{b}$-5.0) : $5.0 < \frac{l}{b} \leq 30$	1,750 : $\frac{l}{b} \leq 4.0$ 1,750-35($\frac{l}{b}$-4.0) : $4.0 < \frac{l}{b} \leq 30$	2,000 : $\frac{l}{b} \leq 4.0$ 2,000-42($\frac{l}{b}$-4.0) : $4.0 < \frac{l}{b} \leq 27$	2,500 : $\frac{l}{b} \leq 5.0$ 2,500-62($\frac{l}{b}$-5.0) : $5.0 < \frac{l}{b} \leq 25$
75 ~ 100			1,950 : $\frac{l}{b} \leq 4.0$ 1,950-40($\frac{l}{b}$-4.0) : $4.0 < \frac{l}{b} \leq 27$	2,450 : $\frac{l}{b} \leq 5.0$ 2,450-60($\frac{l}{b}$-5.0) : $5.0 < \frac{l}{b} \leq 25$
$\frac{A_\omega}{A_c} > 2$	Below 40	1,400 : $\frac{l}{b} \leq \frac{9}{K}$ 1,400-12($K\frac{l}{b}$-9) : $\frac{9}{K} < \frac{l}{b} \leq 30$	1,900 : $\frac{l}{b} \leq \frac{8}{K}$ 1,900-19($K\frac{l}{b}$-8) : $\frac{8}{K} < \frac{l}{b} \leq 30$	2,100 : $\frac{l}{b} \leq \frac{7}{K}$ 2,100-21($K\frac{l}{b}$-7) : $\frac{7}{K} < \frac{l}{b} \leq 27$
40 ~ 75	1,300 : $\frac{l}{b} \leq \frac{10}{K}$ 1,300-11($K\frac{l}{b}$-10) : $\frac{10}{K} < \frac{l}{b} \leq 30$	1,750 : $\frac{l}{b} \leq \frac{8}{K}$ 1,750-17.5($K\frac{l}{b}$-8) : $\frac{8}{K} < \frac{l}{b} \leq 30$	2,000 : $\frac{l}{b} \leq \frac{8}{K}$ 2,000-21($K\frac{l}{b}$-8) : $\frac{8}{K} < \frac{l}{b} \leq 27$	2,500 : $\frac{l}{b} \leq \frac{10}{K}$ 2,500-31($K\frac{l}{b}$-10) : $\frac{10}{K} < \frac{l}{b} \leq 25$
75 ~ 100			1,950 : $\frac{l}{b} \leq \frac{8}{K}$ 1,950-20($K\frac{l}{b}$-8) : $\frac{8}{K} < \frac{l}{b} \leq 27$	2,450 : $\frac{l}{b} \leq \frac{10}{K}$ 2,450-30($K\frac{l}{b}$-10) : $\frac{10}{K} < \frac{l}{b} \leq 25$
Remark	A_ω : Total section area of web (cm2) A_c : Total section area of compression flange (cm2) l : Distance between fixed points of compression flange(cm) b : Width of compression flange(cm) $K=\sqrt{3+\frac{A_\omega}{2A_c}}$			

● 허용전단응력 및 지압응력(kgf/cm²)

응력종류		판 두께 \ 강종	SS400 SM400 SMA41	SM490	SM490Y SM520 SMA50	SM570 SMA58
전 단 응 력		40 이하	800	1,100	1,200	1,500
		40 초과 75 이하	750	1,000	1,150	1,450
		75 초과 100 이하			1,100	1,400
지압응력	강판과 강판사이의 지압응력	40 이하	2,100	2,800	3,100	3,900
		40 초과 75 이하	1,950	2,600	3,000	3,750
		75 초과 100 이하			2,900	3,650
	헬츠 공식으로 산출하는 경우	40 이하	6,000	7,000	-	-
		40 초과 75 이하				
		75 초과 100 이하				

● 양연 지지판의 국부좌굴에 대한 허용응력

강종	강재 판 두께(mm)	국부좌굴에 대한 허용응력(kgf/cm²)
SS400 SM400 SMA41	40 이하	1,400 : $\frac{b}{39.6i} \le t$ $2,200,000\ (\frac{ti}{b})^2$: $\frac{b}{80i} \le t < \frac{b}{39.6i}$
	40 초과 100 이하	1,300 : $\frac{b}{41.1i} \le t$ $2,200,000\ (\frac{ti}{b})^2$: $\frac{b}{80i} \le t < \frac{b}{41.1i}$
SM490	40 이하	1,900 : $\frac{b}{34.0i} \le t$ $2,200,000\ (\frac{ti}{b})^2$: $\frac{b}{80i} \le t < \frac{b}{34.0i}$
	40 초과 100 이하	1,750 : $\frac{b}{35.5i} \le t$ $2,200,000\ (\frac{ti}{b})^2$: $\frac{b}{80i} \le t < \frac{b}{35.5i}$
SM490Y SM520 SMA50	40 이하	2,100 : $\frac{b}{32.4i} \le t$ $2,200,000\ (\frac{ti}{b})^2$: $\frac{b}{80i} \le t < \frac{b}{32.4i}$
	40 초과 75 이하	2,000 : $\frac{b}{32.2i} \le t$ $2,200,000\ (\frac{ti}{b})^2$: $\frac{b}{80i} \le t < \frac{b}{33.2i}$
	75 초과 100 이하	1,950 : $\frac{b}{33.6i} \le t$ $2,200,000\ (\frac{ti}{b})^2$: $\frac{b}{80i} \le t < \frac{b}{33.6i}$
SM570 SMA58	40 이하	2,600 : $\frac{b}{29.1i} \le t$ $2,200,000\ (\frac{ti}{b})^2$: $\frac{b}{80i} \le t < \frac{b}{29.1i}$
	40 초과 75 이하	2,500 : $\frac{b}{29.7i} \le t$ $2,200,000\ (\frac{ti}{b})^2$: $\frac{b}{80i} \le t < \frac{b}{29.7i}$
	75 초과 100 이하	2,450 : $\frac{b}{30.0i} \le t$ $2,200,000\ (\frac{ti}{b})^2$: $\frac{b}{80i} \le t < \frac{b}{30.0i}$

• Allowable shear stress and bearing stress (kgf/cm²)

Stress type / Plate thickness		Steel type	SS400 SM400 SMA41	SM490	SM490Y SM520 SMA50	SM570 SMA58
Shear stress		below 40	800	1,100	1,200	1,500
		40 ~ 75	750	1,000	1,150	1,450
		75 ~ 100			1,100	1,400
Bearing stress	bearing stress between steel plates	below 40	2,100	2,800	3,100	3,900
		40 ~ 75	1,950	2,600	3,000	3,750
		75 ~ 100			2,900	3,650
	In the case of calculating with Hertz formula	below 40	6,000	7,000	-	-
		40 ~ 75				
		75 ~ 100				

• Allowable stress for the local buckling of fixed bearing plate

Steel type	Steel plate thickness(mm)	Allowable stress for the local bucking(kgf/cm²)	
SS400 SM400 SMA41	Below 40	1,400	: $\frac{b}{39.6i} \leq t$
		2,200,000 $(\frac{ti}{b})^2$	: $\frac{b}{80i} \leq t < \frac{b}{39.6i}$
	40 ~ 100	1,300	: $\frac{b}{41.1i} \leq t$
		2,200,000 $(\frac{ti}{b})^2$	: $\frac{b}{80i} \leq t < \frac{b}{41.1i}$
SM490	Below 40	1,900	: $\frac{b}{34.0i} \leq t$
		2,200,000 $(\frac{ti}{b})^2$	: $\frac{b}{80i} \leq t < \frac{b}{34.0i}$
	40 ~ 100	1,750	: $\frac{b}{35.5i} \leq t$
		2,200,000 $(\frac{ti}{b})^2$	: $\frac{b}{80i} \leq t < \frac{b}{35.5i}$
SM490Y SM520 SMA50	Below 40	2,100	: $\frac{b}{32.4i} \leq t$
		2,200,000 $(\frac{ti}{b})^2$	: $\frac{b}{80i} \leq t < \frac{b}{32.4i}$
	40 ~ 75	2,000	: $\frac{b}{32.2i} \leq t$
		2,200,000 $(\frac{ti}{b})^2$	: $\frac{b}{80i} \leq t < \frac{b}{33.2i}$
	75 ~ 100	1,950	: $\frac{b}{33.6i} \leq t$
		2,200,000 $(\frac{ti}{b})^2$	: $\frac{b}{80i} \leq t < \frac{b}{33.6i}$
SM570 SMA58	Below 40	2,600	: $\frac{b}{29.1i} \leq t$
		2,200,000 $(\frac{ti}{b})^2$	: $\frac{b}{80i} \leq t < \frac{b}{29.1i}$
	40 ~ 75	2,500	: $\frac{b}{29.7i} \leq t$
		2,200,000 $(\frac{ti}{b})^2$	: $\frac{b}{80i} \leq t < \frac{b}{29.7i}$
	75 ~ 100	2,450	: $\frac{b}{30.0i} \leq t$
		2,200,000 $(\frac{ti}{b})^2$	: $\frac{b}{80i} \leq t < \frac{b}{30.0i}$

● 자유 돌출판의 국부좌굴에 대한 허용응력

강종	강재 판 두께(mm)	국부좌굴에 대한 허용응력(kgf/cm²)	
SS400 SM400 SMA41	40 이하	1,400 $2{,}400{,}000\ (\frac{t}{b})^2$	: $\frac{b}{13.1} \le t$: $\frac{b}{16} \le t < \frac{b}{13.1}$
	40 초과 100 이하	1,300 $2{,}400{,}000\ (\frac{t}{b})^2$	: $\frac{b}{13.6} \le t$: $\frac{b}{16} \le t < \frac{b}{13.6}$
SM490	40 이하	1,900 $2{,}400{,}000\ (\frac{t}{b})^2$	: $\frac{b}{11.2} \le t$: $\frac{b}{16} \le t < \frac{b}{11.2}$
	40 초과 100 이하	1,750 $2{,}400{,}000\ (\frac{t}{b})^2$	: $\frac{b}{11.7} \le t$: $\frac{b}{16} \le t < \frac{b}{11.7}$
SM490Y SM520 SMA50	40 이하	2,100 $2{,}400{,}000\ (\frac{t}{b})^2$	: $\frac{b}{10.7} \le t$: $\frac{b}{16} \le t < \frac{b}{10.7}$
	40 초과 75 이하	2,000 $2{,}400{,}000\ (\frac{t}{b})^2$	: $\frac{b}{11.0} \le t$: $\frac{b}{16} \le t < \frac{b}{11.0}$
	75 초과 100 이하	1,950 $2{,}400{,}000\ (\frac{t}{b})^2$	: $\frac{b}{11.1} \le t$: $\frac{b}{16} \le t < \frac{b}{11.1}$
SM570 SMA58	40 이하	2,600 $2{,}400{,}000\ (\frac{t}{b})^2$	: $\frac{b}{9.6} \le t$: $\frac{b}{16} \le t < \frac{b}{9.6}$
	40 초과 75 이하	2,500 $2{,}400{,}000\ (\frac{t}{b})^2$	: $\frac{b}{9.8} \le t$: $\frac{b}{16} \le t < \frac{b}{9.8}$
	75 초과 100 이하	2,450 $2{,}400{,}000\ (\frac{t}{b})^2$	: $\frac{b}{9.9} \le t$: $\frac{b}{16} \le t < \frac{b}{9.9}$

• Allowable stress for the local buckling of free overhanging plate

Steel type	Steel plate thickness(mm)	Allowable stress for the local bucking(kgf/cm^2)	
SS400 SM400 SMA41	Below 40	1,400	: $\frac{b}{13.1} \leq t$
		$2{,}400{,}000\ (\frac{t}{b})^2$	: $\frac{b}{16} \leq t < \frac{b}{13.1}$
	40 ~ 100	1,300	: $\frac{b}{13.6} \leq t$
		$2{,}400{,}000\ (\frac{t}{b})^2$	: $\frac{b}{16} \leq t < \frac{b}{13.6}$
SM490	Below 40	1,900	: $\frac{b}{11.2} \leq t$
		$2{,}400{,}000\ (\frac{t}{b})^2$	: $\frac{b}{16} \leq t < \frac{b}{11.2}$
	40 ~ 100	1,750	: $\frac{b}{11.7} \leq t$
		$2{,}400{,}000\ (\frac{t}{b})^2$	: $\frac{b}{16} \leq t < \frac{b}{11.7}$
SM490Y SM520 SMA50	Below 40	2,100	: $\frac{b}{10.7} \leq t$
		$2{,}400{,}000\ (\frac{t}{b})^2$	: $\frac{b}{16} \leq t < \frac{b}{10.7}$
	40 ~ 75	2,000	: $\frac{b}{11.0} \leq t$
		$2{,}400{,}000\ (\frac{t}{b})^2$	: $\frac{b}{16} \leq t < \frac{b}{11.0}$
	75 ~ 100	1,950	: $\frac{b}{11.1} \leq t$
		$2{,}400{,}000\ (\frac{t}{b})^2$	: $\frac{b}{16} \leq t < \frac{b}{11.1}$
SM570 SMA58	Below 40	2,600	: $\frac{b}{9.6} \leq t$
		$2{,}400{,}000\ (\frac{t}{b})^2$	: $\frac{b}{16} \leq t < \frac{b}{9.6}$
	40 ~ 75	2,500	: $\frac{b}{9.8} \leq t$
		$2{,}400{,}000\ (\frac{t}{b})^2$	: $\frac{b}{16} \leq t < \frac{b}{9.8}$
	75 ~ 100	2,450	: $\frac{b}{9.9} \leq t$
		$2{,}400{,}000\ (\frac{t}{b})^2$	: $\frac{b}{16} \leq t < \frac{b}{9.9}$

• 보강된 판의 국부좌굴에 대한 허용응력(kgf/cm²)

강종 / 강재 판 두께 (mm)	SS400 SM400 SMA41	SM490	SM490Y SM520 SMA50	SM570 SMA58
40 이하	1,400 : $\frac{b}{28in} \le t$ 1,400−25($\frac{b}{tin}$ −28) : $\frac{b}{56in} \le t < \frac{b}{28in}$ 2,200,000 ($\frac{tin}{b}$)^2 : $\frac{b}{80in} \le t < \frac{b}{56in}$	1,900 : $\frac{b}{24in} \le t$ 1,900−39($\frac{b}{tin}$ −24) : $\frac{b}{48in} \le t < \frac{b}{24in}$ 2,200,000 ($\frac{tin}{b}$)^2 : $\frac{b}{80in} \le t < \frac{b}{48in}$	2,100 : $\frac{b}{22in} \le t$ 2,100−45($\frac{b}{tin}$ −22) : $\frac{b}{46in} \le t < \frac{b}{22in}$ 2,200,000 ($\frac{tin}{b}$)^2 : $\frac{b}{80in} \le t < \frac{b}{46in}$	2,600 : $\frac{b}{22in} \le t$ 2,600−67($\frac{b}{tin}$ −22) : $\frac{b}{40in} \le t < \frac{b}{22in}$ 2,200,000 ($\frac{tin}{b}$)^2 : $\frac{b}{80in} \le t < \frac{b}{40in}$
45 초과 75 이하	1,300 : $\frac{b}{28in} \le t$ 1,300−22($\frac{b}{tin}$ −28) : $\frac{b}{58in} \le t < \frac{b}{28in}$	1,750 : $\frac{b}{24in} \le t$ 1,750−33($\frac{b}{tin}$ −24) : $\frac{b}{50in} \le t < \frac{b}{24in}$	2,000 : $\frac{b}{22in} \le t$ 2,000−41($\frac{b}{tin}$ −22) : $\frac{b}{46in} \le t < \frac{b}{22in}$ 2,200,000 ($\frac{tin}{b}$)^2 : $\frac{b}{80in} \le t < \frac{b}{46in}$	2,500 : $\frac{b}{22in} \le t$ 2,500−63($\frac{b}{tin}$ −22) : $\frac{b}{40in} \le t < \frac{b}{22in}$ 2,200,000 ($\frac{tin}{b}$)^2 : $\frac{b}{80in} \le t < \frac{b}{40in}$
75 초과 100 이하	2,200,000 ($\frac{tin}{b}$)^2 : $\frac{b}{80in} \le t < \frac{b}{58in}$	2,200,000 ($\frac{tin}{b}$)^2 : $\frac{b}{80in} \le t < \frac{b}{50in}$	1,950 : $\frac{b}{22in} \le t$ 1,950−38($\frac{b}{tin}$ −22) : $\frac{b}{48in} \le t < \frac{b}{22in}$ 2,200,000 ($\frac{tin}{b}$)^2 : $\frac{b}{80in} \le t < \frac{b}{48in}$	2,450 : $\frac{b}{22in} \le t$ 2,450−60($\frac{b}{tin}$ −22) : $\frac{b}{42in} \le t < \frac{b}{22in}$ 2,200,000 ($\frac{tin}{b}$)^2 : $\frac{b}{80in} \le t < \frac{b}{42in}$

• Allowable stress for the local buckling of reinforced plate (kgf/cm^2)

<table>
<tr><th>Steel type / steel plate thickness (mm)</th><th>SS400
SM400
SMA41</th><th>SM490</th><th>SM490Y
SM520
SMA50</th><th>SM570
SMA58</th></tr>
<tr><td>Below 40</td><td>1,400 : $\frac{b}{28in} \le t$
1,400-25($\frac{b}{tin}$ -28)
: $\frac{b}{56in} \le t < \frac{b}{28in}$
2,200,000 $(\frac{tin}{b})^2$
: $\frac{b}{80in} \le t < \frac{b}{56in}$</td><td>1,900 : $\frac{b}{24in} \le t$
1,900-39($\frac{b}{tin}$ -24)
: $\frac{b}{48in} \le t < \frac{b}{24in}$
2,200,000 $(\frac{tin}{b})^2$
: $\frac{b}{80in} \le t < \frac{b}{48in}$</td><td>2,100 : $\frac{b}{22in} \le t$
2,100-45($\frac{b}{tin}$ -22)
: $\frac{b}{46in} \le t < \frac{b}{22in}$
2,200,000 $(\frac{tin}{b})^2$
: $\frac{b}{80in} \le t < \frac{b}{46in}$</td><td>2,600 : $\frac{b}{22in} \le t$
2,600-67($\frac{b}{tin}$ -22)
: $\frac{b}{40in} \le t < \frac{b}{22in}$
2,200,000 $(\frac{tin}{b})^2$
: $\frac{b}{80in} \le t < \frac{b}{40in}$</td></tr>
<tr><td>45 ~75</td><td rowspan="2">1,300 : $\frac{b}{28in} \le t$
1,300-22($\frac{b}{tin}$ -28)
: $\frac{b}{58in} \le t < \frac{b}{28in}$
2,200,000 $(\frac{tin}{b})^2$
: $\frac{b}{80in} \le t < \frac{b}{58in}$</td><td rowspan="2">1,750 : $\frac{b}{24in} \le t$
1,750-33($\frac{b}{tin}$ -24)
: $\frac{b}{50in} \le t < \frac{b}{24in}$
2,200,000 $(\frac{tin}{b})^2$
: $\frac{b}{80in} \le t < \frac{b}{50in}$</td><td>2,000 : $\frac{b}{22in} \le t$
2,000-41($\frac{b}{tin}$ -22)
: $\frac{b}{46in} \le t < \frac{b}{22in}$
2,200,000 $(\frac{tin}{b})^2$
: $\frac{b}{80in} \le t < \frac{b}{46in}$</td><td>2,500 : $\frac{b}{22in} \le t$
2,500-63($\frac{b}{tin}$ -22)
: $\frac{b}{40in} \le t < \frac{b}{22in}$
2,200,000 $(\frac{tin}{b})^2$
: $\frac{b}{80in} \le t < \frac{b}{40in}$</td></tr>
<tr><td>75 ~100</td><td>1,950 : $\frac{b}{22in} \le t$
1,950-38($\frac{b}{tin}$ -22)
: $\frac{b}{48in} \le t < \frac{b}{22in}$
2,200,000 $(\frac{tin}{b})^2$
: $\frac{b}{80in} \le t < \frac{b}{48in}$</td><td>2,450 : $\frac{b}{22in} \le t$
2,450-60($\frac{b}{tin}$ -22)
: $\frac{b}{42in} \le t < \frac{b}{22in}$
2,200,000 $(\frac{tin}{b})^2$
: $\frac{b}{80in} \le t < \frac{b}{42in}$</td></tr>
</table>

3-2-6 하중조합(강도설계법)

① 강도설계법의 주요 하중조합

하중 \ case	고정하중	활하중(충격포함)	원심하중	토압	풍하중	차량에 작용하는 풍하중	제동하중	지진하중	충돌하중	부등침하 크리프 건조수축 온도	부력 양압력 수압 파압
1	1.3	2.15	1.3	1.7	–	–	–	–	–	–	1.3
2	1.3	–	–	1.7	1.3	–	–	–	–	–	1.3
3	1.3	1.3	1.3	1.7	0.65	1.3	1.3	–		–	1.3
4	1.3	1.3	1.3	1.7	–	–	–	–	–	1.3	1.3
5	1.25	–	–	1.65	1.25	–	–	–	–	1.25	1.25
6	1.25	1.25	1.25	1.65	0.625	1.25	1.25	–	–	1.25	1.25
7	1.0	–	–	1.0	–	–	–	1.0	–	–	1.0
8	1.3	1.3	1.3	1.7	–	–	–	–	1.3	–	1.3
9	1.2	–	–	1.55	1.2	–	–	–	1.2	–	1.2

• 주요 하중조합에서 고정하중계수와 토압하중계수를 바꾸어 설계하는 경우

Case	고정하중계수	토압하중계수	
	기둥설계 시 최소 축하중 및 최대 모멘트 또는 최대 편심에 대하여 설계할 경우	라멘구조에 횡토압에 의한 정모멘트를 검토하는 경우	연직토압의 경우
①~④ ⑦~⑧	1.00	0.65	1.50
⑤~⑥	0.95	0.65	1.45
⑨	0.90	0.80	1.40

② 강도감소계수

부재		강도감소계수
휨, 휨 + 축방향 인장	보통 철근 콘크리트	0.85
	프리스트레스트 콘크리트(현장치기)	0.85
축 방 향 인 장		0.85
축방향 압축 휨 + 축방향 압축	나선철근으로 보강된 철근콘크리트	0.75
	그 외의 철근 콘크리트	0.70
전 단 및 비 틀 림		0.80
콘 크 리 트 지 압		0.70

3-2-6 Load Combination (Strength Design Method)

① Major load combination of strength design method

Load / Case	Dead load	Live load (include impact load)	Centrifugal load	Soil pressure	Wind load	Wind load applied to vehicles	braking load	Earth quake load	Collision load	Differential settlement creep Drying shrinkage Temperature	buoyancy uplift water pressure wave pressure
1	1.3	2.15	1.3	1.7	–	–	–	–	–	–	1.3
2	1.3	–	–	1.7	1.3	–	–	–	–	–	1.3
3	1.3	1.3	1.3	1.7	0.65	1.3	1.3	–		–	1.3
4	1.3	1.3	1.3	1.7	–	–	–	–	–	1.3	1.3
5	1.25	–	–	1.65	1.25	–	–	–	–	1.25	1.25
6	1.25	1.25	1.25	1.65	0.625	1.25	1.25	–	–	1.25	1.25
7	1.0	–	–	1.0	–	–	–	1.0	–	–	1.0
8	1.3	1.3	1.3	1.7	–	–	–	–	1.3	–	1.3
9	1.2	–	–	1.55	1.2	–	–	–	1.2	–	1.2

- When design by exchanging dead load coefficient and soil pressure coefficients from major load combination

Case	Dead load coefficient	Soil pressure load coefficient	
	When designing pile for minimum axis load and maximum moment, or maximum eccentric	When positive moment is evaluated for traversing soil pressure of rahmen structure	Vertical soil pressure
①~④ ⑦~⑧	1.00	0.65	1.50
⑤~⑥	0.95	0.65	1.45
⑨	0.90	0.80	1.40

② Strength reduction coefficient

Member		Strength reduction coefficient
bending, bending+axi direction tensile	Normal reinforced concrete	0.85
	Prestressed concrete(concreting in situ)	0.85
Axis direction tensile		0.85
axis direction compression bending+axis direction compression	reinforced concrete reinforced with spiral reinforcement	0.75
	Other type of reinforced concrete	0.70
Shear and torsion		0.80
Concrete bearing		0.70

3-2-7 하중조합(허용응력 설계법)

① 적용대상 구조물 : PSC 구조물, 강구조물, RC 구조물의 사용성 및 피로검토

② 설계하중 조합

• 강교에서의 하중조합 및 증가계수

	하 중 조 합	증가계수
1	주하중 + 주하중에 해당하는 특수하중 + 온도변화의 영향	1.15
2	주하중 + 주하중에 해당하는 특수하중 + 풍하중	1.25
3	주하중 + 주하중에 해당하는 특수하중 + 온도변화의 영향 + 풍하중	1.35
4	주하중 + 주하중에 해당하는 특수하중 + 풍하중 + 제동하중	1.25
5	주하중 + 주하중에 해당하는 특수하중 + 충돌하중	1.70
6	풍하중만 고려할 때	1.20
7	제동하중만 고려할 때	1.20
8	활하중 및 충격 이외의 주하중 + 지진의 영향	1.50
9	가설 시 하중	1.25

• 콘크리트교에서의 하중조합 및 증가계수

	하 중 조 합	증가계수
1	주하중 + 주하중에 해당하는 특수하중 + 온도변화의 영향	1.15
2	주하중 + 주하중에 해당하는 특수하중 + 풍하중	1.25
3	주하중 + 주하중에 해당하는 특수하중 + 온도변화의 영향 + 풍하중	1.35
4	주하중 + 주하중에 해당하는 특수하중 + 풍하중 + 제동하중	1.25
5	주하중 + 주하중에 해당하는 특수하중 + 충돌하중	1.50
6	풍하중만 고려할 때	1.20
7	활하중 및 충격 이외의 주하중 + 지진의 영향	1.33
8	가설 시 하중	1.25

3-2-7 Load Combination (Allowable Stress Design Method)

① Target structures : serviceability and fatigue evaluation of PSC structures, steel structures, RC structures

② Design load combination

• Load combination and increase factor of steel bridge

	Load combination	increase coefficient
1	Main load + Special load corresponding to main load + Effects of temperature change	1.15
2	Main load + Special load corresponding to main load + Wind load	1.25
3	Main load + Special load corresponding to main load + Effects of temperature change + Wind load	1.35
4	Main load + Special load corresponding to main load + Wind load + Braking load	1.25
5	Main load + Special load corresponding to main load + Collision load	1.70
6	When only wind load is considered	1.20
7	When only braking load is considered	1.20
8	Main load besides active load and impact + Effects of Earthquake	1.50
9	Load during erection	1.25

• Load combination and increase factor for concrete bridge

	load combination	increase coefficient
1	Main load + Special load corresponding to main load + Effects of temperature change	1.15
2	Main load + Special load corresponding to main load + Wind load	1.25
3	Main load + Special load corresponding to main load + Effects of temperature change + Wind load	1.35
4	Main load + Special load corresponding to main load + Wind load + Braking load	1.25
5	Main load + Special load corresponding to main load + Collision load	1.50
6	When only wind load is considered	1.20
7	Main load besides active load and impact + Effects of Earthquake	1.33
8	Load during erection	1.25

3-2-8 철근콘크리트 구조 세목

① 최소철근비

항 목	내 용
휨 부 재	$\rho_{min} > 14 / f_y$, $\frac{0.80\sqrt{f_{ck}}}{f_y}$ 중 큰 값 이상
압축부재 (기 둥)	기둥부재의 축 방향 철근은 다음 표에 의함을 원칙으로 한다. 기둥부재의 축 방향 철근 <table><tr><td>항 목</td><td>최 대</td><td>최 소</td></tr><tr><td>철 근 량(cm²)</td><td>0.08A</td><td>0.01A</td></tr><tr><td>철근지름(mm)</td><td></td><td>16</td></tr></table>여기서, A : 기둥의 전 단면적(cm2)
벽 체 (벽식 기둥 포함)	수직철근 $\rho_{min} > 0.0015$ 수평철근 $\rho_{min} > 0.0025$
슬 래 브	$\rho_{min} > 14 / f_y$
확대기초	$\rho_{min} > 14 / f_y$
옹 벽	$\rho_{min} > 0.0025$
건조 수축 및 온도 철근	$f_y \leq 4000\ kgf/cm^2$, $\rho_{min} > 0.002$ $f_y \leq 4000\ kgf/cm^2$, $\rho_{min} > 0.002 \times 4000/f_y$ 단, 어떤 경우에도 철근비는 0.0014 이상이어야 한다.

② 균형철근비와 최대철근비

구 분	균 형 철 근 비(P_b)	최 대 철 근 비(P_{max})
인장철근을 갖는 직사각형 단면	$\rho_b = \overline{\rho_b} = 0.85\ K_1 \frac{f_{ck}}{f_y}(\frac{6,000}{6,000+f_y})$	$\rho_{max} = 0.75\ \overline{\rho_b}$
인장철근만을 갖는 I형 단면과 T형 단면	$\rho_b = \frac{b_0}{b}(\overline{\rho_b} + \rho_1)$ 여기서, $\rho_1 = \frac{A_{sf}}{b_0 d}$ $A_{st} = 0.85\ \frac{f_{ck}}{f_y}(b - b_0)t$ t : 플랜지의 두께 ρ_b : 균형철근비 $\overline{\rho_b}$: 인장철근만 있을 때의 철근비	$\rho_{max} = 0.75\ \frac{b_0}{b}(\overline{\rho_b} + \rho_1)$
압축철근만을 갖는 직사각형 단면	$\rho_b = \overline{\rho_b} + \rho_1 \frac{f'_s}{f_y}$ 여기서, f's = 균형조건에서의 압축철근의 응력 $f'_s = 6,000(1-\frac{d_1}{d}) \cdot (\frac{6,000+f_y}{6,000+f_y}) \leq f_y) \leq f_y$	$\rho_{max} = 0.75\ \overline{\rho_b} + \rho\frac{f_s}{f_y}$

3-2-8 Reinforced Concrete Structure Details

① Minimum reinforcement ratio

Item	Details
bending member	$\rho_{min} > 14 / f_y , \frac{0.80\sqrt{f_{ck}}}{f_y}$ - bigger value
Compression member (Pillar)	Axis direction reinforcement of pillar member shall be as following table in principle Axis direction reinforcement of pillar member Item / Max / Min reinforcement amount(cm^2): 0.08A / 0.01A reinforcement diameter(mm): – / 16 Where, A : Total section area of pillar (cm^2)
Wall (including wall type pillar)	Vertical reinforcement $\rho_{min} > 0.0015$ Horizontal reinforcement $\rho_{min} > 0.0025$
Slab	$\rho_{min} > 14 / f_y$
Expansion foundation	$\rho_{min} > 14 / f_y$
Retaining wall	$\rho_{min} > 0.0025$
Drying shrinkage and temperature reinforcement	$f_y \le 4000\ kgf/cm^2,\ \rho_{min} > 0.002$ $f_y \le 4000\ kgf/cm^2\ ,\ \rho_{min} > 0.002 \times 4000/f_y$ reinforcement ratio shall always be bigger than 0.0014 in any cases.

② Balanced steel ratio and maximum steel ratio

Classification	Balanced steel ratio(Pb)	Maximum steel ratio(Pmax)
Rectangular section with tensile reinforcement	$\rho_b = \overline{\rho_b} = 0.85\ K_1 \frac{f_{ck}}{f_y}(\frac{6,000}{6,000+f_y})$	$\rho_{max} = 0.75\ \overline{\rho_b}$
I-type and T-type section with only tensile reinforcement	$\rho_b = \frac{b_0}{b}(\overline{\rho_b} + \rho_1)$ Where, $\rho_1 = \frac{A_{sf}}{b_0 d}$ $A_{st} = 0.85 \frac{f_{ck}}{f_y}(b - b_0)t$ t : Thickness of flange ρ_b : Balanced steel ratio $\overline{\rho_b}$: steel ratio with only tensile reinforcement	$\rho_{max} = 0.75 \frac{b_0}{b}(\overline{\rho_b} + \rho_1)$
Rectangular section with only compression reinforcement	$\rho_b = \overline{\rho_b} + \rho_1 \frac{f'_s}{f_y}$ Where, f's = Stress of compression reinforcement at balanced condition $f'_s = 6,000(1-\frac{d_1}{d}) \cdot (\frac{6,000+f_y}{6,000+f_y}) \le f_y) \le f_y$	$\rho_{max} = 0.75\ \overline{\rho_b} + \rho\frac{f_s}{f_y}$

③ 철근의 정착길이 및 이음 길이

구 분 \ 직경, 강도		D16			D19			D22			D25			D29		
		240	270	400	240	270	400	240	270	400	240	270	400	240	270	400
정착 길이	일반철근	300	360	300	340	430	350	490	620	510	570	720	590	730	910	750
	상부철근	370	460	380	440	560	460	640	810	660	750	940	770	940	1180	970
이음 길이	일반철근	390	470	390	450	560	460	640	810	670	750	940	770	950	1190	980
	상부철근	490	600	500	580	730	600	840	1060	860	980	1230	1010	1230	1540	1270
설 계 기 준		• 도로교 설계기준 247쪽 식(4.3.8)을 적용하였으며 철근의 중심간격 125mm, 철근의 중심 으로 부터 콘크리트 표면까지의 거리로 나타낸 피복두께는 80mm로 계산된 치수임. • 철근의 중심간격 및 피복두께의 변화에 따른 이음길이의 변화는 도면설계기준 각 구조물별 설계기준에 명기하였음.														

④ 철근 구부리기 최소 반지름 및 연장길이

철 근 직 경		D13	D16	D19	D22	D25	D29	D32
철근 중심의 최소 구부림 반경 ([ ]는 스터럽)		50[40]	60[40]	70	80	90	130	150
라멘구조의 우각부 구부림 반경(10db)		140	170	210	240	270	310	340
주철근 연장길이	90°(4db)	160	200	230	270	310	350	390
	180°(12db)	60	70	80	90	110	120	130
스터럽 또는 띠철근의 연장길이	90°(6db or 12db)	80	100	230	270	310	-	-
	135°(6db)	80	100	120	140	160	-	-

※ 도로교 설계기준(건설교통부, 2003) 4.3.4적용(465쪽)

⑤ 철근 피복두께(철근의 중심으로부터 콘크리트 표면까지의 거리)

구 분			시방 피복(mm)	적용 피복(mm)	비 고
PSC Girder & 바닥판	상 면(바닥판) 및 복부		50	60	
	하면		25	50	
	사재		50	60	
교 대	구 체		지중 : 80	100	무늬거푸집 사용 시도 동일
	날 개 벽	노 출 면	30	50(70)	( )는 무늬거푸집 사용 시
		지 중 면	60	80	
교 각	구 체	육 상 부	80	100	
기 초 (Footing)	직 접 기 초		80	100	

※ 도로교 설계기준(건설교통부, 2003) 4.3.3적용(463쪽)

③ Anchorage length and joint length of connection reinforcement

Classification \ Diameter & strength		D16			D19			D22			D25			D29		
		240	270	400	240	270	400	240	270	400	240	270	400	240	270	400
Anchorage length	Normal reinforcement	300	360	300	340	430	350	490	620	510	570	720	590	730	910	750
	Upper reinforcement	370	460	380	440	560	460	640	810	660	750	940	770	940	1180	970
Joint length	Normal reinforcement	390	470	390	450	560	460	640	810	670	750	940	770	950	1190	980
	Upper reinforcement	490	600	500	580	730	600	840	1060	860	980	1230	1010	1230	1540	1270
Design standard		• Highway bridge design standard page 247, equation (4.3.8) was applied, where center distance of reinforcement is 125mm, cover thickness shown as the distance from the center of reinforcement to concrete surface is 80mm. • The changes of joint length by the changes of center distance of reinforcement and cover thickness is specified at the design standard of each structure in reference to drawing design.														

④ Reinforcement bending minimum diameter and expansion length

Reinforcement diameter		D13	D16	D19	D22	D25	D29	D32
Minimum bending diameter of the center of steel reinforcement ([ ] is stir-up)		50[40]	60[40]	70	80	90	130	150
Coner of opening bending diameter of rigid frame structure(10db)		140	170	210	240	270	310	340
Main reinforcement expansion length	90°(4db)	160	200	230	270	310	350	390
	180°(12db)	60	70	80	90	110	120	130
Expansion length of stir-up	90°(6db or 12db)	80	100	230	270	310	-	-
	135°(6db)	80	100	120	140	160	-	-

※ Highway bridge design standard (MTLM, 2003) 4.3.4 is applied (Page 465)

⑤ Reinforcement cover thickness(distance from the center of reinforcement to the surface of concrete)

Classification			Specified cover (mm)	Applied cover (mm)	Remark
PSC Girder & floor plate	Upper surface(floor slab) and web		50	60	
	Lower surface		25	50	
	diagonal member		50	60	
Abutment	Body		Underground : 80	100	Same when using pattern mold
	Wing wall	Exposed surface	30	50(70)	( ) is when using pattern mold
		Underground surface	60	80	
Pier	Body	Land part	80	100	
Foundation (Footing)	Spread foundation		80	100	

※ Highway bridge design standard (MTLM, 2003) 4.3.3 is applied (Page 463)

3-2-9 사용성 검토

- 강도설계법에 의한 하중계수와 강도에 대하여 설계된 휨 부재는 사용하중 하에서 피로에 대한 규정 및 균열제어, 처짐 등에 대한 사용성 및 내구성을 확보

- 관련기준 : 콘크리트 구조설계기준, 도로교 설계기준

① 균열 검토

계 산 균 열 폭	강재 부식의 환경조건에 따른 허용 균열폭 : Wa(mm)				
	강재의 종 류	건조한 환 경	일반(습윤) 환 경	부식성 환 경	고부식성 환 경
$W = 1.08R f_s \sqrt{d_c A} \times 10^{-5}$ (mm) 여기서, R = 1.2 (보) , 1.35 (슬래브)	철 근	0.006C	0.005C	0.004C	0.0035C
	PS 강재	0.005C	0.004C	-	-

② 활하중에 의한 처짐의 제한

교량의 형식 및 부재		단순지지 및 연속거더	캔틸레버부
철근 콘크리트 슬래브가 있는 강교	10<L≤40m	L^2 / 20,000	L^2 / 12,000
	L > 40m	L / 500	L / 300
기타의 슬래브가 있는 강교		L / 500	L / 300
콘크리트교(자동차 전용도로)		L / 800	L / 300

③ 피로검토

- 콘크리트 구조 : 충격을 포함한 사용 활하중에 의한 철근의 응력범위 및 프리스트레싱 긴장재의 인장응력 변동 범위 해석

- 피로를 고려하지 않아도 되는 철근과 프리스트레싱 긴장재의 응력 범위

철근의 종류	철근의 인장 및 압축응력 범위	PS 긴장재의 위치	PS 긴장재의 인장응력 변동 범위
이형철근 SD30	1,300 kgf/cm^2	연결부 또는 정착부	1,400 kgf/cm^2
이형철근 SD40	1,500 kgf/cm^2	기타 부위	1,600 kgf/cm^2

- 주철근의 방향이 차량진행방향에 직각인 강교의 콘크리트 바닥판에 대해서는 피로응력의 한계를 고려하지 않음.

3-2-9 Serviceability Review

- For bending member designed for strength and load factor by strength design method, serviceability and durability have to be secured for the specification on fatigue, crack, and deflection under the service load.

- Related standard: Concrete structure design standard, Highway bridge design standard

① Crack evaluation

Calculated crack width	Allowable crack with by the steel corrosion environment condition: Wa(mm)				
	Steel type	Dry environment	General (moist) environment	Corrosion environment	Highly corrosive environment
$W = 1.08R f_s \sqrt{d_c A} \times 10^{-5}$ (mm) Where, R = 1.2 (Bream), 1.35 (Slab)	Steel reinforcement	0.006C	0.005C	0.004C	0.0035C
	PS Steel material	0.005C	0.004C	-	-

② Restriction of deflection by live load

Type and member of bridge		Simple bearing and continuous girder	Cantilever part
Steel bridge with reinforcement concrete slab	$10<L\leq40m$	L^2 / 20,000	L^2 / 12,000
	L > 40m	L / 500	L / 300
Steel bridge with other slab		L / 500	L / 300
Concrete bridge (auto-only road)		L / 800	L / 300

③ Fatigue evaluation

- Concrete structure : Analysis of stress range of reinforcement by service live load including impact and tensile stress variation of pre-stressing tendon.

- Stress range of reinforcement, that fatigue doesn't have to be considered and pre-stressing tendon.

Steel reinforcement type	Tensile and compression stress range of steel reinforcement	Location of PS tendon	Tensile stress variation range of PS tendon
Deformed reinforcement SD30	1,300 kgf/cm^2	connecting part or Anchor part	1,400 kgf/cm^2
Deformed reinforcement SD40	1,500 kgf/cm^2	Other part	1,600 kgf/cm^2

- The limit fatigue stress is not considered for the concrete floor slab of steel bridge where the direction of the cast reinforcement is perpendicular to the direction of vehicle movement.

3-2-10 내진설계

① 목　적

- 지진에 의하여 교량이 입는 피해의 정도를 최소화시킬 수 있는 내진성 확보를 위하여 필요한 최소설계 요구 조건을 규정하는데 있음

- 인명피해를 최소화
- 지진 시 교량 부재들의 부분적인 피해는 허용하나 전체적으로 붕괴는 방지
- 지진 시 가능한 한 교량의 기본 기능은 발휘
- 교량의 정상수명 기간 내에 설계 지진력이 발생할 가능성은 희박
- 설계기준은 남한 전역에 적용
- 본 규정을 따르지 않더라도 창의력을 발휘하여 보다 발전된 설계를 할 경우에는 이를 인정

② 내진설계 일반사항

- 설계지반운동은 부지 정지작업이 완료된 지표면에서의 자유장 운동으로 정의한다.
- 국지적인 토질조건, 지질조건과 지표 및 지하 지형이 지반운동에 미치는 영향이 고려되어야 한다.
- 설계지반운동은 흔들림의 세기, 주파수 내용 및 지속시간의 세 가지 측면에서 그 특성이 잘 정의되어야 한다.
- 모든 점에서 똑같이 가지는 것이 합리적일 수 없는 특징을 갖는 교량 건설부지에 대해서는 지반운동의 공간적 변화모델을 사용하여야 한다.

③ 내진설계 흐름도

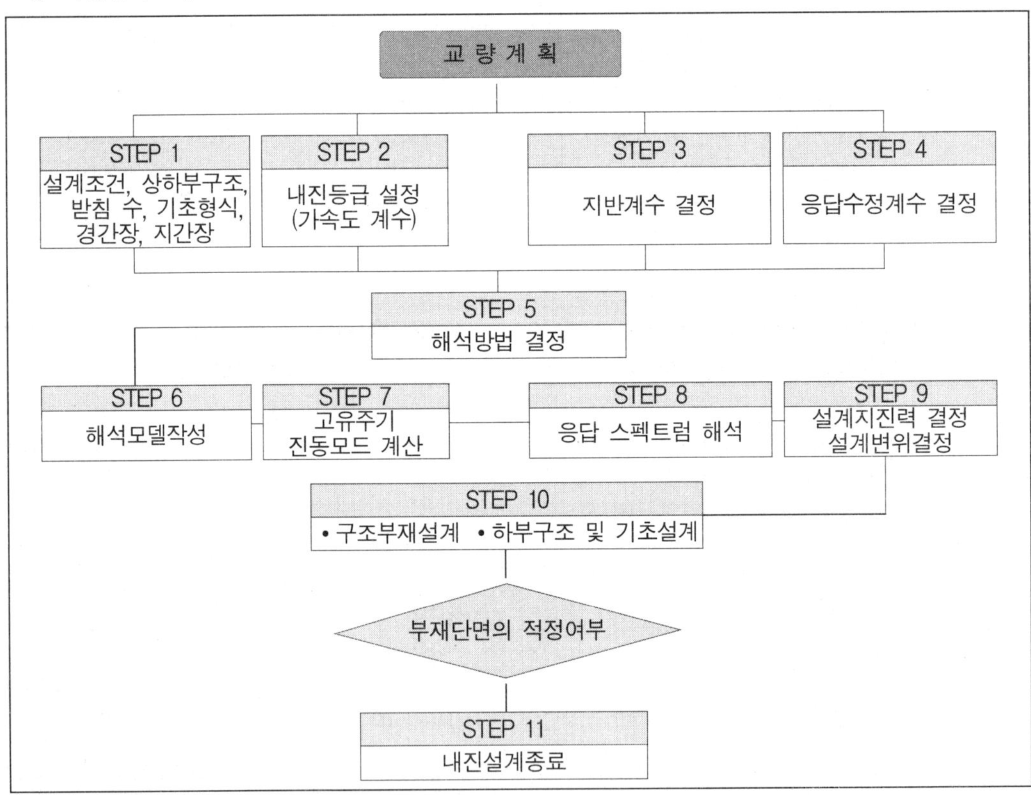

3-2-10 Seismic Design

① Purpose

- The purpose is to specify the minimum design requirement to secure the earthquake-proof function to minimize the damages to the bridge by earthquake.

- To minimize the damages to human life
- Partial damages of member are allowed during the earthquake, but the total collapse must be prevented
- Basic function of bridge should work as much as possible during earthquake
- Possibility of observing the design earthquake load in the lifetime of the bridge is rare
- Design standard shall be applied to entire area of South Korea
- Better design using creativeness might be accepted even though that design doesn't follow the above Standard

② Seismic design general items

- Design ground activity is defined as free field motion on the ground surface when the ground preparation for construction is completed.
- Local soil condition, geological features conditions, and effects of surface and topography to ground motion shall be considered.
- The characteristics of design ground motion shall be well defined in three perspectives - intensity, frequency contents, and duration.
- Considering the limitations of the bridge construction site that not all conditions are equally rational, spatial change model of the ground motion shall be utilized.

③ Seismic design flow chart

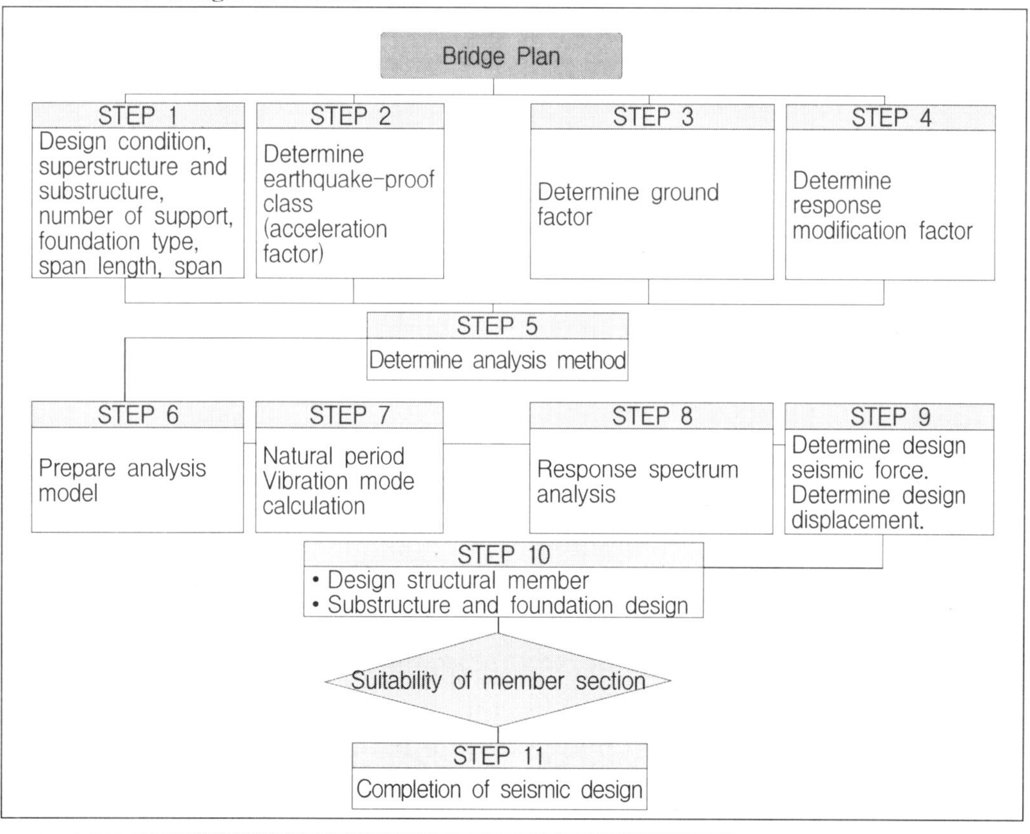

④ 가속도 계수

• 지진구역 구분

지진구역	행 정 구 역	
I	시	서울특별시, 인천광역시, 대전광역시, 부산광역시, 대구광역시, 울산광역시, 광주광역시
	도	경기도, 강원도 남부(1), 충청북도, 충청남도, 경상북도, 경상남도, 전라북도, 전라남도 북동부(2)
II	도	강원도 북부(3), 전라남도 남서부(4), 제주도

주 : (1) 강원도 남부(군, 시) : 영월, 정선, 삼척시, 강릉시, 동해시, 원주시, 태백시

(2) 전라남도 북동부(군, 시) : 장성, 담양, 곡성, 구례, 장흥, 보성, 여천, 화순, 광양시, 나주시, 여천시, 여수시, 순천시

(3) 강원도 북부(군, 시) : 홍천, 철원, 화천, 횡성, 평창, 양구, 인제, 고성, 양양, 춘천시, 속초시

(4) 전라남도 남서부(군, 시) : 무안, 신안, 완도, 영광, 진도, 해남, 영암, 강진, 고흥, 함평, 목포시

(5) 행정구역의 경계를 통과하는 교량의 경우에는 구역계수가 큰 값을 적용

• 지진구역계수 및 위험도 계수

- 지진구역계수 (재현주기 500년에 해당)

지진구역	I	II
구역계수	0.11	0.07

- 위험도 계수

재현주기(년)	500	1000
위험도계수 I	1.0	1.4

⑤ 내진등급과 설계지진수준

내진등급	교 량	설계지진의 평균재현주기
내진 I 등급교	• 고속도로, 자동차전용도로, 특별시도, 광역시도 또는 일반국도상의 교량 • 지방도, 시도 및 군도 중 지역의 방재계획상 필요한 도로에 건설된 교량, 해당도로의 일일계획교통량을 기준으로 판단했을 때 중요한 교량 • 내진 I 등급교가 건설되는 도로 위를 통과하는 고가 교량	1000년
내진 II등급교	• 내진 I 등급교에 속하지 않는 교량	500년

④ Acceleration coefficient

• Earthquake region classification

Earthquake region		Administrative district
I	Si	Seoul, Incheon, Daejeon, Busan, Daegu, Ulsan, Gwangju
	Do	Gyeonggi-do, Southern Gangwon-do(1), Chungcheongbuk-do, Chungcheongnam-do, Gyeongsangbuk-do, Gyeongsangnam-do, Jeonrabuk-do, North-easter Jeonranam-do(2)
II	Do	Northen Gangwon-do(3), South-west Jeonranam-do(4), Jeju-do

Note (1) Southern Gangwon-do (Gun, Si) : Yeongwol, Jeongseon, Samchuk-si, Gangreung-si, Donghae-si, Wonju-si, Taebak-si
(2) North-eastern Jeonranam-do (Gun, Si) : Jangseong, Damyan, Goksung, Gurye, Jangheung, Bosung, Yoecheon, Hwasoon, Kwangyang-si, Naju-si, Yeochun-si, Suncheon-si
(3) Northern Gangwon-do (Gun, Si) : Hongcheon, Cheorwon, Hwacheon, Hengseong, Pyeongchang, Yangu, Inje, Yangyang, Chuncheon-si, Sokcho-si
(4) South-Western (Gun, Si) : Mooan, Sinahn, Wando, Yeonggwang, Jindo, Haenam, Youngam, Gangjin, Goheung, Hampyeong, Mokpo-si
(5) In case the bridge lies between two districts, apply the bigger region coefficients.

• Earthquake region coefficient and risk factor

- Earthquake region coefficient(corresponding to recurrence interval of 500 years)

Earthquake region	I	II
Region coefficient	0.11	0.07

- Risk level coefficient

Repetition cycle (year)	500	1000
Risk level coefficient I	1.0	1.4

⑤ Earthquake-proof class and design seismic force

Earthquake-pro of class	Bridge	Average recurrence interval of Design Earthquake
Earthquake-pro of class I bridge	• Bridges of expressway, free way, Seoul area road, Metropolitan road or general national roads. • Bridges constructed on the roads that require for the disaster-proof plan among the municipal and local roads, and bridges that are considered to be important when judging by the daily planed traffic volume. • Elevated bridge crossing the road that have earthquake-proof class I bridge	1000 years
Earthquake-pro of class II bridge	• Bridges except earthquake-proof class I bridge	500 years

⑤ 지반의 분류

- 지반계수

지반계수	지반종류			
	I	II	III	IV
S	1.0	1.2	1.5	2.0

- 지반의 분류

지반종류	지반종류의 호칭	지표면 아래 30m 토층에 대한 평균값		
		전단파속도(m/s)	표준관입시험(N치[(1)])	비배수전단강도(kPa)
I	경암지반 보통암지반	760 이상	-	-
II	매우 조밀한 토사지반 또는 연암지반	360에서 760	> 50	> 100
III	단단한 토사지반	180에서 360	15에서 50	50에서 100
IV	연약한 토사지반	180 미만	< 15	< 50
V	부지 고유의 특성평가가 요구되는 지반			

주 : (1) 비점착성 토층만을 고려한 평균 N치

⑥ 응답수정계수 및 지진력 조합

- 응답수정계수

적 용 구 조			응답수정계수(R)
하부구조	벽식 교각		2
	철근 콘크리트 말뚝가구(Bent)	1. 수직말뚝만 사용한 경우	3
		2. 한 개 이상의 경사말뚝을 사용한 경우	2
	단주		3
	강재 또는 합성강재와 콘크리트 말뚝가구	1. 수직말뚝만 사용한 경우	5
		2. 한 개 이상의 경사말뚝을 사용한 경우	3
	다주가구		5
연결부분	상부구조와 교대		0.8
	상부구조의 한 지간 내의 신축이음		0.8
	기둥교각 또는 말뚝가구와 캡 빔(Cap Beam) 또는 상부구조		1.0
	기둥 또는 교각과 기초		1.0

⑤ Classification of ground

- Modulus of foundation

Modulus of foundation	Ground type			
	I	II	III	IV
S	1.0	1.2	1.5	2.0

- Classification of ground

Ground type	Name of ground type	Average value of soil layer 30 meter under the surface		
		Shear wave velocity (m/s)	Standard penetration test(N value[(1)])	Undrained shear strength(kPa)
I	Hard rock ground Normal rock ground	Over 760	-	-
II	Very dense soil ground or soft rock ground	360 ~ 760	> 50	> 100
III	Hard soil ground	180 ~ 360	15 ~ 50	50 ~ 100
IV	Soft soil ground	Less than 180	< 15	< 50
V	Ground that requires site specific characteristics evaluation			

Note : (1) Average N value considering only non cohesive soil layer

⑥ Combination of response modification factor and seismic force

- Response modification factor

Application structure			Response modification factor(R)
Sub-structure	Wall type pier		2
	Reinforced Concrete Pile frame(Bent)	1. Using only vertical piles	3
		2. Using more than one inclined pile	2
	Short Column		3
	Steel or composite material and concrete pile frame	1. Using only vertical piles	5
		2. Using more than one inclined pile	3
	Multi-column frame		5
Joints	Superstructure and abutment		0.8
	Elastic joint within one span of superstructure		0.8
	Pillar pier or pile frame and cap beam, or superstructure		1.0
	Pillar or pier, and foundation		1.0

- 직교 지진력의 조합

부재의 각각의 주축에 대한 설계지진력은 지진력을 다음과 같이 조합하여 사용

- 하중 경우 1 : 종방향 축의 해석으로부터 구한 탄성지진력(절대값)에 횡방향 축의 해석으로부터 구한 탄성지진력(절대값)의 30%를 합한 경우
- 하중 경우 2 : 횡방향 축의 해석으로부터 구한 탄성지진력(절대값)에 종방향 축의 해석으로부터 구한 탄성지진력(절대값)의 30%를 합한 경우

⑦ 설계지진력

- 최대하중 = 1.0(D + B + F + H + EM)

여기서, D : 고정하중 B : 부력 F : 유체압 H : 횡토압
EM : 설계지진력 (탄성지진력을 응답수정계수 R로 나눈 값)

- 허용응력 증가

구 조 물 의 종 류	허용응력 증가계수
강 구 조 물	1.50
철근콘크리트 구조물	1.33

⑧ 설계변위

- 최소받침 지지길이

- N = (200+1.67L+6.66H)(1+0.000125Θ2)(mm)

여기서, L : 인접 신축이음부까지 또는 교량단부까지의 거리(m). 다만, 지간 내에 힌지가 있는 경우의 L은 힌지 좌・우측방향의 거리인 L1과 L2의 합

- H : 다음 각 경우에 대한 평균 높이(m)

교 대 : 인접 신축이음부의 교량상부를 지지하는 기둥의 평균 높이
단경간교의 평균높이는 0
기둥 또는 교각 : 기둥 또는 교각의 평균높이
지간 내의 힌지 : 인접하는 양측 기둥 또는 교각의 평균높이

- Θ : 받침선과 교축직각방향의 사잇각(도)

- Combination of perpendicular seismic force

Design seismic force for the main axis of each member shall combine the seismic forces as below

- Load case 1 : When 30% of elastic seismic force (absolute value) taken from the transverse direction analysis is added to the elastic seismic force (absolute value) taken from the longitudinal direction analysis

- Load case 2 : When 30% of elastic seismic force (absolute value) taken from the longitudinal direction analysis is added to the elastic seismic force (absolute value) taken from the traversing direction analysis

⑦ Design Seismic force

- Maximum load = 1.0(D + B + F + H + EM)

Where, D : dead load B : buoyancy F : fluid pressure
H : traversing soil pressure
EM : design seismic force (Elastic seismic force divided by response modification factor, R)

- Increase of allowable stress

Type of structures	Allowable stress increase factor
Steel structures	1.50
reinforced concrete structure	1.33

⑧ Design displacement

- Minimum Bearing support length

- N = (200+1.67L+6.66H)(1+0.000125Θ2)(mm)

Where, L : Distance to elastic joint or bridge end part (m). When there is a hinge inside span, then L is sum of L1 and L2, which are hinge distance in left and right direction.

- H : Average height for each of following cases (m)

Abutment : Average height of pillars that support the bridge upper part of adjacent expansion joint Average height of single span bridge is 0
Pilar or pier : average height of pilar or pier
Hinge within span : Average height of adjacent pillars or pier on both sides

- Θ : Angles between the support line and the perpendicular line to bridge axis(degree)

⑨ 교량의 내진해석방법 선정

구 분	단일모드스펙트럼 해 석 법	다중모드스펙트럼 해 석 법	시간이력해석법	등가하중에 의한 환산해석법
개 요	• 교량의 첫 번째 고유 진동만을 고려하여 해석	• 교량의 최종 응답에 대한 다중모드의 기여 효과를 결정하기 위한 동적 해석	• 가설지역의 지진이력에 의거한 인공지진파를 작성하여 동적 해석	• 단경간교에 적용하는 방법으로 상부구조가 교대의 연결부에 대하여 고정하중 반력에 규정된 가속도 계수를 곱한 값의 수평 관성력을 작용
적 용 기 준	• 질량분포가 일정하고 경간당 강성비가 거의 비슷한 구조물로서 첫 번째 동적 응답을 지배 하는 경우	• 교량의 강성이 종방향 으로 크게 변화하거나, 교량이 비정형적인 경우 또는 실제 지배적인 진동방향으로 쉽게 파악 할 수 없는 경우	• 재료의 비선형성 및 구조물의 기하학적비선형을 고려한 해석 기능 지진하중을 같은 천이영역(Transient Region)에서 구조물의 응답을 정확히 구함 • 각 Mode에 대한 위상차(Phase)를 고려하므로 정밀해석 가능	• 단경간의 경우

3-2-11 교량 상세설계 시 유의사항

① 조사 및 계획업무 결과로 선정된 최적의 구조형식에 따라 상부구조·하부구조·기초구조의 단면을 설정하고, 구조물의 안전성·내구성·사용성 등을 검토한 후 상세설계

② 관련 시방서 및 규정, 지침에 따라 세부사항 검토 후 설계

③ 구조물의 각 단면을 상세하게 선정하고, 관련 설계기준에 규정된 해석방법으로 모델링하여 구조계산에 필요한 단면계수를 산출

④ 주하중 및 부가하중을 선정하여 최대단면력이 산출되도록 하중재하 위치를 결정

⑤ 지진력 계산은 구조물 형식·형태를 고려, 계산방법 결정

⑥ 설계방법에 따른 하중계수, 강도감소계수 등을 적용

⑦ 시공방법 및 순서를 고려하여 구조계산 수행

⑨ Selection of bridge seismic analysis method

Classification	Single mode spectrum analysis method	Multi-mode spectrum analysis method	Time history analysis method	Conversion analysis method by equivalent load
Overview	• Only the first natural frequency of the bridge is considered for the analysis	• Dynamic analysis for determining the contribution effect of multi-mode according to final response of the bridge	• Based on the history of earthquake of the erection sites, synthetic earthquake wave is generated for dynamic analysis	• This is applied to single span bridges, horizontal inertial force is applied from the value of dead load reaction multiplied by specified acceleration coefficients for the connection part of abutment to superstructure
Application standard	• For the structures with constant weight distribution and similar stiffness for each span, and when the first dynamic response is dominant	• When stiffness of the bridge varies largely in longitudinal direction, when the bridge is not standard, or actually dominant vibration direction cannot be easily identified	• This analysis function considers the non-linear material and non-geometric structures, response of the structure be accurately obtained for the earthquake load in the same transient region. • Precise analysis is possible by considering the phase difference of each mode	• In case of single span

3-2-11 Cautions during Bridge Detail Design

① Detail design shall follow the review on safety, durability, and usability of the structures, after determining the section of superstructure, substructure, and foundation in accordance with the optimum structure type selected as a result of investigation and plan.

② Design after reviewing the detail items in accordance with related specification and regulations

③ Section of each structure shall be selected in detail, and generate section coefficients required for the structural calculation by modeling with the specified method in related design standard

④ Loading location shall be determined in order to produce maximum section force by selecting main and editional load.

⑤ Seismic force calculation method shall be determined by considering type and shape of structures.

⑥ Apply load coefficient and strength decrease coefficients according to design method

⑦ Execute structure calculation by considering construction method and sequence

⑧ 산출된 최대단면력(휨모멘트, 축력, 전단력, 비틀림모멘트)에 대한 안전성, 내구성, 사용성 등을 검토·확인

⑨ 상부구조의 처짐, 하부구조의 전도 및 활동, 기초구조의 지지력·변위 등을 검토

⑩ 구조물에 국부적으로 작용하는 하중 및 단면력을 산정하고, 안전성을 확인

⑪ 사장교, 현수교와 같은 특수교량은 풍하중에 대한 안정성을 반드시 검토

⑫ 강도설계 및 피로설계에 따른 상세사항을 관련 기준에 따라 설계

⑬ 하중집중부에 대한 상세설계를 관련 기준에 따라 설계

⑭ 받침부와 상하구조물 연결구조를 상세히 검토하여 설계

⑮ 온도변화에 따른 구조물의 변위 및 구조물 단면력을 상세히 검토하여 설계

⑯ 구조계산에 따른 전산프로그램은 검증된 것을 사용

⑰ 구조계산에 필요한 가시설물을 상세히 검토하여 설계

⑧ Safety, durability, and serviceability of the calculated maximum section force(bending moment, axial force, shear force, torsional moment) shall be reviewed and confirmed.

⑨ Review deflection of superstructure, overturning and sliding of substructure, supporting force and displacement of foundation.

⑩ Calculate the section force and load locally acting the to structure shall be calculated, and the safety shall be confirmed.

⑪ Safety against wind load must be reviewed for the special bridges such as cable strayed bridge and suspension bridge.

⑫ Details on strength design and fatigue design shall be designed in accordance with the related standard.

⑬ Detail design on the load concentrating section shall be designed in accordance with the related standard

⑭ Support part and connection part between superstructure and substructure shall be reviewed in detail and designed.

⑮ Displacement of structure and section force of structure depending on the temperature changes shall be reviewed in detail and designed.

⑯ Proven computer programs shall be used for structural calculation.

⑰ Temporary facilities required for the structural calculation shall be reviewed in detail and designed.

Chapter 11

Pavement

1. 일반사항

1-1 포장의 개요

1-1-1 아스팔트 콘크리트 포장

① 골재를 역청재료(bituminous material)로 결합하여 만든 표층이 있는 포장

② 표층, 중간층, 기층 및 보조기층으로 구성

③ 교통하중에 의해서 발생하는 응력이 포장을 구성하는 각 층에 분포되어 하층으로 갈수록 점차 넓은 면적에 분산시킴

④ 포장의 역학적 거동특성에 초점을 두어 아스팔트 콘크리트 포장을 가요성 포장(flexible pavement)으로 부름

【아스팔트 콘크리트 포장의 구성과 각 층의 명칭】

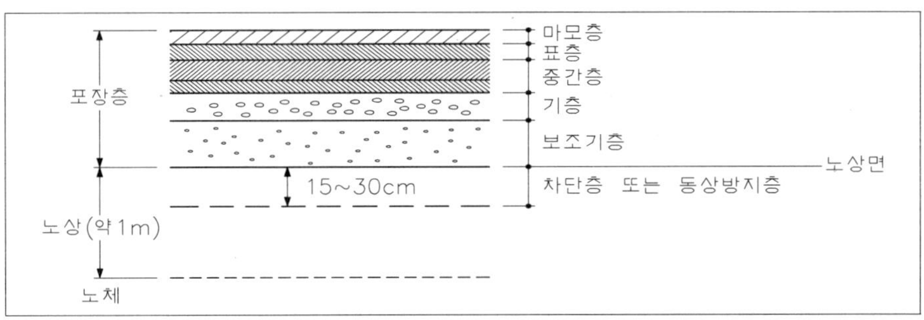

1-1-2 시멘트 콘크리트 포장

① 시멘트 콘크리트 슬래브 및 린 콘크리트, 동상방지층으로 구성

② 대표적인 강성포장(rigid pavement)으로 시멘트 콘크리트 슬래브가 교통하중으로 인한 전단이나 휨에 저항하여 그 슬래브 작용으로 하중을 기초에 넓게 전달

③ 동상방지층의 역할은 슬래브에 균등한 지지력을 주고 투수층으로 동상현상을 방지하고 시멘트 콘크리트 슬래브를 타설하기 위한 안전한 작업기반을 조성

【시멘트 콘크리트 포장의 구성】

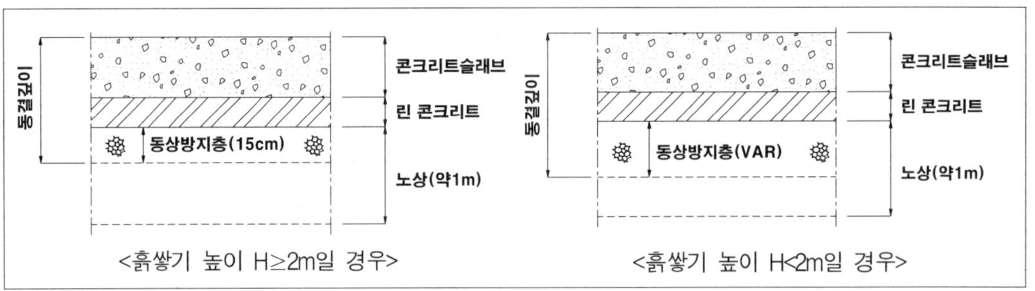

1. General Items

1-1 Pavement Overview

1-1-1 Asphalt Pavement

① Asphalt is outer-layered pavement that aggregates are integrated with bituminous material.

② Asphalt consists of outer surface course, binder course, base course and subbase layer

③ Stress from traffic load is dispersed to each layer of asphalt so that the lower layer would have scattered force in the larger area.

④ Asphalt pavement is called as flexible pavement because of its dynamics characteristics.

【Structure of asphalt pavement and the name of each layer】

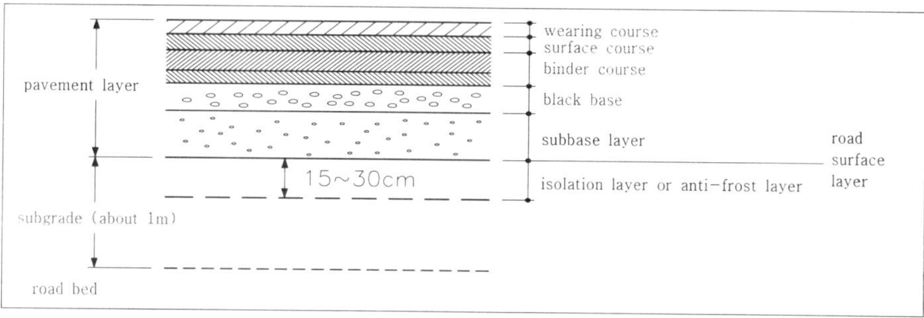

1-1-2 Concrete Pavement

① Composition of the pavement is concrete slab, lean concrete and anti-frost layer.

② Concrete pavement is typical rigid pavement. Cement concrete slab resists the shear or bending force so that the slab disperse the load to the base layer.

③ The role of anti-frost layer is to provide even support for slab, to prevent the freezing in permeable layer and to provide safe base layer for pouring cement concrete slab.

【Structure of concrete pavement】

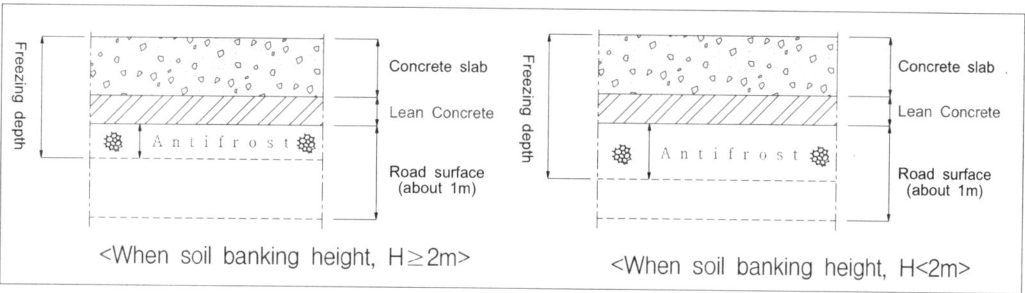

1-2 포장의 구조 및 특징

구 분	아스팔트 콘크리트 포장 (가요성 포장)	시멘트 콘크리트 포장 (강성 포장)
하중전달	• 교통하중을 표층→기층→보조기층→노상으로 확산·분산시켜 하중을 절감하는 형식	• 교통하중을 콘크리트 슬래브가 직접 지지하는 형식
표 층	• 교통하중을 일부 지지하며 하부층으로 전달 • 표면수의 침입을 방지하여 하부층 보호	• 슬래브 자체가 빔(Beam)으로 작용하여 교통하중에 발생되는 응력을 휨 저항으로 지지
기 층	• 입도조정처리 재료 또는 아스팔트 혼합물로 구성 • 표층과 일체가 되어 교통하중에 의한 전단에 저항하여 하중을 분산시켜 보조기층에 전달	-
보조기층	• 압상재료 또는 토사 안정처리 재료 등으로 구성 • 상부층에서 전달된 교통하중을 지지하며 노상으로 전달 • 포장층 내 배수기능 전달 역할 • 미립질의 노상재료가 기층부로 침투하는 것을 방지 • 동결작용의 손상효과를 최소화	• 빈배합 콘크리트 또는 시멘트 및 아스팔트 안정처리 재료로 구성 • 콘크리트 슬래브에 대한 균일한 지지력 확보 • 줄눈부 및 균열부근의 우수 침투처리 및 펌핑현상 방지
구조특성	• 포장층 일체로 하중을 지지하고 노상에 윤하중 분포 • 포장두께는 교통하중과 노상지지력에 의거 설계 • 기층 또는 보조기층에도 큰 응력이 작용 • 반복되는 교통하중에 민감	• 콘크리트 슬래브 자체로 하중 및 온도변화에 대해 지지 • 슬래브에 불규칙한 균열방지를 위해 가로수축줄눈 설치 • 골재 맞물림 작용 및 다웰바를 통해 슬래브간 하중 전달
재료특성	• Asphalt 혼합물 이용 • 쇄석골재 • Asphalt, 굵은골재, 잔골재, 채움재	• cement 혼합물 이용 • 강자갈 또는 쇄석골재 사용 • 시멘트, 물, 굵은골재, 잔골재, 혼화재료, 줄눈재, 양생재

1-2 Structure and Characteristics of Pavement

Classification	Asphalt concrete pavement (flexible pavement)	Cement concrete pavement (Rigid pavement)
Load Transfer	• Traffic load is dispersed through surface course → black base → Subbase layer → subgrade to curtail load concentration 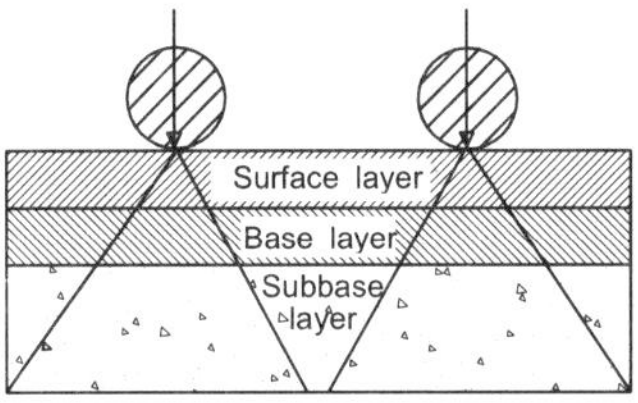 Surface layer / Base layer / Subbase layer	• Traffic load is supported directly by concrete slab 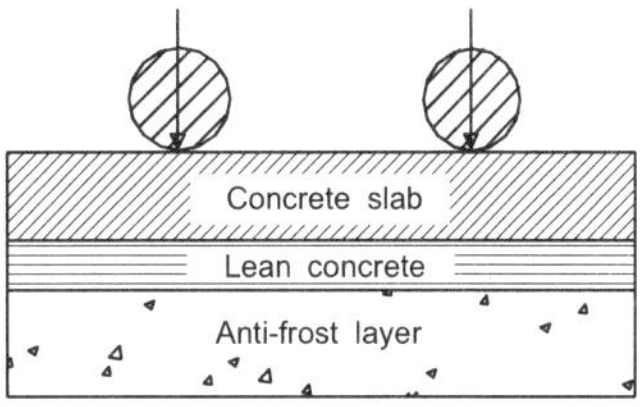 Concrete slab / Lean concrete / Anti-frost layer
Surface course	• Support partial load and transfer the rest to the lower layers • Prevent surface water from penetration to protect lower layers	• Slab itself works as beam to support stress from traffic load with bending resistance
Base course	• Grain adjustment processed material or asphalt mixture • Combined into surface layer to resist share from traffic load so that the load is dispersed into subbase layer	-
Subbase layer	• Forced material or sand stabilizing processed material • Support traffic load transferred from upper layer and transfer some of it to road surface • Drainage within pavement layer • Prevent particles of road surface from penetrating into base layer • Minimize damage from freezing	• Dry lean concrete or cement and asphalt stabilizing processed material • Uniform bearing power against concrete slab • Prevent rainwater from penetration or pumping into jointer or cleave
Structural Characteristics	• Load is supported by entire pavement and dispersed into road surface layer • Thickness of pavement is designed according to traffic load and surface bearing power • Large stress is applied to base course and subbase layer • Sensitive to repetitive traffic load	• Load and temperature change is supported solely by concrete slab • To prevent irregular cleavage, horizontal contraction joints are installed • Load is transferred among slabs through aggregate gearing and dowel-bar
Material Characteristics	• Asphalt mixture • Crashed aggregate • Asphalt, coarse aggregate, fine aggregate and filler	• Cement mixture • River gravel or crashed aggregate • Cement, water, coarse aggregate, fine aggregate, compound, joint and cure materials

구 분	아스팔트 콘크리트 포장	시멘트 콘크리트 포장
종 류	• 내유동성 아스팔트 포장 • 가열 아스팔트 포장	• 무근콘크리트포장 • 연속철근콘크리트포장
시공성	• 신속성 및 간편성 측면에서 유리 • 단계시공 방식에서 유리 • 아스팔트 피니셔 사용 포설	• 줄눈 설치 및 콘크리트 양생으로 다소 불리 • 슬립 폼 페이버 사용 포설
내구성	• 중차량이 많은 도로에서 소성변형 발생	• 중차량에 대한 적응도 양호
유지보수	• 부분 보수 용이 • 유지관리비 고가 • 잦은 보수로 교통소통에 지장 초래 • 보수시기가 늦어지면 큰 하자 발생	• 유지관리비 저렴 (단, 줄눈부의 정기적인 유지보수 필요)
공용성	• 공사 후 즉시 교통개방 • 평탄성 및 승차감 양호 • 소음이 적음 • 중차량에 의한 요철 발생 • 강설 이후 용해 빠름 • 야간 시인성 부족 • 우천 시 반사 발생	• 장기간 양생 필요(보통 포틀랜드 시멘트 사용할 때 14일 이상 소요) • 수축줄눈의 설치로 승차감 불량 • 적설 시 결빙시간이 빠르고, 다소 늦게 녹음 • 회색으로 초기반사도 높음
시험방법	• 마샬시험기를 사용한 역청혼합물의 소성흐름에 대한 저항력 시험 • 골재혼합물의 피막 박리 시험 • 굵은 골재의 밀도 및 흡수율 시험 • 골재의 안정성 시험 • 굵은 골재의 마모시험 • 굵은 골재 중 편장석 함유량 시험	• 슬럼프 시험 • 휨강도 시험 • 단위용적 질량 및 공기량 시험 • 골재 체가름 시험
첨가제 (개질재)	• SBS • SBR Latex • PBS • LDPE • DAMA • CRM • 캠크리트 • 천연 아스팔트 • 폐고무 아스팔트	• AE제 • 감수제 • AE감수제 • 고성능 AE감수제 • 플라이애시 • 팽창재 • 고로슬래그 미분말
적용도로	• 중차량 구성비가 낮은 도로 • 조기 개방이 필요한 도로 • 연약지반에 축조하는 도로 • 저소음이 필요한 도로 • 조기 완공이 필요한 도로 • 교량, 암거, 터널 등 구조물이 많은 구간	• 중차량 구성비가 높은 도로 • 땅깎기 · 흙쌓기 경계부가 많은 도로 • 신설도로

Classification	Asphalt concrete pavement (flexible pavement)	Cement concrete pavement (Rigid pavement)
Type	• Rut resistant asphalt pavement • Hot asphalt pavement	• Plain concrete pavement • Continuously reinforced concrete
Construction easiness	• Advantage in faster and convenient construction • Advantage in staged construction • Install with asphalt finisher	• Disadvantage due to joint installation and concrete cure • Install with slim form paver
Durability	• Requent traffic of heavy vehicles causes plastic deformation	• Ood resistance to heavy vehicles
O&M	• Easy partial repair • O&M cost is high • Frequent repair may induce traffic jam • If repair is not conducted quickly, serious defect may occur	• O&M cost is low (Regular maintenance is required for the joints)
Commonality	• Immediate open to public after construction • Evenness and comfortable to ride • Less noise • Unevenness may occur by heavy vehicle traffic • Rapid dissolve after snowfall • Less night visibility • Reflection may occur during the rain	• Long term cure is required (14 days for regular portland cement) • Contraction joint decrease riding comfort • Rapid freezing when snow. Melting is a bit slow. • Initial reflection is high due to gray color
Test method	• Resistance test on non-associated flow of bituminous mixture using Marshall tester • Membrane peeling test for aggregate mixture • Density and absorptance test for coarse aggregate • Stability test for aggregate • Abrasion test for coarse aggregate • Flat and elongated particles content test for coarse aggregate	• Slump test • Bending intensity test • Unit volume weight and air quantity test • Aggregate sieve test
Additive	• SBS • SBR Latex • PBS • LDPE • DAMA • CRM • Chemcrete • Natural asphalt • Crumb rubber modified asphalt	• Air-entering agent • water reducing agent • AE water reducing agent • high performance AE water reducing agent • Fly ash • Leavening agent • Blast-furnace slag fine grain
Applicable Road	• Road with lower ratio of heavy vehicle • Road that requires immediate operation • Road constructed on the weak foundation • Road that requires minimum noise • Road that requires earlier completion • Road with many structures (bridge, duct, tunnel, etc)	• Road with higher ratio of heavy vehicle • Road with many filling or cutting boundaries • Newly constructed road

1-3 포장의 종류

1-3-1 아스팔트 콘크리트 포장의 종류

① 입도 분포에 의한 분류

- 밀입도(Dense graded)
  - 연속입도로서 골재입자 간의 공간이 가장 작은 상태로 채워지면서 최대밀도를 나타내는 내구성이 큰 입도

- 개립도(Open grade)
  - 합성입도에 있어 2.5mm체 통과량이 5~20% 범위로 구성되는 입도분포

- 갭입도(Gap grade)
  - 내유동성이나 마찰력을 개선할 수 있도록 일부를 불연속으로 하는 입도
  - 합성입도에 있어 2.5mm~0.6mm 또는 5mm~0.6mm의 입경부분이 10% 정도 내의 불연속 입도로 되어있는 것으로 내마모성, 내유동성을 향상시키기 위해 사용

② 개질 및 특수아스팔트 혼합물에 의한 분류

- 개질아스팔트

  아스팔트 콘크리트 포장의 내구성 및 내유동성 향상을 목적으로 일반 아스팔트에 일정량의 개질재(Modifier)를 첨가하여 아스팔트의 물성(점탄성, 감온성 등)을 개선시켜 소성변형과 균열발생 등의 문제점을 극복하기 위한 아스팔트 혼합물

  - 고분자 개질아스팔트(SBS, Styrene Butadiene Styrene)
    - ▸ 공중합체(Copolymer)로서 열가소성 수지계통의 고무계열의 고분자 개질아스팔트
    - ▸ 고온에서는 스프링과 같이 아스팔트의 유동성을 억제하여 소성변형 발생을 최소화시키고, 저온에서는 온도응력을 흡수 완화하여 취성균열을 방지

  - 폐타이어 고무아스팔트(CRM, Crumb Rubber Modified)
    - ▸ 아스팔트, 재활용 타이어 고무, 일정량의 첨가제 혼합체
    - ▸ 고무성분이 혼합체 총 중량의 15% 이상
    - ▸ 고무입자가 가열된 아스팔트 속에서 팽창하도록 충분히 반응된 아스팔트
    - ▸ 점착력(Toughness)과 연화점 개선으로 혼합물의 안전성과 마모방지기능이 향상

1-3 Types of Pavement

1-3-1 Types of Asphalt Pavement

① Classification by grain size distribution

• Dense graded

- Space among aggregate is filled with continuous grain so that the density and resultant durability becomes the largest.

• Open grade

- Mixed grain size distribution where 5~20 percent of grain is passing through 2.5mm sieve

• Gap grade

- Partial gap grade to improve rut resistance and friction.
- Mixed grain size distribution where 10 percent of the grain is gap grade grain with radius of 2.5mm~0.6mm or 5mm~0.6mm to improve durability abration and rut resistance.

② Classification by modified and special asphalt

• Modified asphalt

Modified asphalt is asphalt mixture that certain amount of modifier is added to improve durability and rut resistance of asphalt pavement. The modifier improves property of asphalt (viscoelasticity, temperature susceptibility, etc.) so that the issues of original asphalt such as plastic deformation and cleavage.

- SBS, Styrene Butadiene Styrene
  - ▸ Copolymer, thermoplastic resin, rubber type polymer modified asphalt
  - ▸ In high temperature, minimize thermoplastic deformation through suppressing fluidity of asphalt. In low temperature, prevent brittle crack through absorbing temperature stress.

- CRM, Crumb Rubber Modified
  - ▸ Asphalt, recycled tire rubber, certain amount of additive
  - ▸ Rubber is more than 15 percent of entire mixture weight
  - ▸ Properly reacted asphalt so that rubber particle would be expended in heated asphalt.
  - ▸ Improve safety and abrasion resistance by improving toughness and softening point.

• 특수아스팔트

내소성변형(Rutting) 증대, 불투수층 형성, 휨강성 저항성 증대 등 제한된 특수목적으로 개발된 아스팔트

- 쇄석 매스틱 아스팔트(SMA, Stone Mastic Asphalt)
  - ▶ 일반 밀입도(입도 균등)에 비해 5mm 이상의 굵은 골재함량을 높여 굵은 골재 간의 맞물림작용(Interlocking)에 의해 중차량 하중을 효과적으로 지지함으로써 소성변형 예방에 탁월
  - ▶ 포장수명도 일반포장의 2~3배(10~15년) 정도 예상

- 구스 아스팔트(Guss Asphalt)
  - ▶ 주로 강상판 포장과 같이 불투수성, 휨에 대한 저항성이 요구되는 구간에서 적용

•

1-3-2 콘크리트 포장의 종류

① 무근콘크리트 포장(JCP, Jointed Concrete Pavement)

- 다웰바나 타이바를 제외하고는 일체의 보강이 없는 포장형태
- 일정한 간격으로 줄눈을 형성함으로써 균열의 발생위치를 인위적으로 조절
- 필요에 따라 줄눈부에 다웰바를 사용하여 하중전달을 도움
- 줄눈 이외의 부분에서는 균열발생을 허용하지 않음
- 온도변화 및 건조수축에 의한 콘크리트 슬래브의 움직임을 억제하는 구속력을 줄이기 위하여 콘크리트 슬래브와 보조기층 사이에 분리막을 설치하여 마찰력을 줄임
- 시간경과에 따라 줄눈부위의 파손(단차, 우각부 균열, 펌핑 등)으로 승차감의 저하를 초래

② 철근콘크리트 포장(JRCP, Jointed Reinforced Concrete Pavement)

- 무근콘크리트 포장의 많은 줄눈사용으로 인한 문제점을 감소시키기 위하여 고안된 포장형태
- 줄눈의 개수를 감소시키는(줄눈과 줄눈 간의 간격은 증가) 대신 줄눈 이외의 부분에서 발생되는 균열을 어느 정도 허용
- 발생된 균열들이 과대하게 벌어지는 것을 방지하기 위하여 일정량의 종방향 철근을 사용
- 무근콘크리트 포장(JCP)에 비해 줄눈의 수가 줄어드나 줄눈부위에서 발생하는 문제점을 여전히 내포하고 있음

③ 연속 철근콘크리트 포장(CRCP, Continuously Reinforced Concrete Pavement)

- 횡방향 줄눈을 완전히 제거한 포장

• Special asphalt

Developed for specific purposes such as rutting, impermeability layer, bending rigidity resistance improvement, etc.

- SMA, Stone Mastic Asphalt
  - ▸ Comparing to general grain size distribution, ratio of coarse aggregate (larger than 5mm) is increased so that the interlocking among those coarse aggregate would effectively support load from heavy vehicle and prevent plastic deformation.
  - ▸ The span of pavement life is expected to be more than 2~3 times than general pavement (10~15 years)

- Guss Asphalt:
  - ▸ Apply to area where impermeability and bending rigidity resistance such as steel deck plate pavement are required

1-3-2 Types of Concrete Pavement

① JCP (Jointed Concrete Pavement)

- No reinforcement other than dowel-bar or tie-bar
- Artificially control the cleavage location through installing joint with regular interval
- Dowel-bar is installed in joint if needed
- Do not allow any cleavage other than joint
- Install separate layer between concrete slab and subbase layer to minimize resistance to concrete slab movement by temperature change or drying shrinkage
- As time passes, brakeage of joint causes decreased ride comfort.

② JRCP (Jointed Reinforced Concrete Pavement)

- JRCP is developed to solve the joint problem of JCP
- In exchange for fewer number of joint, allow cleavage outside of joint by certain level
- To prevent cleavage from cracking apart, use vertical reinforcing rod
- Comparing to JCP, JRCP will have fewer joint. However, the underlying problem of joint is not completely solved here yet.

③ CRCP (Continuously Reinforced Concrete Pavement)

- Horizontal joint is removed completely

- 균열의 발생을 허용하고 상당량(콘크리트 단면적의 0.5%~0.8%)의 종방향 철근을 사용하여 균열 틈의 벌어짐을 억제
- 온도변화 및 건조수축에 의한 콘크리트 슬래브의 움직임을 억제하고자 콘크리트 슬래브와 보조기층 사이에 분리막을 사용하지 않음
- 줄눈 미사용에 따른 쾌적한 승차감 및 긴 포장수명으로 인하여 세계적으로 인기

④ 기타 콘크리트 포장(적용빈도 낮음)

- 프리스트레스 콘크리트포장(PCP : Prestressed Concrete Pavement) :
  - 상당한 시공기술이 필요하며, 외국에서도 공항의 활주로 등 일부에서만 적용
- 로울러 다짐 콘크리트 포장(RCCP : Roller Compacted Concrete Pavement)

1-4 포장의 구성

1-4-1 아스팔트 콘크리트 포장의 구성

① 노상

- 포장층의 기초가 되는 흙의 부분으로 포장면 아래 약 1m 두께의 층
- 노상토의 지지력은 평판재하시험 또는 CBR 시험에 의해 판정
- 노상토의 설계CBR이 2이하인 경우 지지력 증가를 위해 연약지반처리 실시

② 차단층

- 배수층 역할을 하는 입상재료 기층과 보조기층 또는 집수시스템(맹암거, 지하배수구 등)이 노상토 침입에 의하여 막히는 것을 보호
- 지하수위를 낮추기 위한 수단으로서 적정 입도와 투수성을 가지는 15~30cm 두께의 선별 입상재료 또는 지오텍스타일(geotextile)을 사용하여 보조기층과 노상면 사이에 설치

③ 동상방지층

- 동결융해 작용에 의한 포장층 파손을 방지하기 위하여 설치
- 빙막 형성을 방지할 수 있는 품질의 재료를 사용

- In exchange for allowing cleavage, large amount of vertical reinforcing rod (0.5% ~0.8% of total concrete cross-section) is used to suppress the further creaking
- Do not install separate layer between concrete slab and subbase layer to minimize resistance to concrete slab movement by temperature change or drying shrinkage
- Popular worldwide due to the lack of joint and long lifespan.

④ Other concrete pavement (less frequently used)

- PCP : Prestressed Concrete Pavement
  - Requires high-end construction technique. Generally used in runway of airport.
- RCCP : Roller Compacted Concrete Pavement

1-4 Structure of Pavement

1-4-1 Structure of Asphalt Pavement

① Subgrade

- This soil layer below the pavement surface with the thickness of one meter provides the base for pavement layers.
- Bearing capacity of the surface earth can be examined through plate loading test or CBR test
- When design CBR is less than 2 for surface earth, perform weak base process to improve bearing capacity

② Filter layer

- Protect grain material layer, subbase layer or water collecting system (mole drainage, underground waterway, etc) which act as drainage from clogged by surface earth.
- Install between subgrade and subbase course using 15~30cm selected grain materials or geotextile to lower the underground water level.

③ Antifrost layer

- Protect pavement from frost damage
- Use material that can prevent frost grazing

④ 보조기층과 기층

- 표층 또는 중간층을 통해 전달되는 교통하중을 분산시켜서 노상에 안전하게 전달하는 역할
- 충분한 지지력을 지니고 내구성이 풍부한 재료로서 필요한 두께로 잘 다져질 수 있는 소요 품질을 갖춘 골재혼합물
- 시멘트안정처리기층은 7일 양생, 1일 수침 일축압축강도를 30kg/cm^2 이상 확보
- 고로슬래그 보조기층용은 수정 CBR 값이 30 이상
- 기층용은 수정 CBR 값이 80 이상, 14일 양생 일축압축강도가 12kg/cm^2 이상

⑤ 중간층과 표층

- 중간층은 기층 위에 놓이므로 기층 윗면의 요철(凹凸)을 보정하고 표층을 통해 가해지는 하중을 균일하게 기층에 전달하는 기능
- 표층은 포장층의 최상부에 위치하여 차륜에 의한 마모와 전단작용에 충분히 저항하고 평탄성을 유지하는 기능
- 미끄럼 저항이 커서 차량이 쾌적한 주행을 할 수 있고, 빗물이 하부층에 침투하지 않는 치밀한 조직을 가질 수 있도록 하는 기능
- 중간층과 표층은 가열식 아스팔트 혼합물을 사용
- 혼합물의 배합설계는 일반적으로 마샬시험법을 적용

1-4-2 시멘트 콘크리트 포장의 구성

① 노상 : 아스팔트 콘크리트 포장과 동일

② 동상방지층

- 슬래브를 지지하며, 균열부에서의 펌핑 현상을 막아주는 중요한 층으로서, 한 층 이상의 다져진 입상재료 또는 안정처리 재료층으로 구성
- 노상과 콘크리트 슬래브 사이에 다음과 같은 목적으로 설치
  - 균등하고, 안정적이며, 영구적인 지지력 제공
  - 노상반력계수(K)의 증대
  - 동결작용에 의한 손상도 극소화
  - 콘크리트 슬래브의 줄눈부, 균열부 그리고 단부에서의 세립토의 펌핑 방지
- 펌핑을 방지하기 위해 특히 아스팔트나 시멘트 등으로 처리된 재료를 사용하는 것이 좋음
  - 입도조정쇄석, 입도조정슬래그, 수경성입도조정슬래그, 시멘트안정처리재료 및 역청안정처리재료 등을 사용

④ Subbase layer and base course

- Disperse traffic load transferred from surface layer or intermediate layer to secure safety of the road surface layer
- Use aggregate mixture with sufficient bearing power and durability
- Cement stability process layer requires 7 days of cure to secure 30kg/cm^2 of daily unconfined compression.
- Adjusted CBR of more than 30 for subbase layer blast furnace slag
- Adjusted CBR of more than 80, 14 days of cure and 12kg/cm^2 of daily unconfined compression for base layer blast furnace slag

⑤ Binder course and surface course

- On top of base layer, binder course adjusts unevenness of base layer and transfers load from surface layer to base layer uniformly.
- On top of all pavement layers, surface course resists abrasion and stress from vehicle wheels and maintains flatness.
- Surface layer enables comfortable riding due to high slide resistance. It also prevent rainwater from penetrating to lower layers through fineness texture.
- Use heat asphalt mixture for binder and surface course
- Adopt Marshall test for mixture design

1-4-2 Structure of Concrete Pavement

① Subgrade : Similar to asphalt pavement

② Antifrost layer

- Antifrost layer is critical layer to support slab and prevent pumping phenomenon. It consists of more than one layer of minced grain material or stability process materials.

- Installed to perform the following roles:
  - Provide uniform and permanent bearing power
  - Increase surface reaction coefficient (K)
  - Minimize damage from frost grazing
  - Prevent pumping of fine-grained earth to joint and cleavage

- It is recommended to use processed material with asphalt and cement to prevent pumping
  - Graded crushed stone, graded slag, cement stability process material and ituminous process material

③ 콘크리트 슬래브

- 시멘트 콘크리트 포장 구조에서 콘크리트 슬래브의 휨 저항에 의하여 통행 교통 하중을 거의 모두 지지하는 기능
- 구성재료는 공용기간동안 받게 되는 교통하중과 환경영향에 의한 손상을 충분히 지지할 수 있는 강도와 내구성을 가지는 것

④ 보강철근, 하중전달장치, 타이바 및 줄눈재

- 온도, 습도 등 환경적 변화, 슬래브 저면마찰, 그리고 시공 시 수화작용에 의하여 발생하는 응력과 균열을 완화시키고 조절하기 위하여 설치

2. 포장형식의 선정

2-1 본선포장형식 선정 절차

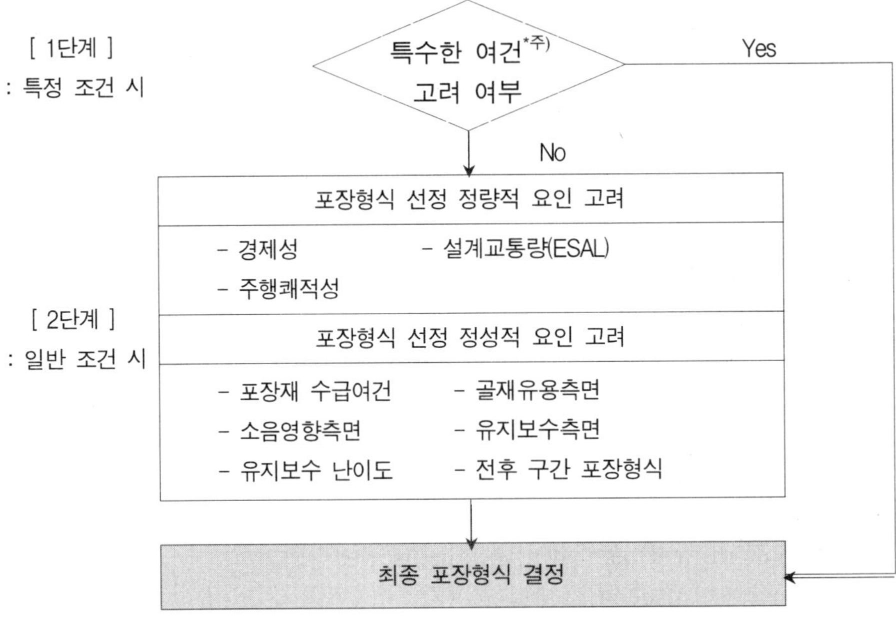

주) 특수한 여건

• 아스팔트 콘크리트 포장	• 시멘트 콘크리트 포장
- 연약지반구간 - 제설재 5ton/km/차로 이상인 지역 ※ 단구간 등 특수한 여건일 경우 2단계 검토가능	- 설계교통량 56백만 ESAL - 연장 500m를 초과하는 터널

※ 위의 요건에 해당하는 경우는 2단계를 거치지 않고 바로 포장형식 확정

③ Concrete Slab

- Concrete slab supports almost all traffic load with bending resistance
- Consider materials with sufficient strength and durability to endure traffic load and environmental impact during the operating period.

④ Reinforce rod, load transfer device, tie-bar and joint

- Install to mitigate or control environmental change (temperature, humidity), slab base friction, stress and cleavage by hydration from construction

2. Selection of Pavement Type

2-1 Main Lane Pavement Type Selection Process

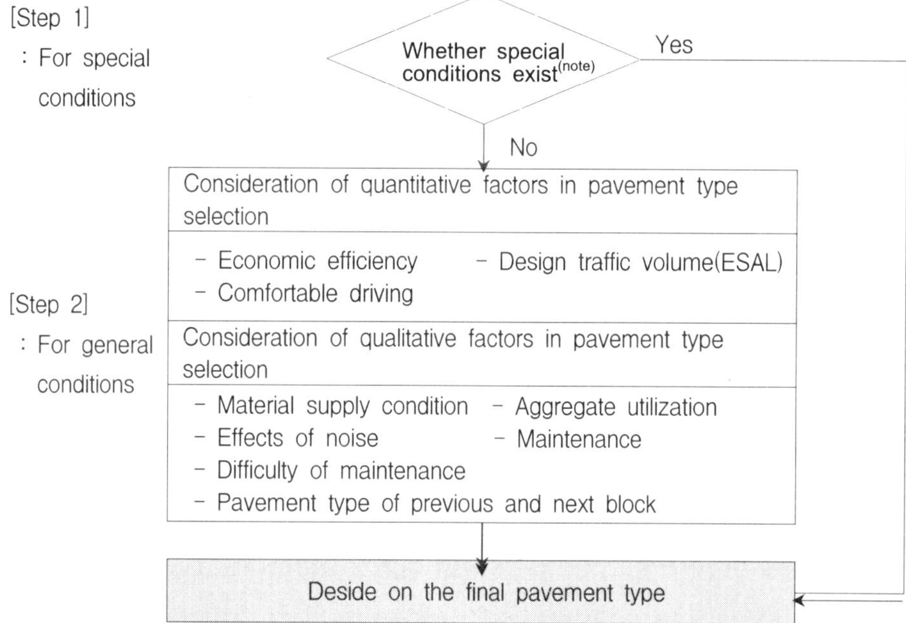

Note) Special conditions

• Asphalt concrete pavement - Soft foundation section - Area with more than 5ton/km/lane of snow removal materials ※ Step 2 can be considered for special conditions such as short section	• Cement concrete pavement - Design traffic rate of 56 million ESAL - Tunnel with the total length over 500m

※ Pavement type is determined without considering step 2 when above conditions are satisfied.

2-2 특수포장형식 결정 시 고려사항

2-2-1 길어깨 포장

① 원활한 교통소통, 안전운행 및 본선포장의 유지보수작업을 위해 설치

② 길어깨포장 두께는 7.5cm(아스콘 표층) 적용

③ 다음 구간에 대해서는 본선포장과 동일하게 보강 조치

- 곡선반경 R≤100m이하 연결로
- 유출연결로 좌·우 길어깨 전 구간
- 연결로 유출부 노즈부터 본선구간 길어깨 20m.
- 도심부 구간 테이퍼 시·종점부 2km
- 가·감속차로 테이퍼 시·종점부 길어깨 30m
- 영업소 테이퍼 시·종점부 길어깨 30m

2-2-2 교면 포장

구 분	고성능콘크리트(30MPa) 노출바닥판	콘크리트계 교면포장	아스팔트계 교면포장
설 치 단 면	마 모 층 50mm 콘크리트교량바닥판 240mm	콘 크 리 트 50mm 콘크리트교량바닥판 240mm	마 모 층 (SMA:10mm) 40mm 고르기층 (SMA:8mm) 40mm 콘크리트교량바닥판 240mm
구 성 및 특 징	• 내구성 및 균열 저항성 양호 • 슬래브 피복 증가 • 침투식 방수제 사용 • 평탄성 불량 • 초기투자비 저렴	• 유동성 및 점착력 증가 • 균열 억제 • 방수효과 우수 • 내구성 증진, 유지관리 양호 • 품질관리, 시공성 다소 불량 • 초기투자비 고가	• 내구성 및 내유동성 탁월 • 방수효과 보통 • 평탄성 양호 • 초기투자비 다소 고가
적 용 구 간	• 염화물 사용과 관계없이 전 지역 교량에 적용 - 평탄성 · 시공성 및 내구성 확보에 불리한 교량은 콘크리트 교면포장 적용	• 일반교량 중 평탄성 및 시공성 확보가 불리한 교량 - 신축이음장치 간격 150m 초과 교량 - 종단 또는 편경사가 3%이상 교량 - 프리캐스트 콘크리트 패널적용 교량 - PC 박스교량 및 특수교량 (현수교, 사장교)	• 강상판형 교량 • 도심지 등 소음민원 예상 교량 • 기타 아스팔트계 교면포장이 필요한 교량

2-2 Considerations for Special Pavement Type Selection

2-2-1 Road Shoulder Pavement

① Install for smooth traffic, safe driving and main lane pavement maintenance
② Thickness for road shoulder pavement should be 7.5cm(Asphalt concrete layer)
③ Apply main lane pavement criteria for the following sections:

- Connecting road with radius less than R≦100m
- All section of both side road shoulder in outflow connection
- 20m of road shoulder from connection outflow nose section to main lane
- Urban area taper beginning / end section 2km
- Accelerating / decelerating lane taper beginning / end section road shoulder 30m
- Operating facility taper beginning / end section road shoulder 30m

2-2-2 Bridge Surface Pavement

Classification	High performance concrete (30Mpa) exposed floor plate	Concrete type bridge surface pavement	Asphalt type bridge surface pavement
Installation section	마 모 층 50mm 콘크리트교량바닥판 240mm	콘 크 리 트 50mm 콘크리트교량바닥판 240mm	마 모 층 (SMA:10mm) 40mm 고르기층 (SMA:8mm) 40mm 콘크리트교량바닥판 240mm
Structure and characteristics	• Good durability and resistance to cracks • Increased slab cover • Use penetration type water proof material • Inferior flatness • Low initial investment	• Increased mobility and cohesion • Crack suppression • Good water proof effect • Increased durability, good maintenance • Slight problems in quality control and constructability • High initial investment	• Excellent durability and resistance to mobility • Normal water proof effects • Good flatness • Slightly high initial investment
Application section	• Applied to entire bridge section regardless of the usage of chloride. - Concrete bridge surface pavement shall be applied for bridges that are difficult to secure flatness, constructability, and durability.	• Bridges that are difficult to secure flatness and constructability among general bridges. - Bridges with flexible joint distance over 150m - Bridges with camber or super elevation over 3% - Bridges with precast concrete panel - PC box bridges or special bridges (suspension bridge, cable-stayed bridge)	• Steel plate type bridges • Bridges in the places where complaints on noise are expected such as downtown area • Other bridges that require asphalt type bridge surface pavement

2-2-3 연결로 포장

① 연결로의 본체 부분의 차도와 측대 포장은 차량 안전성, 포장 시공성과 경제성을 고려하여 적정한 포장형식을 선택

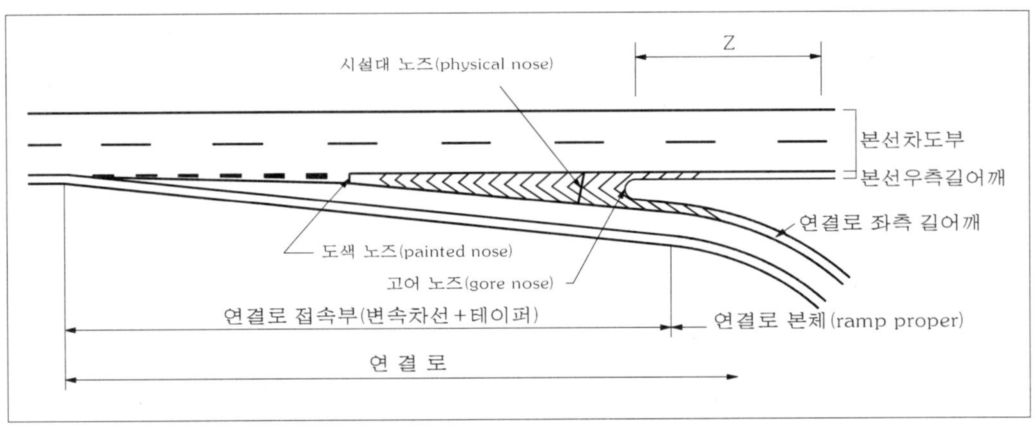

② 연결로의 길어깨포장

- 고속도로 본선 주행차량의 지체를 방지하고자 연결로의 2차로 운용이 가능하도록 진출 연결로의 전폭을 본선포장과 동일시행
- 진입 루프램프에 있어서는 곡선반경 100m 미만의 단곡선과 이에 접속되는 완화구간을 포함하여 차량 진행방향의 길어깨 우측을 본선포장 단면으로 보강

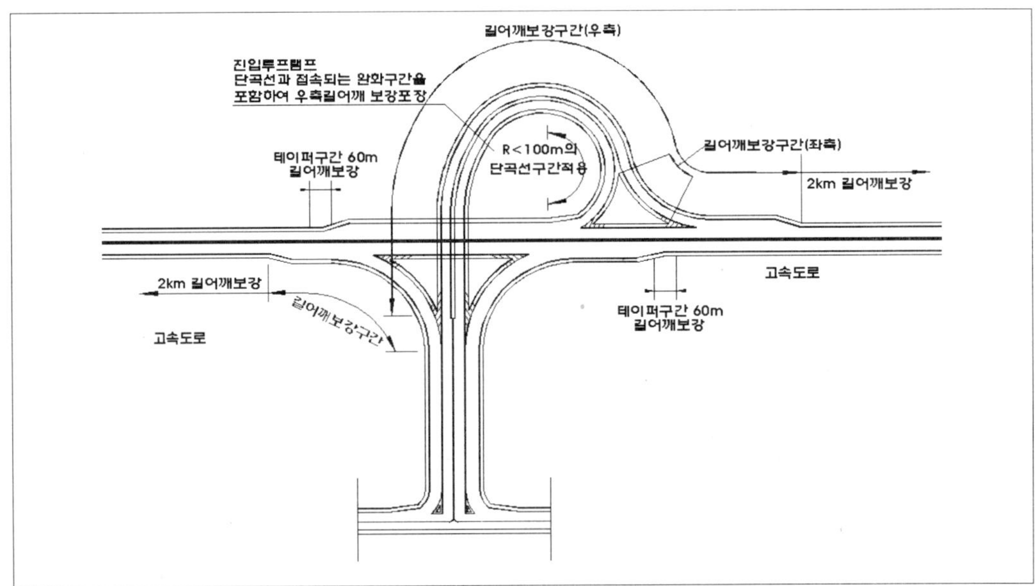

2-2-3 Connection Pavement

① Select pavement that can secure vehicle safety, construction easiness and economic feasibility for driving lane and strip pavement of connection

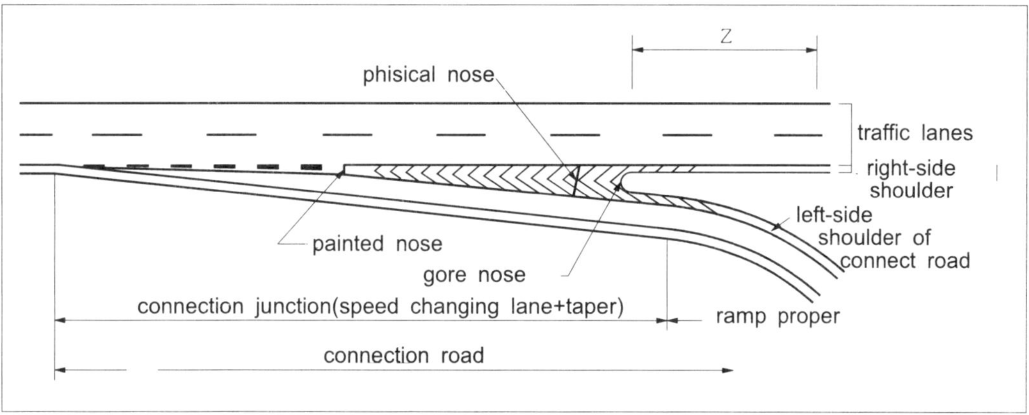

② Road shoulder pavement for connection

- Apply main lane pavement criteria so that the shoulder can be used as the second lane to prevent delay of highway main line vehicles.
- For entering loop lamp, right-side of road shoulder (of vehicle moving direction) including simple curve less than 100m and mitigate section connecting to it should be reinforced with main line pavement section.

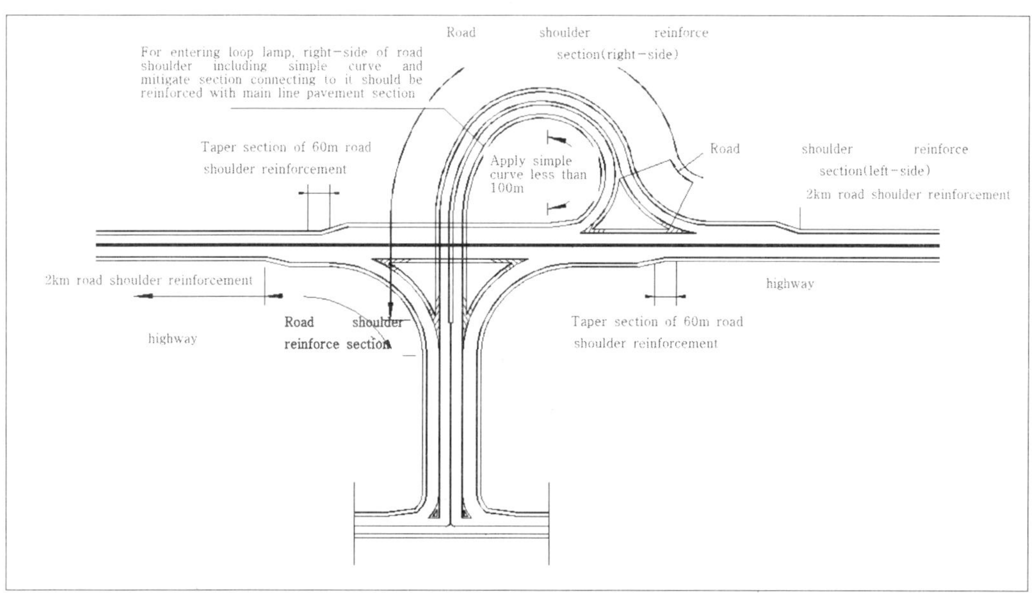

2-2-4 터널 내 포장

① 터널 내의 포장형식은 터널 전후의 토공부 포장과의 연속성, 건설 후 유지관리, 시공성, 저소음 특성 등을 고려하여 적정의 포장형식을 선택

② 터널 내 포장 설계에서 고려할 사항
- 터널 내 포장은 터널 출입부의 일정구간을 제외하고 계절적 온도변화를 적게 받음
- 포장층 내 함수비가 높으므로 내수성 재료가 요구
- 동절기에 포장 표면의 마모 작용이 큼
- 차량 주행 안전성을 위한 명색화 표면이 요구

③ 포장구성
- 시멘트 콘크리트 포장
  - 콘크리트 슬래브 + 시멘트 안정처리 필터층 또는 필터층
- 아스팔트 콘크리트 포장
  - 아스팔트 표층(중간층 포함) + 아스팔트 기층 + 필터층

④ 단면설계
- 시멘트 콘크리트 포장 : 시멘트안정처리 필터층
- 아스팔트 콘크리트 포장 : 필터층
  - 설계 시 두께는 용수량을 고려 200mm 적용(시공 시 용수량에 따라 조정)

⑤ 요철층 보정
- 적용재료 및 두께
  - 시멘트 안정처리 필터층, 두께 100mm
- 공동구 하부
  - 5종 콘크리트, 두께 100mm

2-2-5 영업소 포장

① 영업소 광장부 포장은 무근 콘크리트 포장을, 테이퍼부는 본선 포장과 동일한 형식 및 구조로 함

② 가로수축줄눈의 경우 6m로 설치

③ 품질확보를 위해 기계타설을 원칙으로함

2-2-4 Tunnel Pavement

① Select proper pavement type considering continuity with land-fill pavement, maintenance after completion, construction easiness, noise characteristics.

② Considerations
- Less impact from seasonal temperature change
- Requires durability due to higher water content
- Higher abrasion in winter
- Colored surface for driving safety

③ Pavement structure
- Concrete pavement
  - Concrete slab + Cement stabilize process filer layer
- Asphalt pavement
  - Asphalt surface layer (including binder course) + black base + Filter layer

④ Section design
- Concrete pavement : Cement stabilize process filer layer
- Asphalt pavement : Filter layer
  - Design thickness: 20cm considering water content (Adjust during construction according to water content)

⑤ Unevenness adjustment
- Material and thickness
  - Cement stabilize process filer layer, T=10cm
- Utility-pipe conduit
  - 5 type concrete, T=10cm

2-2-5 Operation Facility Pavement

① Apply JCP for open space part. Apply similar pavement with main lane for taper part.

② Horizontal contraction joint interval: 6m

③ Mechanical casting for quality assurance

④ 광장중앙부(36m, 아일랜드 구간)는 시공성 및 매설시설물(공동구 등)을 감안하여 철근콘크리트(줄눈간격 9m) 적용

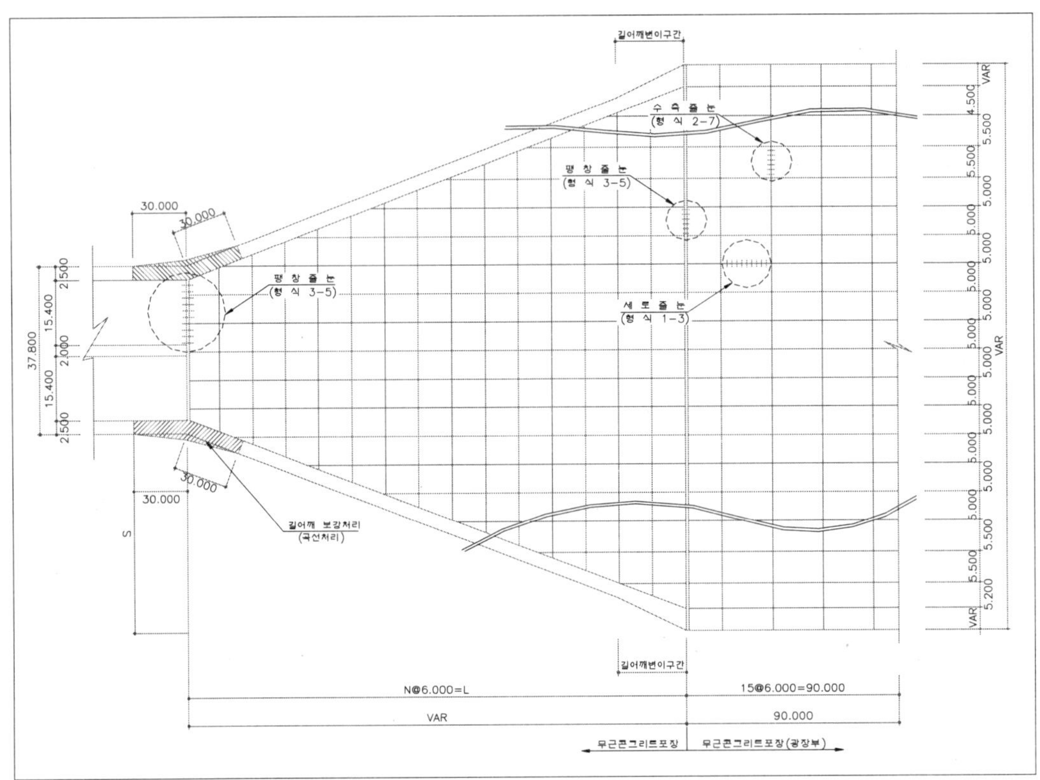

3. 포장의 설계

3-1 아스팔트 콘크리트 포장의 설계방법

3-1-1 아스팔트 콘크리트 포장 설계개념

① 교통하중의 누적손상

- 안전율을 고려하여 최대하중에 대한 저항능력의 확보를 위한 설계가 아니며 설계기간 동안의 누적손상에 대한 저항능력의 확보 설계개념

② 환경하중의 누적손상

- 동결융해 등과 같은 환경에 의한 손상이나 물리적 성질 변화를 고려하여 설계

④ Use reinforced concrete with joint of 9m interval for open space center (36m, island part) considering construction easiness and underground facilities (utility-pipe conduit)

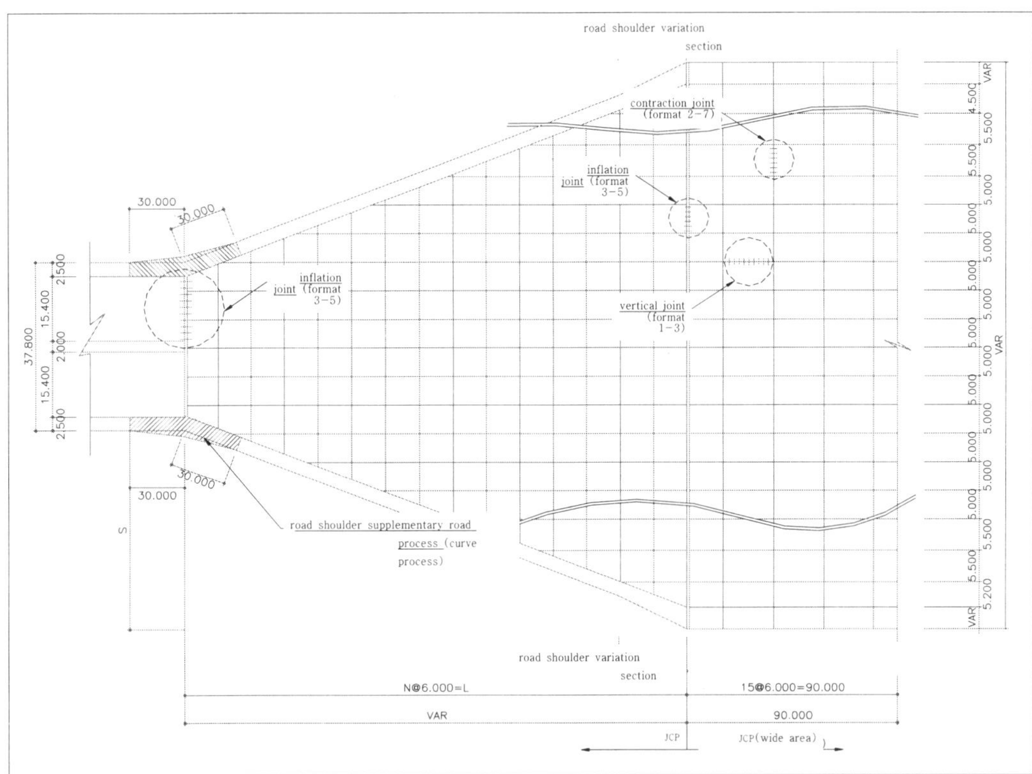

3. Pavement Design

3-1 Design Method for Asphalt Pavement

3-1-1 Design Concept for Asphalt Pavement

① Compounded damage from traffic load

- Consider compounded damage from traffic load for entire operating period to secure proper resistance for safety factor (not resistance for maximum load)

② Compounded damage from environmental load

- Consider physical changes such as frost-melting.

3-1-2 아스팔트 콘크리트 포장 설계법의 종류

국 외	국 내
• 미국 AASHTO 1972 설계법 • 일본 TA 설계법	• 미국 AASHTO 1972 설계법 • 현재 한국형 포장 설계법 개발 중

3-1-3 아스팔트 콘크리트 포장 설계절차

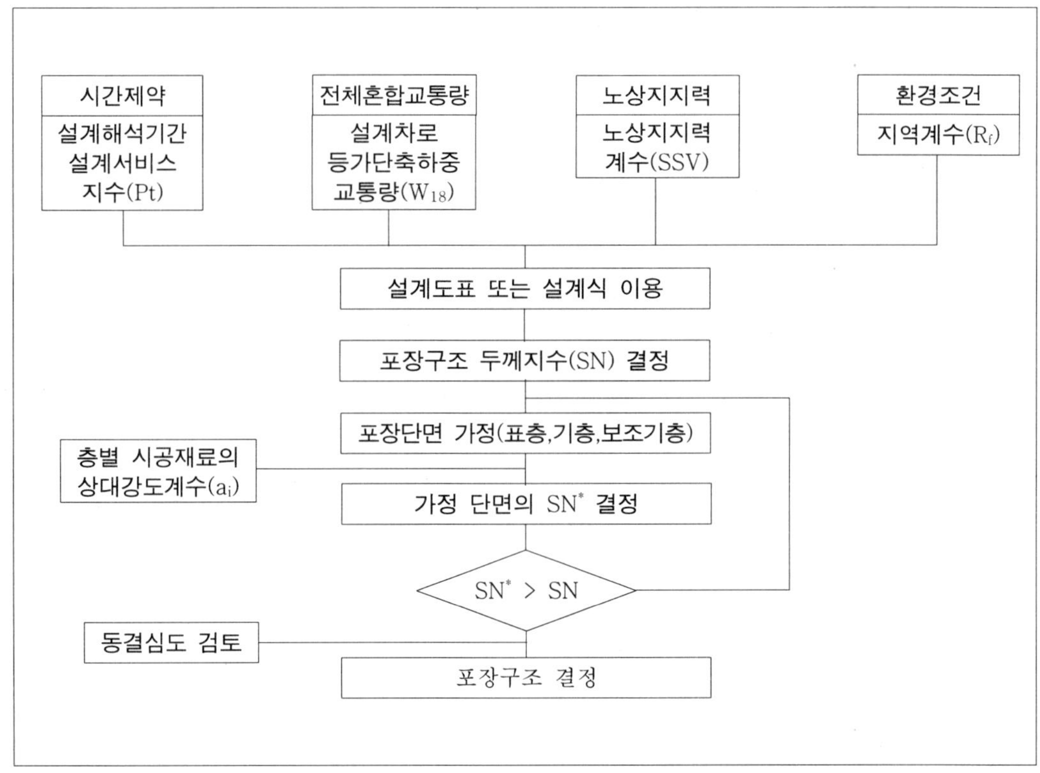

① 설계해석기간 및 설계서비스지수 결정
- 해석기간을 20년으로 설정하여 20년 예측교통량을 산정
- 최종 서비스 지수 (Terminal Serviceability Index, Pt) : 2.5

② 등가단축 교통량 산정
- 등가단축하중(Equivalent Single Axle Load, ESAL)
  - 설계기간의 다양한 총 교통량의 총합을 표현하기 위한 방법
  - 설계대상 도로를 통행하는 예상혼합교통량을 등가단축하중으로 표준화
  - 축하중 8.2ton의 단축(Single Axle)하중

3-1-2 Type of Asphalt Pavement Design Method

International	Domestic
• U.S AASHTO 1972 design method • Japan TA design method	• U.S AASHTO 1972 design method • Domestic design method is in development

3-1-3 Design Process for Asphalt Pavement

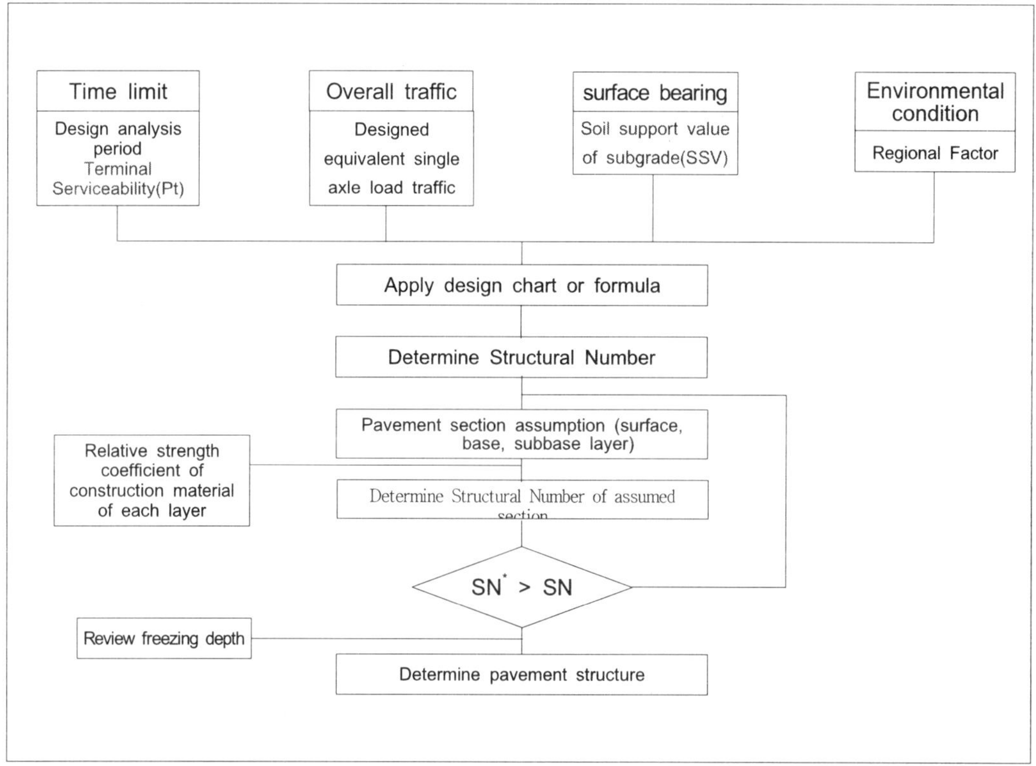

① Design analysis period and Design service point

- Period for analysis: 20 years. Estimate traffic volume for 20 years.
- Terminal Serviceability Index, Pt : 2.5

② Equivalent single axle load traffic

- Equivalent Single Axle Load, ESAL
  - Method to derive total traffic volume from various types of vehicles
  - Standardize estimated mixed traffic into ESAL
  - axle load 8.2ton, Single Axle load

- 등가단축하중 계수 (ESAL Factor, ESALf)

<table>
<tr><th colspan="2" rowspan="2">차 종</th><th rowspan="2">차축구성</th><th colspan="4">전체평균 ESALf</th><th colspan="2">고속도로 평균 ESALf</th><th colspan="2">일반국도+지방도 평균 ESALf</th></tr>
<tr><th colspan="2">가요성</th><th colspan="2">강성</th><th>가요성</th><th>강성</th><th>가요성</th><th>강성</th></tr>
<tr><td colspan="2">승용차</td><td>2A 4T</td><td colspan="2">0.0002</td><td colspan="2">0.0001</td><td>0.0002</td><td>0.0001</td><td>0.0002</td><td>0.0002</td></tr>
<tr><td rowspan="3">버스</td><td rowspan="2">소형</td><td>2A 4T</td><td colspan="2">0.001</td><td colspan="2">0.001</td><td>0.001</td><td>0.0004</td><td>0.001</td><td>0.001</td></tr>
<tr><td>2A 6T</td><td colspan="2">0.001</td><td colspan="2">0.001</td><td>0.001</td><td>0.001</td><td>0.001</td><td>0.001</td></tr>
<tr><td>보통</td><td>2A 6T</td><td colspan="2">0.852</td><td colspan="2">0.839</td><td>1.403</td><td>1.041</td><td>0.762</td><td>0.746</td></tr>
<tr><td rowspan="3">트럭</td><td>소형</td><td>2A 4T</td><td colspan="2">0.004</td><td colspan="2">0.004</td><td>0.015</td><td>0.015</td><td>0.001</td><td>0.001</td></tr>
<tr><td>보통</td><td>2A 4T</td><td>0.613</td><td rowspan="2">1.122</td><td>0.616</td><td rowspan="2">1.638</td><td>0.781</td><td>0.783</td><td>0.602</td><td>0.605</td></tr>
<tr><td>대형</td><td>3A 10T</td><td>2.047</td><td>3.417</td><td>1.472</td><td>2.407</td><td>2.392</td><td>4.022</td></tr>
<tr><td colspan="2" rowspan="3">트랙터+
세미트레일러</td><td>4A 이하</td><td>1.690</td><td rowspan="4">2.130</td><td>2.320</td><td rowspan="4">3.266</td><td>1.734</td><td>2.399</td><td>1.547</td><td>2.068</td></tr>
<tr><td>5A</td><td>1.815</td><td>3.072</td><td>1.292</td><td>2.139</td><td>2.422</td><td>4.156</td></tr>
<tr><td>6A 이상</td><td>0.858</td><td>1.533</td><td>0.471</td><td>0.756</td><td>1.676</td><td>3.173</td></tr>
<tr><td colspan="2">트럭+트레일러</td><td>5A 이하</td><td>3.288</td><td>4.472</td><td>2.977</td><td>4.022</td><td>3.768</td><td>5.165</td></tr>
</table>

- 설계 차로별 교통량 배분 산정

  $DTN = W_{18} \times D_D \times D_L$

  - DTN : 설계차로 교통하중
  - W_{18} : 누가 18kips(8.2ton) 등가 교통하중
  - D_D : 중방향 계수(0.5)
  - D_L : 차로분배계수

일방향 차로 수	분배계수(%)			적 용
	현 행	AASHTO	기존 고속도로 분 석 결 과	
1	100	100	100	100
2	80	80~100	70	80
3	60	60~80	60	60
4	50	50~75	50	50

- ESAL Factor, ESALf

<table>
<tr><th colspan="2" rowspan="2">Type of vehicle</th><th rowspan="2">Axle structure</th><th colspan="4">Total average ESALf</th><th colspan="2">Expressway average ESALf</th><th colspan="2">National road and providential road ESALf</th></tr>
<tr><th colspan="2">Flexibility</th><th colspan="2">Rigidity</th><th>Flexibility</th><th>Rigidity</th><th>Flexibility</th><th>Rigidity</th></tr>
<tr><td colspan="2">Sedan</td><td>2A 4T</td><td colspan="2">0.0002</td><td colspan="2">0.0001</td><td>0.0002</td><td>0.0001</td><td>0.0002</td><td>0.0002</td></tr>
<tr><td rowspan="3">Bus</td><td rowspan="2">Small</td><td>2A 4T</td><td colspan="2">0.001</td><td colspan="2">0.001</td><td>0.001</td><td>0.0004</td><td>0.001</td><td>0.001</td></tr>
<tr><td>2A 6T</td><td colspan="2">0.001</td><td colspan="2">0.001</td><td>0.001</td><td>0.001</td><td>0.001</td><td>0.001</td></tr>
<tr><td>Regular</td><td>2A 6T</td><td colspan="2">0.852</td><td colspan="2">0.839</td><td>1.403</td><td>1.041</td><td>0.762</td><td>0.746</td></tr>
<tr><td rowspan="3">Truck</td><td>Small</td><td>2A 4T</td><td colspan="2">0.004</td><td colspan="2">0.004</td><td>0.015</td><td>0.015</td><td>0.001</td><td>0.001</td></tr>
<tr><td>Regular</td><td>2A 4T</td><td>0.613</td><td rowspan="2">1.122</td><td>0.616</td><td rowspan="2">1.638</td><td>0.781</td><td>0.783</td><td>0.602</td><td>0.605</td></tr>
<tr><td>Large</td><td>3A 10T</td><td>2.047</td><td>3.417</td><td>1.472</td><td>2.407</td><td>2.392</td><td>4.022</td></tr>
<tr><td colspan="2" rowspan="3">Tractor+ Semi-trailer</td><td>Under 4A</td><td>1.690</td><td rowspan="4">2.130</td><td>2.320</td><td rowspan="4">3.266</td><td>1.734</td><td>2.399</td><td>1.547</td><td>2.068</td></tr>
<tr><td>5A</td><td>1.815</td><td>3.072</td><td>1.292</td><td>2.139</td><td>2.422</td><td>4.156</td></tr>
<tr><td>Over 6A</td><td>0.858</td><td>1.533</td><td>0.471</td><td>0.756</td><td>1.676</td><td>3.173</td></tr>
<tr><td colspan="2">Truck+trailer</td><td>Under 5A</td><td>3.288</td><td>4.472</td><td>2.977</td><td>4.022</td><td>3.768</td><td>5.165</td></tr>
</table>

- Traffic distribution for each designed lane

$DTN = W_{18} \times D_D \times D_L$

- DTN : Designed lane traffic load
- W_{18} : Accumulated 18kips(8.2ton) equivalent traffic load
- D_D : Directional distribution (0.5)
- D_L : Lane distribution coefficient

Number of lanes (one-way)	Distribution coefficient (%)			Application
	Current	AASHTO	Analysis result from current highways	
1	100	100	100	100
2	80	80~100	70	80
3	60	60~80	60	60
4	50	50~75	50	50

③ 설계CBR 산정 및 노상지지력 계수 결정

- 설계CBR 산정
  - 시험 조사 시 채취된 노상토에 대하여 실내 CBR 실험으로 산정
    (KS F 2320 노상 지지력비(CBR) 실험 방법)

- 노상지지력 계수 결정
  - 측정 CBR에 의해 산정된 지지력 계수와 노상 다짐 조건에 의한 품질기준인 지지력 계수의 값 중에서 큰 값으로 결정
  - 측정 CBR을 이용하여 "CBR과 k값의 상관계수 도표"를 통한 k 산정
  - 시방에 규정된 노상의 최소 지지력 값인 k_{30} =15kgf/cm^2

④ 지역계수(R_f) 결정

- R.F = 1.5 (대전이남)
- R.F = 2.0 (기타지역)
- R.F = 2.5 (서울이북, 표고 500m 이상 지역)

⑤ 포장재료의 상대강도계수(a) 결정

재 료 종 류	품 질 규 격	상대강도계수	지지력값
아스팔트 표층	안정도 750kg 이상	0.157	
아스팔트 중간층	안정도 500kg 이상	0.145	
아스팔트 기층	안정도 350kg 이상	0.110	
보조기층 (석산 쇄석)	C.B.R 80 이상	0.051	8.5
보조기층 (석산 쇄석)	C.B.R 50 이상	0.048	7.8
보조기층(막자갈, 강모래/자갈)	C.B.R 30 이상	0.043	6.9

⑥ 동결심도 산정

- 동결지수(freezing index)
  - 포장 내의 동결관입깊이를 산정하기 위한 대표적 척도
  - 포장구조와 노상토를 동결시키는 대기온도의 강도와 지속기간(intensity and duration)의 누가영향(cumulative effect)으로 표시
  - 단위는 온도 · 일(℃ · 일, °F · 일)
  - 전국동결지수도에 따라 지점별 동결지수 산정

③ Designed CBR and soil support value

- Designed CBR
  - Interior CBR test with extracted surface soil (KS F 2320 CBR test method)

- Soil support value
  - Higher value between soil support value from measured CBR and quality control value for soil support value from soil compacting condition
  - calculate k vaule from 'coefficient correlation chart between CBR and k value' using measured CBR
  - Minimum soil support value k_{30} =15kgf/cm^2

④ Regional Factor (R_f)

- R.F = 1.5 (South of Daejun)
- R.F = 2.0 (Others)
- R.F = 2.5 (North of Seoul, altitude of 500m and higher)

⑤ Relative strength coefficient for pavement material (a)

Material type	Quality standard	Relative strength coefficient	Support value
Surface course	Stability 750kg and higher	0.157	
Binder course	Stability 500kg and higher	0.145	
Black base	Stability 350kg and higher	0.110	
Subbase layer (quarry crushed aggregate)	C.B.R 80 and higher	0.051	8.5
Subbase layer (quarry crushed aggregate)	C.B.R 50 and higher	0.048	7.8
Subbase layer (gravel, river sand)	C.B.R 30 and higher	0.043	6.9

⑥ Freezing depth

- Freezing index
  - Typical standard to calculate freezing penetration in pavement
  - Indicated as cumulative effect of intensity and duration of atmosphere emperature that freezes pavement structure and surface soil
  - Unit: temperature ・day (℃・day, °F・day)
  - Use national freezing index to calculate the site freezing index

【지역별 동결지수 및 동결기간】

지 역	측후소 지반고(m)	동결지수 (℃·일)	동결기간 (일)	지 역	측후소 지반고(m)	동결지수 (℃·일)	동결기간 (일)
속 초	17.6	181.6	66	합천	32.1	193.0	62
대관령	842.0	873.8	127	거창	224.9	278.2	74
춘 천	74.0	539.0	92	영천	91.3	237.8	64
강 릉	26.0	167.2	57	구미	45.5	278.1	76
서 울	85.5	380.9	80	의성	73.0	425.2	78
인 천	68.9	354.7	78	영덕	40.5	138.8	57
원 주	149.8	613.0	94	문경	172.1	279.4	55
울릉도	221.1	129.3	32	영주	208.0	417.8	77
수 원	36.9	468.4	79	성산포	17.5	-	-
충 주	69.4	528.4	89	고흥	60.0	83.5	49
서 산	26.4	313.2	76	해남	22.1	102.6	49
울 진	49.5	121.6	57	장흥	43.0	130.1	52
청 주	59.0	411.6	78	순천	74.0	179.9	64
대 전	67.2	317.7	68	남원	89.6	272.4	67
추풍령	245.9	303.9	78	정읍	40.5	223.9	61
포 항	2.5	98.5	52	임실	244.0	420.3	86
군 산	26.3	194.9	61	부안	7.0	244.7	61
대 구	57.8	160.9	54	금산	170.7	372.5	77
전 주	51.2	233.5	61	부여	16.0	330.0	74
울 산	31.5	83.6	46	보령	15.1	254.8	76
광 주	73.9	141.4	55	아산	24.5	405.4	78
부 산	69.2	49.6	27	보은	170.0	461.7	76
통 영	25.0	37.4	27	제천	264.4	610.2	91
목 포	36.5	75.6	33	홍천	141.0	635.4	98
여 수	67.0	62.2	31	인제	199.7	614.5	91
완 도	37.5	38.1	26	이천	68.5	511.0	89
제 주	22.0	4.1	3	양평	49.0	619.7	91
남 해	49.8	74.3	38	강화	46.4	486.2	89
거 제	41.5	52.1	39	진주	21.5	132.8	51
산 청	141.8	141.8	49	서귀포	51.9	-	-
밀 양	12.5	180.2	62	철원	154.9	685.0	109

- 설계동결지수
  - 설계동결지수(°F×일) = 동결지수 ± 0.9 × 동결기간 × (표고차/100)
  - 표고차 = 설계구간 최대계획고(m) - 측후소 지반고(m)

- 동결관입 깊이
  - 동결관입깊이와 설계동결지수 상관도표에 의거 산정

【Freezing index and duration per region】

Region	Weather station ground height(m)	Freezing index (℃ · day)	Freezing duration (day)	Region	Weather station ground height(m)	Freezing index (℃ · day)	Freezing duration (day)
Sokcho	17.6	181.6	66	Hapcheon	32.1	193.0	62
Daegyanryung	842.0	873.8	127	Geochang	224.9	278.2	74
Chuncheon	74.0	539.0	92	Youngcheon	91.3	237.8	64
Gangneung	26.0	167.2	57	Gumi	45.5	278.1	76
Seoul	85.5	380.9	80	Uisung	73.0	425.2	78
Incheon	68.9	354.7	78	Youngduk	40.5	138.8	57
Wonju	149.8	613.0	94	Mungyeong	172.1	279.4	55
Ulreungdo	221.1	129.3	32	Youngju	208.0	417.8	77
Suwon	36.9	468.4	79	Seongsanpo	17.5	-	-
Chungju	69.4	528.4	89	Goheung	60.0	83.5	49
Seosan	26.4	313.2	76	Haenam	22.1	102.6	49
Uljin	49.5	121.6	57	Jangheung	43.0	130.1	52
Cheongju	59.0	411.6	78	Suncheon	74.0	179.9	64
Daejeon	67.2	317.7	68	Namwon	89.6	272.4	67
Chupungryeng	245.9	303.9	78	Jungeup	40.5	223.9	61
Pohang	2.5	98.5	52	Imsil	244.0	420.3	86
Gunsan	26.3	194.9	61	Buan	7.0	244.7	61
Daegu	57.8	160.9	54	Gumsan	170.7	372.5	77
Junju	51.2	233.5	61	Buyeo	16.0	330.0	74
Ulsan	31.5	83.6	46	Boryeog	15.1	254.8	76
Gwangju	73.9	141.4	55	Asan	24.5	405.4	78
Busan	69.2	49.6	27	Boeun	170.0	461.7	76
Tongyoung	25.0	37.4	27	Jecheon	264.4	610.2	91
Mokpo	36.5	75.6	33	Hongcheon	141.0	635.4	98
Yeosu	67.0	62.2	31	Injae	199.7	614.5	91
Wando	37.5	38.1	26	Inchon	68.5	511.0	89
Jeju	22.0	4.1	3	Yangpyeong	49.0	619.7	91
Namhae	49.8	74.3	38	Ganghwa	46.4	486.2	89
Geojae	41.5	52.1	39	Jinju	21.5	132.8	51
Sanchung	141.8	141.8	49	Seogyipo	51.9	-	-
Milyang	12.5	180.2	62	Cholwon	154.9	685.0	109

- Designed freezing index
  - Designed freezing index (°F×day) = freezing index ± 0.9 × freezing period × (height difference/100)
  - Height difference = maximum planned height of designing section (m) - Weather station ground height (m)

- Freezing penetration depth
  - Determine based on coefficient correlation chart of freezing penetration depth and designed freezing index

【동결관입깊이와 설계동결지수 상관도표】

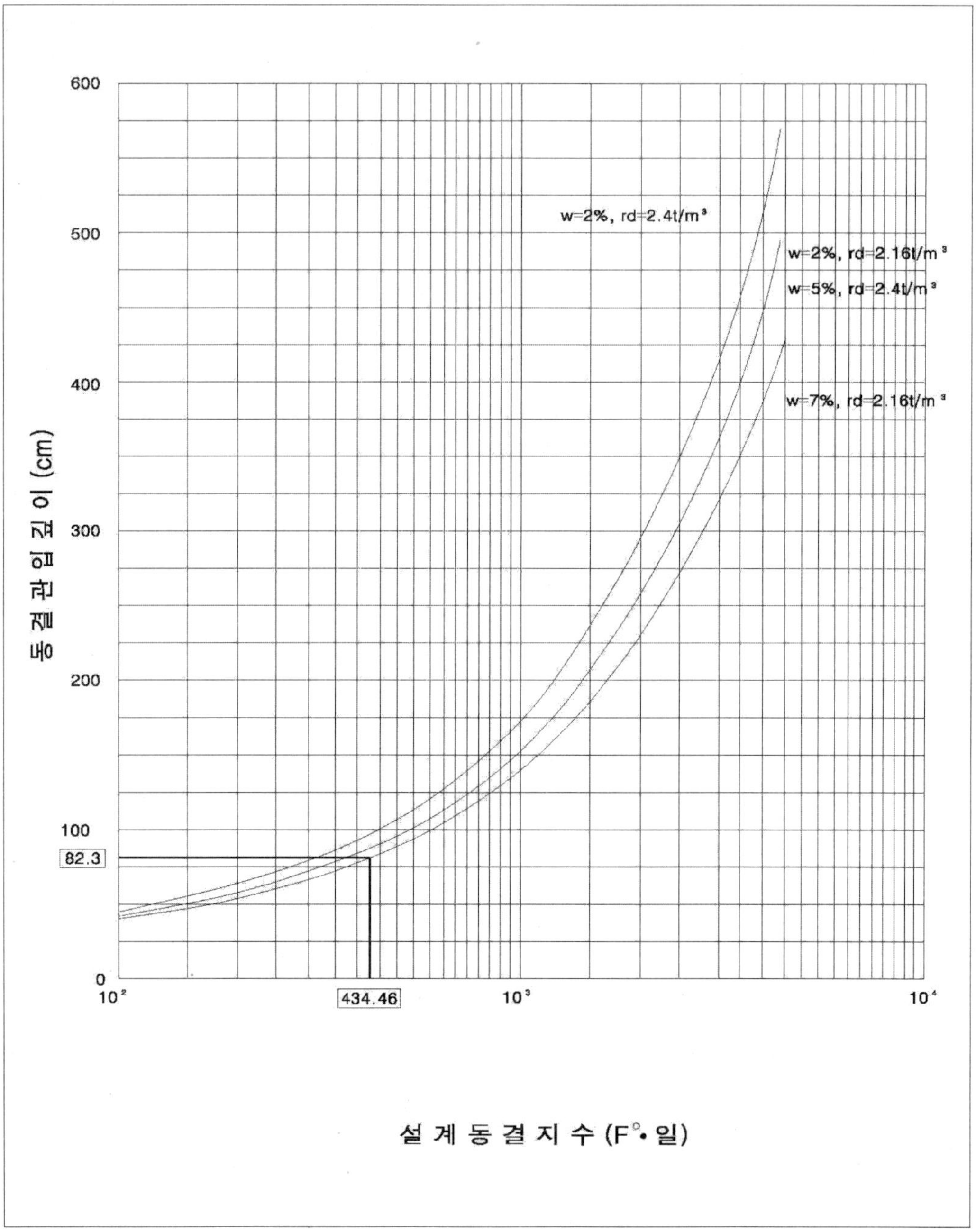

동결관입깊이 (cm)

600
500
400
300
200
100
82.3
0

10^2 434.46 10^3 10^4

w=2%, rd=2.4t/m³
w=2%, rd=2.16t/m³
w=5%, rd=2.4t/m³
w=7%, rd=2.16t/m³

설 계 동 결 지 수 (F°·일)

【Coefficient correlation chart for frost penetration depth and designed freezing index】

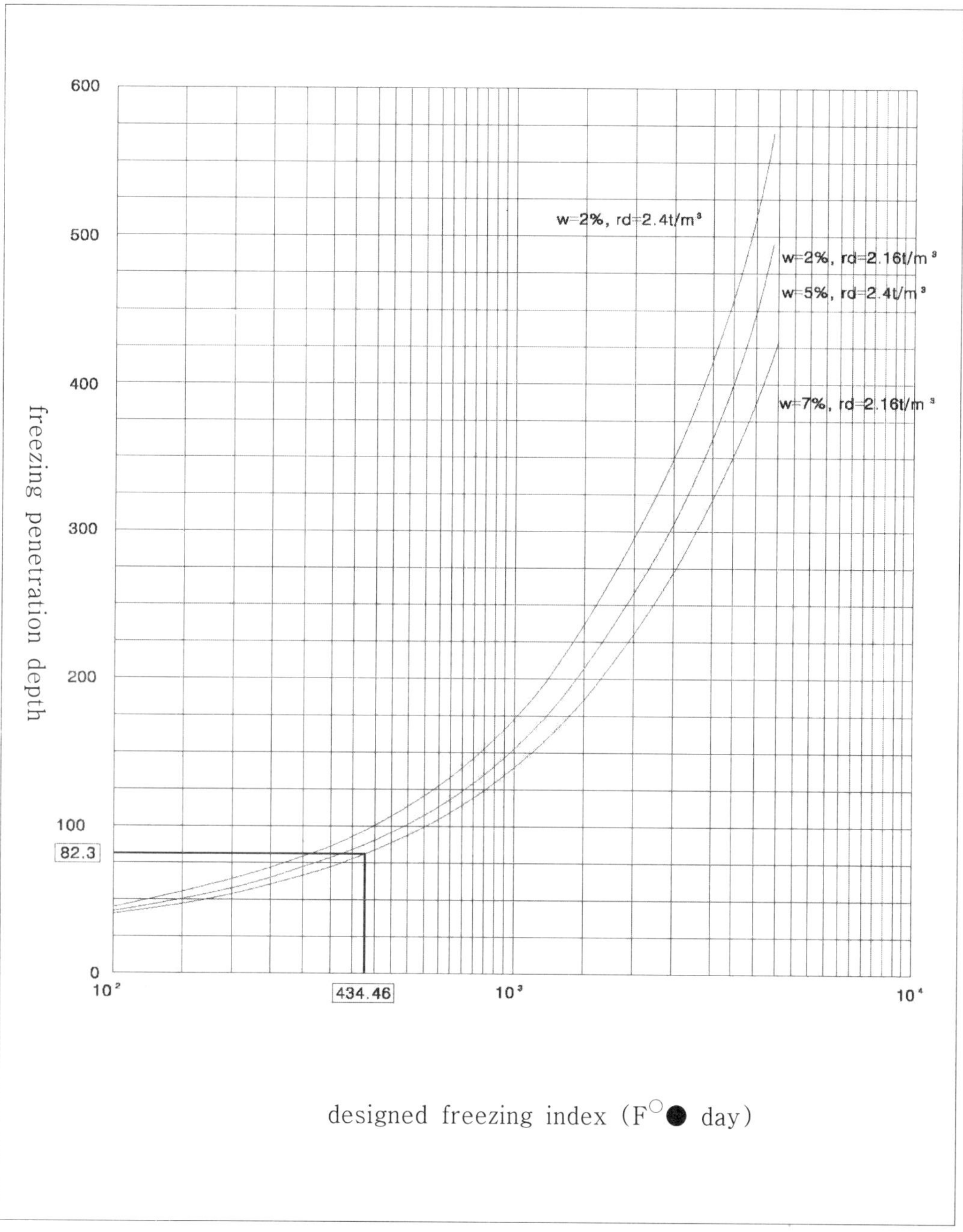

• 동결심도 산정방법의 종류

구 분	개 요
완전방지법	완전방지법은 동결작용으로 인한 포장표면의 변형을 완전히 방지하기 위한 방법으로, 동결심도는 수정 Berggren식에 의거하여 작성한 도표로부터 구함
노상동결관입허용법에 의한 산정	노상동결관입허용법은 동결이 노상 내에 어느 정도 관입되어도 동상으로 인한 피해가 포장파괴를 일으키지 않는 한도 내에서 노상동결을 허용하는 방법으로, 슬래브 두께 P를 가정(30cm)하여 도표로부터 층두께(b)를 산정 a = 콘크리트층 두께 (p) + 비동상재료층 두께(c)

• 동결심도 산정방법 : 노상동결관입허용법을 적용함
- 노상동결관입허용법에 의한 동결심도

$$c = a - p$$

c : 비동결성 재료 치환 최대깊이(cm)
a : 설계 동결관입깊이(cm)
p : 아스팔트 콘크리트 포장의 아스팔트 혼합물층 두께(cm)

- 도표에서 c(cm)의 연직선과 r=2.0선의 만나는 점의 좌측값인 비동결성 재료층의 두께 b(cm)를 얻을 수 있음
- 노상동결관입허용법에 의한 동결심도는 p+b
- 완전방지법에 의한 동결심도는 p+c

⑦ 설계포장 소요두께지수(SN) 산정
• 포장구조의 각 층에서 사용되는 재료의 형태에 따른 적합한 상대강도계수를 사용하여 가요성 포장층의 두께를 환산하기 위하여 교통해석, 노상토의 상태, 환경조건으로부터 유도된 지수

• 경험적 공용방정식에 의한 설계포장 소요두께지수 산정방법

$$\log10(W_{8.2}) = 9.36 \times \log10(SN+1) - 0.20 + \frac{Gt}{0.4 + \frac{1,094}{(SN+1)^{5.19}}}$$

$$- \log10\left(\frac{1}{R_f}\right) + 0.372(SSV - 3.0)$$

- $W_{8.2}$: 설계공용기간 동안의 8.2ton 등가단축하중의 누가통과횟수

• Freezing penetration depth calculation method

Classification	Overview
Complete prevention	complete prevention method prevents surface deformation from freezing entirely. Freezing penetration depth can be estimated by chart based on Berggren equation
Partial allowance for freezing penetration	Allow freezing of surface within the scope that the freezing does not yield pavement damage. Assume slab thickness P (30cm) to calculate layer thickness (b) from chart a = concrete layer thickness (p) + non-freezing material thickness (c)

• Freezing penetration depth calculation : Use partial allowance method

- Freezing penetration depth by partial allowance for freezing penetration

$$c = a - p$$

c : maximum non-freezing material substitution depth (cm)

a : designed freezing penetration depth (cm)

p : asphalt mixture thickness of asphalt pavement (cm)

- From the chart, left-side value of meeting point of vertical line of c(cm) and r=2.0 line is non-freezing material thickness b (cm)
- freezing penetration depth by partial allowance is p+b
- freezing penetration depth by complete prevention is p+c

⑦ Designed pavement required thickness index (SN) calculation

• SN is induced from traffic analysis, soil and environmental conditions to yield flexible pavement thickness using appropriate relative rigidity coefficient according to used material

• SN determination using empirical commonality equation

$$\log10(W_{8.2}) = 9.36 \times \log10(SN + 1) - 0.20 + \frac{Gt}{0.4 + \frac{1,094}{(SN+1)^{5.19}}} - \log10(\frac{1}{R_f}) + 0.372(SSV - 3.0)$$

- $W_{8.2}$: Compounded number of passage of 8.2 ton equivalent single axis load during the designed operation period

- R_f : 지역계수
- SSV : 노상 지지력계수
- SN : 소요 전체 포장층의 구조적 강도를 표시하는 포장두께지수로서, 아스팔트포장의 구성 각 층 두께와 상대강도계수의 곱의 합으로 표시
- $SN = \Sigma a_i \cdot D_i$
- a_i : i번째 층의 상대강도계수
- D_i : i번째 층의 두께(cm)
- $G_t = \log_{10}\left(\dfrac{4.2 - P_t}{4.2 - 1.5}\right)$로 표시되며, 임의시점(최종 서비스지수를 가지는 시점)에 이어 P_t = 1.5인 경우의 최대 서비스 능력 손실량에 대한 서비스 능력의 손실량 비율을 나타내는 함수

④ 설계포장 두께지수(SN^*) 결정

- 각 포장층의 최소두께 값을 기준으로 가정된 포장두께와 상대강도계수를 사용하여 설계포장 두께지수를 결정

- 설계포장 두께지수의 값이 설계포장 소요두께지수보다 크도록 포장두께의 값을 변경하여 적합한 포장두께를 결정

- 설계포장 두께지수 관계식
  $SN^* = a_1D^*_1 + a_2D^*_2 + a_3D^*_3$

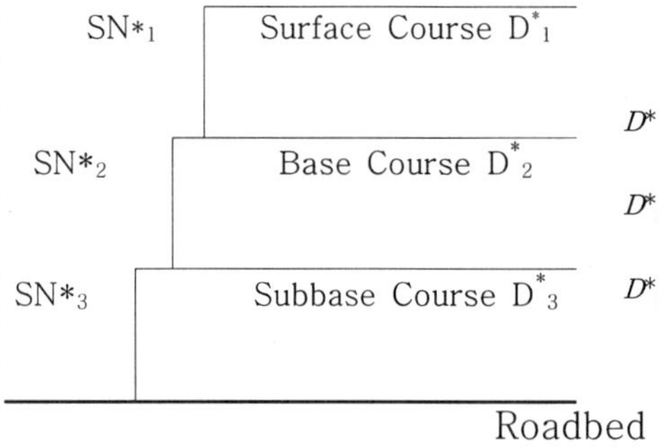

$SN*_1$ Surface Course D^*_1

$SN*_2$ Base Course D^*_2

$SN*_3$ Subbase Course D^*_3

Roadbed

$$D^*_1 \geq \frac{SN^*_1}{a_1} \qquad SN^*_1 = a_1 D^*_1 \geq SN_1$$

$$D^*_2 \geq \frac{SN_2 - SN^*_1}{a_2} \qquad SN^*_1 + SN^*_2 \geq SN_2$$

$$D^*_3 = \frac{SN_3 - (SN^*_1 + SN^*_2)}{a_3}$$

주) D* : 설계두께, SN* : 설계 SN치

- a_1, a_2, a_3 = 포장각층의 상대강도계수
- D^*_1, D^*_2, D^*_3 = 포장각층의 두께

- R_f : Regional factor
- SSV : Soil Support Value
- SN : Structural number to indicate overall structural rigidity. Sum of layer thickness times relative rigidity coefficient
- $SN = \Sigma a_i \cdot D_i$
- a_i : Relative rigidity coefficient of layer i
- D_i : Thickness of layer i (cm)
- $G_t = \log_{10}\left(\frac{4.2 - P_t}{4.2 - 1.5}\right)$ function to indicate serviceability loss ratio of maximum serviceability when P_t=1.5 and certain point of time (when terminal serviceability index is determined)

④ Designed pavement thickness index (SN^*) determination

- Determine SN^* using relative rigidity coefficient assuming minimum thickness for each layer
- Determine thickness of pavement so that SN^* gets larger than SN
- Relation equation of SN^*

  $SN^* = a_1D^*_1 + a_2D^*_2 + a_3D^*_3$

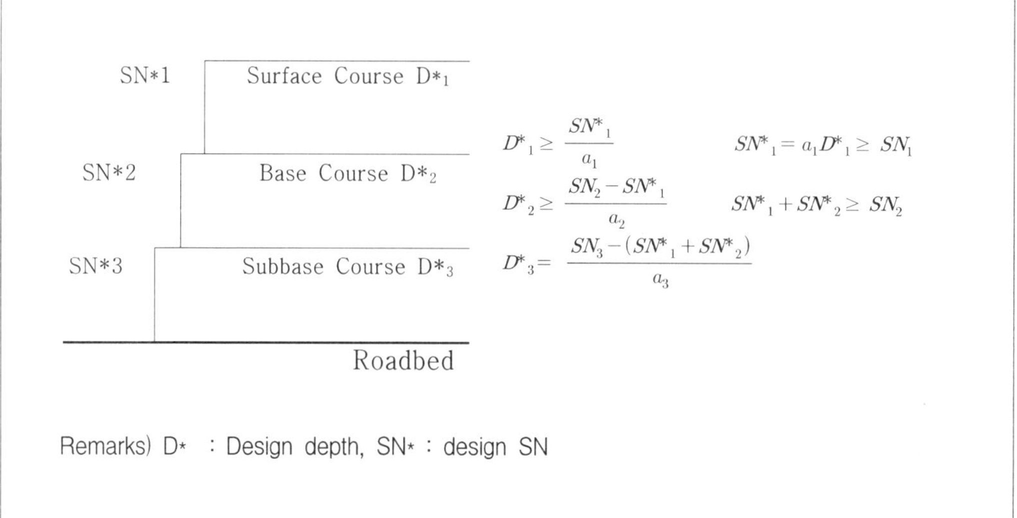

- a_1, a_2, a_3 = Layer cofficients representative of various course
- D^*_1, D^*_2, D^*_3 = Actual thickness of various course

3-2 시멘트 콘크리트 포장의 설계방법

3-2-1 시멘트 콘크리트 포장 설계개념

① 교통하중의 누적손상

- 안전율을 고려하여 최대하중에 대한 저항능력의 확보를 위한 설계가 아니며 설계기간 동안의 누적손상에 대한 저항능력의 확보 설계개념

② 환경하중의 누적손상

- 동결융해 등과 같은 환경에 의한 손상이나 물리적 성질 변화를 고려하여 설계

3-2-2 시멘트 콘크리트 포장 설계법의 종류

국 외	국 내
• 미국 AASHTO 1981 설계법 • 미국 AASHTO 1986 설계법 • 미국 AASHTO 2002 설계법 (적용을 위한 검토 중) • 미국 PCA 설계법 • 일본 도로협회 설계법	• 미국 AASHTO 1986 설계법 + AASHTO 1981(부분적 적용) • 현재 한국형 포장 설계법 개발 중

3-2-3 콘크리트 포장 설계 절차

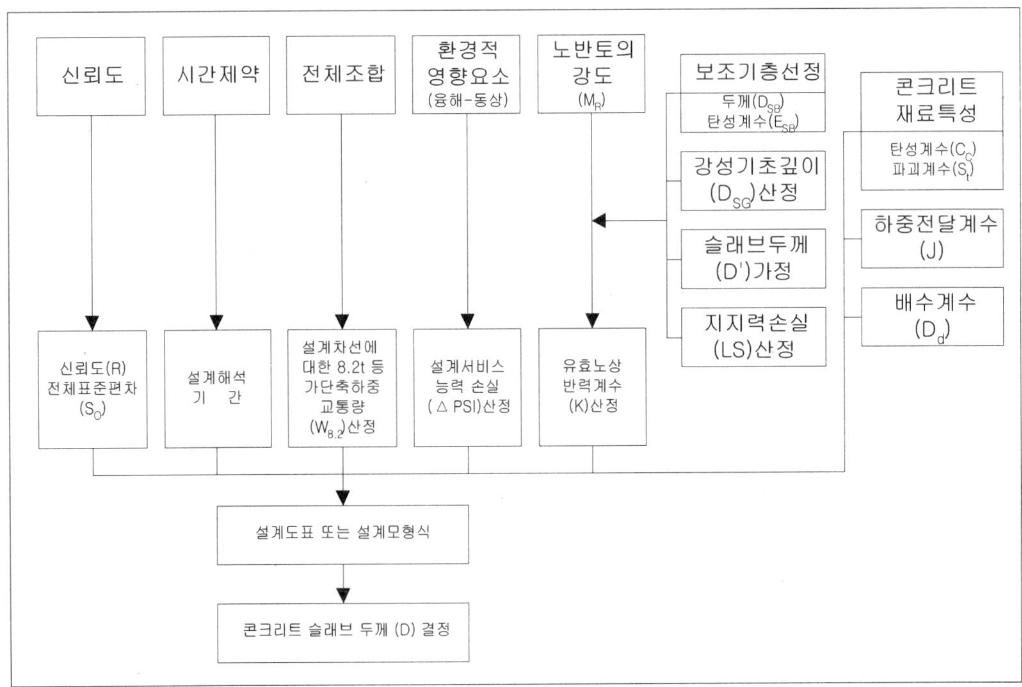

3-2 Design Method for Concrete Pavement

3-2-1 Design Concept for Concrete Pavement

① Compounded damage from traffic load

- Consider compounded damage from traffic load for entire operating period to secure proper resistance for safety factor (not resistance for maximum load)

② Compounded damage from environmental load

- Consider physical changes such as frost-melting.

3-2-2 Types of Concrete Pavement Design Method

International	Domestic
• US AASHTO 1981 design • US AASHTO 1986 design • US AASHTO 2002 design (in review for application) • US PCA design • Japan road association design	• US AASHTO 1986 design + AASHTO 1981(partial application) • Domestic design method is in development

3-2-3 Design Process for Concrete Pavement

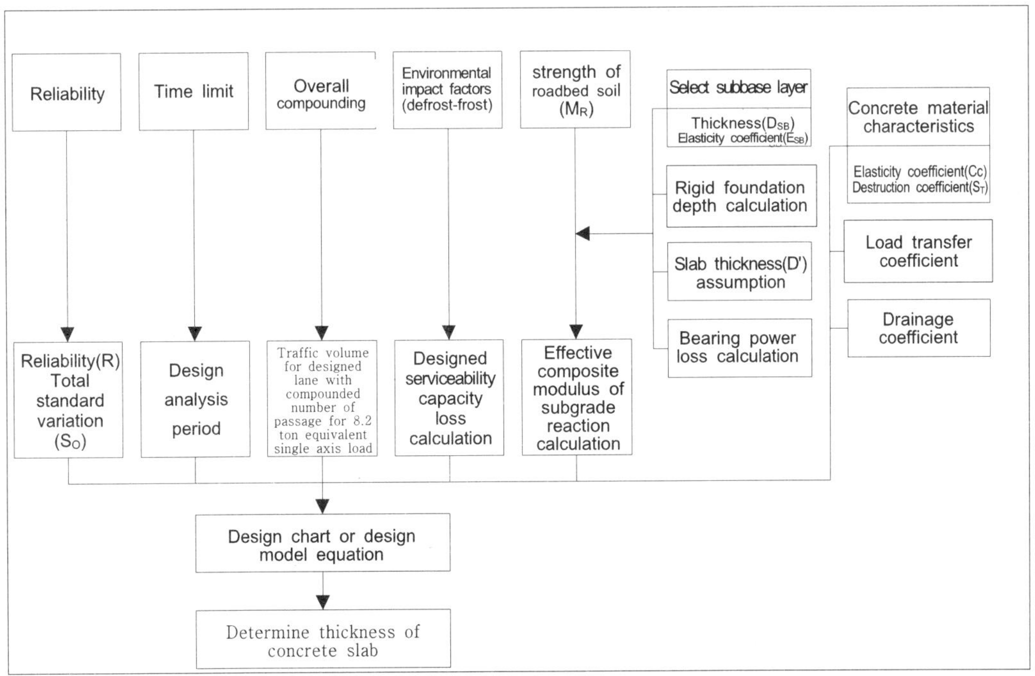

① 설계해석기간 및 설계서비스지수 결정

- 해석기간을 20년으로 설정하여 20년 예측교통량을 산정
- 최종 서비스 지수 (Terminal Serviceability Index, Pt) : 2.5

② 등가단축 교통량 산정(아스팔트 콘크리트 설계 절차와 동일)

③ 설계CBR 산정 및 노상지지력 계수 결정

- 설계CBR 산정
  - 시험 조사 시 채취된 노상토에 대하여 실내 CBR 실험으로 산정 (KS F 2320 노상 지지력비(CBR) 실험 방법)

- 노상지지력 계수 결정
  - 측정 CBR에 의해 산정된 지지력 계수와 노상 다짐 조건에 의한 품질기준인 지지력 계수의 값 중에서 큰 값으로 결정
  - 측정 CBR을 이용하여 "CBR과 k값의 상관계수 도표"를 통한 k 산정
  - 시방에 규정된 노상의 최소 지지력 값인 k30 =15kgf/cm2

④ 노상의 동탄성 계수(MR)

- 노상의 역학적 특성을 대표하는 동탄성 계수
- AASHTO 설계법에서 제시되는 노상 동탄성계수와 지지력 계수의 상관관계 이용
- 상관관계식

$$k = \frac{M_R}{24}$$

- k : 노상지지력계수

④ 합성 노상지지력 계수(Composite modulus of subgrade reaction) 결정

- 노상 위에 설치되는 동상방지층, 린 콘크리트 기층의 기여효과를 고려한 최종적인 포장 슬래브 하단의 지지력 계수

- 기본방정식

$$\ln k = -2.807 + 0.1253\times(\ln D_{SB})2 + 1.062\times\ln M_R - 0.1282\times\ln D_{SB}\times\ln E_{SB}$$
$$-0.4114\times\ln D_{SB} - 0.0581\times\ln E_{SB} - 0.1371\times\ln D_{SB}\times\ln M_R$$

- k : 합성노상반력계수(pci)
- D_{SB} : 상부층(동상방지층, 린콘크리트층)의 두께(inch)
- E_{SB} : 상부층(동상방지층, 린콘크리트층)의 탄성계수(psi)
- M_R : 노상의 동탄성 계수(psi)

① Design analysis period and Design serviceability index

- Period for analysis: 20 years. Estimate traffic volume for 20 years.
- Terminal Serviceability Index, P_t : 2.5

② Calculation of equivalent single axix traffic volume (similar to asphalt pavement design process)

③ Designed CBR and soil support value

- Designed CBR
  - Interior CBR test with extracted surface soil (KS F 2320 CBR test method)
- Soil support value
  - Higher value between soil support value from measured CBR and quality control value for soil support value from soil compacting condition
  - Calculate k value from 'coefficient correlation chart between CBR and k value' using measured CBR
  - Minimum soil support value k_{30} =15kgf/cm^2

④ Dynamic elasticity coefficient of soil surface layer (M_R)

- The coefficient represents dynamic characteristics of soil surface layer
- Use correlation equation suggested in AASHTO design
- The correlation equation

$$k = \frac{M_R}{24}$$

  - k : Modulus of subgrade reaction

④ Composite modulus of subgrade reaction (M_R)

- Final bearing power of paved slab's lower part considering contribution of anti-frost layer, lean concrete layer
- Basic equation

$$\ln k = -2.807 + 0.1253\times(\ln D_{SB})2 + 1.062\times\ln M_R - 0.1282\times\ln D_{SB}\times\ln E_{SB} - 0.4114\times\ln D_{SB} - 0.0581\times\ln E_{SB} - 0.1371\times\ln D_{SB}\times\ln_{MR}$$

  - k : composite modulus of subgrade reaction (pci)
  - D_{SB} : upper layer thickness (anti-frost layer, lean concrete layer) (inch)
  - E_{SB} : upper layer elasticity coefficient (anti-frost layer, lean concrete layer) (psi)
  - M_R : dynamic elasticity coefficient of soil surgrade layer (psi)

⑤ 하중전달계수(J) 결정

- 줄눈, 균열 또는 포장과 길어깨 사이 등 콘크리트포장 슬래브의 불연속 지점에서 하중전달의 능력을 나타내는 계수 (J=2.9)

⑥ 동결심도 산정(아스팔트 콘크리트 포장 설계절차와 동일)

- 동결심도 산정방법 : 노성동결관입허용법을 적용함
  - 노상동결관입허용법에 의한 동결심도

$$c = a - p$$

c : 비동결성 재료 치환 최대깊이(cm)
a : 설계 동결관입깊이(cm)
p : 시멘트 콘크리트 포장의 슬래브두께(cm)

- 노상동결관입허용법에 의한 설계비동결성재료층 두께결정도표(아스팔트콘크리트포장 설계절차 참고)에서 c(cm)의 연직선과 r=2.0선의 만나는 점의 좌측값인 비동결성 재료층의 두께 b(cm)를 얻을 수 있음

- 노상동결관입허용법에 의한 동결심도는 p+b

⑥ 시멘트 콘크리트 포장 두께의 결정

- 경험적 공용방정식에 의하여 반복계산방법에 따라 두께를 산정

$$\log W_{8.2} = 7.35 \times \log_{10}(D+1) - 0.06 + \frac{G_t}{1 + \dfrac{1.624 \times 10^7}{(D+1)^{8.46}}}$$
$$+ (4.22 - 0.32P_t)\log_{10}\left[\frac{S_c' \times (D^{0.75} - 1.132)}{215.63 \times J \times \left(D^{0.75} - \dfrac{18.42}{(\frac{E_c}{K_c})^{0.25}}\right)}\right]$$

- $W_{8.2}$: 설계기간 동안의 8.2톤 등가단축하중(ESAL)의 설계차로 당 누가 통과 횟수
- P_t : 설계최종서비스지수
- J : 시멘트 콘크리트 포장형식에 따른 연속성을 반영하는 하중전달계수
- E_c : 콘크리트 슬래브 설계탄성계수(psi)
- K_c : 콘크리트 슬래브와 접속면에서의 노상반력계수(pci)
- D : 포장슬래브두께(inch)
- S_c' : 콘크리트슬래브의 설계휨강도(psi)
- G_t : $\log_{10}\left[\dfrac{4.5 - P_t}{4.5 - 1.5}\right]$로 표시되며, 임의 시점(최종 서비스지수를 가지는 시점)에 이어 P_t=1.5인 경우의 최대 서비스능력 손실량에 대한 서비스능력의 손실량 비율을 나타내는 함수

⑤ Load transfer coefficient (J)

- load transfer capacity of non-continuum point of concrete pavement slab such as joint, cleavage, between pavement and road shoulder (J=2.9)

⑥ Freeze penetration depth (similar to asphalt pavement)

- Freeze penetration depth calculation : Use partial allowance method
  - Freeze penetration depth by partial allowance for freezing penetration

$$c = a - p$$

c : maximum non-freezing material substitution depth (cm)
a : designed freezing penetration depth (cm)
p : asphalt mixture thickness of asphalt pavement (cm)

- From the chart below, left-side value of meeting point of vertical line of c(cm) and r=2.0 line is non-freezing material thickness b (cm)

- freezing penetration depth by partial allowance is p+b

⑥ Concrete pavement thickness

- Repetitive calculation using empirical commonality equation

$$\log W_{8.2} = 7.35 \times \log_{10}(D+1) - 0.06 + \frac{G_t}{1 + \frac{1.624 \times 10^7}{(D+1)^{8.46}}}$$
$$+ (4.22 - 0.32P_t)\log_{10}\left[\frac{S_c' \times (D^{0.75} - 1.132)}{215.63 \times J \times \left(D^{0.75} - \frac{18.42}{\left(\frac{E_c}{K_c}\right)^{0.25}}\right)}\right]$$

- $W_{8.2}$: compounded number of passage of 8.2 ton equivalent single axis load during the designed operation period
- P_t : Designed terminal serviceability index
- J : load transfer coefficient reflecting concrete pavement continuity
- E_c : concrete slab design elasticity coefficient (psi)
- K_c : soil surface reaction coefficient between concrete slab and soil surface (pci)
- D : Thickness of paved slab (inch)
- S_c' : designed bending strength of concrete slab (psi)
- G_t : $\log_{10}\left[\frac{4.5 - P_t}{4.5 - 1.5}\right]$ function to indicate serviceability loss ratio of maximum serviceability when P_t=1.5 and certain point of time (when final serviceability index is determined)

3-3 동결심도를 고려한 포장단면 결정

아스팔트 콘크리트 포장 및 시멘트 콘크리트 포장의 설계절차에 의하여 도출된 결과와 동결심도를 고려하여 최종단면을 결정한다. 흙쌓기 구간에서는 노상재료가 양호할 경우 동상이 발생되지 않으므로, 흙쌓기 높이 2m 이상 구간에서는 동상방지층을 생략한다.

3-3-1 동상방지층 삭제구간

① 흙쌓기 높이 2m 이상 구간이 50m 이상

② 흙쌓기 높이 2m 이상이 많고, 부분적으로 흙쌓기 높이 2m 미만인 구간

3-3-2 동상방지층 설치구간

① 흙쌓기 높이 2m 이상 구간이 많고 부분적으로 흙쌓기 높이 2m 이상 구간이 존재(2m 미만 구간 연장이 30m 미만)

② 흙쌓기 높이 2m 미만 구간과 흙쌓기 높이 2m 이상 구간이 계속 반복되며 각 연장이 30m 미만인 경우

③ 한쪽깎기 한쪽쌓기 구간

3-3 Pavement Section Determination Considering Freezing Penetration Depth

Determine final section considering various results from asphalt and concrete pavement design process and freezing penetration depth. For land-fill area with fine soil material, freezing is avoidable so that anti-frost layer can be omitted for land-fill of 2m and higher.

3-3-1 Anti-frost Omission Area

① 50m or longer section with land-fill of 2m and higher

② Most of the area has land-fill of 2m and above while there are some small area with land-fill of less than 2m

3-3-2 Anti-frost Installation Area

① Most of the area has land-fill of less than 2m while there are some small area with land-fill of 2m and above (less than 30m)

② Land-fill of less than 2m and land-fill of 2m and above appears one after another with the length of less than 30m for each

③ Cutting and embankment area

Chapter 12
Traffic Safety Facilities

1. 일반사항

1-1 교통안전시설의 정의

- 도로교통의 안전하고 원활한 소통을 확보하고 도로의 미비한 구조상태를 보완하며 도로이용자의 안전을 도모하기 위하여 설치하는 시설

1-2 교통안전시설 설계 시 고려사항

- 안전시설물 설치 시 시인성이 극대화되며, 경제성이 확보 가능한 지점에 설치

- 시설물은 통일성이 유지되도록 설치하여 운전자가 판단에 오해가 발생하지 않도록 함

- 주·야간 또는 주행 중인 차량에서도 시설물을 인지할 수 있도록 양호한 시인성을 확보

- 도로이용자의 행동특성을 배려하여 설치할 수 있도록 도로이용자의 관점에서 설치

2. 교통관리 안전시설

2-1 교통안전표지

2-1-1 개　　요

① 도로이용자에게 일관성 있고 통일된 방법으로 각종 필요한 정보를 제공

② 교통의 안전과 원활한 소통을 도모하고 도로구조와 도로시설물을 보호하기 위한 시설

2-1-2 교통안전표지의 구분 및 종류

① 본표지 : 단독으로 주의, 규제, 지시의 의미를 전달

- 주의표지
  - 도로상태가 위험하거나 도로 또는 그 부근에 위험물이 있는 경우에 필요한 안전조치와 예비동작을 할 수 있도록 이를 도로이용자에게 알리는 표지
  - 도로상태 예고, 노면 또는 연도상황 예고, 기상상황 예고, 기타주의 예고로 구분

- 규제표지
  - 도로교통의 안전을 위하여 각종 제한이나 금지 등의 규제를 하는 경우에 이를 도로이용자에게 알리는 표지
  - 통행금지, 통행제한, 금지사항으로 구분

1. General Considerations

1-1 Definition of Traffic Safety Facility

- Facilities to install for securing smooth flow of traffic, complementing insufficient states of road structures, and improving the safety of road users

1-2 Considerations in the Designing Traffic Safety Facility

- Safety facilities shall have maximum visibility during installation and they shall be installed where economical efficiency is validated.

- Facilities shall be installed consistently so that drivers do not misunderstand in their judgement.

- Facilities shall have good visibility so that they can be recognized during days, nights, or while driving.

- Facilities shall be installed from the view point of road users considering their behaviour characteristics.

2. Traffic Control Safety Facility

2-1 Traffic Safety Sign

2-1-1 Overview

① Facilities to provide the various required information to users in a consistent and unified way.

② Facilities to enable smooth flow of traffic and safe driving, and to protect the road structures and the other road facility

2-1-2 Classification and Type of Traffic Safety Sign

① Main sign : To indicate warning, restriction, and instruction independently

- Warning sign
  - Sign that indicates dangerous road condition or when there are hazard materials on or around the road so that drivers can take the safety precaution
  - Notice of road condition, road surface, roadside condition, weather condition, or other warnings

- Restriction sign
  - Sign indicates various restriction or prohibitions for the safety to drivers
  - Classified to no passing, limited passing, and restricted actions.

② 보조표지

- 본표지와 함께 설치하여야 하며, 본표지의 부연설명·보완 또는 첨부기능을 수행

- 거리, 구역, 구간, 일자, 시간, 본표지 설명, 해제, 기타사항으로 구분

2-1-3 설계기준

① 설치위치

- 정주식
  - 노측에 표지판을 설치할 때에는 차도 측단으로부터 차도 측단의 바깥쪽에 설치함을 원칙
  - 고속도로(자동차 전용도로)는 0.5m 이상, 일반도로는 0.25m 이상 확보하여 바깥쪽에 설치
  - 중앙분리대, 교통섬의 경우 분리대 끝으로부터 0.25m 이상 내측에 설치

- 내민식, 문형식
  - 일반적으로 표지판이 도로 진행방향 차로의 중앙에 위치하게끔 설치

- 부착식
  - 부착되는 도로시설에 따름

② 설치간격

- 주의, 규제 및 지시표지가 가리키는 표시하는 도로의 구역, 구간 또는 지점으로부터 시인거리 및 교통상황을 적응하는데 필요한 예비동작시간이 확보될 수 있도록 적절한 간격으로 중복하여 설치

- 규제와 지시표지는 규제 또는 지시의 시작 시점과 끝 지점에 설치

- 규제 또는 지시가 있는 도로에서 구간 내에 교차로가 있을 때는 교차하는 도로 폭, 교통량 등에 따라 교차로 부근에 규제 및 지시표지를 중복하여 설치

- 표지와 노면표지를 병설할 경우, 표지의 설치간격을 1.5배까지 연장할 수 있음

- 교통안전표지는 주행속도, 곡선반지름, 시거 등의 도로조건 및 도로이용자의 행동특성을 고려하여 설치간격을 결정

- 도로구조 및 교통상황을 고려하여 표지의 설치간격을 연장 또는 축소할 수 있음

② Auxiliary sign

- This must be installed along with main sign that provides additional description, complement, or supplement

- Classified to distance, zone, date, time, main sign description, release, and others

2-1-3 Design Standard

① Installation position

- Pole type
  - When installed on the side of the road, pole shall be installed between the edges of the road and the outer edge of the road side as a principle.
  - Pole shall be installed with over 50 cm of clearance for expressway (driveway) and over 25 cm for general road from the edge of the road.
  - In case of median strip or traffic island, pole shall be installed over 25 cm inside from the edge of the strip.

- Extrusion type, Door type
  - In general, the sign shall be installed so that the sign is positioned in the middle of the road of forward moving direction.

- Attached type
  - Depending on the road structure to attach

② Installation distance

- From the point in zone, section, or place that requires warning, restriction, or instruction, signs shall be installed repetitively with the proper distance so that drivers have enough time to recognize the traffic status and react accordingly.

- Restriction and instruction signs shall be installed at the beginning point and end point of the restriction or instruction zone.

- When there is an intersection within the restricted or instructed zone, restriction or instruction signs shall be installed repetitively around the intersection depending on width and traffic volume of crossing road.

- When signs and road surface signs are installed together, the installation distance can be extended by up to 1.5 times.

- Traffic safety signs shall be installed with the appropriate installation distance considering various road conditions including driving speed, and sight distance and characteristics of behaviour of drivers.

- Installation distance of signs can be extended or shortened depending on the road and traffic conditions.

③ 설치높이

- 정주식은 1.60m를 원칙으로 하되 운전자의 눈 높이와 표지판 훼손 등을 고려하여 설치장소나 종류, 내용에 따라 1.00~2.10m를 기준으로 적용

- 내민식과 문형식은 시설한계기준인 4.50m에 시공오차, 자재의 굴절, 노면 덧씌우기 등에 대한 여유 폭(0.50m)를 더하여 상단 5.00m를 기준으로 적용

- 적설지역에 있어서는 제설방법을 검토하고 적설량을 고려하여 필요에 따라 설치높이를 결정

2-2 도로표지

2-2-1 개　요

① 도로교통의 안전과 원활한 소통을 도모하고 도로 구조를 보존하기 위한 시설
② 도로정보의 전달을 목적으로 하는 시설

2-2-2 도로표지의 구분 및 종류

① 경계표지
- 도·시·군·읍 또는 면 사이의 행정구역의 경계를 나타내는 표지
  - 도계표지 등

② 이정표지
- 목표지까지의 거리를 나타내는 표지
  - 1지명, 2지명 이정표지 등

③ 방향표지
- 방향 또는 방면을 나타내는 표지
  - 방향예고표지와 방향표지 등

④ 노선표지
- 주행노선 또는 분기노선을 나타내는 표지
  - 분기점 등

④ 기타표지
- 상기에 언급한 이외의 표지
  - 터널표지, 양보차로표지, 유도표지 등

③ Installation height

- The height of pole type shall be 160cm as a general rule, and it shall be applied flexibly between 100 and 210 cm depending on the type and details of the signs considering height of drivers' vision and possibility for damages.

- The height of extrusion type and door type shall be 500 cm as a reference with 450 cm, which is a limit by construction regulation, and additional margin (50 cm) for construction error, bending of the material, and over pavement.

- In snowfall area, snow removal method shall be reviewed and the installation height shall be determined depending on the need by considering the amount of snowfall.

2-2 Road Sign

2-2-1 Overview

① Facilities to enable safety of road traffic, smooth flow, and preserving road structures.

② Facilities for the purpose of notifying road information

2-2-2 Classification and Type of Road Signs

① Boundary signs

- Signs indicates the district boundary of do, si, gun, and eup.
  - Do (province) boundary signs, etc

② Milestone signs

- Signs to indicate the distance to the destination
  - Milestone sign for one place name, two place name, etc

③ Direction signs

- Signs indicating directions
  - Advanced direction sign, direction sign, etc

④ Route signs

- Signs indicating driving routes and branch routes
  - Branch point sign, etc

④ Miscellaneous signs

- Signs for the purpose other than mentioned above
  - Tunnel signs, yield lane signs, guide signs, etc

2-2-3 설계기준

① 설치위치

- 단주식, 복주식은 고속도로에서 표지판의 차도를 향한 끝부분이 도로 가장자리로부터 1.00m 벗어난 지점에 설치

- 편지식, 현수식, 문형식 표지판은 일반적으로 표지판이 도로 진행방향 차로의 중앙에 위치하도록 설치

- 부착식 표지판은 부착되는 도로시설에 따름

② 설치높이

- 단주식, 복주식은 2.00m를 원칙으로 하되 운전자의 눈 높이와 표지판 훼손 등을 고려하여 설치장소나 종류, 내용에 따라 2.00~2.50m를 기준

- 내민식과 문형식은 시설한계 기준인 4.50m에 시공오차, 자재의 굴절, 노면 덧씌우기 등에 대한 여유 폭(0.50m)를 더하여 상단 5.00m를 기준

- 적설지역에 있어서는 제설방법을 검토하고 적설량을 고려하여 필요에 따라 설치높이를 결정

③ 설치각도

- 차량진행방향과 직각 또는 10도 이내

- 편지식 및 문형식의 경우 종단경사를 고려하여 연직면으로부터 하향으로 기울여 설치
  - 오르막인 경우 (0%~+4%)
    - ▶ 표지판을 하향으로 약 3° 기울여 차량 전조등 빛이 표지판에 수직 입사하도록 설치
    - ▶ 오르막 종단경사별 표지판 각도변화(Θ)

종단경사(%) / 직선거리(m)	0	1	2	3	4
250m	1.32°	1.90°	2.47°	3.04°	3.48°

2-2-3 Design Standard

① Installation location

- For single pole and double-pole type installation in expressway, the edge of sign board toward the road shall be 100 cm far from the edge of the road

- Branch type, suspension type, and door type signs shall be installed so that the sign boards are located in the middle of lanes in driving direction in general.

- Attached type signs shall be installed depending on the road facilities to be attached to.

② Installation height

- The height shall be 200 cm for single pole and double-pole type, but 200~250 cm shall be applied depending on the location, type, and details considering the height of drivers' vision and possibility of damages.

- The height for extrusion type and door type shall be 500 cm as a reference with 450 cm, which is a limit by construction regulation, and additional margin (50 cm) for construction error, bending of the material, and over pavement.

- In snowfall area, snow removal method shall be reviewed and the installation height shall be determined depending on the need by considering the amount snowfall.

③ Installation angle

- Perpendicular to the moving direction of cars or withing 10 degrees

- For shaped type and door type, signs shall be installed tilted downward from the vertical plane considering longitudinal slope
  - In case of uphill (0%~+4%)
    - ▸ Signs shall be installed downward titled by 3° so that the headlight of cars illuminates the board perpendicularly.
    - ▸ Angles of signs depending on the uphill longitudinal slope(Θ)

longitudinal slope(%) / distance in straight line(m)	0	1	2	3	4
250m	1.32°	1.90°	2.47°	3.04°	3.48°

- 내리막인 경우 (0%~-4%)
  - ▸ 표지판을 현행 수직방향으로 설치 시 차량 전조등 빛이 표지판에 수직입사
  - ▸ 내리막 종단경사별 표지판 각도변화(Θ)은 아래와 같다.

종단경사(%) / 직선거리(m)	-4	-2	-2	-1
250m	-0.97°	0.18°	0.18°	0.75°

④ 기타

- 표지판 지주 보호시설

구 분	안내 및 내민식 표지판	문형식 표지판
땅깎기부	L2측구 + 되메우기 (완화구간 4m+L2구간 12m)	L2측구 + 되메우기 (완화구간 4m+L2구간 12m)
흙쌓기부	1단 가드레일 (L=12m)	2단 가드레일 (L=12m)

- In case of downhill (0%~-4%)
  - ▸ Signs shall be installed vertically so that the headlight from cars illuminates the board perpendicularly.
  - ▸ Angles of signs depending on the downhill longitudinal slope(Θ).

longitudinal slope(%) / distance in straight line(m)	-4	-2	-2	-1
250m	-0.97°	0.18°	0.18°	0.75°

④ Miscellaneous

- Facilities to protect sign poles

Classification	Guide and extrusion signs	Door type signs
Cutting parts	L2 side ditch + post- filling (transition zone 4m+L2 zone 12m)	L2 side ditch + post-filling (transition zone 4m+L2 zone 12m
Banking parts	1 stage guard rail (L=12m)	2 stage guard rail (L=12m)

2-3 노면표시

2-3-1 개요

① 도로교통의 안전을 위하여 각종 주의, 규제, 지시 등의 내용을 노면에 기호, 문자 또는 선으로 도로이용자에게 알리는 표시

② 안전표지의 보완 또는 독자적으로 도로이용자에게 규제 또는 지시의 정보를 전달

2-3-2 노면표시의 구분

① 선과 색채의 의미
- 실선 : 제한을 뜻하며, 2중 실선은 최대의 제한을 의미
- 점선 : 유도 및 융통성 있는 제한을 의미
- 백색 : 동일한 방향의 교통류 분리 및 경계표시
- 황색 : 반대방향의 교통류 분리, 제한 및 지시표시
- 청색 : 지정방향의 교통류 분리표시

② 규제표시
- 도로의 교통안전, 소통 및 도로 구조 보존과 관련한 각종 제한, 금지 등의 규제 내용을 전달하는 표시

③ 지시표시
- 도로의 교통안전, 소통 및 도로 구조 보존과 관련한 도로의 통행방법, 통행구분 등의 지시내용을 전달하는 표시

2-3-3 노면표시의 종류

구 분		종 류
규제표시	선규제	중앙선, 차선, 길가장자리 구역선, 진로변경제한선, 버스전용차선, 유턴구역선, 유도선
	통행방법규제	우회전 및 좌회전 금지, 직진금지, 좌·우회전금지, 유턴금지, 속도제한, 서행, 일시정지, 양보
	정차주차규제	주차금지, 정차금지지대
	장애물규제	노상장애물, 안전지대
지시표시	주차방법지시	평행주차, 직각주차, 경사주차
	유도지시	유도선
	횡단지시	횡단보도예고, 정지선, 횡단보도, 자전거횡단도, 자전거전용도로
	방향 및 방면지시	진행방향표시, 비보호좌회전
기 타		어린이보호구역, 좌회전전용차로 및 테이퍼 길이

2-3 Road Surface Signs

2-3-1 Overview

① Markings that indicate various warnings, restrictions, and instructions for traffic safety by symbols or letters on road surface

② To notify the information of restriction or instruction to drivers as either independently or complementing to safety signs.

2-3-2 Classification of Road Signs

① Meaning of lines and colors
- Solid line : means restriction, and double solid means maximum restriction
- Dotted line : means guide or flexible restriction
- White : Separation of traffic flow in the same direction
- Yellow : Separation between the traffic flow in opposite direction
- Blue : Separation of traffic flow in the designated direction

② Restriction signs
- Signs to notify restriction details of various restriction and prohibition in relation to traffic safety of roads and preservation of the road structures

③ Instruction signs
- Signs to notify various instruction details including passing method and passing classification in relation to traffic safety of roads and preservation of the road structures.

2-3-3 Type of Road Surface Signs

Classification		Type
Restriction signs	Line restriction	Center line, lane markings, road edge zone line, lane change restriction line, bus only lane, u-turn line, guide line
	Travel method restriction	no right turn or left turn, no driving forward, no turns, no u-turns, speed limit, slow, stop, yield
	Stopping and parking restriction	No parking, no stopping or parking
	Obstacle restriction	Road obstacles, safety zone
Instruction signs	Parking method instruction	Parallel parking, perpendicular parking, oblique parking
	Guide instruction	Guide line
	Crossing instruction	Advanced crossing sign, stop line, pedestrian crossing, bicycle crossing, bicycle only road
	Direction instruction	Direction sign, unprotected left turn
Others		Children protection zone, lanes for left turn only, length of taper

2-3-4 설계기준

① 형식선정

- 교통사고 취약구간 및 우천 시 시인성이 요구되는 구간은 차선의 기본 기능 외의 부가기능을 가진 차선도색재료 적용 가능(돌출형 차선 및 수막방지형 차선재료 등)

- 제설작업으로 인하여 비정상적인 마모가 발생한 구간은 일반차선으로 시인성 확보

- 가로등과 같은 조명시설 등이 구비되어 있는 휴게소 광장부(주차구획선 등)의 차선도색은 시인성보다는 경제성 측면에서 상온형 도색재료를 사용

- I.C, JCT, 영업소 부분은 가로등이 설치되어 있고 본선에 비하여 통행량이 적으므로 본선교통량 기준에 따라 가열형 및 상온형 도색재료를 사용(단, 가·감속차선은 내마모형으로 시행)

구 분		A등급	S등급	비고
적용 도색재	중앙선, 길어깨선	상온형	가열형	
	구분선	가열형	내마모형	
	버스전용차선	가열형	내마모형	
	휴게소 광장부	상온형		
	노면문자, 기호	일반 융착식		
	▪ 가·감속차선 ▪ 본선 영업소	내마모형		

※ 내마모형 : 실내마모시험 통과재료
A 등 급 : 차로별 일교통량 10,000대 미만
S 등 급 : 차로별 일교통량 10,000대 이상
일반차선도색 : 가열형 차선도색(Hot paint), 상온형 차선도색(Cold paint)
내마모형 차선도색 : 수용성차선도색, 변성아크릴 차선도색, 돌출형(2액형) 차선도색, 적설형(2액형) 차선도색

2-3-4 Design Standard

① Type selection

- For frequent traffic accident area and where visual clearance is required during raining, additional materials for marking lines that have extra functions can be applied in addition to the basic function of marking lines (projecting lines or line materials for preventing hydroplaning, etc)

- To secure visual clearance as general lanes for the zone where abnormal wear was generated by snow plow or other works.

- For line markings rest area (parking spot marking, etc) where street lights or other illumination facilities are installed, economic efficiency shall be considered over visual clearance so that cold paint materials can be used.

- For IC, JCT, and toll gate area where street lights are installed and traffic volume is less than main roads, heating type or cold type paint materials shall be used depending on the traffic volume of the main roads.(However, acceleration and deceleration lanes shall be done with wear-free type)

<table>
<tr><th colspan="2">Classification</th><th>A class</th><th>S class</th><th>Remark</th></tr>
<tr><td rowspan="6">Applied paint material</td><td>Center line, shoulder line</td><td>Cold type</td><td>Heating type</td><td rowspan="6"></td></tr>
<tr><td>Division line</td><td>Heating type</td><td>Wear-proof type</td></tr>
<tr><td>Bus only lanes</td><td>Heating type</td><td>Wear-proof type</td></tr>
<tr><td>Rest area square</td><td colspan="2">Cold type</td></tr>
<tr><td>Road surface letters, symbols</td><td colspan="2">General fusion type</td></tr>
<tr><td>▪ Acceleration/ deceleration lane
▪ Main road toll gate</td><td colspan="2">Wear-proof type</td></tr>
</table>

※ Wear-proof type: Materials that passed the indoor wear test
A class : Daily traffic per lane is lower than 10,000 cars
S class : Daily traffic per lane is over 10,000 cars
General lines paint : Heating type line paint(Hot paint), cold type line paint (Cold paint)
Wear proof line paint : Water soluble line paint, denaturated acrylic line paint, projection type (two solution type) line paint, snowfall type (two solution type) line paint

② 설치기준

- 중앙선
  - 황색실선(1개)
    - ▶ 어떠한 경우에도 넘거나 침범할 수 없는 중앙선

  - 황색점선
    - ▶ 왕복 2차로(편도 1차로) 도로에서 반대방향의 교통에 주의하면서 도로 양측으로 넘어갈 수 있음을 표시하는 중앙선으로서, 반드시 앞지르기 시거가 확보된 구간에 설치 (양방향 추월선)

  - 황색실선 및 점선
    - ▶ 왕복 2차로(편도 1차로) 도로에서 점선이 있는 한쪽에서만 앞지르기가 허용되며 반대방향에서는 앞지르기를 금지시키는 중앙선(일방향 추월선)

  - 황색실선의 복선
    - ▶ 어떠한 도로나 어떤 경우에도 넘거나 침범할 수 없는 절대적인 중앙선으로 황색실선(1개)와 같은 의미이나 더 넓은 도로에서 사용되며 그 의미를 강조하고, 편도 2차로 이상의 도로와 중앙분리대가 없는 고속도로에 설치

- 차선
  - 백색점선
    - ▶ 동일방향의 교통에 주의하면서 차로를 변경할 수 있는 선

  - 백색실선
    - ▶ 차로를 변경할 수 없는 선으로서 교량, 커브길, 오르막길의 정상부분, 터널 내, 교차로의 정지선 부근에 설치 (진로변경제한선)

- 버스전용차선
  - 청색실선
    - ▶ 지정된 시간에 버스전용차로로 운영되며, 차로를 변경할 수 없음을 표시하는 버스전용차선

  - 청색점선
    - ▶ 지정된 시간에 버스전용차로로 운영되며, 동일방향의 교통에 주의하면서 차로를 변경할 필요가 있음을 표시하는 버스전용차선

② Installation standard

- Center line
  - Yellow solid line (single line)
    - ▸ Center line that cars should not touch or cross over in any cases
  - Yellow dotted line
    - ▸ In two way with two lanes (one lane each way), this line indicates that drivers can driver over the other side to pass while watching the traffic on the other side, and passing must be done only when passing visual distance is secured. (Passing from both sides)
  - Yellow sold line and dotted line
    - ▸ In two ways with two lanes (one lane each way), passing is allowed from the dotted line side, and passing is not allowed from the opposite side. (One way passing)
  - Yellow solid dual lines
    - ▸ These are absolute center lines so that driver cannot drive over the lines in any road and any condition, and it has the same meaning as yellow sold line (single line), but these lines are used in wide roads and emphasize the meaning. These lines are installed in the roads with more than two lanes each side and express way with no median strip.
- Lane dividing line
  - White dotted line
    - ▸ Changing lane is allowed while watching the traffic in the same direction
  - White solid line
    - ▸ Changing lane is not allowed, and this line is installed in bridge, curve, top of uphill, inside tunnel, and around stop line of crossing. (Lane change limit line)
- Bus only lane
  - Blue solid line
    - ▸ Only buses are allowed in specified time, and changing lane is not allowed.
  - Blue dotted line
    - ▸ Only buses are allowed in specified time, and changing lane is allowed while watching the traffics in the same direction

• 길가장자리 구역선
- 길어깨와 차로를 구분하여 차량주행 가능지역을 표시하는 백색실선
- 불량한 시계조건 하에서는 시각적 안내와 유도를 할 수 있도록 표시하는 백색실선

• 진로변경제한선
- 도로가 분리 또는 합류되는 구간 또는 장소 내의 필요한 지점에 설치되는 백색실선
- 백색점선 및 실선
  ▸ 점선측 차로에서 진로변경은 허용되나, 실선측 차로에서 진로변경은 금지를 표시하는 선
- 실선(1개 또는 복선)
  ▸ 차로를 변경할 수 없는 선

• 노상장애물
- 노상에 장애물을 표시하는 것으로 적절한 안전표지와 함께 설치한다.
- 사선방향은 차량의 진행방향에서 좌측으로 45°의 경사를 이루게 한다.
  ▸ 황색 : 도로 중앙에 장애물이 있을 때 설치
  ▸ 백색 : 동일방향으로 진행하는 도로에 장애물이 있을 때 설치

• 유도선
- 평면교차로가 넓거나 언덕 위에 있어서 진입하고자 하는 차로를 찾기 힘들 때(특히 야간) 또는 좌회전하는 거리가 너무 먼 경우 중앙선 또는 차선을 연장하거나 좌회전방향을 유도하는 선

• 진행방향 표시
- 차량의 진행방향을 화살표와 문자로 표시하는 것

• 안전지대
- 차량이 진입하지 못하도록 황색으로 표시하는 것으로 광장 또는 노폭이 넓은 도로의 중앙지대에 설치

- Shoulder zone line
  - This line divides the shoulder and car lane, and white solid line is used to indicate that driving is allowed
  - White solid line is used to indicate and guide visually during bad vision condition

- Lane change restriction line
  - White solid line installed in branching/merging zone or other necessary location
  - White solid and dotted line
    - ▸ Changing lane is allowed from the dotted line side, but it is not allowed from the solid line side.
  - Solid line (single or dual line)
    - ▸ Line indicates that changing lane is not allowed.

- Obstacles on road
  - Marking indicates the obstacles on road and it is installed with appropriate safety signs.
  - Oblique direction shall have 45° angle from the left of the car moving direction.
    - ▸ Yellow line : installed when there are obstacles in the middle of the road
    - ▸ White line : installed when there are obstacles on the road in the same direction

- Guide line
  - Line that is installed by extending center line or lane dividing line when it is difficult to find the entrance lane because plane intersection is wide or at the top of the hill (especially at night) or when the left turn distance is too long.

- Moving direction marking
  - Moving direction of vehicles are shown with arrow or letters.

- Safety zone
  - Zone that are made with yellow line so that cars cannot enter, and it is installed in the middle area of square or wide road.

③ 특수구간 설치기준
- 왕복 4차로 이상의 터널 진입 전 100m와 터널내부는 백색실선으로 도색

- 교량구간은 백색실선을 적용하여 진로변경 제한과 앞지르기 금지를 원칙으로 하며 다음의 교량에 대해서는 교통류의 원활한 흐름을 유도하기 위하여 좌·우 진로변경이 가능한 백색점선을 설치
  - 연장이 짧은 소교량
  - 출입시설 감속차로 전방에 있는 교량
  - 출입시설 사이가 교량으로만 이루어진 경우
  - 엇갈림 구간 내 교량
  - 기타 오르막차로 구간 내 교량 등 진로변경이 필요한 구간

3. 방호안전시설

3-1 방호울타리

3-1-1 개　요

① 주행 중 진행 방향을 잘못 잡은 차량이 길 밖 또는 대향차로 등으로 이탈하는 것을 방지

② 차량이 구조물과의 직접적인 충돌을 방지하여 차량 탑승자 및 차량, 보행자 또는 도로변의 주요 시설을 안전하게 보호하기 위하여 설치하는 시설

3-1-2 방호울타리의 종류

① 가요성 방호울타리
- 차량이 충돌할 때 다소의 변형이 수반되면서 충격에너지를 흡수

- 보(beam)형 방호울타리, 케이블(cable)형 방호울타리 등

② 강성 방호울타리
- 구조물의 변형에 의한 충격 흡수보다는 차량의 복귀를 목적으로 설치

- 콘크리트 벽형(뉴저지형, F형 등)

③ Installation standard of special zone

- White solid line shall be installed in 100 meters before entering tunnel and inside tunnel for roads with more than four lanes both ways.

- White solid line shall be installed in the bridge zone to restrict the lane change and passing as a general rule, and white dotted line that allow left or right lane change can be installed for the following bridge zone to enable smooth flow of traffic.
  - Small bridge that has short length
  - Bridge in front of deceleration lane for exit and entry facilities
  - When there are only bridges between exit and entry facilities.
  - Bridge inside crossing lane zone
  - Other miscellaneous zone that requires lane changes such as bridge within up hill lane

3. Protection Safety Facility

3-1 Protection Fence

3-1-1 Overview

① Facilities installed to prevent the cars of moving in wrong direction to outside of road or opposite direction lane.

② Facilities installed to prevent direction collision between cars and structures to protect passengers, pedestrians, and major facilities around the road safely.

3-1-2 Type of Protection Fence

① Flexible (bendable) protection fence

- To absorb the energy of impact with slight deformation when colliding by cars

- Beam type protection fence and cable type protection fence

② Hard protection fence

- The installation purpose is to keep the car inside the fence rather than to absorbing impact energy by deformation

- Concrete wall type (New Jersey type, F type, etc)

3-1-3 설계기준

① 대상구간

- 노측에 설치하는 경우
  - 자연상태의 토사로 된 산의 비탈면 경사구간
  - 도로변에 철도가 인접하고 있는 구간
  - 도로 폭 및 선형 등과의 관련으로 위험한 구간
  - 구조물과의 관련으로 필요한 구간

- 분리대에 설치하는 경우
  - 4차로 이상인 고속국도 및 자동차 전용도로 구간
  - 일반국도 구간 중 신호교차로의 간격이 짧아 단부처리가 어려운 구간 등 불가피하게 설치하지 못하는 곳을 제외한 전 구간
  - 지방지역의 도로에서 선형조건이 위험하여 설치가 필요하다가 인정되는 구간
  - 도시 내 도로에 있어서는 주행 속도가 높거나 중앙선 침범이 우려되는 위험한 구간 또는 불법 U턴 등을 막기 위하여 설치가 필요한 구간

- 강성방호울타리를 설치하는 경우
  - 주로 구조물이 많은 도로에서 차도 밖으로 벗어나려는 차량이 2차 사고를 일으킬 가능성이 많은 장소
  - 노측의 위험도가 높은 곳 등에서 완충 효과는 다소 저하되더라도 차도 밖으로 벗어남을 방지하는 것이 절대 필요하다고 인정되는 구간

3-1-3 Design Standard

① Target zone

- In case of installation in road side
  - Slope zone of mountain consists of natural state soil.
  - Where railroad is near to road side
  - Dangerous zone in relation to road width and alignment
  - Where fence is necessary in relation to structures

- In case of installation in median strip
  - For express way with more than four lanes or high way zone
  - Entire national road except the place where installation cannot be possible such as difficult end-treatment zone because of short distance between signalized intersections.
  - Certain zones in local roads where it is considered to be necessary because of danger due to alignment condition.
  - Dangerous zone in city area where driving speed is high or there is a possibility to cross the center line, or where fence is necessary to prevent illegal U-turn.

- In case of installation of hard protection fence
  - This type of fence is installed where break away cars are likely to cause secondary accident in the road where there are many structure around.
  - This type of fence is installed where it is considered to be absolutely necessary to keep the cars inside the fence even with the cost of low buffering effects such as highly dangerous road side area.

② 방호울타리 설치기준

<table>
<tr><th colspan="3">등 급</th><th>SB1</th><th>SB2</th><th>SB3</th><th>SB4</th><th>SB5</th><th>SB6</th><th>SB7</th></tr>
<tr><th colspan="3">기준 충격도(kJ)</th><th>60</th><th>90</th><th>130</th><th>160</th><th>230</th><th>420</th><th>600</th></tr>
<tr><td rowspan="7">적용도로 구분</td><td colspan="2">• 저속구간
(설계속도 60km/시 미만인 집산도로, 국지도로)</td><td>◎</td><td>○</td><td></td><td></td><td></td><td></td><td></td></tr>
<tr><td rowspan="3">• 일반구간
(60km/시 이상 ~80km/시 인 주간선 도로, 보조 간선도로)</td><td>- 기본 등급</td><td></td><td>◎</td><td>○</td><td></td><td></td><td></td><td></td></tr>
<tr><td>- 중앙분리대, 교량 구간 및 노측위험도가 큰 구간</td><td></td><td></td><td></td><td>◎</td><td>○</td><td></td><td></td></tr>
<tr><td>- 도로가 타 도로와 교차되는 등 특수구간
- 특수 중차량 통행이 많은 구간</td><td></td><td></td><td></td><td></td><td>◎</td><td>○</td><td></td></tr>
<tr><td rowspan="3">• 고속구간
(100km/시 이상인 고속도로 및 자동차전용 도로)</td><td>- 기본 등급</td><td></td><td></td><td>◎</td><td>○</td><td></td><td></td><td></td></tr>
<tr><td>- 중앙분리대, 교량구간 및 노측위험도가 큰 구간</td><td></td><td></td><td></td><td></td><td>◎</td><td>○</td><td></td></tr>
<tr><td>- 도로가 타 도로와 교차되는 등 특수구간
- 특수 중차량 통행이 많은 구간</td><td></td><td></td><td></td><td></td><td></td><td>◎</td><td>○</td></tr>
</table>

※ ◎표시는 일반적으로 추천하는 등급
○표시는 도로여건이나 시설물 개발 수준에 따라 사용이 권장되는 등급

③ 설치형식(노측용)

• 본선 구간
- 지주 : ○ 형, 가드레일 : Thrie-Beam, 블록아웃시스템 : 연결대

• 높은 흙쌓기 구간
- 지주 : ○ 형, 가드레일 : Thrie-Beam, 블록아웃시스템 : 연결대

• IC 구간
- 지주 : ○ 형, 가드레일 : Thrie-Beam

• 적설지역의 높은 흙쌓기부 및 차량방호시설로 인하여 제설작업이 어려운 길어깨 구간
- 신설구간 : 국내 사용 중인 방호시설로서 적정 등급(SB3) 이상 개방형 가드레일 또는 이와 유사한 형식 적용
- 기존구간 : 기존 스틸파이프를 가드레일 또는 개방형 가드레일로 보완, 시설

② Installation standard of protection fence

<table>
<tr><th colspan="3">Class</th><th>SB1</th><th>SB2</th><th>SB3</th><th>SB4</th><th>SB5</th><th>SB6</th><th>SB7</th></tr>
<tr><th colspan="3">Standard degree of impact(kJ)</th><th>60</th><th>90</th><th>130</th><th>160</th><th>230</th><th>420</th><th>600</th></tr>
<tr><td rowspan="7">Classification of applied road</td><td colspan="2">• Low speed zone (merge/branch road or local road with lower 60km/hour of design speed)</td><td>◎</td><td>○</td><td></td><td></td><td></td><td></td><td></td></tr>
<tr><td rowspan="3">• General zone (Main route road between 60km/hr and 80km/hr, auxiliary main road)</td><td>-Basic class</td><td></td><td>◎</td><td>○</td><td></td><td></td><td></td><td></td></tr>
<tr><td>-Median strip, bridge zone, and other high road side danger zone</td><td></td><td></td><td></td><td>◎</td><td>○</td><td></td><td></td></tr>
<tr><td>-Special zone that crosses with other road
-Lot of special heavy duty vehicles traffic zone</td><td></td><td></td><td></td><td></td><td>◎</td><td>○</td><td></td></tr>
<tr><td rowspan="3">• High speed zone(Express way and high way over 100km/hr)</td><td>-Basic class</td><td></td><td></td><td>◎</td><td>○</td><td></td><td></td><td></td></tr>
<tr><td>-Median strip, bridge zone, and other high road side danger zone</td><td></td><td></td><td></td><td></td><td>◎</td><td>○</td><td></td></tr>
<tr><td>-Special zone that crosses with other road
-Lot of special heavy duty vehicles traffic zone</td><td></td><td></td><td></td><td></td><td></td><td>◎</td><td>○</td></tr>
</table>

※ ◎ marking is for generally recommended class
○ marking is for recommended class according to the road condition or facilities development level

③ Installation type(for road side usage)

- Main route zone
  - Pillar : ○ type, guard rail: Thrie-Beam, block out system: connection block
- High soil banking area
  - Pillar : ○ type, guard rail: Thrie-Beam, block out system: connection block
- IC zone
  - Pillar : ○ type / guard rail : Thrie-Beam
- High soil banking parts of snowfall area and shoulder zone where snow plow is difficult because of vehicle protection facilities
  - Newly constructing roads : Among the protection facilities used domestically, open type guard rail over appropriate class (SB3) or similar type shall be applied.
  - Existing roads : Basic steel pipes are used for the installation and complementing of guard rail or open type guard rail

구 분	설치형식	
본선 일반구간	원형지주 + 3W-BEAM	120~150 4% 810 350 1.300 1 : 1.5 지주보강재 Φ 125x4.3x700 지주 Φ 139.8x4.5x2080
본선 높은 흙쌓기 구간	원형지주 + 3W-BEAM(2단)	2500 1200 1300 350 50 85 165 139 460 1210 200 3.2 350 100 600
I.C 일반구간	원형지주 + 3W-BEAM	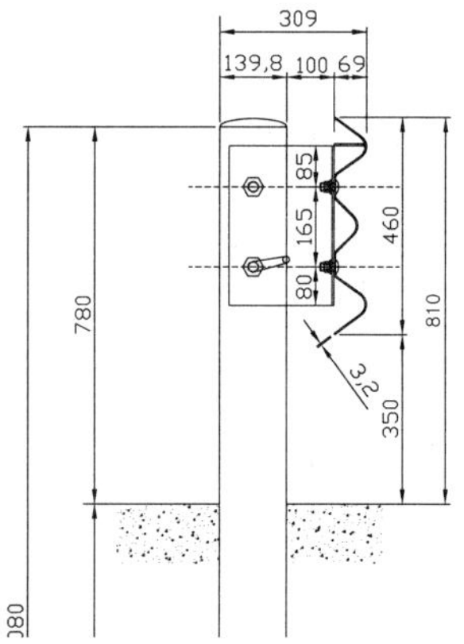 309 139.8 100 69 85 165 80 460 810 780 3.2 350

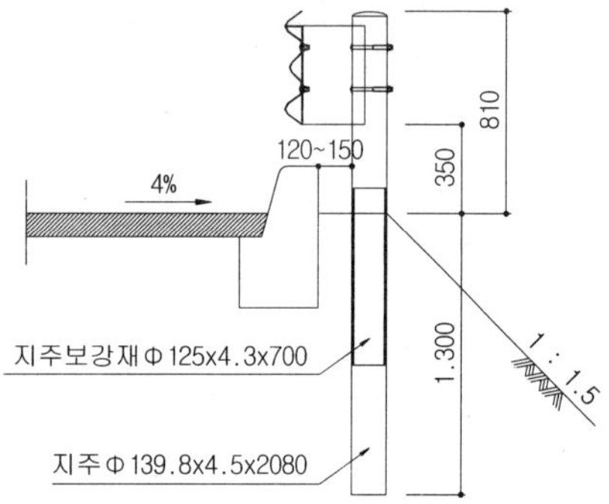

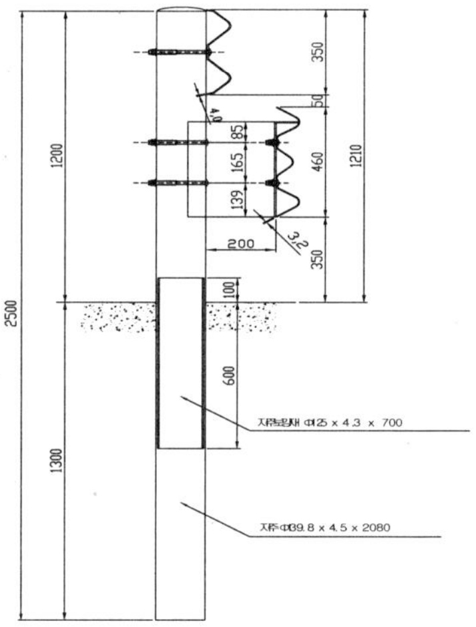

Classification	Installation type	
Main route General zone	Circular pillar + 3W-BEAM	810 350 120~150 4% 1.300 1 : 1.5 pilla support material Φ 125x4.3x700 pillaΦ 139.8x4.5x2080
Main router High soil banking zone	Circular pillar + 3W-BEAM(2 stages)	350 50 85 165 139 460 1210 1200 200 350 100 2500 600 1300 pilla support material ϕ125×4.3×700 pilla ϕ139.8×4.5×2080
I C General zone	pillar Circular pillar + 3W-BEAM	309 139.8 100 69 85 165 80 460 810 780 3.2 350 380

④ 설치간격

- 노측용
  - 가드레일은 부득이한 경우를 제외하고는 연속적으로 설치
  - 가드레일의 지주는 지면에 수직으로 설치
    - ▸ 본선구간, 높은 흙쌓기 구간은 지주간격을 2m로 적용
    - ▸ C구간, SB1 등급 중앙분리대는 지주간격을 4m로 적용
  - 방호울타리의 바람직한 설치 최소길이는 100m이고, 부득이 설치길이를 줄이는 경우에도 60m 이상 적용

- 분리대용(방호벽)
  - 중앙분리대 방호벽은 부득이한 경우를 제외하고는 연속적으로 설치
  - 중앙분리대 설치 시 팽창조인트는 30.0m 마다 설치
  - 중앙분리대 설치 시 수축조인트는 6.0m 마다 설치
  - 중앙분리대의 개구부 단부처리는 8.0m로 적용
  - 교량의 신축이음부는 중앙분리대 및 난간부에 덮개를 설치
  - 콘크리트 포장슬래브 및 중앙분리대 기초의 가로수축줄눈과 일치되도록 설치

- 교량의 경우 신축이음장치 설치부에 신축 유간과 동일하게 완전 절단

⑤ 설치높이

- 노측용

구 분		보	
		상 단	하 단
설치 기준	본선 구간	길어깨 가장자리 포장면에서 0.81m	길어깨 가장자리 포장면에서 0.35m
	높은 흙쌓기 구간	길어깨 가장자리 포장면에서 1.20m	길어깨 가장자리 포장면에서 0.35m
	I.C 구간	길어깨 가장자리 포장면에서 0.78m	길어깨 가장자리 포장면에서 0.35m
	SB1 등급 중앙분리대	길어깨 가장자리 포장면에서 0.80m	길어깨 가장자리 포장면에서 0.35m

④ Installation distance

- For road side usage
  - Guard rail shall be installed continuously unless it cannot be avoided
  - Pillars of guard rail shall be installed vertically over the ground
    - ▸ The pillar distance shall be 2 m for main route zone and high soil banking zone
    - ▸ The pillar distance shall be 4 m for IC zone, SB1 class median strip
  - The minimum appropriate installation length of protection fence is 100m, and over 60 m shall be applied when the distance is to be reduced unavoidably.

- For median strip (protection wall)
  - Protection wall for median strip shall be installed continuously unless it cannot be avoided
  - Expansion joint shall be installed at every 30.0 m when installing center protection wall
  - Contraction joint shall be installed at every 6.0 m when installing center protection wall
  - End treatment for opening part of center protection wall shall be 8.0 m
  - Cover shall be installed at the median strip and the railing of the flexible joints of bridges.
  - The protection wall shall be installed in line with horizontal contraction scale mark of concrete pavement slab and the base of median strip.
  - In case of bridge, protection wall shall be cut completely exactly same as flexible joint clearance.

⑤ Installation height

- For road side usage

Classification		Beam	
		Upper side	Lower side
Installation standard	Main route zone	0.81 m from the shoulder edge pavement	0.35 m from the shoulder edge pavement
	High soil banking zone	0.12 m from the shoulder edge pavement	0.35 m from the shoulder edge pavement
	I.C zone	0.78 m from the shoulder edge pavement	0.35 m from the shoulder edge pavement
	SB1 class median strip	0.80 m from the shoulder edge pavement	0.35 m from the shoulder edge pavement

- 분리대용
  - 와이어메쉬를 보강한 H=1.27m를 원칙으로 적용

3-2 충격흡수시설

3-2-1 개　요

① 차량의 충돌 시 차량의 충격에너지를 흡수하여 차량을 정지하도록 하거나, 방향을 교정하여 안전하게 본래의 주행차로로 복원시켜 주는 역할

② 위험물의 위치를 미리 알려주어 운전자가 주행속도를 줄이거나 차로변경 등의 적절한 운행을 취할 수 있도록 하는 역할

3-2-2 충격흡수시설의 구분 및 종류

① 용도에 따른 구분
- 일반적 충격흡수시설
  - 모래채움통, 폐타이어 조합체 등
- 단부처리용 충격흡수시설
  - CAT, Brakemaster, ET-2000, ADIEM 등
- 특수목적용 충격흡수시설
  - 트럭부착용 등

② 에너지 흡수방법에 따른 구분
- 관성형
  - 운동량 보존의 법칙을 적용한 형식으로 차량의 충격에너지를 시설로 전이시켜 차량을 정지시킴

- 비관성형
  - 시설의 변형을 통하여 충격에너지를 흡수하는 원리를 적용

③ 기능상 차이에 따른 구분
- 주행 복귀형(복원형)
  - 차량 충돌 시 차량을 본래의 주행차로로 복원시켜, 주행의 연속성이 유지되도록 하는 기능을 수행

- For median strip
  - Installation of H=1.27m reinforced by wire mesh shall be applied as a general rule

3-2 Crash Cushion Systems

3-2-1 Overview

① To recover the vehicles to the orignal driving lanes safely by stopping or changing the direction by absorbing the impact energy when cars collide with the facility

② To notify the dangerous object in advance so that drivers can react properly such as reducing speed or changing lane.

3-2-2 Classification and Type of Crash Cushion Systems

① Classification by usage

- General types
  - Sand filled box, structures with scrapped tires, etc

- End treatment
  - CAT, Brakemaster, ET-2000, ADIEM, etc

- Special purpose
  - For attachment to truck, etc

② Classification by method of energy absorption method

- Inertia type
  - To stop the cars by transferring impact energy to the facilities in accordance with the law of conservation of momentum.

- Non-inertia type
  - To use the principle of absorbing impact energy through the deformation of facilities

③ Classification by the difference in function

- Return to driving direction type (recover type)
  - To execute the function to keep the continuity of driving direction by recovering the vehicle to driving direction during impact.

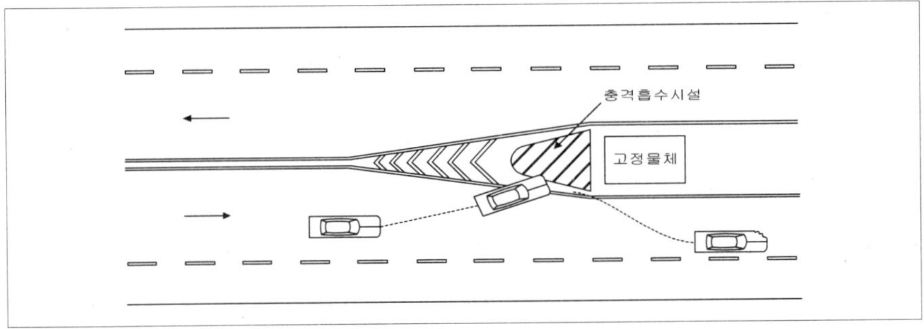

- 주행 비복귀형(비복원형)
  - 차량 충돌 시 차량의 충격에너지를 점진적으로 감소시켜, 차량을 안전하게 정지하도록 하는 기능을 수행

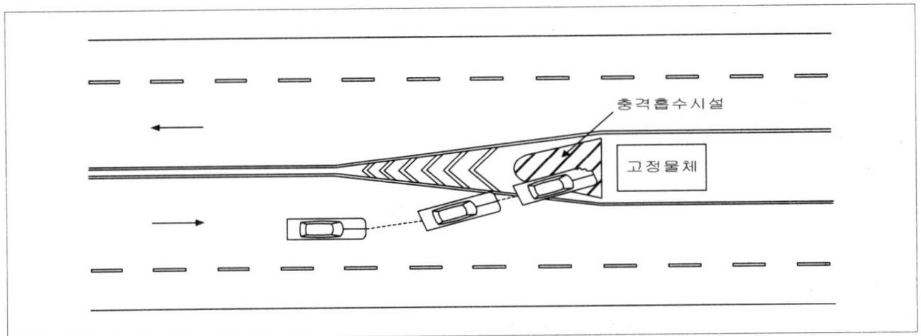

3-2-3 설계기준

① 대상구간

- 요금소 전면 (Hi-pass 부스 전면 및 본선상의 개방식 요금소 아일랜드 전면)

- 연결로 출구 분기점 (분기부시설이 강성구조물(방호벽, 옹벽 등)로 설치된 경우)

- 본선 요금소 전면, 연결로 출구 분기부가 내리막 경사에 위치하면서 대형표지판, 가로등 지주 등 강성시설물 전면과 충돌 시 위험하다고 판단되는 지점

- 대형사고가 빈번하거나 도로 기하구조 여건상 설치가 필요한 지점

- 기타 유형(교각, 교대 앞, 방호울타리 단부, 터널 옹벽 전방 등)에서는 가드레일, 방호벽(PE 방호벽 포함)등의 차량방호시설을 설치

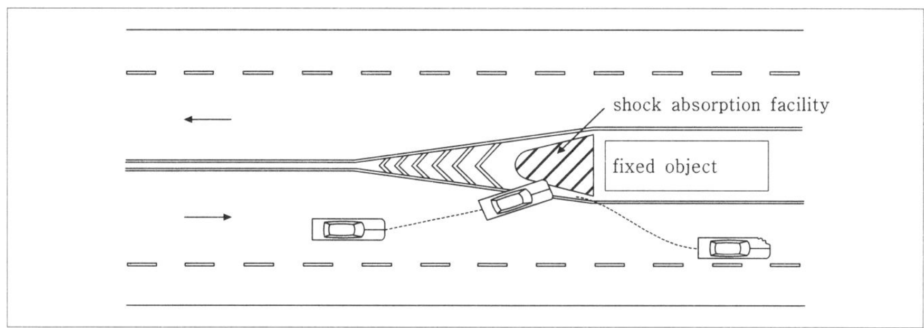

- Non return to driving direction type (non recovery type)
  - To execute the function to stop the vehicle safely by reducing the impact energy gradually during impact

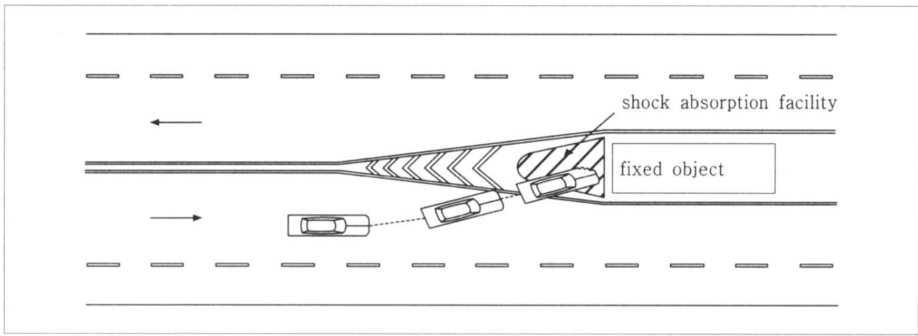

3-2-2 Design Standard

① Installation zone

- Front of toll gate (Front of Hi-pass booth and front of open type island toll booth in main route)

- Branch point of connecting road exit. (when branch part facilities are installed with hard structures (protection wall, barrier, etc)

- Places where it is determined to be dangerous to collide such as front of main route toll booth and front of hard facilities such as large signs as street lights where connecting road branch exit is located in the downhill slope

- Where large scale accidents are frequent or shock absorption is necessary due to the geometric condition of the road

- Protection facilities such as guard rail, protection wall (including PE protection wall) shall be installed for miscellaneous type (front of pier and abutment, end part of protection fence, front of tunnel retainment wall, etc)

② 형식선정

- 설치등급결정
  - 연결로 출구 분기부

등 급	충돌속도	비 고
CC2	80km/hr	본선 분기부
CC3	100km/hr	본선 분기부

  - 요금소

구 분	하이패스 전용(혼용)차로		일반차로
	차단기 미설치	차단기 설치	
본선 영업소	CC2급	CC1급	CC1급
I.C 영업소	CC1급	CC1급	CC1급

- 시설선정
  - 설치장소의 도로교통조건
  - 설치장소의 길이와 폭
  - 충격흡수시설의 수행도
  - 경제성(초기 설치비, 유지관리비 등)
  - 가드레일, 방호벽 등의 대안을 면밀히 검토 후 적용시설 선정

③ 설치공간

- 충격흡수시설의 수행도를 보장하기 위해서는 도로나 구조물의 설계단계에서 충격흡수시설의 설치를 위한 충분한 여유 공간의 확보가 필요

- 설치 공간은 충격흡수시설의 거동과 구성부재의 비산정도 및 충돌 후 차량의 거동 등을 파악하여 후속차량, 인근차로 또는 대향차로의 주행차량에 영향을 미치지 않도록 설치 형식과 변형거리, 탈출거리 등급 선정에 주의

3-3 긴급제동시설

3-3-1 개 요

① 긴급제동시설이란 자동차가 주행 중에 제동장치가 고장날 경우 자동차의 도로이탈 및 충돌사고를 방지하는 시설

② 승객 및 차제에 대한 손상을 최소화하기 위하여 산지부의 내리막 급경사가 연속된 곳에 설치

② Type selection

- Determination of installation class
  - Branch part of connecting road exit

Class	Impact speed	Remark
CC2	80km/hr	Main route branch part
CC3	100km/hr	Main route branch part

  - Toll gate

Classification	Hi-pass only (and cash) lane		General road
	Gate not installed	Gate installed	
Main router toll gate	CC2 class	CC1 class	CC1 class
IC toll gate	CC1 class	CC1 class	CC1 class

- Facility selection
  - Road traffic condition at installation location
  - Length and width of installation location
  - Execution performance of shock absorption facility
  - Economic efficiency (initial installation cost, maintenance cost, etc)
  - Alterative solution such as guard rail and protection wall shall be carefully reviewed and select the applied facilities

③ Installation space

- In order to secure the crash absorption performance, it is necessary to secure sufficient rooms for the installation of crash cushion systems absorption facilities during the design stage of road and structures.

- Care must be taken on selecting installation type, deformation distance, and escape distance class so that the adjacent road and cars in opposite direction are not affected by understanding the behaviour of shock absorption facility, degree of scatter for composition materials, and behaviour of vehicles after impact.

3-3 Emergency Escape Ramps

3-3-1 Overview

① Emergency escape ramp is to keep the cars on the road and to prevent collision accidents when the brake devices are malfunctioning while driving.

② In order to minimize the damages to passengers and vehicle bodies, the facility shall be installed where steep downhill is continuous in mountain area.

3-3-2 긴급제동시설의 구분 및 종류

① 부설재료에 따른 구분

- 모래더미형식
  - 모래더미의 기울기와 모래에 의하여 제동력이 발휘되는 방식

- 골재부설형식
  - 하향경사방식 및 수평경사방식
    ▶ 골재부설층에 의한 구동저항에 의존
  - 상향경사방식
    ▶ 중력에 의한 제동효과와 골재부설층에 의한 구동저항에 의존

3-3-3 설계기준

① 대상구간

- 산지부의 급경사로 인하여 긴급제동시설이 필요하다고 인정되는 구간

- 하향 종단경사 3% 이상 구간이 5km 이상 연속되는 경우

② 형식선정

- 지형적인 특징과 운전자의 안전 등을 고려할 때, 골재부설에 의한 형식(상향경사방식)을 원칙으로 적용(지형여건상 상향경사방식이 곤란한 경우 수평경사방식 적용가능)

- 골재부설연장을 최대한 확보토록 하는 형식(골재부설구간+이탈방지둑) 적용을 원칙으로 적용

- 현장여건상 정지거리(골재부설층구간) 확보가 어려운 장소에는 부족연장분에 대하여 감속보조시설을 설치하는 형식(골재부설구간+감속원통+이탈방지둑)으로 세밀한 검토 분석 후 적용

③ 설계차량 진입속도

- 고장차량의 최악의 상황고려 및 선형취약구간에 대한 주행차량의 절대적 안전성 확보를 위하여 진입속도를 130km/h를 적용

3-3-2 Classification and Type of Emergency Escape Ramps

① Classification by installation materials

- Sand bank type
  - To stop the cars the slant of sand bank and sand

- Aggregate installation type
  - Downward slope type and horizontal slope type
    - ▸ Relying on the braking effects by aggregate installation layer
  - Upward slope type
    - ▸ Relying on the brake effect by gravity and by aggregate installation layer

3-3-3 Design Standard

① Installation zone

- Zones where the emergency escape ramp is considered to be necessary due to the steep slope in mountain area

- Where downward longitudinal slope zone of over 3% continues for more than 5 km

② Type selection

- Aggregate material installation type (downward slope type) shall be applied as a general rule considering the topological characteristics and safety of drivers. (Horizontal slope type can be applied if upward slope type cannot be installed due to topological condition)

- The type to secure maximum extension of aggregate installation (aggregate installation zone + de-lane prevention dam) shall be applied as a general rule

- For the places where the stop distance (aggregate installation layer) cannot be archived due to the site condition, auxiliary deceleration facility type (aggregate installation layer + deceleration cylinder + de-lane prevention dam) shall be carefully reviewed and applied.

③ Design vehicle entering speed

- Entrance speed of 130km/h shall be applied in order to secure absolute safety of the vehicles in vulnerable alignment zone considering the worst case condition of the malfunctioning vehicles.

④ 진입각

- 긴급제동시설의 연결로는 가능한 한 직선으로 설계하는 것이 바람직

- 본선과 연결로의 진입각은 최소화하여야 하며 5도 이하가 바람직

⑤ 연결로

- 연결로의 폭은 최소 7.8m 이상, 9~12m 가 바람직

- 경제적인 연장 산정에 따른 별도의 감속시설 설치가 불가피함에 따라 도표에 의한 적정 감속길이 선정이 가능한 펜실베니아주 추천식을 사용하여 골재부설길이를 산정

⑥ 감속시설

- 감속원통
  - 감속속도의 산출결과에 따라 결정 후 설치
    - ▸ A형 : 20~50km/h, 감속원통 4열, 4.3m X 7.0m
    - ▸ B형 : 50~70km/h, 감속원통 5열, 5.4m X 7.0m
    - ▸ C형 : 20~50km/h, 감속원통 7열, 7.6m X 7.0m

- 이탈방지둑
  - 골재부설구간의 깊이가 충분히 확보되었을 경우에는 설치를 하지 않는 것이 바람직
  - 설치 시에는 1V : 1.5H의 경사(상단폭:0.5m, 하단폭:5m, 높이:2.5m)의 2열 구조로 설치
  - 흙쌓기부는 도로 외측 2개면에 설치하며, 땅깎기부는 미설치

⑦ 골재규격

구 분		규 격	비 고
골재부설층	10%	13~19mm의 세척골재 둥근모양의 자잘	
	15%		
	20%		
	25%		
감속원통채움재	A형	강자갈	0.63m2 이상
	B형		
	C형		
이탈방지둑		골재부설구간 부설재료와 동일	

④ Entrance angle

- It is desirable to design emergency ecape ramps as linear as possible.

- The angle of the connection to main route shall be at minimum and angle of less than 5 degree is desirable.

⑤ IC ramp

- Minimum width of 7.8m and 9~12m is desirable for IC ramp

- When the installation of additional deceleration facilities are unavoidable in consideration of economic efficiency, calculate the aggregate installation length by the recommendation-type of State of Pennsylvania, with which appropriate length can be calculated by tables

⑥ Deceleration facilities

- Deceleration cylinder
  - Install after determining according to the calculation result of deceleration speed
    - ▸ A type : 20~50km/h, four rows of deceleration cylinder, 4.3m X 7.0m
    - ▸ B type : 50~70km/h, five rows of deceleration cylinder, 5.4m X 7.0m
    - ▸ C type : 20~50km/h, seven rows of deceleration cylinder, 7.6m X 7.0m

- Break-away prevention dam
  - It is not desirable to install when enough depth of aggregate material zone is not secured.
  - Install in two rows structure with the slope of 1V:1.5H (width of upper part:0.5m, width of lower part:5m, height:2.5m) when installing.
  - Install two surfaces outside the road for soil banking part and no installation of cutting part

⑦ Aggregate specification

Classification		Specification	Remark
Aggregate installation layer	10%	Washed aggregate of 13~19mm round shape gravels	
	15%		
	20%		
	25%		
Deceleration cylinder filling material	A type	Hard gravels	Over 0.63m2
	B type		
	C type		
De-lane prevention dam		Same as installation material of aggregate installation zone	

4. 시인성 증진시설

4-1 시선유도표지

4-1-1 개　요

① 주·야간에 직선 및 곡선 구간에서 운전자에게 전방의 도로선형이나 기하조건이 변화되는 상황을 반사체를 사용하여 안내하여 줌으로써 안전하고 원활한 차량주행을 유도하는 시설물

② 노면표시를 보조하여 운전자에게 도로 선형을 인식할 수 있는 정보를 제공

4-1-2 설계기준

① 대상구간

- 자동차 전용도로에는 원칙적으로 전 구간에 연속적으로 설치

- 도로조명시설이 있는 경우와 안개등 및 태양열 시선유도기를 설치하여 야간 운전자의 시선유도가 가능한 구간에는 설치를 생략할 수도 있음

② 형식선정

- 반사체의 형상은 합성수지 또는 유리알 및 반사지를 사용한 원형을 원칙

- 반사체의 크기는 반사기의 유효 반사면적의 내경이 100mm를 표준

- 반사체의 색상은 도로 진행방향에서 왼쪽이나 중앙분리대측에는 노란색, 오른쪽에는 하얀색을 사용을 원칙(일방향 연결로 좌측길어깨에 설치되는 시선유도표지의 색상은 차선색상과 동일한 백색 사용)

- 반사체는 단면형을 원칙(왕복2차로인 경우는 양면형 사용)

③ 설치위치

- 길어깨 가장자리로부터 0~2.00m 되는 곳에 지형에 맞게 연속적으로 설치

- 부착위치에 따라 지면위에 설치하는 토공용, 가드레일용, 옹벽용, 교량난간 부착용, 중앙분리대용등으로 분류하고, 전후구간의 시선유도를 고려하여 연속적으로 설치

④ 설치높이

- 노측은 길어깨 가장자리 면에서 반사체 중심까지 1.00m로 함

- 중앙분리대측은 중앙분리대 기초면에서 반사체 중심까지 0.90m로 함

- 시선유도표지가 방호울타리, 옹벽 등에 설치되는 경우에도 전·후 구간에서 사용하는 동일 종류의 반사체를 가능한 동일 높이에 설치하도록 하여야 함

4. Visibility Enhancement Facility

4-1 Delineators

4-1-1 Overview

① Facilities to guide the safe and smooth car driving by guiding road alignment or changes of geometric condition in front using reflecting objects in linear or circular road zone in day and night.

② To provide the information that driver can recognize the road alignment in combination with road surface signs

4-1-2 Design Standard

① Installation zone

- The facilities must be installed in entire zone of highway as a general rule.
- Installation can be skipped where road lighting facilities are installed or where driver vision can be guided during night time with the installation of fog lights and solar vision guiding facilities

② Type selection

- The shape of reflecting materials shall be round type using synthetic resin, glass balls, or reflecting papers as a general rule
- The standard inner diameter of effect reflection surface of 100mm for the size of reflecting object.
- The colors of reflecting object position on the left of the moving direction or toward the median strip shall be yellow, and right side shall be white as a general rule. (vision guide signs installed on the left shoulder of the one way connecting road shall be white which is same an lane dividing line.)
- Reflecting object shall be single sided type as a general rule (Both-sided type is used for two lanes both ways)

③ Installation point

- To install with continuity and appropriately for given topography at 0~200 cm far from the edge of the shoulder.
- They are classified as for soil work, for guard rail, for barrier, for attaching to bridge rail, and for median strip depending on the installation location, and installation shall be continuous considering the vision guiding for front and rear zone

④ Installation height

- The height shall be 100cm from the edge of the shoulder to the center of the reflecting object for road side installation.
- The distance between the base of the median strip to the center of the reflecting object shall be 90 cm for the median strip installation.
- Even when vision guiding signs are installed in protection fence or barrier, same type of reflecting object used in the rear and front zone shall be installed at the same height as much as possible.

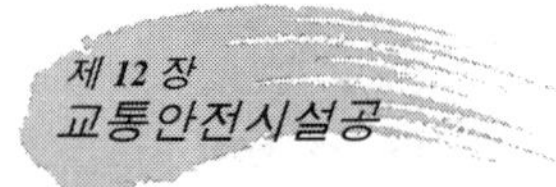

【시선유도표지 설치방법】

구 분	설 치 방 법
토 공 부	VAR, 1,000, 600, 1 : 1.5
가 드 레 일 가드 케이블	VAR, 1,000, 1 : 1.5
L2 측 구	900
중앙 분리대	900
교량방호벽	900

⑤ 설치간격

- 도로의 곡선반지름에 따라 최대 48m 간격으로 설치(인터체인지, 휴게소 등의 연결로 및 변속차로에는 최대 24m 간격으로 설치)
- 차로 수나 차도 폭이 변화하는 구간에 시선유도표지를 설치할 경우에는 그 변화를 정확하게 알릴 수 있는 설치간격이 되도록 하여야 함
- 시선유도표지를 왼쪽 길어깨에만 설치할 경우의 설치간격은 오른쪽에 설치하는 방법과 같은 방법을 적용

【Installation method of vision guide signs】

Classification	Installation method
Soil work parts	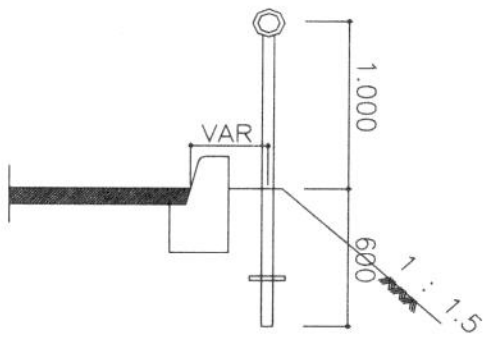 VAR, 1.000, 600, 1 : 1.5
Guard rail Guard cable	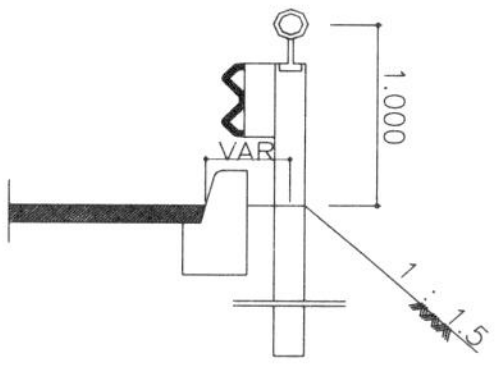 VAR, 1.000, 1 : 1.5
L2 side ditch	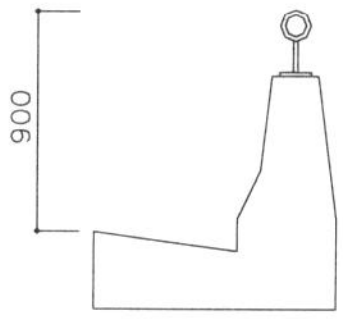 900
Median strip	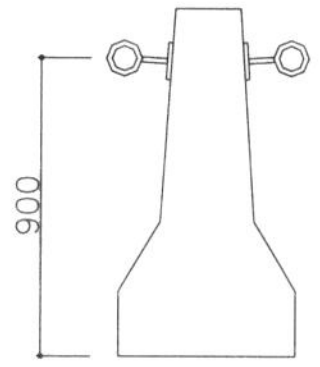 900
Bridge protection wall	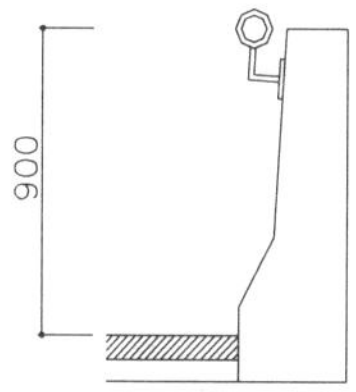 900

⑤ Installation distance

- Install at maximum 48 m distance depending on the radius of the road (install at maximum 24 m distance for connecting road to IC and rest area and changing speed lanes)

- When installing vision guide signs where number of lanes or width of lanes are changed, installation distance shall be set to notify the changes accurately.

- When installing the vision guide signs in only left shoulder, same method for installation on the right side shall be applied.

⑥ 기타

- 설계속도별 곡선반지름이 갈매기표지 설치대상 곡선반지름 이하인 구간은 곡선부 바깥쪽에 갈매기 표지를 안쪽에는 시선유도표지를 설치(갈매기표지와 시선유도표지의 중복설치 배제)

- 설치각도는 자동차의 진행 방향에 대하여 직각으로 설치

- 곡선 반지름이 작은 구간 등 진행 방향에 대하여 직각으로 설치 시 반사 성능이 약할 경우에는 주행 조사 등에 의하여 설치 각도를 변경

4-2 갈매기표지

4-2-1 개　　요

① 급한 평면곡선부 등 시거가 불량한 장소에서 도로의 선형 및 굴곡 정도를 갈매기 기호체를 사용하여 운전자가 명확히 알 수 있도록 하는 시설물

② 갈매기표지는 도로 및 지형조건에 따라 시선유도표지보다 더욱 효과적인 시선유도를 제공할 수 있음

4-2-2 설계기준

① 대상구간

- 도로의 선형이 급격하게 변하는 구간, 공사구간 또는 사고 많은 지점 등과 같이 운전자에게 도로의 상황에 관한 사전정보 제공이 특별히 강조되는 구간

- 대상구간 곡선부 바깥쪽에 설치하며, 안쪽에는 시선유도표지를 설치하고 갈매기표지 설치구간에 시선유도표지는 설치하지 않음(안쪽에 설치하는 시선유도 표지는 시선유도표지 설치기준에 의거하여 설치)

② 형식선정

- 표준형
  - 양방향 2차로의 일반도로 또는 양방향 분리 4차로(편도 2차로) 도로에 설치

- 소형
  - 교통량이 1일 400대 이하의 소교통량 도로 또는 1차로의 연결로 교량 등 도로구조물에 의하여 표준형의 설치공간이 충분치 않아 표지판의 손상을 가져다 줄 수 있는 곳에 적용

⑥ Miscellaneous

- When the radius for each of design speed is lower than the radius that requires sea gull sign, install sea gull sign in the outer side of the curved part and vision guide signs on the inner side of the curve. (Avoid the overlapping installation of sea gull signs and vision guide signs.)

- Installation angle shall be perpendicular to the vehicle moving direction.

- When installing perpendicular to the moving direction including small radius zone, change the installation angle through the driving test if the reflection performance is not good enough.

4-2 Chevron Sign

4-2-1 Overview

① Facilities to clearly notify drivers of the alignment and degree of curve of road where steep horizontal curve area or limited vision area using chevron symbols.

② Chevron signs can provide more efficient vision guidance than vision guide signs depending on the road and topographic condition.

4-2-2 Design Standard

① Target installation zone

- Installation zones are where advance information on the road status provision to drivers is specially emphasized such as zone with rapid change of alignment, construction zone, or frequent accident zone.

- Sign shall be installed at outer side of curve, and vision guide signal installed at inner side or curve. Vision guide signs shall not be installed where chevron signs are installed. (Vision guide signs installed at inner side of the curve shall be installed in accordance with the vision guide sign installation standard)

② Type selection

- Standard type
  - To install in general road of both-way two lanes or both-way separated four lanes (one-way two lanes)

- Small size type
  - To install this type in light traffic road with the traffic volume of less than 400 cars per day, or where damages to the signs are expected because of the limited installation space such as connecting road and bridge of one lane road.

- 대형
  - 편도 3차로 이상의 자동차 전용도로에 적용, 중앙분리대 설치 등으로 편도 2차로이거나 확대 설치 시 설치공간이 충분치 않을 경우 표준형 적용

③ 설치위치

- 차도 시설한계의 바깥쪽 가장 가까운 곳에 설치
- 일반적으로 길어깨 가장자리로부터 0~2.00m 되는 곳에 지형에 맞게 설치

④ 설치간격

- 도로의 곡선반지름에 따라 설치

곡선반지름(m)	설치간격(m)	곡선반지름(m)	설치간격(m)
50 이하	8	246 ~ 320	25
51 ~ 80	12	321 ~ 405	30
81 ~125	15	406 ~ 500	35
126 ~180	20	501 ~ 650	38
181 ~245	22	651 ~ 900	45

- 해당 곡선부에만 설치하되, 진입부 전이구간에 곡선부 설치간격의 2배를 취한 위치에 1개소 추가 설치
- 갈매기표지 설치지점 200m 전방에 우로 굽은 도로표지 및 좌로 굽은 도로표지를 병행하여 설치
- 입체교차시설의 연결로 곡선 외측부에서는 곡선시점에서부터 4개만 곡선반지름별 설치간격에 따라 설치하며, 그 외는 필요에 따라 시선유도표지를 설치

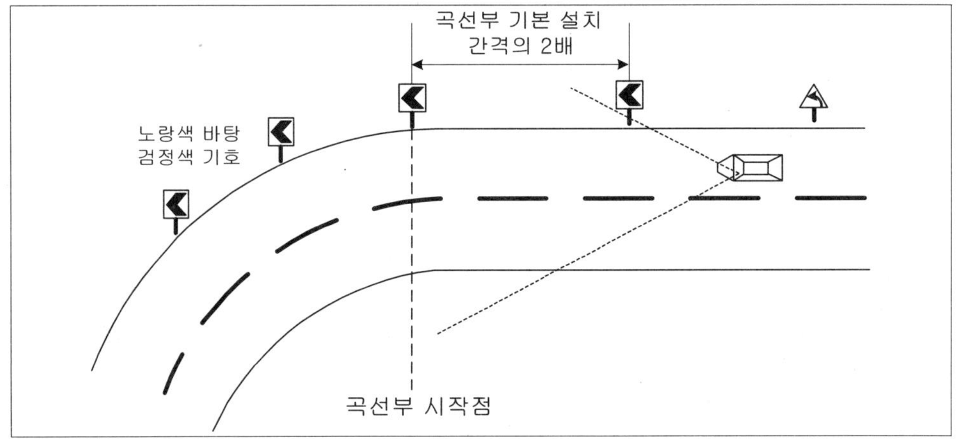

- Big size type
  - To install in high way of more than three lanes each way. Standard type shall be applied for 2 lanes each way or not enough installation room for extended installation due to the installation of median strip.

③ Installation location

- To install at the nearest position to the outside of the road facility limit.
- To install at 0~200 cm from the edge of the shoulder and appropriately from the given topographic condition in general.

④ Installation distance

- Installation depending on the radius of the road

Radius(m)	Installation distance(m)	Radius (m)	Installation distance(m)
Lower than 50	8	246 ~ 320	25
51 ~ 80	12	321 ~ 405	30
81 ~125	15	406 ~ 500	35
126 ~180	20	501 ~ 650	38
181 ~245	22	651 ~ 900	45

- To install at the corresponding curve part only, and to install one additional sign at the transition zone of the entrance area with the twice the distance of the curved part installation distance.

- Road signs for left curve or right curve shall be installed together at 200 m ahead of the chevron sign installation position.

- In the outside of the connection curved ramp of interchange, only four signs shall be installed according to the installation distance for corresponding radius, and additional vision guide signs shall be installed if necessary.

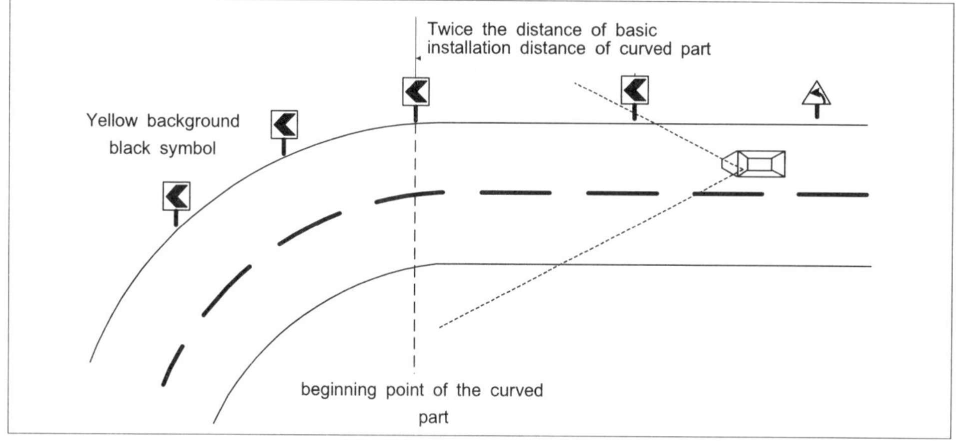

⑤ 설치높이

- 본선은 길어깨 가장자리면에서 표지하단까지 1.30m, 연결로의 경우는 바깥쪽 길어깨 가장자리 면에서 표지 하단까지 1.20m로 하여 설치

- 중앙분리대측은 현장여건에 따라 조정 설치

⑥ 기타

- 2차로 도로에서는 양면형으로 하고, 중앙분리대로 분리된 4차로 이상 도로에서는 단면형으로 설치

- 설치각도는 자동차의 진행방향에 대하여 직각 또는 10 이내에서 설치하되, 표지의 시인성과 자동차의 진행방향을 고려한 주행조사 등에 의하여 설치 각도를 변경 설치할 수 있음

4-3 표지병

4-3-1 개　　요

① 야간 및 악천후 시 노면표시의 기능을 수행

② 자동차 타이어와의 접촉음을 통하여 운전자에게 경고의 의미 전달

4-3-2 설계기준

① 대상구간

- 도로의 중앙선, 차선 경계선, 전용차선, 노상장애물, 안전지대 등 노면표시의 기능을 보완할 필요가 있는 곳에 설치

- 급곡선부, 터널, 차로의 감소, 분리 또는 합류 구간, 통행로의 변경 구간, 교통섬, 인터체인지 고어지역, 좌회전차로를 포함한 2차로 도로, 물리적으로 분리되지 않은 다차로 도로, 도로폭이 좁은 교량 등 선형 유도 또는 도로환경 변화에 대한 운전자의 인식을 높일 필요가 있는 구간에서 도로교통여건에 적합하게 설치

- 횡단보도 및 교차로 정지선 등 표지병의 설치로 인하여 안전주행을 해칠 우려가 있는 지점에는 설치불가

- 강설량이 많은 지역에서는 제설 작업의 지장 여부를 검토하여 설치

⑤ Installation height

- The distance from the edge surface of the shoulder to the bottom of the sign shall be 130 cm for main route, and that shall be 120 cm for connecting road.

- The height for median strip side shall be adjusted depending on the site condition.

⑥ Miscellaneous

- Double sided signs shall be installed in two lanes road, and single sided signs shall be installed in roads with more than four lanes divided by median strip.

- Installation angle shall be perpendicular or within 10 degrees to the vehicle moving direction, and the installation angle can be changed by the driving test considering visibility and vehicle moving direction.

4-3 Raised Pavement Marker

4-3-1 Overview

① To execution the function of road surface sign during night time or bad weather condition

② To warn the drivers with the noise from the contact with vehicle tires.

4-3-2 Design Standard

① Installation zone

- To install where it is necessary to complement the function of road surface sign such as median strip, lane dividing line, dedicated lane, on-road obstacle, and safety zone.

- To install appropriately for the road traffic condition where there is a need to guide the alignment and to raise the recognition of drivers about the changes in road environment such as sharp curve, tunnel, reduction in number of lanes, merge and branch zone, driving lane changing zone, traffic island, interchange gore zone, two lanes road including left turn lane, not physically separated multi-lane road, and bridge with narrow road width.

- It is not allowed to install raised pavement marker where the installation could disturb the safe driving such as pedestrian crossing or stopping line of crossing.

- Whether signs are causing any trouble to snow plow in heavy snowfall area shall be reviewed before the installation.

② 설치높이

- 표지병 몸체의 밑면(하단 지면)부터 윗면까지의 높이는 최대 30mm 이하

- 표지병과 타이어의 마찰이 빈번한 곳에서는 높이가 20mm를 넘지 않는 것이 좋으며, 가급적 지주 없는 부착식 표지병 사용을 권장

- 중앙선이나 안전지대 등 자동차의 출입을 금지하는 곳에서 자동차의 진입 시 운전자에게 경고의 의미를 전달하기 위한 부가적인 기능을 발휘할 수 있도록 표지병의 높이를 높이는 경우에도 최대 30mm로 제한

③ 설치간격

- 표지병 최소 설치간격

구 분		설치간격	비 고
직선부	시가지도로	1N(8m)	· 공학적 판단에 의하여 조정 가능
	지방도로	1N(13m)	· 공학적 판단에 의하여 조정 가능
	전용도로	1N(20m)	· 공학적 판단에 의하여 조정 가능
	편도1차로	N/2	· 간격은 도로구분별로 달리 적용
곡선부		N/4~N/2	· 반지름의 크기에 따라 공학적 판단 하에 설치
진출·입 연결로 고어부		N/4	· 미국 FHWA 기준 적용
교차로 좌회전 차로		N/2	· 미국 FHWA 기준 적용

- 곡선부에서는 최소설치간격 기준을 따르되, 기하구조상 시계에 장애가 있을 때에는 연속적으로 4개 이상이 보일 수 있도록 설치

② Installation height

- Maximum height from the bottom of the road line delineator (lower ground surface) and to the upper side shall be 30 mm.

- It is desirable to limit the height to 20 mm where the contact between tires and road line delineators are frequent, and it is recommended to install the road delineators type without pillar as much as possible.

- Even when height is elevated for the additional functions to inform the drivers of warning messages when vehicles enter the prohibited zone such as median strip or safety zone, the maximum height shall be limited to 30 mm.

③ Installation distance

- Minimum installation distance between raised pavement markers

Classification		Installation distance	Remark
Linear part	City road	1N(8m)	· Possible to adjust based on engineering judgement
	Local road	1N(13m)	· Possible to adjust based on engineering judgement
	Dedicated road	1N(20m)	· Possible to adjust based on engineering judgement
	One lane each way	N/2	· Different distance by road classification shall be applied
Curved part		N/4-N/2	· Install under the engineering judgement depending on the radius size
Entrance/Exit connecting road gore part		N/4	· Apply American FHWA standard
Left turn lane of intersection		N/2	· Apply American FHWA standard

- Minimum installation distance standard shall be followed in curved road, and at least four continuous signs shall be visible when there is a difficulty in visibility due to geometric structure.

5. 기타 안전시설

5-1 미끄럼 방지시설

5-1-1 개　요

① 미끄럼 방지시설이란 노면의 미끄럼 저항이 낮아진 곳, 도로의 평면 및 종단 선형이 불량한 곳 등에서 포장면의 미끄럼 저항력을 높여 주는 시설

② 자동차의 제동 거리를 짧게 하기 위한 목적으로 설치되는 시설

5-1-2 미끄럼 방지시설의 구분 및 종류

① 표면에 신재료를 추가하는 방식에 따른 구분

- 개립도 마찰층 (OGFC : Open-Graded Friction Course)
  - 플랜트 믹스로 실코팅하는 공법

- 슬러리실
  - 상온에서 유화 아스팔트, 잔골재, 석분, 물 등을 혼합한 유동체인 슬러리 혼합물을 6~10mm 정도 포장면에 포설하는 공법

- 수지계 표면처리
  - 포장면에 에폭시 수지를 도포한 후 마찰계수가 큰 경질 골재를 살포하여 고착시키는 방법

② 표면의 재료를 제거하는 방식에 따른 구분

- 그루빙
  - 다이아몬드 날 또는 텅스텐 카바이드 드럼 등을 여러 개 부착시킨 그루빙 기계로 포장층에 홈을 내어 우천 시 수막현상(hydroplaning)을 억제하거나 노면과 타이어의 마찰 저항을 개선하기 위하여 실시하는 미끄럼 개량 공법
  - 그루빙의 방향으로는 주로 종방향과 횡방향으로 나눌 수 있으며, 종방향의 경우 횡미끄럼에 대하여 효과가 있으므로 곡선구간에 적합하다. 또한 횡방향의 경우 제동 정지 거리의 단축, 수막 현상 억제, 배수 경로 제공, 조면조직 회복 등에 효과가 있으며, 급경사·교차로 등에도 적합

- 숏 블라스팅
  - 블라스터(blaster)라 불리는 다량의 쇠구슬을 고압으로 노면에 연속 타격하여 조면 조직을 회복시키는 방법
  - 도로 포장면의 미끄럼 증진용이나 시멘트 콘크리트 포장 덧씌우기 층의 접착력 증진을 위하여 사용

5. Other Safety Facilities

5-1 Slip Preventing Facilities

5-1-1 Overview

① Facilities to improve the slip resistance of pavement surface in place where slip resistance becomes low and horizontal and longitudinal alignment are bad

② Facilities to install for the purpose of shortening the braking distance of vehicles

5-1-2 Classification and Type of Slip Preventing Facilities

① Classification by the method of adding new materials to the surface

- OGFC (Open-Graded Friction Course)
  - Process method of seal coating using plant mix

- Slurry seal
  - Process method of paving 6~10 mm of slurry fluid mixture over the pavement surface which is a mixture of emulsified asphalt, fine aggregate, stone power, and water.

- Surface treatment with resin
  - Process method to put and fix the hard aggregate that has big friction coefficient after spraying epoxy resin on the pavement surface.

② Classification by the method of removing surface materials

- Grooving
  - Slip enhancement process method to suppress hydroplaning during raining or to improve the friction resistance between tires and road surface by making grooves with the grooving machine that is attached with multiple diamond blades or tungsten carbide drums.
  - Grooving direction can be divided into later and longitudinal direction, and longitudinal direction has effects of slip in transverse direction so it is suitable in curved zone. transverse grooving is suitable for steep slope or intersection since it has effects on shortening braking distance, suppressing hydroplaning, provision of water drain path, and recovery of rough surface structure.

- Shot blasting
  - Process method to recover rough surface structure by continually hitting road surface in high pressure with large number of iron beads called blaster.
  - To use for improving the adhesion force of the overcoated concrete pavement layer or for improving the slip behaviour of pavement surface

• 노면 평삭
  - 포장 노면을 전체적으로 약간 깎아내는 방법으로 조면 조직을 회복시키는 공법

5-1-3 설계기준

① 대상구간

• 기존의 노면 마찰계수가 도로교통 조건에 부합되지 않아 위험한 구간

• 도로의 전·후 선형의 연속성이 이루어지지 않아 주행속도의 차이가 20km/h 이상인 구간의 변화구간

• 기타 사고발생의 위험이 높아 설치효과가 있다고 인정되는 구간
  - 선형이 양호한 직선구간 사이에 위치한 급곡선 구간
  - 종단경사가 급한 내리막길에 위치한 곡선 등의 굴곡도로
  - 곡선부와 낭떠러지가 복합된 선형 불량 구간
  - 배수시설 불량으로 강우 시 마찰력이 크게 감소하는 지점
  - 철도 건널목 진입부의 선형 불량으로 사고 위험이 많은 지점

② 형식선정

• 아스팔트 콘크리트 포장
  - 개립도 마찰층
  - 슬러리실
  - 노면 평삭
  - 수지계 표면처리

• 시멘트 콘크리트 포장
  - 그루빙
  - 숏 블라스팅
  - 노면 평삭

③ 직선구간 설치방법

• 일반적인 직선구간

- Road surface flat cutting
  - Process method to recover rough surface structure by cutting entire pavement surface slightly.

5-1-3 Design Standard

① Installation zone

- Dangerous zone where the existing road surface friction coefficient doesn't meet the road traffic condition.

- The speed changing zone of over 20km/h difference because there is no continuity between front and rear alignment.

- And other area that is considered to have high installation effects because of high change of traffic accidents
  - Sharp curve zone located between highly linear zone
  - Curved road like curve located at the downhill road with steep longitudinal slope.
  - Bad alignment zone where curved parts and cliff are combined
  - Where friction force decreases largely during raining because of bad water drain facilities.
  - Bad alignment in the entrance of railroad crossing that causes frequent accidents

② Type selection

- Asphalt concrete pavement
  - OGFC (Open-Graded Friction Course)
  - Slurry seal
  - Road surface flat cutting
  - Surface treatment with resin

- Cement concrete pavement
  - Grooving
  - Shot blasting
  - Road surface flat cutting

③ Installation method in linear section

- General linear section

- 마찰력이 최소 마찰계수 이하인 구간 전체의 마찰력을 개선
- 미끄럼 마찰의 개선이 주 목적인 경우 전면처리 방식을 추천
- 미끄럼 마찰력은 기준에 부합하나, 다른 이유로 인하여 사고가 발생하는 경우에는 직선구간 중 문제 구간을 세부적으로 분석 검토하여 해당 구간에만 최소한의 길이로 미끄럼방지포장을 설치

• 교차로 또는 횡단보도 접근부
- 설계속도로부터 여유 있게 정지할 수 있는 거리를 최소 길이로 적용
- 문제구간에는 전면처리 미끄럼방지포장을 적용
- 문제구간의 전방에 운전자의 주의를 환기시키기 위하여 필요에 따라 이격식을 적용할 수 있음
- 이격식 시점은 전면처리 구간 전방에 인지반응시간 2.5초를 고려한 거리로부터 하며, 설치구간은 시점부터 1초간 주행하는 거리를 같은 식으로 산정하여 설치

【이격식 미끄럼방지포장 설치방법】

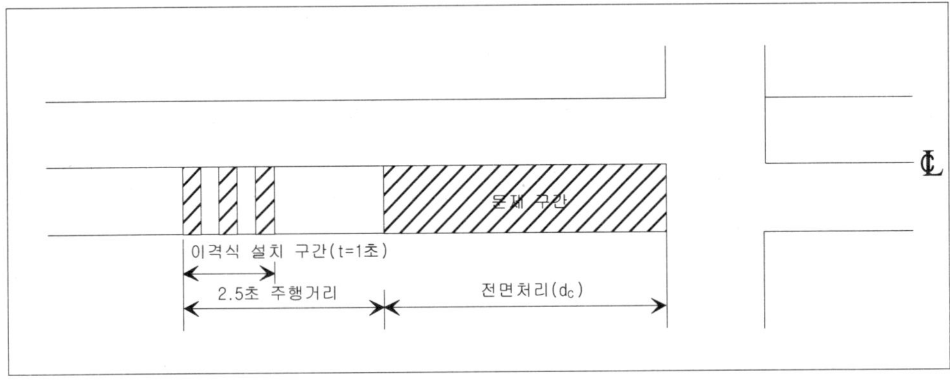

• 5% 이상의 내리막 경사가 100m 이상인 곳
- 내리막 경사 전체에 전면처리 미끄럼방지포장을 설치하는 것이 바람직
- 부득이한 경우 내리막 종단경사의 시점 5% 이상의 경사가 되는 지점으로부터 100m 내려간 지점에서 내리막 경사가 끝나는 지점까지 도로환경 조건을 고려하여 필요하다고 판단되는 길이를 최소 길이로 하여 미끄럼방지포장을 설치

- To improve the friction force of entire zone where friction coefficients is lower than minimum friction coefficient.
- Entire surface treatment method is recommended if the improvement of the slip friction is the main purpose.
- For frequent accident area while slip friction meets the requirement, problematic sections among the alignment sections shall be reviewed in detail and install slip preventing pavement at minimum required length in only corresponding section.

• Zone approaching intersection or pedestrian crossing
  - The distance that vehicle can stop from the designed speed with margin shall be applied as minimum distance.
  - Slip preventing pavement shall be applied to entire surface for problematic zone
  - Stripe type can be installed in front of the problematic area if necessary to remind the drivers.
  - The beginning point of the stripe type shall be set considering the recognition reaction time of 2.5 seconds in front of entire surface treatment zone, and installation zone shall be set and installed using the same method for one second driving distance.

【Installation method of stripe type slip preventing pavement】

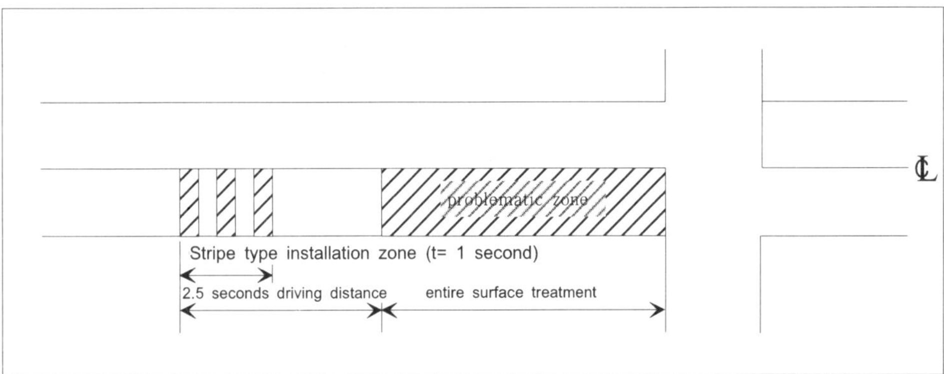

• Where downhill slope of over 5% is longer than 100 m
  - It is desirable to install total surface slip preventing pavement on entire down hill slope surface.
  - If unavoidable, install slip preventing pavement on the minimum determined distance considering road environment condition from 100 meters down from the beginning of the longitudinal downhill slope over 5% until the end of down hill slope.

- 노면 미끄럼과 관계없이 과속만이 문제가 될 경우는 미끄럼방지포장보다 감속유도시설의 설치를 고려하는 것이 경제적

【미끄럼방지포장의 최소 설치길이】

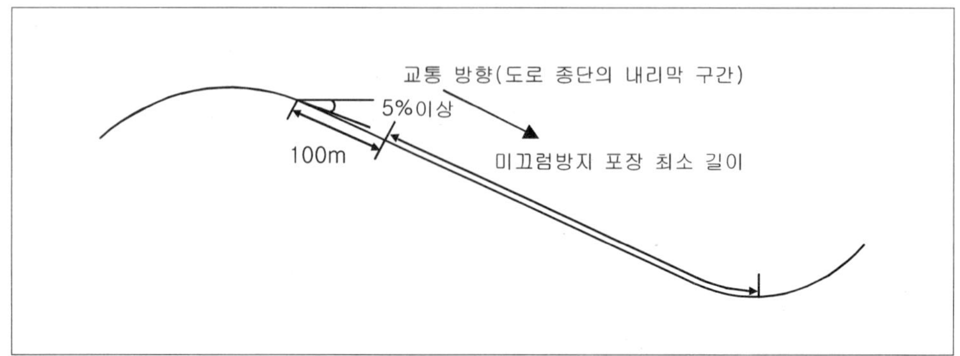

④ 곡선구간 설치방법

- 완화구간이 있는 곡선구간
  - 설치 대상이 되는 전 구간에 걸쳐 전면처리 미끄럼방지포장을 설치하며, 완화구간이 있는 경우 진입부 완화구간에도 전면처리
  - 완화구간이 없거나 불분명한 경우에는 원곡선 구간에만, 완화구간만으로 된 곡선구간에는 완화구간에 전면처리

【완화구간이 있는 곡선구간에서의 미끄럼방지포장 설치】

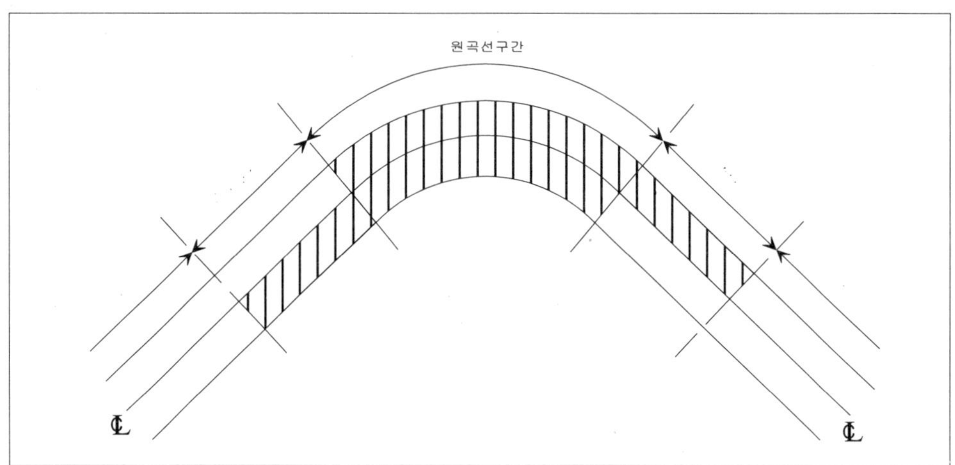

- 원곡선만으로 된 곡선구간
  - 원곡선만으로 된 곡선구간에는 곡선부 전체에 전면처리 미끄럼방지포장을 설치
  - 곡선구간의 진입부인 직선구간에는 운전자의 주의를 환기시킬 목적으로 필요에 따라 이격식을 적용 가능

- If over speeding is a problem without the concern of road surface slip, then it is more economical to consider installing speed reduction guide facility rather than slip preventing pavement.

【Minimum installation distance of slip preventing pavement】

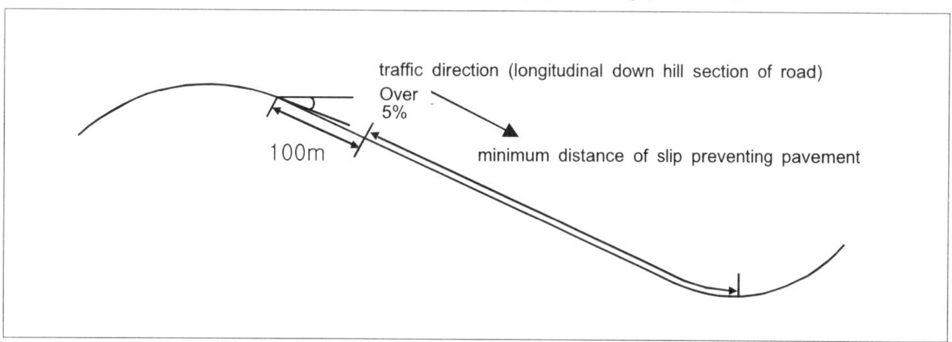

④ Curve section installation method

• Curve zone with transition curve
  - To install the entire surface treatment of slip preventing pavement across the entire section of the target installation road and the entire surface of transition section of entrance part shall be treated if it exists.
  - If there is no transition section or boundary is unclear, only circular curve section shall be treated. For curved section consisting of only transition section, entire surface of transition section shall be treated.

【Installation of slip preventing pavement in curve section with transition section】

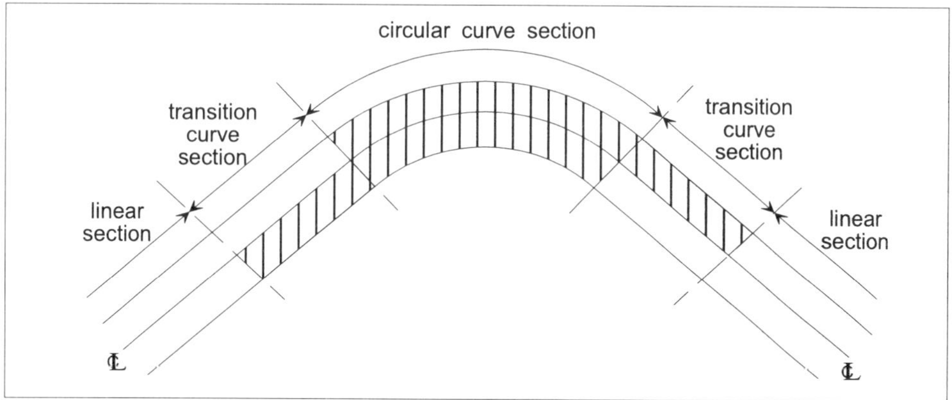

• Curve section consisting of circular curve only
  - To install slip preventing pavement on entire surface of curved parts for curved section consisting on circular curve only.
  - Stripe type can be applied depending on the need of reminding drivers in linear section that is entrance part of curved section.

【원곡선구간에서의 미끄럼방지포장 설치】

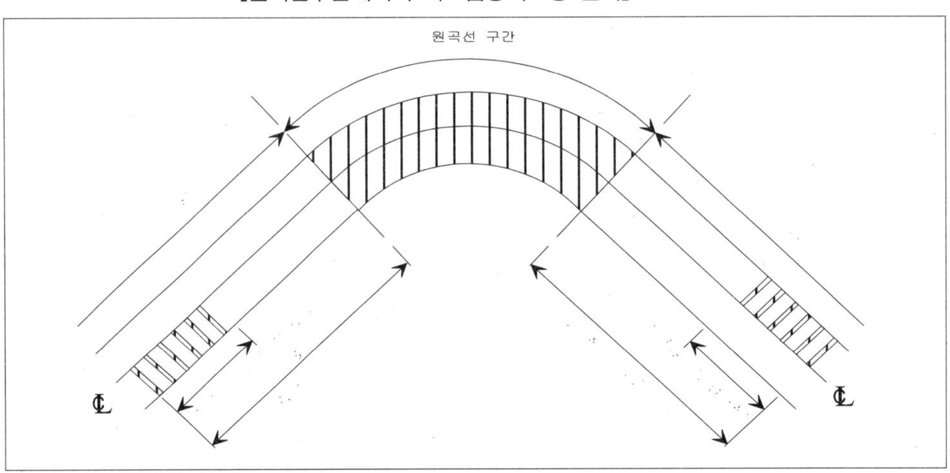

【원곡선구간에서의 이격식 미끄럼방지 포장 설치길이】

주행속도(km/h)	이격식처리방식	s(m)	d(m)
60	1 : 3 방식	40	15
70		50	20
80	3 : 6 방식	55	22
90		60	25
100		70	30

- I.C 및 JCT 구간의 Loop 연결로 포장(콘크리트)의 곡선구간
  - 시공방향별 특성을 고려하여 종방향 그루빙 처리를 원칙으로 적용 (길어깨부는 제외)

【Installation of slip preventing pavement in circular curve section】

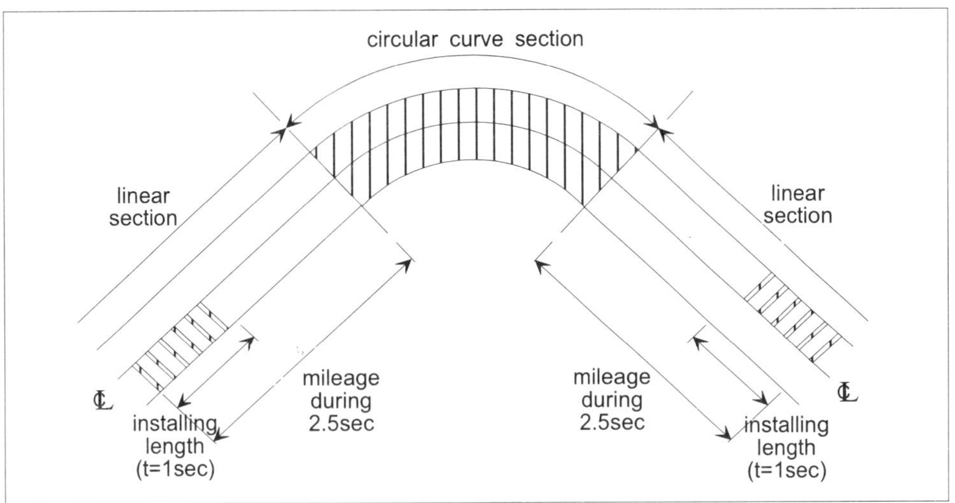

【Installation distance of separation type slip preventing pavement in circular curve section】

Driving speed(km/h)	Separation type process method	s(m)	d(m)
60	1 : 3 type	40	15
70		50	20
80	3 : 6 type	55	22
90		60	25
100		70	30

- Curve section of loop connecting road pavement (concrete) of I.C and JCT
  - Longitudinal grooving shall be applied considering the characteristics of constructions direction as a general rule (excluding shoulder part)

【시공방향별 특성에 따른 그루빙 처리방법】

구 분	종방향 그루빙 (Longitudinal Grooving)	횡방향 그루빙 (Transverse Grooving)
개 요	• 차량 진행 방향으로 시공	• 차량 진행 직각방향으로 시공
효 과	• 급제동 시 제동거리 단축 • 시선유도, 조향성 증대 • 곡선구간 횡방향 미끄럼방지 • 우천 시 배수촉진 • 타이어 소음 진동억제 • 동결노면의 수막분리로 노면빙설현상 억제	• 미끄럼 저항성 증대로 제동거리 단축 • 우천 시 배수촉진 • 소음·진동발생으로 운전자 경고효과 있음 • 동결노면의 수막분리로 노면 빙설현상 억제
장 · 단점	• 한 차로씩 시공하므로 교통차단최소화 • 연속작업으로 시공성 우수 • 홈을 따라 진행하려는 특성으로 인하여 불쾌감 유발	• 마모된 타이닝 대안으로서, 종방향 마찰저항 우수 • 전 차로 폭을 대상으로 시공하므로 교통 차단 단속 작업으로 시공성 불량
적 용	• 곡선부, 교량, 내리막, 횡풍이 심한 구간	• 영업소 전방 경고구간, 과속방지구간, 졸음방지구간 등

【시공방법특성에 따른 그루빙 처리 방법】

구 분	건식 그루빙	습식 그루빙
개 요	• 그루빙 작업 시 집진장치 사용	• 그루빙 작업 시 냉각수 사용
효 과	• 물을 사용하지 않으므로 부대장비 필요 없음 • 오탁수가 발생하지 않으므로 환경친화적 • 작업공간 축소 • 블레이드 팁 마모가 심함	• 블레이드 팁 마모가 적음 • 절단 시 발생열에 의한 포장영향이 적음 • 오탁수 처리를 위하여 별도의 살수차, 청소차 등 부대장비 필요
장 · 단점	• 한 차로씩 시공하므로 교통차단최소화 • 연속작업으로 시공성 우수 • 홈을 따라 진행하려는 특성으로 인하여 불쾌감 유발	• 마모된 타이닝 대안으로 종방향 마찰저항 우수 • 전 차로 폭을 대상으로 시공하므로 교통 차단 단속 작업으로 시공성 불량
적 용	• 상수원 보호구역, 동절기 시공 등	• 도심지, 하절기 시공 등

【Process method of grooving for each construction direction】

Classification	Longitudinal Grooving	Transverse Grooving
Overview	• Construction in vehicle moving direction	• Construction in perpendicular to vehicle moving direction
Effects	• Shortening braking distance in emergency braking • Improving vision guide and steerability • Preventing transverse slip in curve section • Enhancing water drain during raining • Suppressing tire noise • Suppressing road freezing symptom by separating hydroplaning from frozen surface	• Shortening of braking distance by increasing slip resistance • Enhancing water drainage during raining • Warning effects to drivers by noise and vibration effects • Suppressing road freezing symptom by separating hydroplaning from frozen surface
Merits and demerits	• Minimum traffic blocking because only one vehicle is used for construction • Excellent construction workability due to continuous work • Unpleasant feeling of tendency to move along the groove	• Excellent longitudinal friction resistance as an alternative to worn out tining • Poor workability because entire lanes have to be blocked
Application	• Curve part, bridge, down hill, strong transverse wind area	• Front of tollgate, warning zone, over speed preventing zone, zone to prevent doze off, etc

【Grooving process method for each construction method】

Classification	Dry grooving	Wet grooving
Overview	• To use dust collector while grooving	• To use cooling water while grooving
Effects	• Additional equipment are not necessary because waters are not used • Environment friendly because not pollution water is generated • Minimum work area • Wear level of blade tip is severe	• Wear level of blade tip is small • Effects to pavement due to cutting heat is small • Additional water spraying truck and cleaning truck will be required for processing pollution water
Merits and demerits	• Minimum traffic blocking because construction is done for each lane • Excellent construction workability due to continuous work • Unpleasant feeling of tendency to move along the groove	• Excellent longitudinal friction resistance as an alternative to worn out tining • Poor workability because entire lanes have to be blocked
Application	• Water source protection area, winter season construction, etc	• City area, summer season construction, etc

5-2 노면요철포장

5-2-1 개　　요

잠재적인 위험을 지니고 있는 구간의 노면에 인위적인 요철을 만들어 차량이 이를 통과할 때 타이어에서 발생하는 마찰음과 차체의 진동을 통하여 운전자의 경각심을 높임으로써 차량이 안전하게 주행할 수 있도록 유도하는 시설

5-2-2 노면요철포장의 종류

① 절삭형(Milled Type)

- 포장면을 지름 600mm의 회전드럼에 부착된 절삭장비로 깎아서 만드는 방식

- 설치방법이 용이하고 소음 및 진동효과가 큼

② 다짐형(Rolled Type)

- 아스팔트 콘크리트 포장에 사용되며 원형(또는 사각형)의 강봉을 길이 방향으로 잘라 부착한 롤러를 사용하여 고온의 아스팔트 표면을 다짐으로써 반원형 모양의 홈을 만드는 방식

③ 틀형(Formed Type)

- 콘크리트 포장에 사용되며 콘크리트 타설 및 표면 마무리 작업 후 길어깨 표면을 일정 모양의 틀로 눌러서 만드는 방식

④ 부착형(Raised Type)

5-2-3 설계기준

① 대상구간

- 신규구간
  - 도시부 고속도로는 가급적 적용 제외
  - 방음벽 설치구간은 전후 100m 구간 포함하여 설치 제외
  - 출입시설 구간은 진입연결로 및 진출연결로 전후 200m 구간 포함하여 설치 제외 (길어깨 주행가능성 고려)
  - 구조물(터널, 교량)구간은 설치 제외

- 기존도로 구간
  - 운전자의 부주의, 지루함, 피로 및 졸음운전에 의한 차로이탈사고 다발구간에 대한 사업우선순위 선정 후 자체 운영계획을 수립 시행
  - 신규구간의 일반적인 적용기준 준용

5-2 Rumble Strip

5-2-1 Overview

① Facility to lead the safe driving by increasing the warning level through the vibration and noise generated from tires by producing artificial unevenness of the road surface in potential danger zone

5-2-2 Type of Road Surface Rumble Strip

① Milled Type

- Method to cut pavement surface with cutting equipment attached to the revolving drum with the diameter of 600mm.

- Easy installation method and big noise and vibration effects

② Rolled Type

- To use the roller which is used for asphalt pavement by cutting the circular (or rectangular) shape steel rod in longitudinal direction to produce the grooves of half circle shape by rolling hi temperature asphalt surface

③ Formed Type

- Method to press the shoulder surface using certain shape of frame after installation and surface finishing of concrete pavement.

④ Raised Type

5-2-3 Design Standard

① Installation zone

- New constructing road
  - City highways shall be avoided as much as possible
  - Installation shall be excluded noise barrier installation zone including before and after 100m length
  - At Interchange zones, installation shall be excluded including before and after 200m length. (Considering driving in shoulder)
  - Structure section(tunnel, bridge) shall be excluded from the installation.

- Existing road
  - To set up internal operating plan and execute after selecting project priorities for the frequent de-lane accidents section by careless driving, boring, and doze off.
  - General standard for new constructing road shall be applied as well.

② 형식선정

- 형식선정 시 주요 고려사항
  - 소음, 진동효과
  - 시공성(선형유지, 온도관리, 장비운용 등의 시공편의성)
  - 효과지속성(형상유지, 내구성)
  - 연도주민의 생활환경 침해여부 등

- 일반적으로 절삭형(Milled Type)이 바람직
  - 아스팔트 콘크리트 포장 : 절삭형 (Milled Type)
  - 시멘트 콘크리트 포장 : 전압형 (Rolled Type)

- 도시지역 및 취락지 등에 불가피하게 설치 시는 전압형(Rolled Type)을 제한적으로 선정

③ 설치위치

- 길어깨 폭, 포장여건을 감안하여 차로를 이탈한 차량이 복귀할 수 있는 공간 확보를 위해서 가능한 길어깨 차선에 가깝게 설치
  - 시멘트 콘크리트 포장 : 길어깨 차선 우측 끝에서 100mm 이격하여 설치
  - 아스팔트 콘크리트 포장 : 길어깨 차선 우측 끝에서 100mm 이격하여 설치

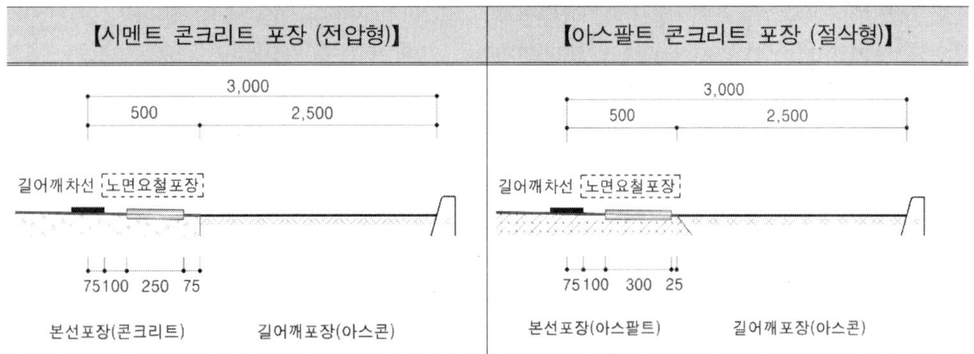

- 설치간격은 연속형과 단속형의 두 가지 형태로 구분되며 연속으로 설치하는 것을 원칙

② Type selection

- Major considerations during type selection
  - Noise and vibration effects
  - Workability (Level of convenience in construction including keeping alignment, temperature control, and equipment operation)
  - Durability of effects (shape maintainability, durability)
  - Whether construction could cause any trouble to living environment of nearby residents

- Milled Type shall be preferred in general
  - Asphalt pavement: Milled Type
  - Concrete pavement: Rolled Type

- When installing in city area or settlement area unavoidably, rolled type shall be used within limitation.

③ Installation location

- To install near to shoulder lane as much as possible to secure the room for cars to recover back to the lane considering shoulder width and pavement condition.
  - Concrete pavement: to install at 100 mm apart from rightmost end of shoulder lane
  - Asphalt pavement: to install at 100 mm apart from rightmost end of shoulder lane

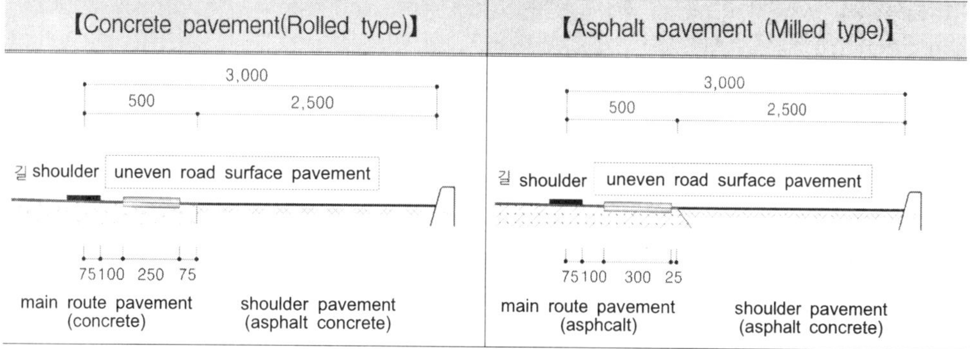

- Installation distance is classified into continuous type and intermittent type, and continuous type shall be installed as a general rule.

- 적절한 안내표지판을 함께 설치

- 고속도로 요금소 전방 노면요철포장은 동 시설물의 주의환기 등을 통하여 감속을 유도하는 안전시설물임을 감안하여 본선 요금소에만 설치

5-3 낙석방지시설

5-3-1 개　요

① 도로 절개면의 낙석, 토사붕괴 등으로 인한 교통 장애, 도로 구조물의 손상, 재산 및 인명상의 손실을 예방하기 위하여 설치하는 구조물

5-3-2 낙석방지시설의 구분 및 종류

① 낙석보강공법

- 낙석 발생이 예상되는 암편을 사전에 제거하거나 절개면에 고정시키는 공법

- 절취공법, 면정리, 록앵커, 록볼트, 콘크리트 버팀벽, 콘크리트 블록공법, 숏크리트, 배수공법(지표수, 지하수), 식생공법 등

② 낙석보호공법

- 절개면으로부터 낙하하는 암편이 도로로 유입됨을 방지하기 위하여 도로변이나 도로시설 위에 설치하는 공법

- 낙석방지망
  - 강우나 풍화, 나무뿌리의 작용 등에 의하여 불연속면의 이완이 심화되어 낙석 발생 가능성이 있는 부분을 철제망으로 덮어 낙석을 예방하는 공법
  - 비포켓식 낙석방지망, 포켓식 낙석방지망

- 낙석방지울타리(Rockfall Protection Fence)
  - 지주와 지주 사이를 와이어 로프와 철망 등으로 연결하여 절개면으로부터 떨어지는 낙석의 에너지를 흡수함으로써 낙석으로 인한 피해를 막는 시설

- 낙석방지옹벽(Barrier)
  - 벽면 배후에 설치한 모래방석 등으로 낙석의 충격력을 분산 감소시켜 낙석을 방호하는 옹벽형 구조물

- To be installed together with appropriate guide signs

- Road surface uneven pavement in front of highway toll gate shall be installed only at the main route toll gate since it is safety facility to lead the speed reduction by reminding drivers

5-3 Rockfall Preventing Facility

5-3-1 Overview

① Structures to prevent traffic blocking, and damages to properties and people caused by falling rocks from road cutting surface or washout

5-3-2 Classification and Type of Rockfall Preventing Facility

① Rock fall reinforcing method

- Method to remove or fix the piece of rocks to cutting surface where rockfall is expected from.

- Cutting method, surface cleaning, lock anchor, lock bolt, concrete support, concrete block method, shotcrete, water drain method (surface, underground water), gardening method, etc

② Rockfall protection method

- Method of installation over road side or on the road facilities to prevent the rocks or stones falling into road from cutting surface.

- Rockfall preventing mesh
  - Method to prevent rockfall preventing metal mesh over the area where rockfall is probable because of severe deterioration of the discontinuous surface by the effects of rainfall, wind, and roots of trees.
  - Non pocket type rock falling preventing mesh, pocket type rock falling preventing mesh

- Rockfall protection Fence
  - Facility to prevent the damage by rockfall by absorbing the energy of rockfall from the cutting surface by connecting between pillars with wire rope or metal mesh.

- Rockfall preventing barrier
  - Barrier type structure to prevent rockfall by distributing and reducing the impact from rockfall by sand cushion installed in rear side of wall surface

- 피암터널(Rock Shed)
  - 절개면 상부에서 발생된 낙석을 도로 바깥쪽으로 이동하게 하여 낙석에 의한 피해를 방지하는 시설

5-3-3 설계기준

① 대상구간

- 낙석방지망
  - 발파암이 10m 이상 분포하는 땅깎기부
  - 발파암고 10m 미만 및 리핑암으로만 구성된 땅깎기부와 소단부 낙석방지 울타리 상단부는 대상에서 제외

- 낙석방지울타리
  - 높이 15m 미만의 땅깎기부나 암층이 리핑암으로만 구성된 높이 15m 미만의 비교적 낮은 땅깎기부에서 낙석방지시설이 필요한 구간
  - 리핑암으로만 구성된 암이 높이 15m 이상 분포하고 낙석방지시설이 필요한 구간이나 발파·리핑암의 혼합고가 15m 미만인 땅깎기부로서 낙석방지시설이 필요한 구간
  - 발파·리핑암의 혼합고가 15m 이상인 땅깎기부로서 낙석방지시설이 필요한 구간
  - 발파암이 높이 30m 이상 분포하는 땅깎기부로서 낙석방지시설이 필요한 구간이나, 리핑암으로만 구성된 암이 높이 40m 이상 분포하고 낙석방지설이 필요한 구간

② 형식선정

- 낙석방지망
  - 절취면의 형상 및 위치에 따라 용도를 고려하여 TYPE-1 및 TYPE-2로 구분하여 적용

구 분	적 용
상단부에 예기치 못한 낙석 방호 및 점검원 안전 확보가 요구되는 절취부	TYPE-1 : 상부지주(보강)
상단부 낙석 및 유지관리 시 안전사고 우려가 없는 구간	TYPE-2 : 상부지주(비보강)

  - 절취 후 표면이 고르게 처리된 절개면에서는 비포켓식 낙석방지망 적용

- 낙석방지울타리

- Rock Shed
  - Facility to prevent damages of rockfall by moving fallen rocks from the cutting surface to the outside the road

5-3-3 Design Standard

① Installation zone

- Rockfall preventing mesh
  - Cut-slope section where blasting rocks are distributer over 10m
  - Except cut-slope section consisting of ripping rock or blasting rock with the height less than 10m and upper part of slope berm installed rockfall preventing fence

- Rockfall preventing fence
  - To install where rockfall preventing facility is necessary at the section of cut-slope that the height of slope is less than 15m relatively consisted of only ripping rocks
  - To install where only ripping rocks with the height over 15m are distributed and rockfall preventing facility is necessary and cut-slope section where blasting and ripping rock are mixed with the height less than 15m and rockfall preventing facility is necessary.
  - Cut-slope section where blasting and ripping rock are mixed with the height over 15m and rockfall preventing facility is necessary.
  - Cut-slope section with blasting rocks with the height ver 30m and where rockfall preventing facility is necessary, and where only ripping rocks with the height over 40m are distributed and rockfall preventing facility is necessary.

② Type selection

- Rockfall prevention mesh
  - Classified and applied into TYPE-1 and TYPE-2 considering the usage depending on the shape and position of the cutting surface

Classification	Application
Cut-slope section where protection from unexpected rockfall at upper side and safety of monitoring person are required	TYPE-1 : Upper pillar (reinforcement)
Where there is no concern of upper part rockfall and safety accident during maintenance	TYPE-2 : Upper pillar (non-reinforcement)

  - Non pocket type rockfall preventing mesh shall be applied to the cutting surface with even processed surface after cutting.

- Rockfall preventing fence

구 분	대상구간	낙석방지시설
형식-1	• 높이 15m 미만의 땅깎기부나 암층이 리핑암으로만 구성된 높이 15m 미만의 비교적 낮은 땅깎기부에서 낙석방지시설이 필요한 구간	프리캐스트 방호벽
형식-2	• 리핑암으로만 구성된 암이 높이 15m 이상 분포하고, 낙석방지시설이 필요한 구간	낙석방지울타리 (H=2.0m)
	• 발파·리핑암의 혼합고가 15m 미만인 땅깎기부로서 낙석방지시설이 필요한 구간	
형식-3	• 발파·리핑암의 혼합고가 15m 이상인 땅깎기부로서 낙석방지시설이 필요한 구간	낙석방지울타리 (H=2.5m)
형식-4	• 발파암이 높이 30m 이상 분포하는 땅깎기부로서 낙석방지시설이 필요한 구간	소단부 낙석방지울타리 추가설치 (H=2.0m)
	• 리핑암으로만 구성된 암이 높이 40m 이상 분포하고 낙석방지설이 필요한 구간	

③ 설치위치

- 낙석방지망
  - 낙석방지울타리가 있는 경우
    - ▸ 낙석방지울타리 상단에서부터 설치하는 것으로 설계적용
    - ▸ 시공 시 암질 및 절리상태와 비탈경사(1:1 이상)등을 종합적으로 검토하여 조정

  - 낙석방지울타리가 없는 경우
    - ▸ 땅깎기부 하단에서 1m 이격되도록 설치

- 낙석방지울타리
  - 낙석방지울타리가 있는 경우
    - ▸ 낙석방지울타리의 위치는 콘크리트 옹벽(L2측구)을 설치하고 방지울타리를 설치
    - ▸ 뒤채움은 하지 않는 것을 원칙

Classification	Target installation zone	Rockfall preventing facility
Type -1	• Rockfall preventing facility is necessary in soil bank part with the height of less than 15m or relatively low soil bank part where rocks distribution are with only ripping rocks with the height of less than 15 m	Precast protection wall
Type-2	• Where only ripping rocks are distributed with the height of over 15m and rockfall preventing facility is necessary	Rockfall preventing fence (H=2.0m)
	• Soil bank part where blast and ripping rocks are distributed as mixed with the height of less than 15 m and rockfall preventing facility is necessary	
Type-3	• Soil bank part where blast and ripping rocks are distributed as mixed with the height of over 15 m and rockfall preventing facility is necessary	Rockfall preventing fence (H=2.5m)
Type-4	• Soil bank part where blast rocks with the height of over 30m are distributed and rockfall preventing facility is necessary	Small end part additional installation of rockfall preventing fence (H=2.0m)
	• Where rocks consisting of ripping rocks only with the height of over 40m are distributed and rockfall preventing facility is necessary	

③ Installation location

- Rockfall preventing mesh
  - Where there is rockfall preventing fence
    - ▸ Installation from the upper end of the rockfall preventing fence shall be applied to design
    - ▸ Rock quality, joint statues and slope grade shall be reviewed comprehensively and the design shall be adjusted.
  - When there is not rockfall preventing fence
    - ▸ Install 1m apart from the bottom of cutting area
- Rockfall preventing fence
  - Where there is rockfall preventing fence
    - ▸ Concrete barrier (L2 side ditch) shall be applied at where rockfall preventing fence is to be installed and rockfall preventing fence shall be installed.
    - ▸ Backfill shall not be done as a general rule

④ 설치간격

- 낙석방지울타리은 유지관리 및 케이블 장력 강화를 위하여 100m 마다 단부지주를 설치하는 것을 원칙

- 설치연장에 따라 60~100m 마다 등간격으로 단부지주를 조정

- 낙석방지울타리의 배면이나 땅깎기 비탈면의 유지관리를 위하여 출입문을 설치

④ Installation space

- End pillar shall be installed at every 100 meters for maintenance and strengthening the tension of cable when installing rockfall preventing fence as a general rule.

- End pillar installation distance shall be adjusted to equal space of 60~100m depending on the length of installation

- Access door shall be installed to maintain rear side of rack fall preventing fence or cutting slope area.

Chapter 13

Subsidiary Facilities

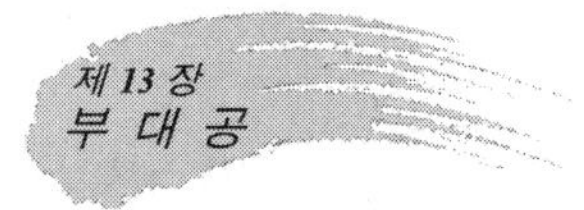

1. 일반사항

1-1 부대시설물의 정의

도로의 기하구조상 본선 이외에 설치되어 도로이용자에게 도로를 효율적으로 이용할 수 있도록 편의를 도모하고 안전성과 쾌적성 제공 및 사고에 대한 피해를 최대한 줄일 수 있도록 설치되는 시설물

1-2 부대시설물 설계 시 고려사항

- 운전자가 쉽게 인식할 수 있도록 곡선반지름이 작은 구간이나 급경사 구간의 설치를 피하며 양호한 시인성에 가장 적합한 장소에 설치

- 부대시설간의 거리는 너무 가깝거나 또는 너무 멀리 설치되지 않도록 적정 간격을 유지하여 설치

- 부대시설 내 차량이 시설물로 안전하게 진입할 수 있도록 유도 안내의 적정성을 가져야 함

2. 부대시설의 종류 및 설계기준

2-1 휴게시설

2-1-1 개 요

① 출입이 제한된 도로에서 장시간의 연속주행으로 인한 운전자의 생리적 욕구 및 피로를 해소

② 자동자의 급유, 정비 등의 필요한 서비스를 제공

2-1-2 휴게시설의 종류

① 휴게소

- 사람과 자동차가 필요로 하는 서비스를 거의 완벽하게 제공할 수 있는 휴게시설

- 변속차로, 주차장, 녹지, 화장실, 주유소, 차량정비소, 식당, 매점 등으로 구성

② 간이휴게소

- 운전자의 생리적 욕구를 해소하고, 짧은 시간 내에 차의 점검·정비 등이 이루어질 수 있는 최소한도의 필요시설이 갖추어진 휴게시설

- 변속차로, 주차장, 녹지, 화장실 등으로 구성

1. General Consideration

1-1 Definition for Subsidiary Facilities

The facilities installed outside of the main lane of the road to provide convenience and safety of the drivers who use the road and minimize the damages from any possible accidents.

1-2 Consideration in Designing Subsidiary Facilities

- Avoid installing in steep slope or corner so that drivers can easily recognize the facilities.

- Maintain proper intervals between subsidiary facilities.

- For the safe entry to the facilities, the guide should induce vehicles properly.

2. Types of subsidiary facilities

2-1 Resting Facility

2-1-1 Overview

① Facility to resolve driver's fatigue and other physiological requirements from continuous driving in the road.

② Facility to provide refueling and maintenance services.

2-1-2 Types of Resting Facility

① Resting Area

- Resting facility to provide almost all required services for drivers and vehicles.

- Resting area includes the following facilities: speed changing lane, parking lot, green area, rest rooms, gas station, vehicle service center, food court and convenient store.

② Simple Resting Area

- Resting area with minimum facilities (rest rooms, facilities for self-maintenance, etc

- Simple resting area includes only the following facilities :
  speed changing lane, parking lot, green area, rest rooms.

2-1-3 휴게시설의 설치기준

① 휴게소의 입지조건

- 자연환경 조건
  - 자연경관 훼손 최소화 가능

- 교통, 기술적 조건
  - 휴게소 설치구간 본선선형 조건 만족
  - 타 시설물과의 이격거리 만족

- 기타 조건
  - 주변 장래계획 등을 고려

② 휴게소의 설치간격

【휴게소 상호간 설치간격】

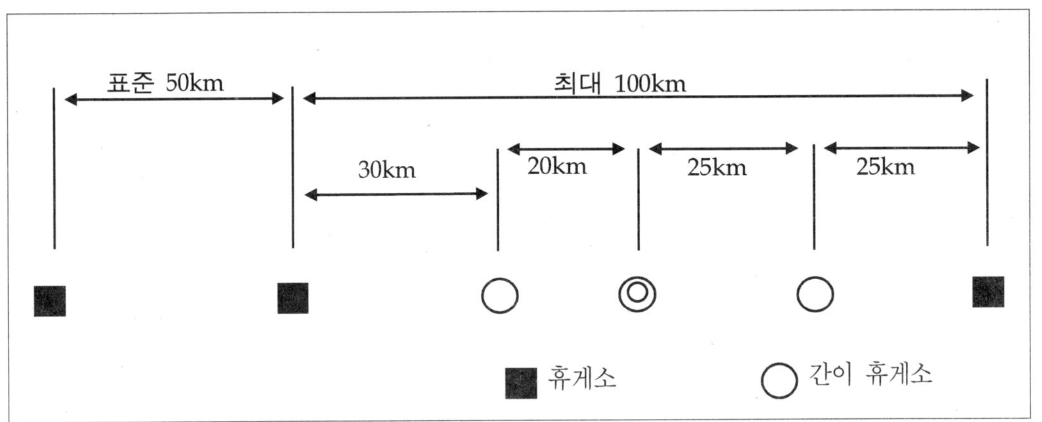

구 분	표준간격 (km)	최대간격 (km)
모든 휴게시설 상호간	25	50
휴게소 상호간	50	100
주유소 상호간	50	60

③ 휴게소의 형식선정

- 휴게소 형식은 분리식 외향형을 원칙으로 적용

- 주유소의 위치는 출구형 또는 중앙형 적용

2-1-3 Installation Standard

① Conditions of location

- Natural condition
  - Minimize the damage of natural scenary

- Technical condition
  - Satisfy shape condition for resting area installation section.
  - Satisfy interval condition from other facilities

- Other condition
  - Consider future plan of adjacent area

② Installing interval for resting area

【Intervals among resting areas】

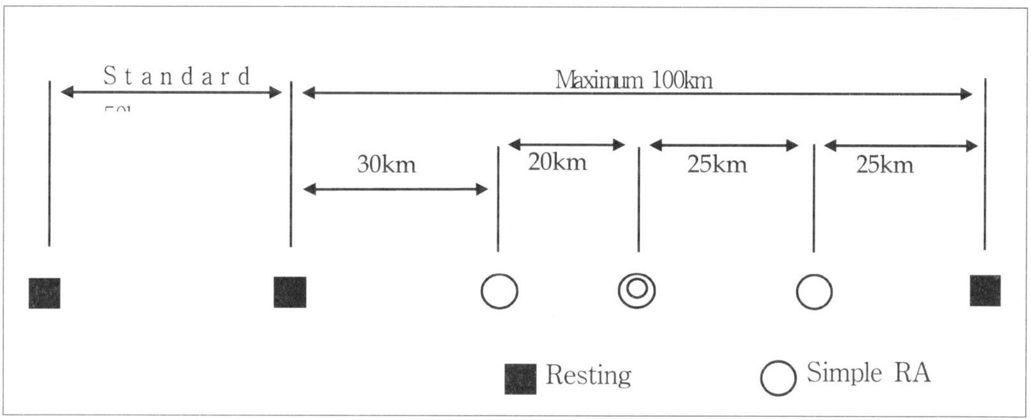

Classification	Standard Interval (km)	Maximum Interval (km)
Between all resting areas	25	50
Between resting areas	50	100
Between gas stations	50	60

③ Selection of resting area type

- In principle, resting area should be the following type: separated and extrovert

- For the location of gas station, apply exit or center type.

<table>
<tr><th>분 류 기 준</th><th>형 식</th><th>내 용</th></tr>
<tr><td rowspan="2">주차장 위치별</td><td>분리식</td><td>주차장이 본선 차도를 기준으로 양쪽에 분리되어 배치된 형식</td></tr>
<tr><td>집약식</td><td>주차장이 본선의 한 쪽 또는 상·하행선 양 차도의 중앙에 모여서 배치된 형식</td></tr>
<tr><td rowspan="2">식당 위치별</td><td>외향형</td><td>식당이 유출부 차도의 바깥쪽에 배치된 형식</td></tr>
<tr><td>내향형</td><td>식당이 유출부 차도의 안쪽에 배치된 형식</td></tr>
<tr><td rowspan="3">주유소 위치별</td><td>입구형</td><td>주유소가 휴게시설의 입구에 배치된 형식</td></tr>
<tr><td>중앙형</td><td>주유소가 휴게시설의 중앙에 배치된 형식</td></tr>
<tr><td>출구형</td><td>주유소가 휴게시설의 출구에 배치된 형식</td></tr>
</table>

③ 휴게소의 규모 산정

- 휴게시설의 전체규모는 각 구성 요소의 규모를 합산하여 구하는 것을 원칙으로 하며, 각 구성요소의 규모는 계획 교통량에서 산정된 주차면 수를 기준으로 하여 결정

- 휴게시설의 규모는 공용기간을 10년으로 하여 결정되고, 각 구성요소는 단계 건설을 고려하여 계획

- 휴게시설 전체규모
  - 주차장 + 건축물 부지면적 + 녹지 등 기타면적

④ 주차장 면적 산출

- 면적산출방법
  - Σ[{차종별 교통량×이용율×(혼잡율/회전율)}×차종별 주차면적]

- 차종별 교통량은 정기 교통량 조사 결과를 인용

- 이용율, 혼잡율, 회전율 : 최근(실제) 조사 결과를 인용
  - 이용율(%) : 휴게소 이용차량대수(대/일)/본선교통량(대/일)
  - 혼잡율(%) : 가장 혼잡시간 이용대수(대/시간)/일이용대수(대/일)×100
  - 회전율(회) : 1시간(시)/평균주차시간(시)

- 차종별 1대당 주차면적은 부지형태에 따라 표준 제원을 적절히 적용

Standard for classification	Type	Remarks
Location of parking lot	Separated	Parking lots are located in both side of main lane.
	Consolidated	Parking lot is located in ether side of the lane or in the center of the main lane.
Location of food-court	Extrovert	Foodcourt is located in outside of the driving lane.
	Introvert	Foodcourt is located in inside of the driving lane.
Location of gas station	Entrance	Gas station is located in the entering of the resting area
	Center	Gas station is located in the center of the resting area
	Exit	Gas station is located in the exiting of the resting area

③ Scale Calculation of Resting Area

- Scale of the resting area is the sum of all consisting facilities. Scale of each consisting facility is decided by the standard using the parking space calculated in traffic volume planning.

- Scale of the resting area should be determined based on 10 year usage assumption. Each consisting facility should be planned with staged construction.

- Overall scale of resting area
  - Size of parking lot + Size of building site + Size of others(green area, etc)

④ Calculation of Parking Lot Space

- Method for calculation
  - Σ[{traffic volume by type of vehicle × rate of usage × (rate of crowdedness / turnover)} × parking space by type of vehicle]

- Use periodic traffic survey result for traffic volume by type of vehicle

- Use recent (actual) survey result for the following variables:
  - Rate of usage (%) : number of vehicles using resting area parking lot (vehicles/day) / main lane traffic volume (vehicles/day)
  - Rate of crowdedness (%) : number of vehicles in most crowded time (vehicles/hour) / number of vehicles using the parking lot per day (vehicles/day) × 100
  - Turnover (times) : 1 hour / average parking time (hour)

- Use standard specification for parking space by type of vehicle

⑤ 목표 교통량 구분 적용

- 면적산출방법
  - 영업시설(매장, 편의점 등) : 공용(도로개통) 후 5년
  - 화장실, 기전설비시설 등 : 공용(도로개통) 후 10년

⑤ 서비스 계수

연평균일교통량 Q (양방향 : 대/일)	서비스계수
0 < Q ≤ 25,000	1.40
25,000 < Q ≤ 5,0000	$1.65 - Q \times 10^{-5}$
50,000 < Q	1.15

⑥ 건축물 부지면적 산출기준

- 주차용량 기준

구 분	주 차 대 수	비 고
소 형	100 ~ 150 미만	
중 형	150대 이상 ~ 200 미만	
대 형	200대 이상 ~ 250	

- 건축물 부지 면적 산정

구 분	계	휴 게 소	주 유 소	충 전 소	
대 형	15,700	11,000(220×50m)	3,200(80×40m)	1,500(50×30m)	건물+부대공간(전면 DECK, 파고라, 운동시설공간, 쓰레기장, 진입로 등)
중 형	10,440	6,840(180×38m)	2,100(60×35m)	1,500(50×30m)	
소 형	8,940	6,840(180×38m)	2,100(60×35m)		

- 녹지 등 기타면적 : 가・감속차로+녹지+기타면적=3,000㎡+주차장 면적의 2배

가・감속차로	진출·입 속도에 따른 면적으로 본선속도가 120km/h인 경우 약3,000㎡ 소요	감속테이퍼 90m, 감속차로 175m, 가속차로 470m, 가속테이퍼 90m ※휴게소 노즈부 속도는 40km/h적용
환경보존녹지, 이용녹지, 분리녹지	주차면적의 1.0배 이상으로 설계 - 본선분리녹지는 교통안전 등을 고려, 최소 15m 이상의 폭원 확보	세계보건기부(WHO) 권장 도시지역 1인당 최소 생활녹지 면적기준(9㎡이상)적용
기타면적	휴게소 형태특성상 주차장면적의 약 1.0배에 해당	주차부지, 건물부지, 녹지, 가감속차로 이외의 부지

⑤ Target traffic volume application

- Method for size calculation
  - Operation facilities (stores): 5 years after road opening
  - rest rooms, electricity/mechanic facilities: 10 years after road opening

⑤ Service coefficient

Daily Traffic (yearly average) Q (Both way: vehicles per day)	Service coefficient
0 < Q ≤ 25,000	1.40
25,000 < Q ≤ 5,0000	$1.65 - Q \times 10^{-5}$
50,000 < Q	1.15

⑥ Standard for building area calculation

- Parking capacity standard

Classification	Parking Space	Remarks
Small	100 ~ less than 150	
Medium	150 and more ~ less than 200	
Large	200 and more ~ 250	

- Building area calculation

Classification	Total	Resting Area	Gas Station	Charging Station	Building+Subsidiary Area (Front DECK, pagora, exercise facility area, dumping ground, entrance, etc.)
Large	15,700	11,000(220×50m)	3,200(80×40m)	1,500(50×30m)	
Mid	10,440	6,840(180×38m)	2,100(60×35m)	1,500(50×30m)	
Small	8,940	6,840(180×38m)	2,100(60×35m)		

- Other areas including green area : speed changing lane + green area + other areas = 3,000㎡ + twice of parking space

Speed changing lane	3,000㎡ is required when speed of main lane is 120km/h	Decelerating taper 90m, decelerating lane 175m, accelerating lane 470m, accelerating taper 90m ※Speed in resting area nose: 40km/h
Natural preservation green area, green area for usage, separate green area	More than 1.0 time of parking area - separating green area should be more than 15m for the safety.	WHO recommended life green area per person in urbanized area (9㎡)
Others	About 1.0 time of parking area	Other than parking, building, green area and speed changing lane.

2-2 체인탈착장

2-2-1 개 요

① 적설 한랭지 등에서 주행하고 있는 차량이 도로 노면의 결빙에 따라 체인을 설치하거나 제거하기 위한 공간

② 체인탈착장은 적설 한랭지 등에서 주행하는 차량이 도로 노면의 결빙에 따라 체인을 설치하거나 제거하기 위한 장소를 제공

③ 차량이 결빙된 노면 위를 미끄러지면서 발생하는 사고를 미연에 방지

2-2-2 체인탈착장의 종류

① 대규모 체인탈착장

- 도로를 이용하는 전 차량을 대상으로 의무적으로 체인탈착을 행할 필요가 있는 경우에 적합

② 소규모 체인탈착장

- 도로 이용자가 임의로 사용하는 것을 목적
- 길어깨를 확폭한 정도의 간이구조

2-2-3 설치기준

① 설치구간

- 대규모 체인탈착장
  - 강설조건이 급격히 변하는 곳으로 적설지와 비적설지의 경계, 산악지 진입 전
  - 상향경사가 4% 이상인 급경사 구간의 전방에 설치
  - 길고 큰 터널이나 터널연속구간의 연장이 10km 이상 되는 구간의 전후
  - 적설지에서 출입시설의 일반도로 접속 부근

- 소규모 체인탈착장
  - 노선의 기상이 급변하는 지역을 통과하는 경우
  - 대규모의 체인탈착장을 보완하는 위치
  - 터널길이가 2km를 넘는 터널의 양쪽 입구에 설치

② 설치방법

- 대규모 체인탈착장
  - 설계교통량, 체인의 설치 제거를 필요로 하는 차의 비율, 체인의 설치 제거에 필요한 시간 및 대형차의 비율을 고려

2-2 Facilities for Chain Install and Removal

2-2-1 Overview

① Area required for vehicles driving in snowing and icing area to install or remove chain.

② Facility for chain install and removal provides the area so that the driver can install or remove chain in order to cope with the icing condition of the road.

③ By providing the area, sliding accidents can be prevented.

2-2-2 Types

① Large scale facility
- Applicable for the case when all the vehicle using the road should install chain compulsorily.

② Small scale facility
- Applicable when the driver can decide to install / remove the chain of his own accord.
- Simple structure with little wider than usual road shoulder.

2-2-3 Installation Standard

① Installing location
- Large scale
  - Area with sudden changes for snowing: boundary between snowy area and non-snowy area or entrance for mountainous area.
  - In front of stiff upper hill (on-ward slope more than 4%)
  - Front and back of large, long tunnel or continuous tunnel area more than 10km.
  - In snowy area, around the connection of facilities and main road

- Small scale
  - Area with rapid weather change
  - Area supplementing large scale chain install facility
  - Both end of tunnel longer than 2km

② Installing method
- Large scale facility
  - Consider designed traffic volume, ratio of vehicles requiring chain install /removal, time required for chain install/removal, ratio of large sized vehicles.

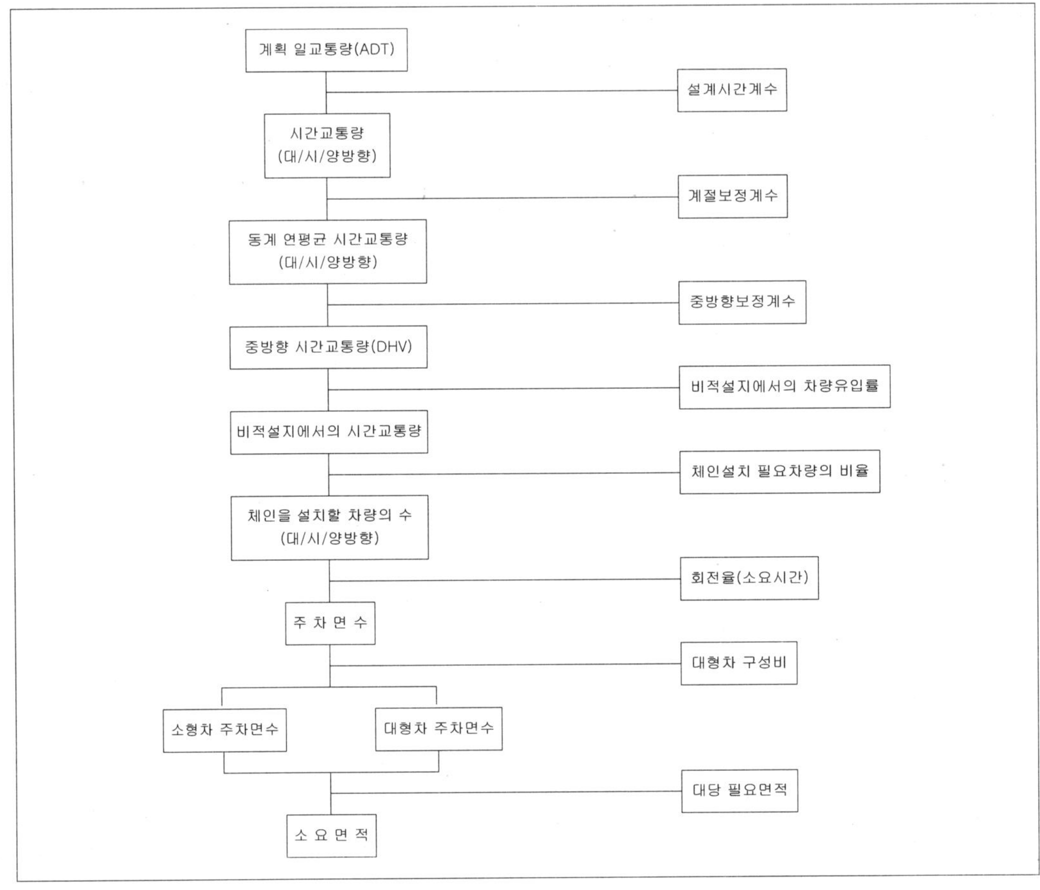

- 소규모 체인탈착장
  - 원칙적으로 우측 길어깨를 확폭하여 확보하며, 평행식 주차방식으로 함
  - 폭은 5m, 테이퍼 길이는 20m 정도로 함
  - 터널 입구에 체인탈착장을 설치할 경우에 지형적인 제약으로 우측 길어깨의 확폭이 곤란할 경우에는 중앙분리대의 공간을 확폭하여 이용 가능
  - 본선 주행차가 유출입 차량을 알아볼 수 있도록 시거의 확보, 표지설치 등의 안전대책에 충분히 유의

③ 기타사항

- 체인탈착장에는 조명설비를 설치

- 체인탈착장의 경사는 주차차량의 종방향으로 2%, 횡방향으로 3% 이하로 하고, 배수에 충분한 주의를 기울여야 함

- 체인탈착장으로 사용되는 부분은 포장을 하고, 교통섬은 원칙적으로 미설치

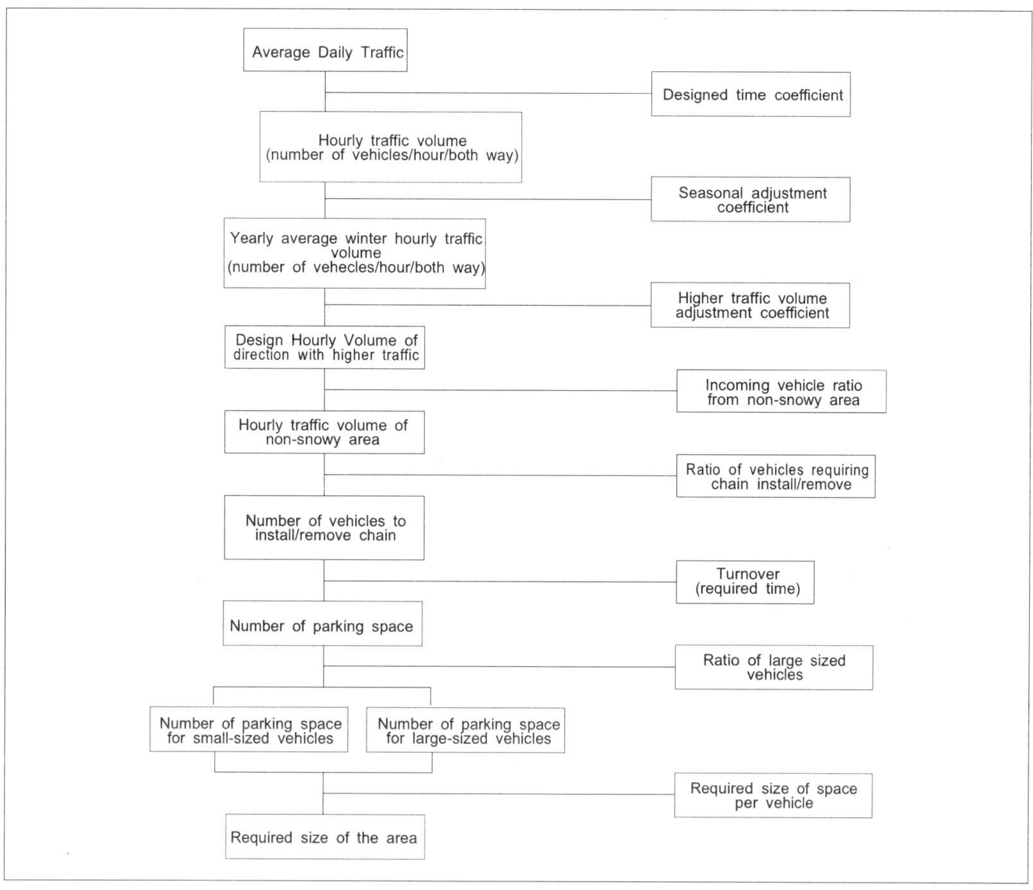

- Small size facility
  - Extend right-side road shoulder with parallel parking
  - Width: 5m, Taper length: 20m
  - In case of installing chain install facility in front of tunnel (unable to extend road shoulder), extend median strip
  - Secure the field of vision and install proper signs so that the drivers in the main lane can recognize the entering / exiting vehicles

③ Others

- Install lights in the facility

- Slope in the facility should remain less than 2% per vertical section and less than 3% per cross section with consideration of drainage

- Pave the surface. Do not install traffic island

- 살수, 소설(掃雪) 등의 융설 시설은 설치하지 않는 것이 바람직
- 대규모 체인탈착장에는 화장실을 설치하는 것이 바람직
- 체인탈착장이 겨울철에만 이용될 경우에는 유지관리면을 고려해서 화장실은 정화시설을 하지 않은 간이형을 원칙으로 적용
- 간이휴게소로 이용할 수 있는 경우에는 휴게소의 기준을 따름

2-3 비상주차대

2-3-1 개 요

① 우측 길어깨의 폭이 협소한 장소에서 고장난 차량이 본선 차도에서 벗어나 대피할 수 있도록 제공하는 장소

② 고장차량을 본선에서 벗어나게 함으로써 본선의 교통용량의 감소 방지

③ 사고 예방 기능

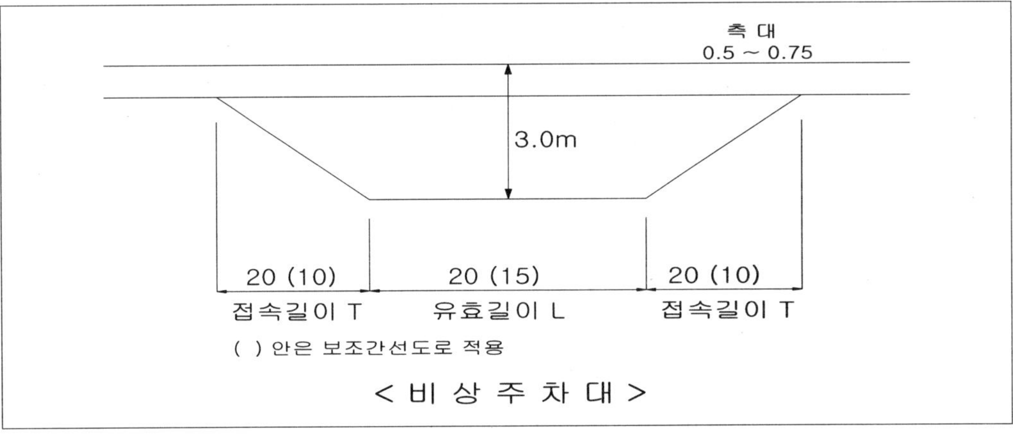

< 비 상 주 차 대 >

2-3-2 설치장소

① 장대교, 터널 등에서는 길어깨 폭이 2.0m 미만이면서 구조물의 연장이 1,000m미만 일 때에는 그 구조물 전후의 토공구간에 설치

② 오르막 차로 구간에 대하여는 토공, 교량부에 준하여 설치

③ 토공구간에서는 표준설치간격에 의거하여 용지취득이 용이한 곳으로 선정

④ 한쪽깎기한쪽쌓기구간이나 구조물 설치구간은 피하는 것이 바람직

- Snow-melting facilities (water sprinkling, etc) are not recommended to install

- It is recommended to provide rest rooms for large scale facilities

- When the facility is used only in winter season, design rest rooms without sewage system.

- Follow standard for simple rest area when the facility can be used for rest area.

2-3 Emergency Parking

2-3-1 Overview

① Area for emergency shelter when a vehicle is out of order when width of road shoulder is limited

② Prevent traffic volume decrease by removing defected vehicle from the main lane.

③ Prevent accidents

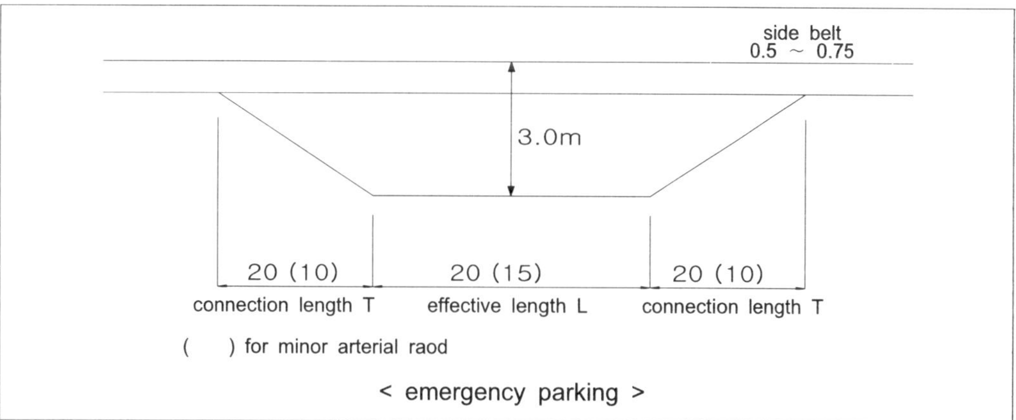

< emergency parking >

2-3-2 Installing location

① Install in land-fill area in front of / at the end of long bridge or tunnel when the structure is shorter than 1,000m and has road shoulder with width of less than 2.0m

② Follow the similar criteria for upper-hill section

③ Follow the standard installing interval and select area easy to purchase for land-fill area

④ Avoid installing in cutting and embankment section or area with other facilities

2-3-3 설치기준

① 설치간격

- 신설되는 고속도로의 경우 길어깨 폭을 3m 이상으로 설치하고 있으므로 비상주차대 설치는 일부구조물을 제외하고 특별히 고려할 필요가 없으나 길어깨 폭이 2.5m 미만인 고속도로에서는 비상주차대를 반드시 설치하여야 함

도로 구분	설치간격(m)	비 고
고속도로	750	
지방지역 일반도로	750	

② 접속길이 및 유효길이

- 접속길이는 본선에서 비상주차대로 진입한 후 본선과 평행이 될 때까지 필요한 거리로 20~30m를 표준으로 함

- 비상주차대 유효길이는 일반적으로 승용차 3대, 세미트레일러연결차 1대가 주차 가능한 20m를 최소길이로 하고, 설치장소 및 위치 등 제반여건을 고려하여 30m로 설치함이 바람직

- 터널구간의 비상주차대는 유효길이 20m와 접속길이(2개소) 10m, 총 30m의 길이로 설치

- 교량구간에서는 경간장을 고려하여 가급적 하나의 경간 내에서 설치

도 로 구 분	접 속 길 이 (m)	유 효 길 이 (m)
고 속 도 로	20 ~ 30	20 ~ 30
도 시 고 속 도 로	20	20
주 간 선 도 로	20	20
보 조 간 선 도로	10	15

③ 폭원

- 비상주차대의 폭원은 3.0m로 하고, 측대가 있는 경우에는 측대를 포함한 폭원으로 함

④ 비상주차대 표지판

- 양방향 4차로에서는 비상주차대 감속차로 시점에서 200m 전방에 설치하고, 양방향 2차로 구간에서는 100m 전방에 설치

2-3-3 Installation Standard

① Installing interval

- As the width of road shoulder is larger than 3m for newly constructing highways, it is not required to consider installing emergency parking except for some specific sections. For the highways with road shoulder of smaller than 2.5m, emergency parking must be installed.

Road classification	Installing interval (m)	Remarks
Highway	750	
Provincial surface street	750	

② Connecting length and effective length

- Connecting length means the length required for a vehicle to be parallel to the main lane after entering the parking area. The standard is 20~30m

- The effective length of the emergency parking should be more than 20m (parking space for three regular-sized vehicles and one semi-trailer vehicle.) Recommended effective length is 30m.

- For emergency parking within tunnel, it is recommended to have a length of 30m (effective length of 20m and connecting length of 10m)

- For bridge section, install in single span. (Consider length of the span)

Road classification	Connecting length (m)	Effective length (m)
Highway	20 ~ 30	20 ~ 30
Expressway	20	20
Main Arterial Road	20	20
Minor Arterial Road	10	15

③ Specification for width

- Width of emergency parking should be 3.0m (including strip)

④ Sign for emergency parking

- For four lane section, install a sign in 200m ahead of speed change lane. For two lane section, install a sign in 100m ahead of speed change lane.

2-4 버스정류장

2-4-1 개 요

- 노선버스가 통행하는 고속국도 및 자동차 전용도로에서 노선버스가 승객의 승강을 위하여 전용으로 이용하는 시설물

- 통과 차로로부터 분리되어 버스의 감속, 대기 및 가속에 필요한 장소를 제공

- 버스 이용자의 편의를 제공

2-4-2 버스정류장의 종류

① 버스정류장 (BUS BAY)
- 버스승객의 승강을 위하여 본선차로에서 분리하여 설치된 띠 모양의 공간

② 버스정류소(BUS STOP)
- 버스승객의 승강을 위하여 본선의 외측차로를 그대로 이용할 경우

③ 간이버스정류장
- 버스승객의 승강을 위하여 일반도로의 왕복 2차로 도로에서 본선차로에서 분리하여 최소한의 목적을 달성하기 위하여 설치된 공간

2-4-3 설치장소

① 고속도로, 도시고속도로, 주 간선도로에 설치

② 보조 간선도로로서, 특히 본선의 교통류가 버스정차로 인하여 혼란이 야기될 우려가 있는 경우에 설치

③ 그 외의 경우라도 버스정류소를 설치했을 때 그 도로의 교통용량이 설계교통량에 비하여 부족할 경우에 설치

④ 고속도로, 도시고속도로, 주 간선도로와 같이 주행속도가 높고 교통류의 혼란과 그로 인한 사고 발생의 위험이 예상되는 도로에는 모두 본선에서 분리한 정차대를 설치하고 보조 간선도로라도 교통류의 혼란이 예상되는 경우에 설치

2-4-4 설치기준

① 단독버스정류장
- 평면선형
  - 가능한 큰 곡선반지름을 사용하여 원활한 주행을 도모, 본선의 평면선형에 따라서는 원활하지 못한 선형이 되기 쉬우므로 가능한 한 큰 곡선반지름을 사용하여 원활한 주행을 도모

  - 버스 정류장에서 사용하는 원곡선 반지름은 버스 정류장의 구성요소에 따라 아래 표의 값 이상으로 함

2-4 Bus Stop

2-4-1 Overview

- Facility used for bus passengers boarding in the expressways and motorways where routing buses are operating

- Provide area for bus to decelerate, stop and accelerate

- Provide convenience for bus passengers.

2-4-2 Types

① BUS BAY
- Strip shaped area separated from main lane for bus passengers to board.

② BUS STOP
- Outside lane used for passenger boarding

③ Simple bus bay
- Minimum required area for passengers boarding separated from two lane main road

2-4-3 Installing Location

① Highway, expressway and main arterial road

② Minor arterial road where passengers boarding can induce crowdedness.

③ For other cases, install bus bay when the traffic volume of the road cannot meet the designed traffic volume if bus stop is installed.

④ Install bus bay for high speed road with risk of crowdedness and accident (not only for highway, expressway and main arterial road, but also for the minor arterial road if needed)

2-4-4 Installation Standard

① Independent bus bay
- Linear shape of plane section
  - Apply as large curvature radius as possible to facilitate smooth traffic movement

  - For curvature radius, secure minimum curvature radius shown below:

【최소 원곡선 반지름】

본선 설계속도(km/h)	120	100	80	60
주감속차로	500m	500m	400m	200m
보조감속차로	500m	400m	300m	200m
정 차 로	400m	300m	250m	200m
보조가속차로	400m	300m	250m	200m
주가속차로	500m	400m	300m	200m

- 종단선형
  - 보조 감속 차로와 보조 가속 차로의 종단경사는 5.0% 이하
  - 정차로의 종단경사는 원칙적으로 2.0% 이하(지형상황에 따라 3.0%까지 허용)

【종단곡선 변화비율】

구 분	종단곡선 종류	최소 종단곡선 비율(m/%)
노 즈 부 근	볼 록 형	20
	오 목 형	15
기 타 구 간	볼 록 형	15
	오 목 형	10

- 본선선형의 최소기준

본 선 설계속도(km/h)		120	100	80	60	50
평면곡선반지름(m)		1000	700	400	200	150
종단경사(%)		2.0	3.0	4.0	5.0	6.0
종단곡선 변화비율(m/%)	볼록형	170	100	45	20	12
	오목형	60	45	30	15	10

- 버스정류장의 구성 및 제원
  - 감속부, 정류부, 가속부로 구성
  - 감속부는 변이구간(L1), 주 감속차로(L2), 보조 감속차로(L3)로 구성

【Minimum Curvature Radius】

Main lane Designed Speed (km/h)	120	100	80	60
Main decelerate lane	500m	500m	400m	200m
Minor accelerate lane	500m	400m	300m	200m
Stop lane	400m	300m	250m	200m
Minor accelerate lane	400m	300m	250m	200m
Main accelerate lane	500m	400m	300m	200m

- Linear shape of vertical section
  - Vertical slope for minor decelerate lane and minor accelerate lane should be less than 5.0 percent
  - Vertical slope for stop lane should be less than 2.0 percent (in some cases with specific topology condition, less than 3.0 percent)

【Vertical curve change ratio】

Classification	Type	Minimum vertical curve ratio (m/%)
Near Nose	Convex	20
	Concave	15
Others	Convex	15
	Concave	10

- Minimum standard for main lane linear shape

Main lane designed speed (km/h)		120	100	80	60	50
Plane curvature radius(m)		1000	700	400	200	150
Vertical slope(%)		2.0	3.0	4.0	5.0	6.0
Vertical curvature change ratio (m/%)	Convex	170	100	45	20	12
	Concave	60	45	30	15	10

- Structure and specification of bus stop
  - Consist of decelerate section, stop section and accelerate section
  - Decelerate section is divided into changing section (L1), main decelerate lane (L2) and minor decelerate lane (L3)

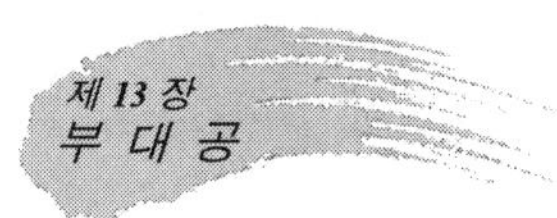

- 가속부는 보조 가속차로(L5), 주 가속차로(L6), 변이구간(L7)으로 구성

【버스정류장의 구성】

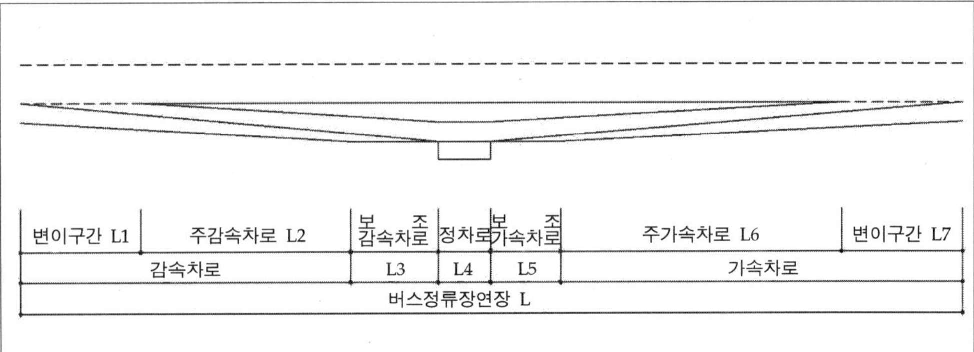

【버스정류장의 제원】

구 분	설계속도(km/hr)	120	100	80	비고
감속부	변이구간 길이(m)	70	60	50	
	주 감속차로 길이(m)	120	100	90	
	감속차로 길이(m)	190	160	140	
	보조 감속차로 길이(m)	50(40)	50(40)	50(40)	
정차로	정차로 길이(m)	30(24)	30(24)	30(24)	
가속부	보조 가속차로 길이(m)	40(30)	40(30)	40(30)	
	주 가속차로 길이(m)	160	130	110	직접식
		220	190	120	평행식
	변이구간 길이(m)	70	60	50	
	가속차로 길이(m)	230	190	160	직접식
		290	250	170	평행식
버스정류장 길이(m)		540	470	420	직접식
		600	530	430	평행식

※ ( )안의 수치는 제반여건 등을 감안한 최소 설치 길이임

- Accelerate section is divided into minor accelerate lane (L5), main accelerate lane (L6) and changing section (L7)

【Structure of Bus Stop】

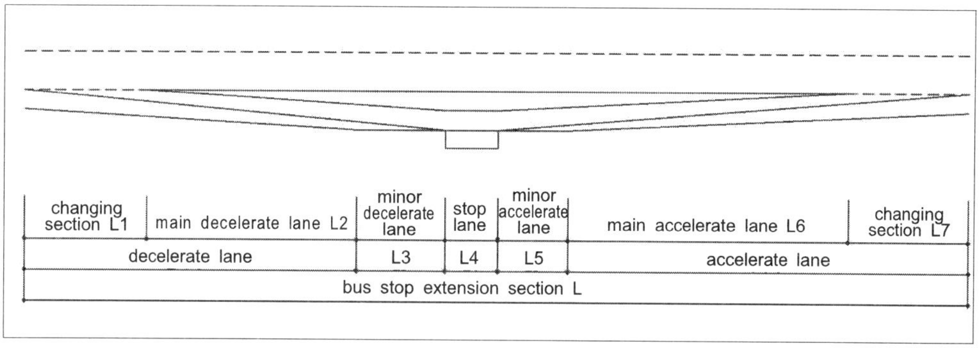

changing section L1	main decelerate lane L2	minor decelerate lane	stop lane	minor accelerate lane	main accelerate lane L6	changing section L7
decelerate lane		L3	L4	L5	accelerate lane	
bus stop extension section L						

【Specification for Bus Stop】

Classification \ Designed speed(km/hr)		120	100	80	Remarks
Decelerate section	Changing section (m)	70	60	50	
	Main decelerate lane (m)	120	100	90	
	decelerate lane (m)	190	160	140	
	Minor decelerate lane (m)	50(40)	50(40)	50(40)	
Stop section	Stop lane (m)	30(24)	30(24)	30(24)	
Accelerate section	Minor accelerate lane (m)	40(30)	40(30)	40(30)	
	Main accelerate lane (m)	160	130	110	Direct
		220	190	120	Parallel
	Changing section (m)	70	60	50	
	Accelerate lane (m)	230	190	160	Direct
		290	250	170	Parallel
Total length of bus bay (m)		540	470	420	Direct
		600	530	430	Parallel

※ Numbers in parenthesis are minimum length considering various matters.

② 출입시설에 병설하는 버스정류장

- 형식선정
  - 유출입 영업소를 분리시켜 그 사이에 설치
  - 영업소의 앞이나 뒤에 설치
  - 가감속차로를 출입시설과 공용(共用)하여 본선에 설치
  - 버스 정류장을 출입시설에 떨어뜨려 본선에 설치
  - 영업소의 외측에 설치

- 평면선형
  - 정차로가 설치된 곳의 최소 평면곡선반지름은 30m로 적용
  - 버스차로의 최소 평면곡선반지름은 15m로 적용

- 종단선형
  - 정차로가 설치된 곳의 최대 종단경사는 2.0%로 적용 지형 조건을 감안하여 부득이한 경우 3%까지 허용

- 횡단경사는 2.0%로 적용

- 정차로의 폭
  - 정차한 버스를 앞질러 갈수 있도록 5.6m로 적용

- 정차로의 길이
  - 두 대의 버스가 정차할 수 있도록 30m를 표준으로 하며, 본선의 설계속도가 80km/h 이하인 경우 20m까지 줄일 수 있음

- 버스차로의 횡단면 구성은 해당 출입시설의 일방향 일차로 연결로의 횡단면 구성과 동일

③ 휴게시설에 병설하는 버스정류장

- 평면선형
  - 정차로가 설치된 곳의 최소 평면곡선반지름은 500m 적용
  - 가감속차로의 일부분을 빠져나가는 차로와 공용하는 경우에 버스 차로의 최소 평면곡선반지름은 안전을 고려하여 15m로 할 수 있으며, 이 경우 정차로의 최소 평면곡선반지름은 30m로 적용

- 종단선형
  - 정차로가 설치된 곳의 최대 종단경사는 2.0%로 적용(지형조건을 감안하여 부득이한 경우 3%까지 허용)
  - 정차로가 설치되는 곳의 최소 종단곡선 변화율은 볼록형일 경우 15m/%, 오목형일 경우 10m/%로 적용

② Annexed bus stop (with entrance / exit operation facilities)

- Type selection
  - Separate entrance and exit operation facilities to install bus stop in between.
  - Install bus stop in front of the operation facility or in the rear of the facility
  - Share decelerate / accelerate lane with entrance / exit facilities and install those lane within the main lane
  - Seperate bus stop from entrrance / exit facility and install it within the main lane
  - Install bus stop in the external side of the operation facility

- Linear shape of plane section
  - Minimum curvature radius is 30m for stop lane area
  - Minimum curvature radius is 15m for bus lane

- Linear shape of vertical section
  - Maximum vertical slope is 2.0 percent for stop lane area (if unavoidable, maximum 3.0 percent is allowed)

- Cross-section slope is 2.0 percent

- Width of stop lane
  - 5.6m to enable passing ahead stoped bus

- Length of stop lane
  - 30m to allow two buses in the stop lane. It is possible to reduce to 20m when the designed speed is less than 80km/h

- Cross-sectional structure for bus lane is identical to the one-way connecting lane of corresponding entrance / exit facility

③ Annexed bus stop (with resting area)

- Linear shape of plane section
  - Minimum curvature radius 500m for stop lane area
  - When accelerate / decelerate lane is shared with exit lane, minimum curvature radius for bus lane can be 15m for safety reason. In this case, minimum curvature radius for stop lane should be 30m.

- Linear shape of vertical section
  - Maximum vertical slope is 2.0 percent for stop lane area (if unavoidable, maximum 3.0 percent is allowed)
  - Minimum vertical curvature change ratio is 15m/% for convex type and 10m/% for concave type for stop lane area

- 횡단경사는 2.0%로 적용(정차로)

- 정차로의 폭
  - 노선버스 전용의 차로를 설치하는 경우 5.6m로 적용
  - 연결로 또는 통과 차로에 설치하는 경우 3.6m로 적용

2-5 긴급연락시설

2-5-1 개 요

- 고속도로나 일반도로 중 차량의 진출·입이 제한되는 도로에서 차량의 사고나 고장 시 긴급히 연락할 수 있도록 설치하는 시설

- 주행속도가 높은 도로에서 사고나 고장 차량의 방치로 야기될 수 있는 위험을 신속한 연락으로 사전에 예방하는 기능

2-5-2 긴급연락시설의 종류

① 경찰 또는 소방관서와 직결된 경우

- 공중전화를 부가 설치하여 필요한 대상자와 통화가 가능토록 함

② 도로관리청의 교환에 연결된 경우

- 긴급전화를 도로 관리청의 교환에 연결하여 필요한 당사자와 통화할 수 있도록 함

- 수화기를 드는 것만으로 통화가 가능할 수 있도록 장치

- 도로관리청의 교환에서는 발신자의 위치를 자동으로 확인할 수 있도록 함

2-5-3 대상구간

① 자동차 전용도로 본선구간

② 일반 공중전화가 없는 휴게소, 간이휴게소

③ 터널 내부

2-5-4 설치기준

① 설치간격

- 고속도로 본선구간에서는 2km 간격으로 설치

- 터널내부에서는 200m 간격으로 설치

- 기타 구간에서는 1km 간격으로 설치

- 가능한 한 버스정류장이나 비상주차대에 설치하도록 그 설치간격에 유의

- Cross-section slope is 2.0 percent (stop lane)

- Width of stop lane
  - 5.6m for dedicated lane for routing bus
  - 3.6m for connecting lane or through lane

2-5 Emergency Communication Facility

2-5-1 Overview

- Facilities installed in highway or other road where entrance or exit is limited to provide emergency communication related to vehicle breakage or traffic accident.

- Prevent potential risk of leaving mal-functioning or accident vehicles in the high-speed road by enabling fast communication to related agencies.

2-5-2 Types of Emergency Communication Facilities

① Direct connection to police or fire department
- Additionally install public telephone to provide communication to related agencies

② Connected to exchanger of road management agency
- Exchanger is connecting the call to related agencies

- Instant connection through holding up the receiver

- The exchanger should recognize the location of caller automatically

2-5-3 Installing Location

① Main lane of Motorway

② Resting area without public telephone

③ Inside tunnel

2-5-4 Installation Standard

① Installing interval
- Every 2km for highway main lane

- Every 200m inside tunnel

- Every 1km for other areas

- If possible, install the facility within bus bay or emergency parking facility

② 설치 시 유의사항

- 가능한 한 버스 정류장이나 비상주차대에 설치
- 긴급전화의 설치 장소에는 도로표지 429 「긴급신고표지」를 설치
- 야간에는 확인이 용이하도록 표지를 반사체로 하는 외에 가능하면 설치 장소 내·외부에는 조명시설을 하는 것이 바람직

2-6 방음시설

2-6-1 개　요

- 차량에서 발생하는 음을 차단 또는 흡수하거나 음의 회절에 의하여 소음을 감소
- 도로 인접 지역의 생활환경을 보전하는 목적으로 설치

2-6-2 방음시설의 구분

① 음향성능상의 원리에 따른 구분

- 반사형
- 흡음형
- 간섭형
- 공명형 등

② 사용재료에 따른 구분

- 칼라형
- 투명형
- 금속형
- 목재형
- PVC형
- 콘크리트형 등

② 방음판 형상에 따른 구분

- 직립형
- 경사형
- 꺾임형

② Considerations

- If possible, install the facility within bus bay or emergency parking facility
- Install road sign 429 「Emergency Call Sign」
- Sign should be made of reflexible material to be recognized night-time. It is recommended to provide lights around the facility.

2-6 Soundproof Facility

2-6-1 Overview

- Facilities to arrest or absorb noise from the vehicles
- To preserve quality of life of the area near the road

2-6-2 Classification of Soundproof Facility

① Classification by theoretical ground

- Reflection
- Absorption
- Interference
- Echo

② Classification by used material

- Colored
- Transparent
- Metallic
- Wood
- PVC
- Concrete

③ Classification by shape of soundproof plate

- Upright
- Sloped
- Folded

2-6-3 방음시설의 종류

① 콘크리트(반사형)

② 금속판(반사형)

③ 투명 방음벽(반사형)

④ 방음터널

- 회절에 의한 감쇄효과를 이용한 방음벽 사용으로 목표치를 달성할 수 없는 경우 차음성능이 큰 재료로 도로를 터널모양으로 덮음으로써 소음을 차폐하는 방법

⑤ 방음림 및 방음둑

- 주거밀집지역 등 경관적·시각적 요소를 중시하는 일부지역에서 사용

2-6-4 설치대상구간

① 4km 이상의 도로신설 또는 2차로 이상으로서 10km 이상의 도로 확장 구간에 대하여는 환경영향평가를 실시한 후 그 결과와 보호대상구역의 용도조사를 실시하고 주민의견을 수렴한 결과에 필요한 장소에 설치

② 주택, 학교, 병원, 도서관, 휴양시설의 주변지역 등 조용한 환경을 요하는 지역(이하 "보호대상지역") 중 소음의 영향을 크게 받는 지역으로 상주 인구밀도, 학생 수, 병상 수 등이 많고 소음이 환경기준을 초과하여 소음문제가 발생하거나 발생할 우려가 큰 지역부터 우선하여 설치

③ 예측소음도가 환경정책기본법상 소음환경기준치를 상회하는 지역에 설치

④ 환경영향평가 시 설치 제외된 구간이라도 현장 여건상 필요하다고 판단되는 곳은 설치여부를 재검토하여 반영

2-6-5 형식별 선정기준

① 반사형

- 일반구간 적용

② 흡음형

- 도로의 좌우에 방음벽을 설치하여야 하는 경우(평행방음벽 구간)

- 폭과 높이의 비율이 1:20 이상인 경우는 반사형 적용

③ 투명형

- 조망, 일조, 채광 등이 요구될 경우

- 불투명 방음벽 설치 시 결빙예상지역

- 불투명 방음벽 설치로 인한 I.C 및 교차로 진입 시 시거가 불량하여 시거의 확보가 필요한 구간

2-6-3 Type of Soundproof Facilities

① Concrete (reflection)

② Metal plate (reflection)

③ Transparent soundproof wall (reflection)

④ Soundproof tunnel
- Method of covering the road in the tunnel-shape with material of high soundproof function to screen noise when it is impossible to reach the target value by using soundproof wall utilizing reduction effect through diffraction

⑤ Soundproof forest / soundproof bank
- Used in certain areas where scenic and visual factors are concerned such as residence congested area

2-6-4 Installing Location

① Conduct environmental impact assessment for new road construction longer than 4km or for two lane or above road extension longer than 10km. In addition to the result of assessment, conduct land-use assessment for the preservation area and collect opinions from residents. Sum up the result of all the research and install the facility as required.

② Soundproof facilities should be installed in the residence, hospital, library, recreation facilities and their adjacent area (hereinafter "preservation area"). The priority can be assigned to the areas where noise from the road exceeds the environmental standard and the residents can be highly influenced by the noise. Other considerations include high population density, large number of students, or large number of patient beds.

③ Area where predicted noise exceeds the noise standard of environmental law.

④ Re-assess requirement and install the facility even when the area is neglected by environmental assessment if required.

2-6-5 Selection Standard by Type

① Reflection
- apply to general area

② Absorption
- Apply to area where both side of the road require soundproof wall (Parallel wall section)

- Apply reflection type when ratio of width and height is more than 1:20

③ Transparent
- apply when view, sunshine, lighting is required

- apply when freezing is expected with regular wall

- apply to IC and crossroad area where the field of vision is required

- 환경영향평가상 경관확보를 필요로 하는 구간

- 불투명 방음벽 설치 시 전파장애가 예상되는 난청지역

④ 칼라형
- 대도시 주변 대단위 밀집지역 및 종합병원 등 요양시설이 위치한 지역으로 연장이 200m 이상 설치하여야 할 지역

④ 방음둑
- 원지반과 도로계획고와의 고저차가 적은 지형으로 고속도로 용지편입이 적어 토지이용을 극대화할 수 있는 지역

- 편입용지 비용이 적은 지역

- 현장 발생 사토를 활용할 수 있거나 인근에 토사를 이용할 수 있는 지역

- 경관, 도시미관, 환경적 측면 등의 자연환경요소와 연계되어야 하는 지역

【방음시설의 형식 예】

흡음형	반사형

투명형	칼라형

방음터널	방음둑

- Landscape area according to environmental impact assessment

- Fringe area where regular wall may interfere radio wave

④ Colored
- Area with length more than 200m and adjacent to metropolitan area with medical facilities such as general hospital are located

⑤ Soundproof bank
- Area where height difference between original ground and designed road height so that land-use can be maximized

- Area where purchased land is relatively small

- Area where can obtain nearby soil or soil from the construction can be used

- Area aesthetically linked to nearby environment

【Examples of Soundproof Facilities】

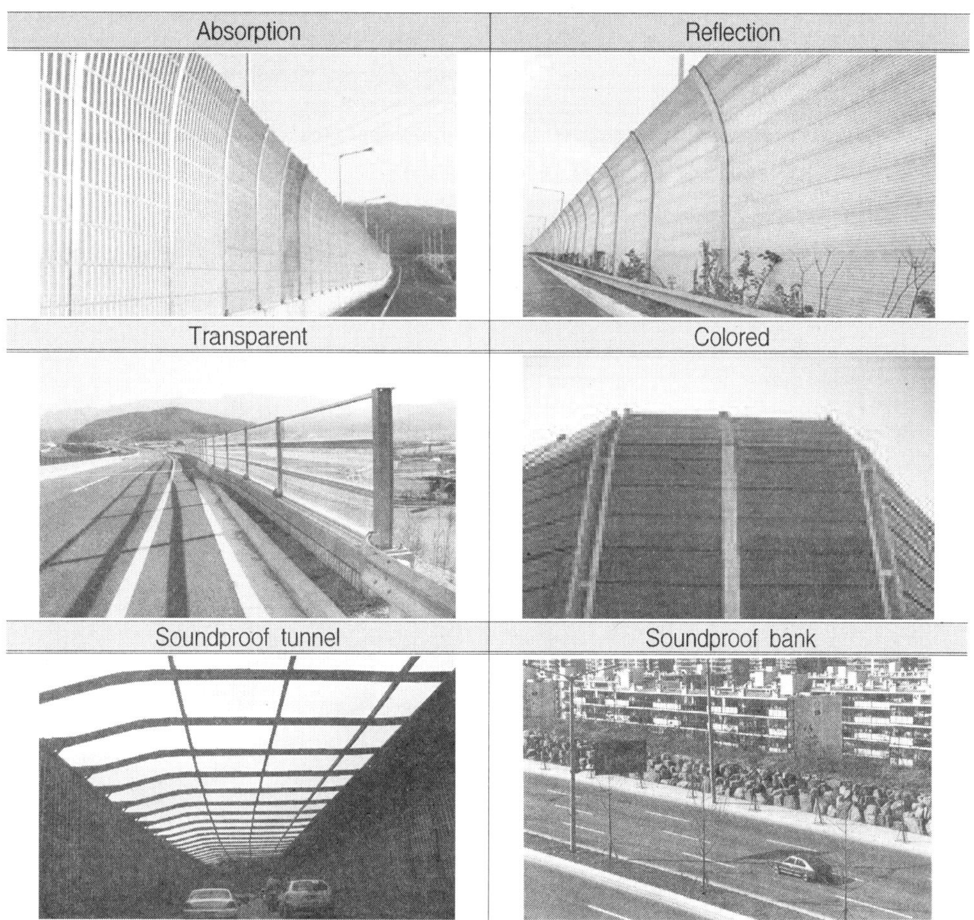

2-6-6 설치기준

① 설치위치

- 흙쌓기부
  - 가드레일과 함께 사용할 경우 길어깨 단으로부터 1.5m 이상 떨어져 설치
  - 토사 방음둑과 겸용하여 설치하는 경우 길어깨 단으로부터 3m 이상 이격

- 땅깎기부
  - 길어깨의 외측에 방음벽을 시공하는데 필요로 하는 폭의 위치에 설치

- 구조물부
  - 콘크리트 벽인 경우에는 벽 상단에 직접 설치하여도 좋으며, 현장 상황에 따라 가장 적절한 위치에 설치

② 설치높이

- 방음벽의 설치위치를 결정하고 소음도를 설계목표수준까지 감쇠시키기 위하여 필요한 높이를 결정

- 방음벽의 소요높이 계산에서 필요 감쇠치(dB)와 음원의 대표 주파수(Hz)를 설정하고 경로차에 의한 식을 이용하여 방음벽의 높이를 결정

③ 설치길이

- 방음벽의 길이는 방음벽 측단으로 입사하는 음의 영향을 고려하여 설계목표를 충분히 달성할 수 있는 길이를 가져야 함

- 방음벽의 설치길이는 직접음 감쇠치에 의하여 결정

④ 경관 및 미관설계

- 시행시기 : 실시설계 단계에서 환경영향평가 초안이 작성된 후

- 시행방법
  - 지역특성과 토사 또는 암 사토 등 현장여건을 종합적으로 분석하여 재질, 형태, 색상 등의 변화를 검토
  - 다양한 디자인을 구상하여 미관설계를 수행하여 의무적으로 방음시설 미관설계 평가회를 실시

2-6-6 Installation Standard

① Installing location

- Land-fill section
  - When used together with guard rail, install the facility 1.5m away from the edge of road shoulder.
  - When used together with land-fill soundproof bank, install the facility 3.0m away from the edge of road shoulder.

- Land-cut section
  - Proper location external to the road shoulder

- Structure section
  - Install to the top of the wall if the wall is made of concrete. Determine most proper area according to the site condition.

② Installing height

- Determine height to decrement the noise to designed level
- Determine required decrementing noise value (dB) and representative frequency (Hz) so that the height can be calculated by the path difference formula.

③ Installing length

- Consider noise incidence from the side of the soundproof wall when determining the length of the facility.

- The length can be determined by the direct sound decrement value.

④ Landscape and Aesthetic design

- Time : After drafting environmental impact assessment in detailed design stage.
- Method
  - Comprehensively analyze site conditions such as earth and rock status so that the material, shape and color can be properly changed.

  - Plan various design alternatives for landscaping requirement. Landscape review for soundproof facility is obligatory.

⑤ 기타사항

- 기초부와 최하단 방음판 사이에는 옥외 기후에도 내구성이 우수한 재료 및 모르타르, 발포 고무판 등의 자재로 밀폐하여 음의 누출을 방지

- 방음벽의 교차부분 또는 방음벽 밑 부분이나 방음벽과 나란히 배수로를 설치하는 등 도로의 배수흐름을 방해하지 않도록 하여야 함

- 방음벽의 보호를 위하여 도로여건에 따라 필요한 경우에 한하여 방호책을 설치

2-7 생태이동통로

2-7-1 개 요

- 도로건설로 인하여 서식지가 절단되는 것을 막기 위하여 야생동물이 지나는 길을 인공적으로 만든 것을 의미함
- 도로건설에 따른 야생동물의 로드킬(Road Kill) 방지, 야생동물의 이동로 제공, 야생동물 서식지로의 이동, 생태계의 연속성 유지 등의 역할

2-7-2 생태이동통로의 종류

① 터널형 이동통로

- 암거형 통로
  - 도로통과 지역 중 흙쌓기에 의하여 농수로, 늪지, 개울 등이 단절되는 경우 설치
  - 소형동물의 이동이 만은 지역의 골짜기를 따라 설치
  - 동물전용(최소규격 2.5m×2.5m)과 수로겸용 암거로 구분

- 파이프형 통로
  - 작은 소택지를 따라 이동통로 연결
  - 동물전용과 수로겸용 파이프형 이동통로로 구분

- 하천통과 교량
  - 비교적 폭이 넓은 하천에 조성되는 교량 하부 공간을 복합적으로 활용
  - 물, 초지 등의 공간이 공존할 수 있도록 고려

② 육교형 이동통로

- 도로건설로 인하여 생태계 단절이 예상되는 곳에 설치

- 육교형 통로 설치위치는 관련전문가의 자문을 받아 기존에 야생동물이 이용하던 자연통로에 설치하는 것이 바람직

- 대규모 땅깎기 발생에 따른 구조물의 과다 등으로 경제성·시공성 측면에서 문제 발생 시에는 유도휀스, 유도식재 등의 설치로 설치위치를 변경 할 수 있음

⑤ Others

- Use highly durable material such as mortar, foamed rubber plate that can endure harsh exterior conditions between foundation section and lowest soundproof plate to prevent the noise leakage.

- Install drainage in the intersection of the wall or under the wall parallel to the wall so that the wall would not prevent drainage of the road.

- If required, install guard-rail to protect the wall.

2-7 Eco-Corridor

2-7-1 Overview

- Eco-corridor is artificial wildlife passage to prevent wild habitat from division by the road construction.

- The facility can prevent road kill of wildlife and provide wildlife passage so that the ecosystem can maintain its continuity.

2-7-2 Types of Eco-Corridor

① Tunnel type

- Duct type
  - Install in the areas where land-fill cuts agricultural waterway, wetland, or streamlet
  - Install along the valley where there are lots of movement of small wildlife.
  - Dedicated for wildlife (minimum 2.5m×2.5m) type or waterway combined type.

- Pipe type
  - Connect passage along small marshland
  - Dedicated for wildlife type or waterway combined type.

- Under-Bridge type
  - Use the lower section of the bridge over relatively large scale watercourse.
  - Consider coexistence of water and grassland.

② Viaduct type

- Install in the areas where road construction may cut ecosystem continuity.

- It is recommended to consult to the relevant professional so that the viaduct can be installed to the original pathway of wildlife.

- If there is economic or constructional issue of excessive structure due to large scale land-cut, installation location can be modified with inducing fence or inducing planting.

- 육교형 통로 설치 시 중앙부의 폭은 30m를 최소(지형적인 여건에 따라 최소폭을 30m 이하로 할 수 있음)로 하고, 입·출구부는 넓게 하여 야생동물의 이동을 자연스럽게 유도

- 생태적으로 보전가치가 우수하고 넓은 면적이 단절되거나 대형동물 출현이 많은 곳 등에 주로 설치

2-7-3 설치기준

① 위치 결정 시 고려사항

- 주요 대상동물의 서식지 특성 파악

- 기존 이동로의 위치 및 개수와 훼손 가능성 파악

- 주변 서식지와의 생태적 연속성 및 동물이동의 장애물 존재여부 파악

② 규모 결정 시 고려사항

- 연결대상 서식지간 거리는 가능한 짧고 직선을 유지

- 통로의 길이가 길수록 폭은 넓게 적용

③ 보조시설

- 현지여건에 따라 적합한 형태의 보조시설(토양, 초본류, 식재, 선반 설치, 서식처와 피난처 제공, 경사로 설치, 유도휀스, 나무그루터기벽, 동물출현포지판 등)을 설치

- 생태통로 입구 주변은 야생동물과 파충류의 이동 시 은폐를 위하여 인근 자생수목과 유사한 수목을 식재하여 수림을 조성

- 농로 및 임도는 사람의 왕래가 적으므로 야생동물의 이동통로을 겸용하여 설치 가능

- 야생동물이 자유롭게 오르내릴 수 있도록 경사로 설치 및 적정 간격의 이동로 설치

2-8 오탁방지망

2-8-1 개　요

- 하천을 통과하는 공사 시행 시 부유 토사의 확산을 방지하기 위하여 설치

- The width of the center should be more than 30m (if unavoidable due to topology condition, less than 30m can be accepted.) Entrance and exit section should be wider to induce wildlife movement.

- Install in the area with high preservation value, large division of ecosystem or area with frequent appearance of large animals.

2-7-3 Installation Standard

① Considerations for location

- Research the characteristics of main wildlife habitat.
- Assess the location of existing passage and possible damage from the construction.
- Consider ecosystem continuity of nearby habitat and existence of barrier for wildlife movement.

② Consideration for size

- The length of the corridor should be as short as possible and remain straight.

- If the length gets wider, width should be increased accordingly.

③ Subsidiary facilities

- According to the site condition, provide proper subsidiary facilities (earth, herb, planting, shelf, habitat, shelter, slope, inducing fence, wood stubble wall and proper signage, etc.)

- Near the corridor entrance, provide plantage similar to nearby native plants so that the wildlife and reptiles can hide themselves.

- Agricultural road or forest road can be used as eco-corridor as the road is used by human less frequently.

- Consider slope and interval of corridor to facilitate smooth movement of wildlife.

2-8 Taint Prevention Net

2-8-1 Overview

- Facility to prevent diffusion of floating earth and sand during the road construction.

2-8-2 설치기준

① 설치높이

- 수심의 50%로 설계 적용

- 현장여건을 고려하여 부유토사 및 폐수의 확산에 의한 수생태계 보존에 필요한 설치범위를 결정하여야 함

② 현장적용

- 일반 하천의 경우 앵커는 콘크리트 블록형으로 설계 적용하고, 공사시행 시 현장여건에 맞는 앵커를 설치

- 앵커블록의 설치가 불리한 소규모 하천의 경우 좌·우측에 로프를 연결하여 고정하고, 하부체인으로 앵커역할이 가능한 형식으로 설계 적용

2-9 터널 유출수 관리시설

2-9-1 개 요

- 터널 공사 시 배출되는 지하수 또는 유출수가 하천이나 호소에 직접 유입되지 않도록 처리하는 시설

2-9-2 설치기준

① 터널 폐수처리시설

- 운영기간은 수질기준 만족 시까지로 하며, 기간변경에 따른 인건비·약품비·슬러지처리비·운전동력사용비 등이 정산 가능하도록 단가 구성

- 설계 중 터널조사 및 건설 중 터널굴착 이후 유출수량(지하수+작업수)에 따라 처리시설규모를 변경

② 터널 타일 세척수 처리시설

- 터널 세척수와 지하수 배수관로를 구분하여 터널 타일 세척수를 별도 집수처리

- 타일청소 시의 세척수 사용량을 기준으로 소요 규모를 산정하여 설치

2-10 노면오염수 저감시설

2-10-1 개 요

- 비점오염원에서 하천으로 유입되는 오염물질을 저감하기 위한 시설

2-8-2 Installiation standard

① Installing height

- 50 percent of the water depth.

- Consider site condition to determine installing location required to preserve water ecosystem and prevent tainted water diffusion.

② Site application

- For regular watercourses, design anchor to be concrete convex type. When constructed, install the anchor properly according to the site condition.

- For small streams where installing concrete convex anchor is not applicable, connect rope in the both side and design lower part chain to become anchor.

2-9 Tunnel Drainage Management Facility

2-9-1 Overview

- Facility to manage subterranean water and other outflow water from tunnel construction so that the water would not directly come into other watercourse.

2-9-2 Installiation standard

① Tunnel waste water management facility

- Facility should be operated until satisfying water quality standard. Unit cost should be consist of labor cost, chemical cost, sludge treatment cost and power usage cost so that each item can be settled up independently.

- Change the size of the facility according to the tunnel research and the water volume after the tunnel excavation.

② Tunnel tile cleaning water management facility

- Separate tunnel cleaning water drainage and subterranean water drainage.

- Based on water usage for cleaning tiles, calculate the required size of facility.

2-10 Surface Polluted Water Reduction Facility

2-10-1 Overview

- Facility to reduce inflow of pollutant from non-point pollution source to watercourses.

2-10-2 오염물질의 특징 및 종류

구 분	점오염원	비점오염원
배출원	공장, 가정하수, 분뇨처리장, 축산농가 등	대지, 도로, 논, 밭, 임야, 대기 중의 오염물질 등
특 징	• 인위적 • 배출지점이 특정적, 명확 • 관거를 통하여 한 지점(주로 처리장)으로 집중적 배출 • 자연적 요인에 영향을 적게 받아 연중 배출량의 차이가 일정함 • 모으기 용이하고 처리효율 높음	• 인위적 및 자연적 • 배출지점이 불특정, 불명확 • 회석, 확산되면서 넓은 지역으로 배출 • 자연적 요인에 따른 배출량의 변화가 심하여 예측 곤란함 • 모으기 어렵고 처리효율이 일정하지 않음

2-10-3 비점오염원 저감시설의 종류

① 자연형 시설

- 저류시설
  - 강우유출수를 저류하여 침전 등에 의하여 비점오염물질을 저감하는 시설로 저류지(연못), 저류조 등을 포함

저 류 지	저 류 조

- 인공습지
  - 침전 · 여과 · 흡착 · 미생물 분해, 식생식물에 의한 정화 등 자연상태의 습지가 보유하고 있는 정화능력을 인위적으로 향상시켜 비점오염물질을 저감하는 시설

2-10-2 Characteristics and Types of Pollutants

Classification	Point pollution source	Non-point pollution source
Source	Factories, households, night-soil treatment, stock raising farm, etc.	Land, road, rice field, garden, woodland, the atmosphere, etc.
Characteristics	• Artificial • Source is distinctive • Discharged to a certain spot through the pipes. • Less influenced by natural circumstances. Discharged volume is regular for all season. • Easy to gather and higher management efficiency	• Artificial and natural • Source is not distinctive • Dispersed to wider area • Hard to predict as highly influenced by natural conditions. • Hard to gather and management efficiency varies

2-10-3 Type of Non-point Pollution Mitigation Facilities

① Natural type

- Undercurrent
  - Gather discharged rainwater and mitigate pollution through settlement such as pond and undercurrent tank, etc.

- Artificial wetland
  - Improves natural purification function of natural wetland to mitigate pollutant

- 침투시설
  - 강우유출수를 지하로 침투시켜 토양의 여과·흡착작용에 따라 비점오염물질을 저감하는 시설로, 유공(有孔)포장·투수성포장·침투조·침투저류지·침투도랑 등을 포함

지역 내 침투시설	침투성 도로

- 식생형시설
  - 토양의 여과·흡착 및 식물의 흡착작용으로 비점오염물질을 저감함과 동시에 동·식물 서식공간을 제공하면서 녹지경관으로 기능

식생수로 설치	식생여과대

② 장치형시설

- M-Filter
  - 오염된 물속의 부유물질과 용존물질을 제거하는 기술로서 복개구조물 내에 상소하대공극 필터층, 투수격자망 및 월류수여과기를 수직으로 결합하여 초기우수를 분리차집하여 필터층 상부 저류공간에 저장하고 모래층과 자갈층을 통과하면서 부유물질과 용존유기물질을 분해하는 장치

- Penetration
  - Penetrate rainfall discharge water into underground to mitigate pollutant such as poled pavement, permeable pavement, penetration tank, penetration pond and penetration drain, etc.

- Planting
  - Mitigate pollutants through soil filtration and absorption. Can be used as green area and environmental plant habitat

② Mechanical

- M-Filter
  - Remove floating particles and dissolved pollutants through combining poled filter layer, coffer net, and overflow percolator vertically. Collect and separate initial rainfall water and contain the water in the upper tank to disintegrate pollutants passing sand layer and gravel layer.

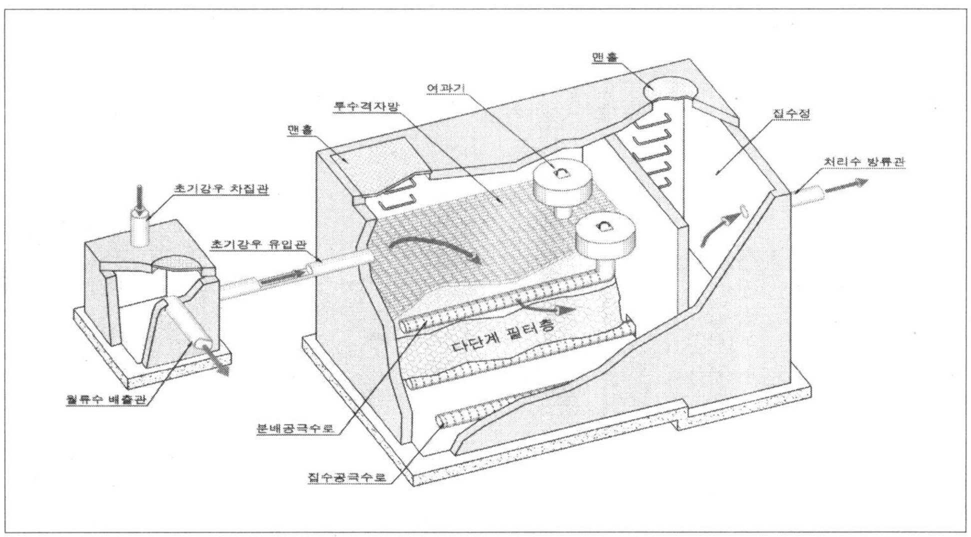

- Storm-Filter
  - 콘크리트 구조물 내부에 카트리지형 여과기를 다수 설치하여 부유물질과 용존물질을 제거하며 미처리된 우수는 월류 방류시키는 장치

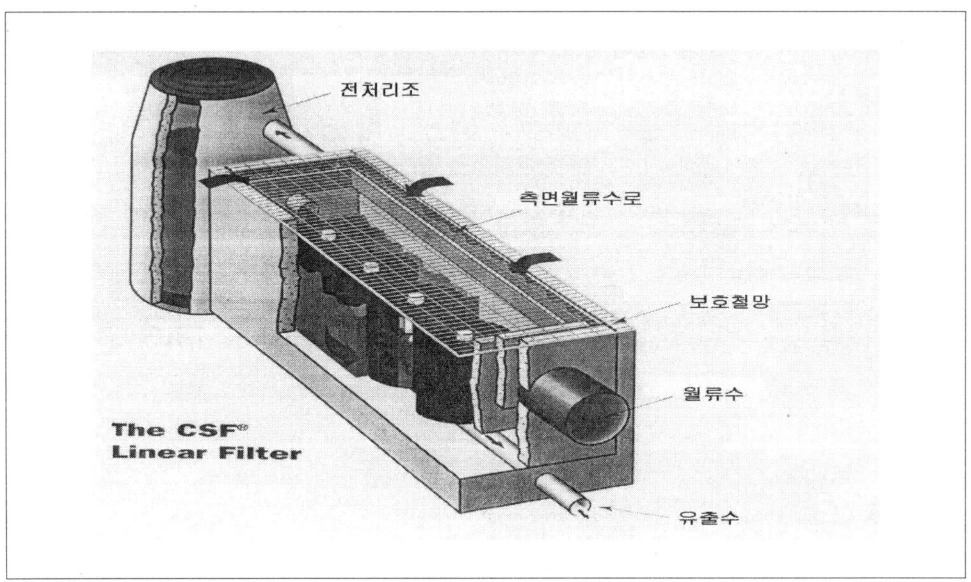

- 에코탱크
  - 수두 차 및 유입수의 관성력을 이용하여 무동력으로 처리가 이루어지며, 초기우수만 선택적으로 처리 가능(소요부지 적음)

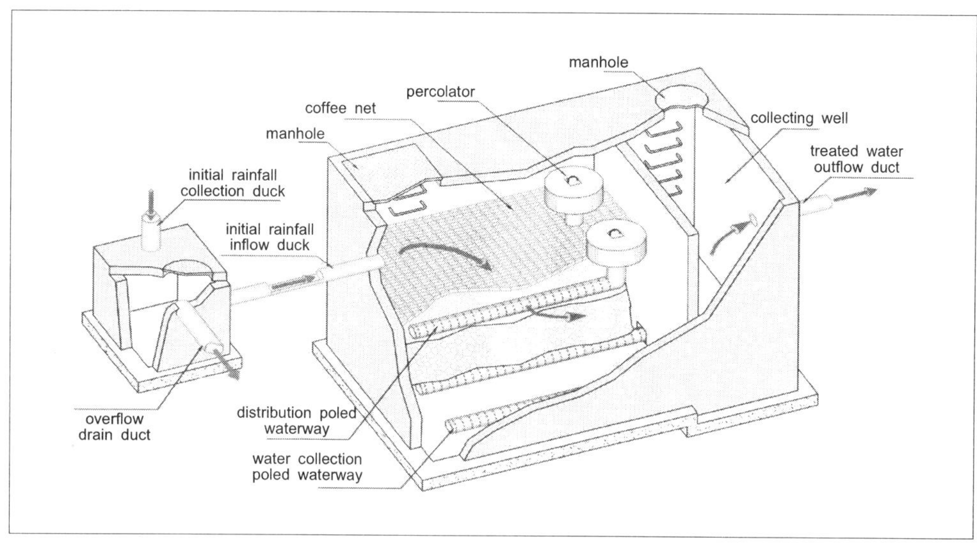

- Storm-Filter
  - Install a large number of cartridge-type percolators inside the concrete structure to remove floating particles and dissolved pollutants and to discharge untreated rainwater

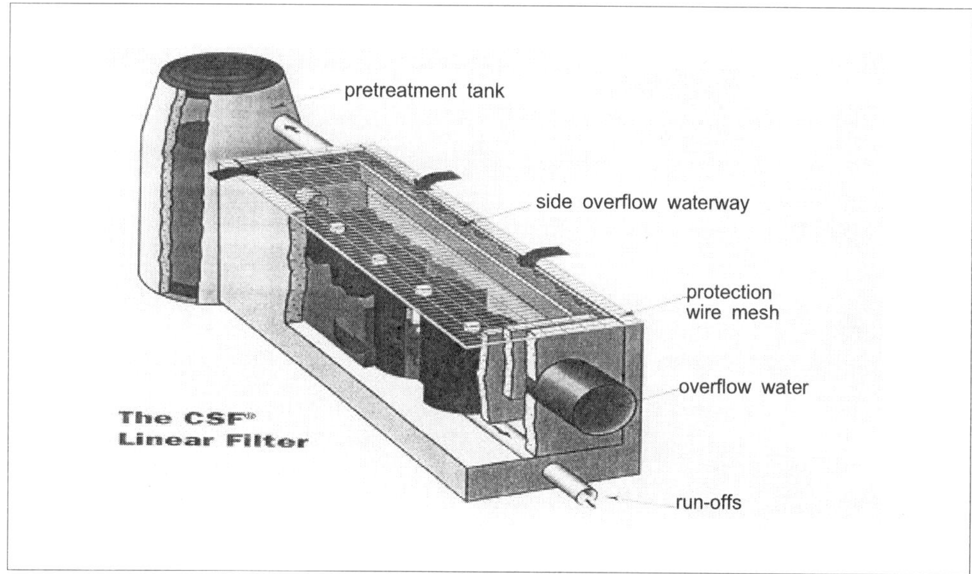

- Eco tank
  - No power supply is required as initial rainfall water can be purified by water head difference and inertial force (smaller area required)

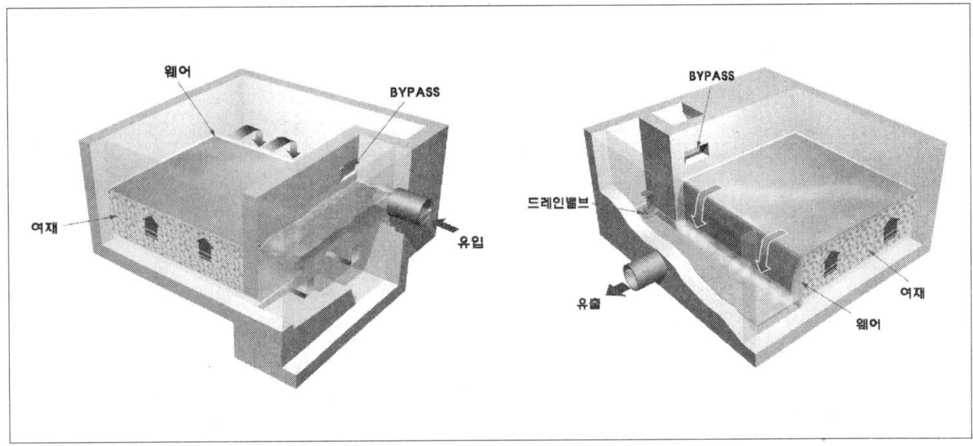

2-10-4 설치기준

① 설치지역

- 각종 법령에 의하여 수질 보호 및 개선이 요구되는 하천 및 호소 인근 지역 (수질오염총량관리제 대상지역, 특별대책지역, 상수원 보호구역, 수변구역 및 호소수질 보전구역 등)

- 강우 시 노면에서 수계로 비점오염물질이 직접 유입되어 수질오염의 원인으로 작용하는 지역

- 비점오염원에 의한 수질오염 민원 및 분쟁이 발생하는 지역

- 상수원보호구역과 수변구역 경계에서 1km 이내 지역

- 환경영향평가, 관계기관, 지역주민 등과 협의 지역

② 용량산정

- 자연형(저류형, 침투형, 식생형 등)
  - WQV(Water Quality Volume) = (PI)/1,000 × (Rv) × (A)
    ※ PI : 초기우수(mm), Rv : 유출계수, A : 배수면적(㎡)

- 초기우수(PI)는 80% 확률강우사상, 최소 7.4mm 적용

- 유출계수(Rv)는 "하수도 시설기준의 토지이용별 기초 유출계수"의 도로 표면형태 0.9 적용

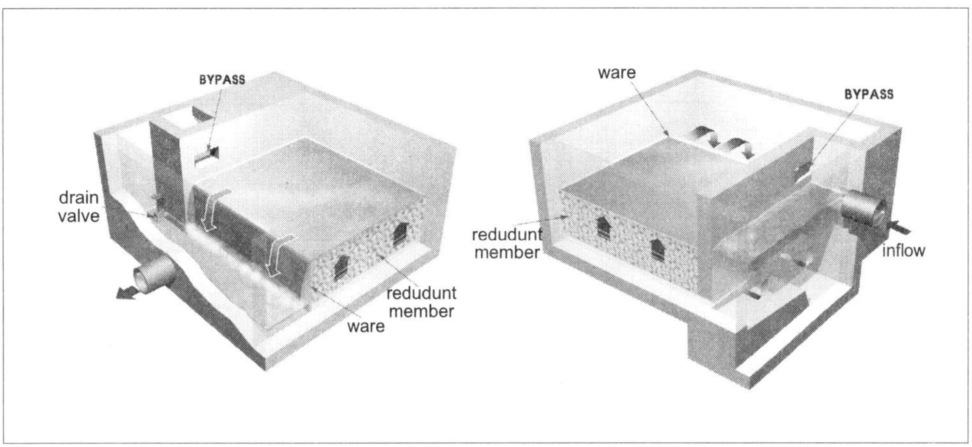

2-10-4 Installation Standard

① Installing location

- Area near to watercourses where water quality should be preserved and improved according to relevant regulations (water quality management area, area subject to special measure, source water preservation area and waterside preservation area, etc)

- Area where non-point pollution can inflow from road surface to watercourse directly when it rains

- Area subject to civil appeal due to water pollution from non-point pollution sources

- Area within 1km from water source preservation area and waterside preservation area

- Area designated by consultation of environmental assessment, relevant agencies and nearby civilian

② Capacity calculation

- Natural type (undercurrent, penetration and planting, etc)
  - WQV(Water Quality Volume) = (PI)/1,000 × (Rv) × (A)
    - ※ PI : Initial rain water (mm), Rv : discharge coefficient, A : drainage area(m^2)

- Apply minimum 7.4mm for PI according to 80% precipitation probability history

- Apply Road surface condition 0.9 for Rv according to "sewage facility standard for land use base discharge coefficient"

- 장치형
  - Q(계획 우수유출량, ㎥/초) = 1/360 C·I·A
    ※ C : 유출계수, I : 초기우수(mm/h), A : 배수면적(ha)
  - 초기우수(I)는 80% 확률강우강도, 유달시간(t) 내의 평균강우 강도(mm/h), 최소 5mm/h 적용
  - 유출계수는 "하수도 시설기준의 토지이용별 기초 유출계수"의 도로 표면형태 0.9 적용.
  - 여유율은 미적용

③ 설치형식

- 노면수량, 처리효율, 유지관리 등을 고려하여 자연형의 형식(저류형, 침투형, 식생형)으로 적용

- 저류형은 설치개소가 최소화 되도록 통합체계 배수계획을 수립

- 장치형은 자연형 설치 및 오염수 처리효율 확보 곤란 시 적용

2-11 폐수 및 오수처리시설

2-11-1 폐수처리시설

① Batch Plant, 골재생산시설 등에서 발생하는 폐수는 높은 SS 농도와 pH 성분을 함유하고 있으므로 저감시설을 설치

② 적절한 폐수처리시설 설치 · 운영 후 방류하는 방안이나 처리 전 또는 처리 후 재이용 하는 방안 등을 고려하여 폐수처리 계획을 수립

2-11-2 오수처리시설

① 현장사무소 및 건설현장 근무인원에 의하여 배출되는 오수

- 오수처리시설과 간이화장실을 설치 · 처리

- 방류수 수질은 하천수질환경기준, 호소수질환경기준 등을 고려하여 "오수·분뇨 및 축산폐수의 처리에 관한 법률 시행규칙 제9조"의 오수처리시설 등의 방류수 수질기준에 준하여 계획

- Mechanical
  - Q(Designed rainwater discharge, m^3/sec) = 1/360 C·I·A
    ※ C : Discharge coefficient, I : Initial rainwater (mm/h), A : drainage area (ha)
  - Apply minimum 5mm/h for PI according to 80% precipitation probability history and average rainfall intensity (mm/h) within time of concentration
  - Apply Road surface condition 0.9 for Rv according to "sewage facility standard for land use base discharge coefficient"
  - Do not apply rate of allowance

③ Installing form

- Apply natural type considering surface water content, processing efficiency and maintenance requirement

- Consolidated drainage plan should be prepared to minimize installing locations for underflow type

- Apply machine type when it is hard to secure required process efficiency with natural type

2-11 Waste Water & Polluted Water Management Facility

2-11-1 Waste Water Management Facility

① Dedicated wastewater management facility is required for Batch Plant, aggregation production facility as the wastewater of those facility has high rate of SS density and pH

② Prepare wastewater management plan with considerations on discharging or reusing after the process

2-11-2 Polluted Water Management Facility

① Polluted water from site office and site workers

- Install and operate polluted water management facility and temporary restroom

- For water quality of discharged water, consider watercourse quality standard according to "article 9 from law related to management of polluted water and wastewater"

② 휴게소 운영에 의하여 배출되는 오수

- 오수의 적정처리는 하수처리구역을 고려하여 계획을 수립하고, 가능한 한 공공처리시설로 유입·처리

- 오수를 공공처리시설로 유입시키는 것이 불가능할 경우 자체 오수처리시설을 설치

2-12 영업시설

2-12-1 개 요

- 유료도로를 건설하는 경우 통행요금을 징수하기 위하여 설치하는 시설

- 도로구조의 보전과 통행차량의 안전을 도모하기 위하여 영업소에 축중계를 설치하여 과적차량의 통행을 제한

2-12-2 영업시설의 종류 및 구분

① 설치장소에 따른 구분

- 본선영업소
  - 본선상에 설치되는 영업소

- 인터체인지 영업소
  - 인터체인지 내에 설치되는 영업소

② 통행요금징수체계에 따른 구분

- 전선균일 요금제
  - 일반 유료도로에서 많이 채택되는 방식으로서 입구 또는 출구에만 영업소 설치

- 구간별 균일 요금제
  - 고속도로 본선상에 30~40km마다 영업소 광장 설치
  - 이용자가 영업소를 통과할 때마다 일정액을 지불

- 완전한 구간별 요금제
  - 이용자가 고속도로 진입 시 통행거리에 따른 요금을 지불하고 통행권을 받아 출구에서 통행권을 반납하는 방식

③ 통행료의 징수방법에 따른 구분

- 폐쇄식 영업방식
  - 고속도로를 진입하여 톨게이트 입구에서 통행권을 발급받고 톨게이트 출구에서 요금을 지불하는 영업방식

② Polluted water from resting area operation

- Prepare polluted water management plan with consideration of sewage process area. If possible, connect to public management facility

- Install independent management system if connecting to public facility is not feasible

2-12 Operation Facility

2-12-1 Overview

- Facility to collect fee for toll road

- Install weight indicator to operation facility to restrict passage of overloaded vehicles and secure road safety

2-12-2 Type and Classification of Operation Facility

① Classification by location

- Main lane operational facility
  - Installed in the main lane

- Interchange operational facility
  - Installed in the interchange

② Classification by toll fee charging system

- Flat fee for all lane
  - Operation facility is installed only on exit or entrance. Used for common toll road frequently.

- Flat fee by section
  - Install open space for operational facility on highway main lane
  - Drivers pay certain amount of toll fee everytime they pass through the toll

- Complete section fee
  - Drivers pay toll fee according to destination in the highway entrance toll and return the ticket in the exit toll

③ Classification by toll fee gathering system

- Closed type
  - Drivers receive ticket in the entering toll and pay the fee in the exit toll

- 개방식 영업방식
  - 고속도로 본선상에 톨게이트를 설치하고, 통과차량에 한하여 일률적으로 통행료를 회수하는 영업방식
  - 통행요금을 평균주행거리에 따라 차종별로 차등 징수

- NTCS 영업방식
  - 개방식 또는 폐쇄식 영업체계에 적용할 수 있는 자동요금징수 시스템
  - 차량탑재장치와 영업소와이 데이터 통신방식에 의하여 차량이 무정차로 요금을 지불하도록 하는 시스템

2-12-3 영업소 차로 수 설계기준

① 소요 차로 수 산정

구 분	내 용
교 통 량	• 첨두설계시간교통량(PDDHV)을 사용 - PDDHV=(AADT×D×K)/PHF - 공용 개시 후 10년 교통량으로 산정
평균 서비스시간	• 요금징수에 필요한 서비스시간을 말함 - 구간별 요금제(폐쇄식) : 입구 6초, 출구 13초 - 균일 요금제(개방식) : 입·출구 8초
서비스 기준	• 평균 대기차량 대수, 원칙적으로 3대를 기준 (부득이한 경우 5대)

② 축중기 차로 수 산정

- 고속도로 영업소에는 1개 이상의 차로에 고정식 축중기를 설치

- 폐쇄식에는 이용차량의 진입차로에, 개방식에는 양방향 차로에 고정식 축중기 설치

- 고정식 축중기는 이용차량 진입방향의 맨 우측 차로부터 설치

- 소요 차로 수 산정기준

차 로 수 / 화물교통량	1	2	3	4	5
시간 교통량	300	850	1,420	2,000	2,590
일평균 교통량	4,000	11,300	18,900	26,700	34,500

- Open type
  - Toll gate in the main lane of highway gathers the flat fee from passing drivers
  - Toll fee is graded by vehicle type according to the mean mileage.

- NTCS type
  - Automated toll fee gathering system which can be applied to both open and closed type
  - Non-stop charging system through data communication between device in vehicle and operation facility

2-12-3 Design Standard for Operation Facility Lanes

① Calculation for required lanes

Classification	Content
Volume of Traffic	• Peak design hour volume (PDDHV) - PDDHV=(AADT×D×K)/PHF - traffic volume after 10 years from opening
Average Service Time	• Service time required to collect fee - Sectional (closed) : entrance 6 sec, exit 13 sec - Flat (open) : 8 sec for both
Service Standard	• Average waiting vehicles : 3 (5 vehicles if unavoidable)

② Calculation for weight indicator lanes

- Install fixed weight indicator to more than one lane for highway

- Install fixed weight indicator to entering lane for closed type, to both lanes for open type

- Install fixed weight indicator from the very right side of the entering lane.

- Required lane determination standard

# of lanes / Volume of freight traffic	1	2	3	4	5
Hourly traffic	300	850	1,420	2,000	2,590
Daily average traffic	4,000	11,300	18,900	26,700	34,500

※ 화물교통량 : 일방향 화물차량 교통량 적용
※ 시간당 교통량 : 영업소 차로 수 산정방법 적용
- 평균서비스시간 : 6초
- 평균대기차량대수 : 1.0대 기준
※ 일평균 교통량 : 시간당 교통량/설계시간계수(K)

③ 하이패스 차로 수 산정

- 이용차종
  - 승용차(경차 포함), 버스(소형, 보통), 2톤 이하 화물차(탑차, 밴차)

- 차로 수
  - 기본 1차로
  - 특별한 경우 1차로 추가
    (3만대 이상, 분리형 영업소, 다수 진출입로 등)

- 차로위치
  - 하이패스 차로 설치위치는 좌측차로를 원칙(영업소 환경 및 교통특성 등에 따라 위치조정 가능)
  - 본선형 진출·입램프 200m 이내 : 방향별 교통량을 고려 선정
    (추가차로는 중앙차로 고려)
  - I.C형 진출입램프 150m 이내 : 중앙차로 고려
  - 신호교차로 : 중앙차로 고려
  - 대형차량용 확폭 차로(B=3.5m)가 있는 경우 : 확폭 차로 고려

④ 왕복 공용 차로 수

- 출구 및 입구의 게이트가 분리되어 있는 경우 중앙부분의 차로 수를 왕복 공용으로 사용하는 것으로 하며, 이 때 왕복 공용으로 사용하는 차로수는 3차로까지로 함

- 초기년도에는 원칙적으로 왕복 공용 차로는 고려하지 않음

⑤ 차로 및 부스의 구성

- 영업소 내의 차로폭은 3.00m로 한다. 단 1방향에 대해서 최우측(진행 방향에 대하여)의 1차로는 폭을 3.50m로 하고, 축중기 설치 차로는 4.20m로 적용

- 하이패스차로는 이용차량의 교통특성을 감안하고 주행안전성을 고려 폭원을 3.5m 적용하고, 부득이한 경우는 3.0m를 적용

- 1방향에 대한 필요차로가 1차로인 경우에도 최소 2차로를 확보하는 것으로 하고 각각의 좌측에 교통섬, 부스를 설치(우측차로를 폭 3.5m로 하여 예비차로로 적용)

※ Freight traffic volume: unidirectional freight vehicle traffic volume
※ Hourly volume: Apply operation facility lane determination standard
- Average service time: 6 sec
- Average waiting vehicles : 1.0
※ Daily average traffic volume: Hourly volume/Design hour coefficient (K)

③ Calculation for number of Highpass lanes

- Type of applicable vehicle
  - Sedan (include mini-sedan), bus (small, regular), Freight vehicle under 2 ton (Tower, Van)

- Number of lanes
  - One lane
  - If required, add one lane
    (over 30,000 vehicles, separated operation facility and many gates)

- Lane location
  - Highpass lane should be very left of the lane (Subject to change according to the operation facility condition and traffic characteristics)
  - Within 200m from entrance ramp : Consider traffic volume
    (Consider central lanes for any required addition)
  - Within 150m from IC type entrance ramp : consider central lanes
  - Signaled crossroad : consider central lanes
  - When there is extended width lane (B=3.5m) for heavy vehicles :
    consider extended width lane

④ Shared lane for both way traffic

- When entrance and exit gates are separated, central lanes are used jointly. Number of jointly used lanes should be no more than three lanes.

- In initial year, do not consider both way traffic

⑤ Structure of lanes and booths

- Lane width in operation facility is 3.00m. For one direction, the very right lane should be 3.50m and weight indicator lane should be 4.20m
- Highpass lane should be 3.5m to secure driving safety. (3.0m if unavoidable)
- Secure at least two lanes even when only one lane is required for the direction. Install traffic island booth on the left side (3.5m for right side lane for allowance)

【영업소의 요금소 구간 횡단구성】

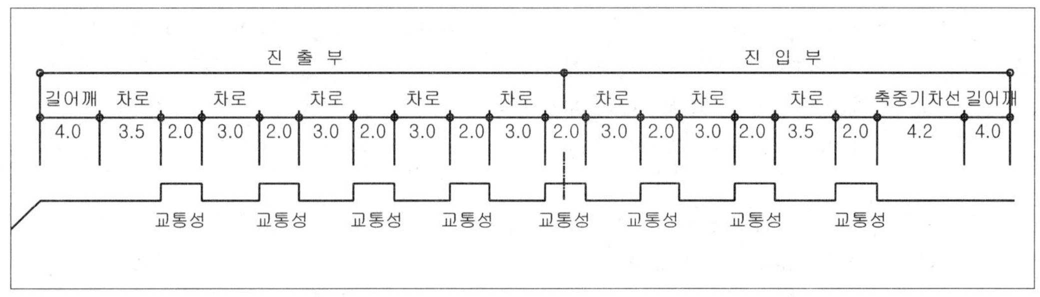

2-12-4 영업소 광장부 설계기준

① 평면선형

- 본선영업소의 경우 곡선반지름을 최소 1,500m 이상 또는 인터체인지를 설치하는 곳의 본선의 기준에 따름
- 인터체인지 영업소의 경우 최소 곡선반지름 200m

② 종단선형

- 영업소에서의 종단경사는 원칙적으로 2% 이하, 특별한 경우에도 3.0% 이하로 적용
- 적용 범위는 영업소 중심선의 전·후 각각 최소 90m
- 아일랜드 등에 의하여 횡배수가 차단되므로 요금소 차로부의 원활한 종배수를 위하여 최소 종단경사를 0.7%(부득이한 경우 0.5%) 이상으로 적용
- 영업소 광장을 설치하는 부분의 종단곡선 변화비율은 본선 영업소의 경우에는 인터체인지를 설치할 곳의 본선의 기준에 따름
- 인터체인지 영업소의 경우에는 원칙적으로 10m/% 이상, 특별한 경우에도 7m/% 이상으로 적용

③ 횡단경사

- 횡단경사는 표준 1.5%, 최대 2.0%로 적용

④ 광장부 연장

- 본선 영업소의 경우 영업소 중심선의 전·후 각각 90m를 원칙으로 적용
- 인터체인지 영업소의 경우 입구측은 영업소 중심선의 진입부 90m, 진출부 60m로 한다. 또한 출구측은 진입부 60m, 진출부 60m를 원칙으로 적용

⑤ 광장부 접속율

- 본선영업소의 경우 진출·입부 1/10 이하로 적용
- 인터체인지 영업소의 경우 진출·입부 1/5 이하로 적용

【Horizontal structure of operation facility toll section】

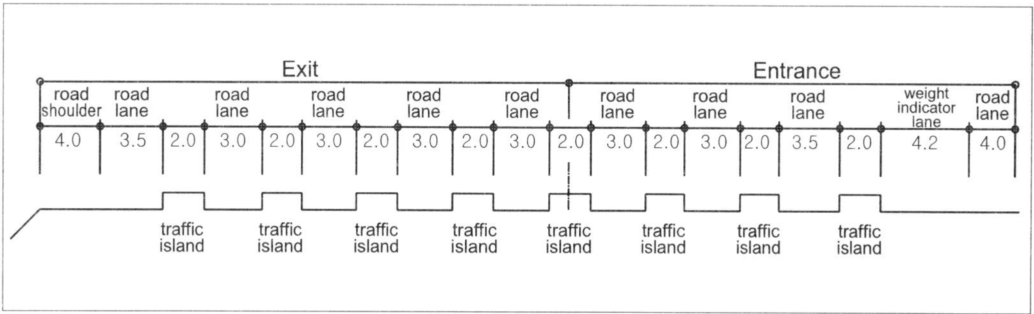

2-12-4 Design Standard for Open Space of Operation Facility

① Linear shape of plane section

- Minimum curvature radius of 1,500m for main lane operation facility or follow main lane standard for interchange construction

- Minimum curvature radius of 200m for interchange operation facility

② Linear shape of vertical section

- Vertical slope should be less than 2 percent. Less than 3 percent for special occasion.

- Applicable to front and back minimum 90m from center of operation facility

- Minimum vertical slope 0.7 percent (0.5 percent if unavoidable) for smooth drainage as horizontal drainage is blocked by traffic islands

- For terminal curvature change ratio of main lane operation facility, follow main lane standard of interchange construction

- For interchange operation facility, apply 10m/% and above (7m/% and above for special occasion)

③ Cross section slope

- Horizontal slope standard: 1.5%, (maximum 2.0%)

④ Open space length

- Applicable to front and back 90m from center of operation facility

- For interchange operation facility
  - Entrance: entrance 90m and exit 60m from the center
  - Exit: entrance 60m and exit 90m from the center

⑤ Open space connection ratio

- For main lane facility, less than 1/10 for entrance and exit
- For interchange facility, less than 1/5 for entrance and exit

- 본선 영업소의 경우 확장 계획 시 인근시설물 및 지장물 저촉, 관계기관 협의 곤란 등으로 현 기준 적용이 곤란할 경우 본선 및 영업소 차로 수를 고려한 최소 접속연장 340m~360m 이상 확보

⑥ 광장부 접속부
- 본선 영업소의 경우 자동차의 주행에 무리가 없도록 R=400m로 곡선처리
- 인터체인지 영업소의 경우 R=100m로 곡선처리

⑦ 축중기 차로
- 축중기 설치 차로의 평탄포장구간은 자동통행권 발행기 기준으로 45m 적용
- 종단조정 구간의 최대 종단경사는 5% 이하(부득이한 경우 8% 이하) 최대 28.5m로 적용

⑧ 포장
- 광장부와 테이퍼부는 무근 시멘트 콘크리트 포장(줄눈간격 6m)을 원칙
- 단, 광장중앙부 (36m, 아일랜드구간)은 시공성 및 매설시설물(공동구 등)을 감안하여 철근 시멘트 콘크리트 포장(줄눈간격 9m)을 적용
- 품질확보를 위하여 기계타설을 원칙으로 적용

2-13 방풍시설

2-13-1 개 요

- 산악지대나 해안지역을 통과하는 경우 강풍으로 인한 차량의 주행안전성을 확보하기 위하여 설치하는 시설
- 차량손상, 도로시설물 파손 및 인명손실 방지
- 사고의 방지를 통한 물류비용의 절감 및 경제적 이익 향상

2-13-2 설치기준

① 위험구간 판정기준
- 위험풍속은 10분 평균풍속 25m/sec를 기준으로 산정하고, 위험풍속 산정 시 해당하는 재현기간은 1/6년으로 적용
- 단, 계곡사이의 교량과 같이 풍속이 0인 터널을 빠져나오자마자 바로 높은 풍속의 바람을 접하게 되는 구간의 경우에 위험풍속은 3초 평균풍속 25m/s를 기준으로 적용

- Secure minimum connection length of 340~360m if extension is limited by nearby structure or difficulty in consultation with relevant agencies

⑥ Open space connecting section
- For main lane facility, curvature R=400m for smooth traffic

- For interchange facility, R=100m

⑦ Weight indicator lanes
- Flat pavement for weight indicator should be 45m from automated ticket issue machine

- Maximum vertical slope for vertical adjustment section (maximum 28.5m) should be less than 5 percent (8% for special occasion)

⑧ Pavement
- Cement concrete pavement (joint interval 6m) for open space and taper

- For center of open space (36m, island section), apply reinforced cement concrete pavement (joint interval 9m) considering construction easiness and underground facilities (utility conduit)

- Machine cast to secure construction quality

2-13 Wind-proof Facility

2-13-1 Overview

- Facility to secure driving stability in high-wind mountainous area or seashore area

- Prevent vehicle and road facility damage and life threat

- Improve economic benefit through preventing logistics accident

2-13-2 Installation Standard

① Risk area judgement standard
- Dangerous wind speed : 25m/sec for 10 minutes mean value. 1/6 recurrence year base

- For rapid wind speed changing area such as opening from valley, dangerous wind speed: 25m/s for 3 seconds mean value.

③ 방풍벽 설계

- 효율적인 방풍벽 설계를 위하여 고려되어야 할 변수들을 결정하기 위하여 풍동실험을 수행
- 일반적으로 방풍벽의 재료로는 철망이나 유공절판강판 또는 팽창메탈(expanded metal)등이 주로 사용
- 풍하중 감소와 시계의 확보 나아가서 미관 및 유지관리성을 고려하여 다공성방풍벽의 공극률을 포함한 형식선정을 고려

② 위험지역 선정기준

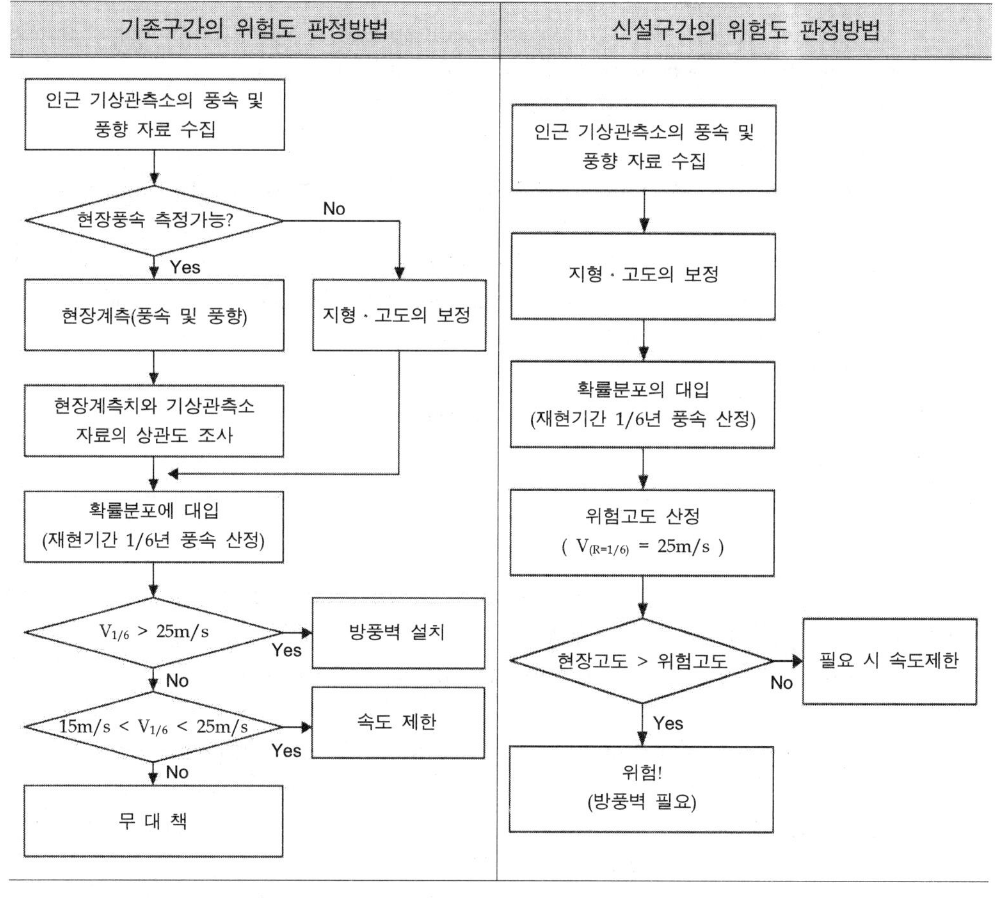

③ Windproof wall design

- Conduct wind dynamic test to decide variation to consider for efficient windproof wall design
- Use holed steel plate, steel net or expanded metal for windproof wall
- Consider types and forms including aperture rate of holed windproof wall to minimize wind road and improve landscape and maintenance easiness.

② Risk area determination standard

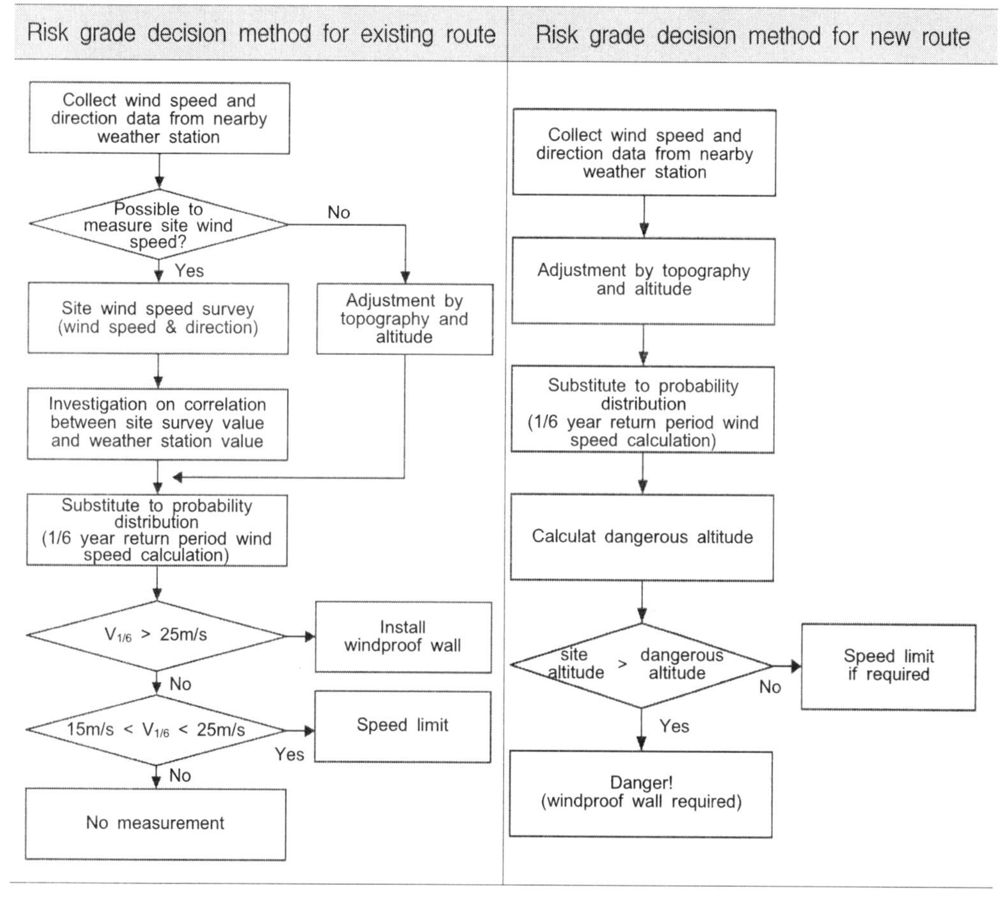

2-14 긴급구난시설

2-14-1 개　　요

- 고속도로에서 긴급상황(돌발상황) 발생으로 이용차량의 대피가 필요한 경우, 교통전환 및 회차를 위한 개구부 및 응급구난시설의 연결을 위한 진출·입로 등의 시설

- 긴급 상황 시 신속한 대처로 대국민 고객 서비스 증대

2-14-2 긴급구난시설의 종류

① 중앙분리대 개구부

- 중앙분리대에 회차 및 차로변경을 위하여 중앙분리대를 개폐할 수 있는 시설

② 터널 입·출구부 개구부

- 터널 화재 시 대피 및 차량회전을 위하여 터널 전·후 구간에 설치하는 시설

③ 긴급용 개구부

- 고속도로를 횡단하는 하급도로가 있는 경우 병원 및 소방서에 연결되도록 설치하는 진출·입로

④ 제설 작업용 개구부

- 제설작업용 차량의 진출입을 위하여 인터체인지 인접지역에 외부와 연결되도록 설치

2-14-3 설치기준

- 설치간격

<table>
<tr><th colspan="2">구　분</th><th>설 치 간 격</th><th>비 고</th></tr>
<tr><td rowspan="2">중앙분리대 개구부</td><td>형식-1</td><td>5.0km</td><td></td></tr>
<tr><td>형식-2</td><td>2.5km
(형식 1의 중간에 설치)</td><td></td></tr>
<tr><td rowspan="2">터널 입·출구부</td><td>형식-1</td><td>터널의 설치로 선형이 분리되는 경우</td><td></td></tr>
<tr><td>형식-2</td><td>터널과 터널, 장대교량의 연속설치로 선형이 완전 분리된 경우</td><td></td></tr>
<tr><td rowspan="2">긴급 및 제설용 간이진출입시설</td><td>하급도로가 있는 경우</td><td>출입시설과 출입시설 사이에서 교차하는 하급도로(양방향 2차로 이상)와 접속 가능구간에 1개소 설치를 원칙
※ 현장여건을 고려 조정 가능</td><td></td></tr>
<tr><td>하급도로가 없는 경우</td><td>출입시설과 출입시설 사이에 설치 가능한 교량구간을 이용하여 1개소 설치를 원칙
※ 현장여건을 고려 조정 가능</td><td></td></tr>
</table>

2-14 Emergency Rescue Facility

2-14-1 Overview

- Opening and exit facility for traffic switchover required in emergency situation in highway

- Fast reaction to emergency to promote quality of public service

2-14-2 Types of Emergency Rescue Facility

① Bisection opening

- Opening and closing facility for median strip to facilitate returning or lane changing

② Tunnel entrance / exit opening

- Installed in front and back of tunnel to facilitate evacuation and return of vehicles in case of fire

③ Emergency opening

- Entrance / exit to facilitate emergency connection from highway to overlapped and crossing regular motorway

④ Snow removal opening

- Connection to external area to provide entrance / exit for snow removal vehicles

2-14-3 Design Standard

① Installing interval

<table>
<tr><th colspan="2">Classification</th><th>Interval</th><th>Remark</th></tr>
<tr><td rowspan="2">Bisection opening</td><td>Type-1</td><td>5.0km</td><td></td></tr>
<tr><td>Type-2</td><td>2.5km
(in the middle of type 1)</td><td></td></tr>
<tr><td rowspan="2">Tunnel entrance / exit</td><td>Type-1</td><td>when lanes are divided in the tunnel</td><td></td></tr>
<tr><td>Type-2</td><td>complete division of lanes by continuous tunnels or bridges</td><td></td></tr>
<tr><td rowspan="2">Emergency and Snow removal entrance / exit</td><td>minor road exists</td><td>between entrance facilities, install one exit to connect crossing minor road (more than 2 lanes bidirectional)
※ Subject to be adjusted by site situation</td><td></td></tr>
<tr><td>minor road not exist</td><td>between entrance facilities, install one exit using applicable bridge section
※ Subject to be adjusted by site situation</td><td></td></tr>
</table>

Chapter 14

Tunnelling

1. 터널계획

1-1 선형계획

1-1-1 평면선형

① 평면선형은 사용목적 및 시공측면에서 가급적 직선으로 계획하고, 부득이한 경우는 터널 내 정지시거를 고려하여 큰 평면곡선반지름을 적용

② 편압이 예상되는 비탈면부 및 습곡지역, 애추(talus) 분포지역, 용수대, 단층 및 파쇄대지역 등은 피하는 것이 바람직

③ 터널 갱구위치는 주변 환경과 조화되고 가능한 한 자연 비탈면에 직교되도록 선정

1-1-2 종단선형

① 터널의 종단경사는 주행의 안정성, 환기의 효율성, 방재 설비, 배수 및 시공능률 등을 고려하여 가급적 완만하게 계획(최소 0.3%~0.5% 이상)

② 장대터널의 경우 환기를 고려하여 2~3% 이내를 적용하고, 환기 등을 감안하여 가능한 한 일방향 경사로 계획하는 것이 바람직

1-2 터널 단면계획

1-2-1 터널 내공단면 선정흐름도

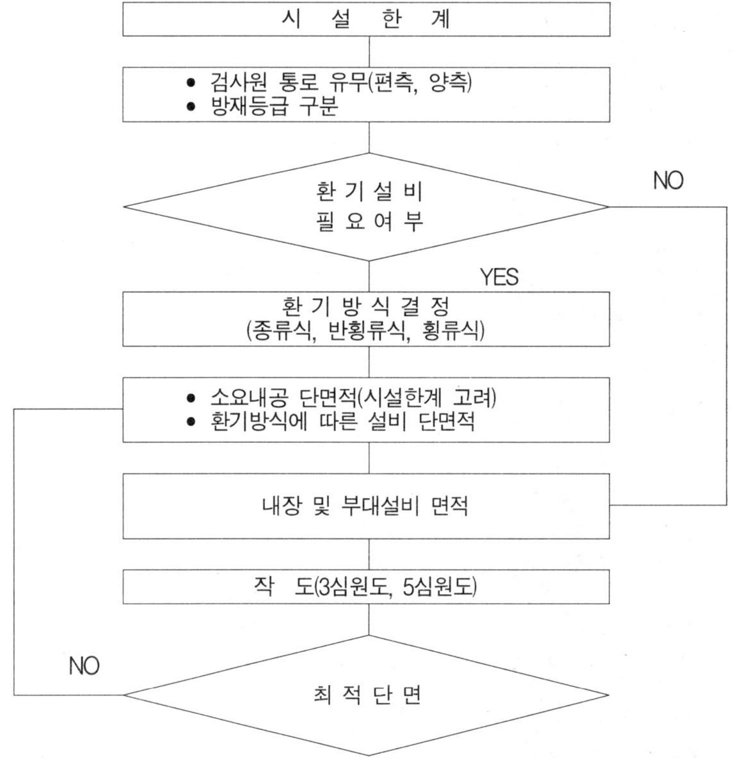

1. Tunnel Planning

1-1 Linear Planning

1-1-1 Linear Shape of Plane Section

① Design linear shape to be as straight as possible. If unavoidable, apply large curvature radius considering stop sight distance.

② Avoid areas where differential pressure is expected such as slope, fold, earth dislocation, talus, water belt, stratum, cracked surface, etc.

③ Tunnel entrance should be located in the place where the entrance matches with nearby natural scenary and crosses at the right angle with natural slope

1-1-2 Linear Shape of Vertical Section

① Design slope as gentle as possible considering driving stability, ventilation, disaster prevention facilities, drainage, construction efficiency, etc (minimum 0.3%~0.5%)

② For very long tunnel, apply 2~3% considering ventilation. If possible, let the direction of slope to be unidirectional

1-2 Tunnel Section Planning

1-2-1 Flowchart for Tunnel Section Planning

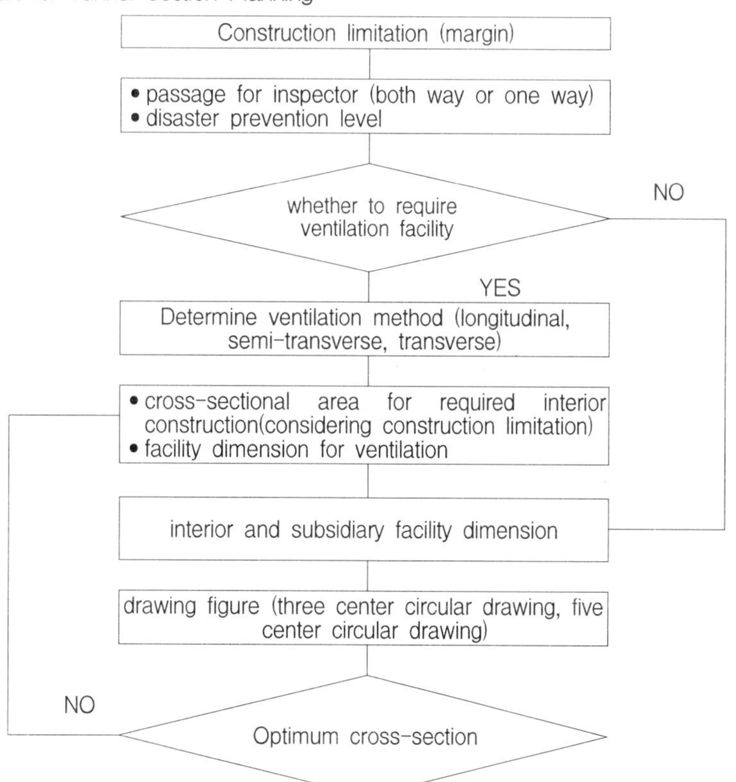

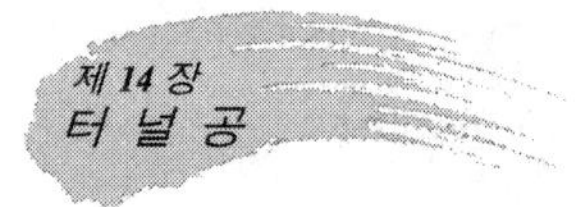

1-2-2 터널 내공단면의 구성요소

① 터널의 횡단구성

- 터널의 횡단은 차로폭, 측방 여유폭, 시설대로 구성
- 방재설비와 계측기구(교통감지 카메라 등) 등을 설치할 수 있는 시설대를 확보.

② 도로 폭원 및 시설한계

- 차량통과 높이는 장래의 유지보수 및 시공오차를 감안하여 4.8m 를 확보
- 조명등은 터널 단면의 양측 우각부에 설치하며, 시공오차 및 조명시설 설치에 따른 여유폭 0.30m 확보

- 각종 케이블 등의 설치를 위한 공동구 설치

【 2차로 도로터널의 내공단면 선정(예) 】

시설한계	터널 내공 여유공간	제트팬 이격거리
시 설 한 계 / 터널중심선 / 2차로기준 / 4.800 / 1.000 3.600 3.600 2.500	강섬유보강 숏크리트 / 콘크리트 라이닝 / 800 / 1.000	0.5D 설치한계선 / 시설한계
		주행차로 검사원 통로
		50 이상 / 2.000 / 750
		공동구 크기

1-2-2 Structure of Interior Cross-section of Tunnel

① Horizontal structure of tunnel

- Driving lane width, side space width, facility belt
- Secure facility belt to install disaster prevention facilities and measuring facilities (traffic detection camera, etc)

② Road width specification and construction limitation

- Vehicle passage height should be 4.8m considering construction error and future maintenance
- Light should be installed in superior angle of both side of the tunnel. Secure 30cm of additional space for construction error and facility installment
- Install utility-pipe conduit for cable installment

【 Cross-section example for two lane tunnel】

Facility limitation (margin)	Allowance space	Jet-fan estrangement distance
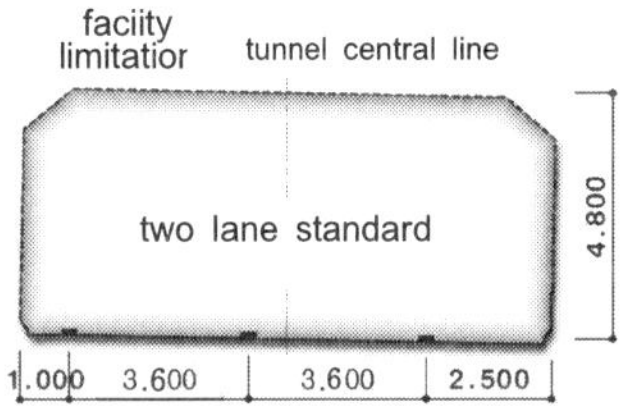 faciity limitatior tunnel central line two lane standard 4.800 1.000 3.600 3.600 2.500	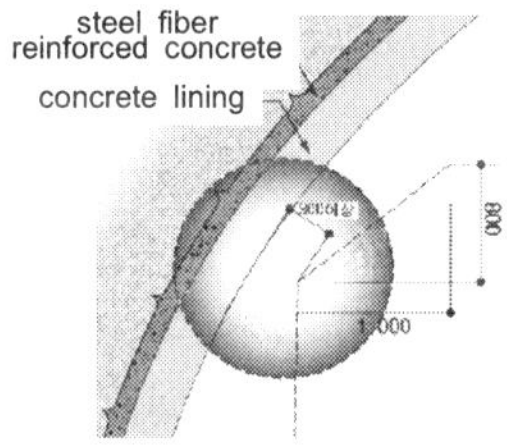 steel fiber reinforced concrete concrete lining 800 1.000	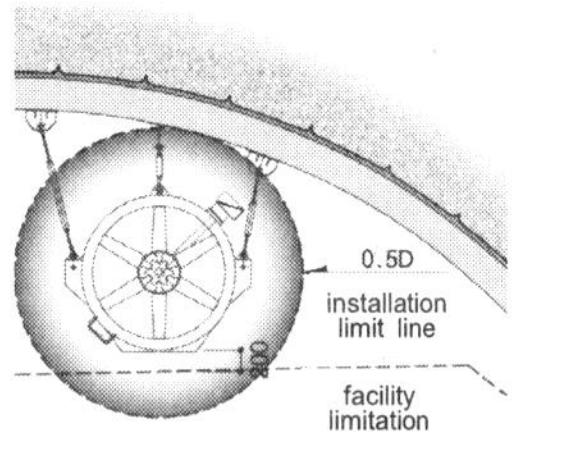 0.5D installation limit line 200 facility limitation
		Driving lane inspector passage
		50 10 2.000 750
		Utility pipe conduit

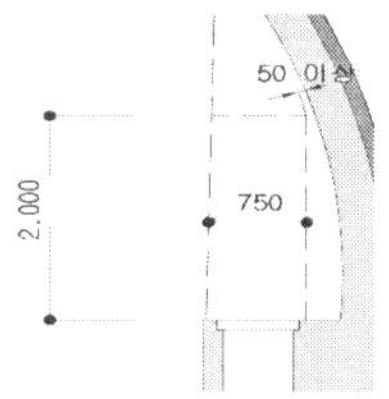

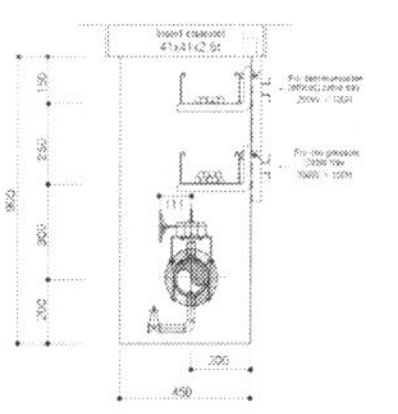

③ 포장공

- 터널 내 포장은 시멘트 콘크리트 포장이 기본
- 표 층 : 0.30m, 보조기층 : 0.15m, 필터층 : 0.20m (갱구부 50m는 동결심도 고려 조정)

④ 수압처리 방식

배수식	비배수식

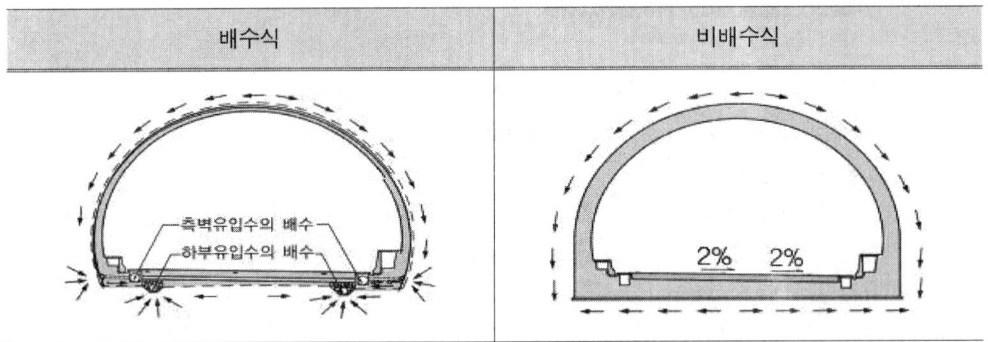

⑤ 배수공

- 터널의 아치부, 측벽부에 설치된 시트 방수재의 배면을 타고 온 침투수는 측벽하단 종배수관(고밀도폴리에틸렌유공관 ø100 mm)을 설치하여 배수구로 유도
- 터널 내 세정오수 및 비상 시 사용되는 소화수, 기타 수처리는 일정한 횡단경사에 의해 배수구(스파이럴씸닥트관 ø300 mm)로 유도
- 바닥 슬래브 배면에 발생되는 양수압을 배제시키기 위하여 집수관(유공관, ø200 mm)을 설치
- 공동구로 유입되는 오수처리를 위해 P.V.C 관(ø50 mm)을 C.T.C 50m마다 연결

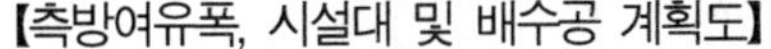
【측방여유폭, 시설대 및 배수공 계획도】

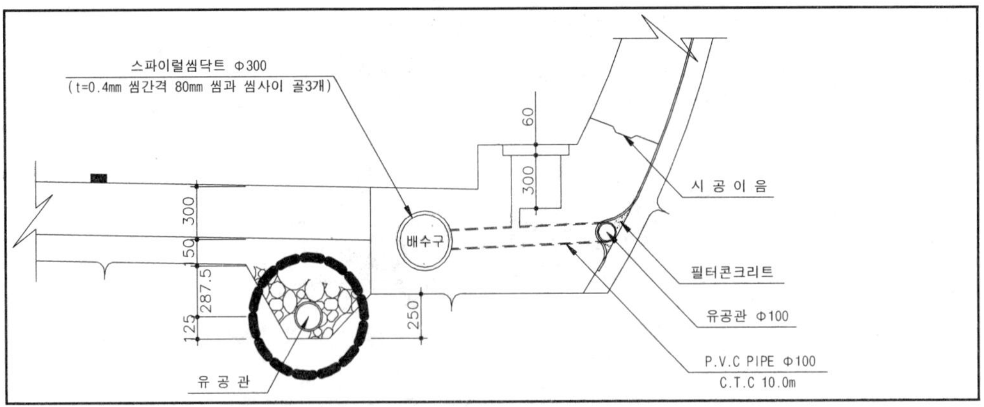

③ Pavement

- Concrete pavement for inside tunnel
- Surface layer : 30 cm, subbase layer: 15 cm, filter layer: 20 cm (consider freezing depth of pit mouth 50m)

④ Water pressure process method

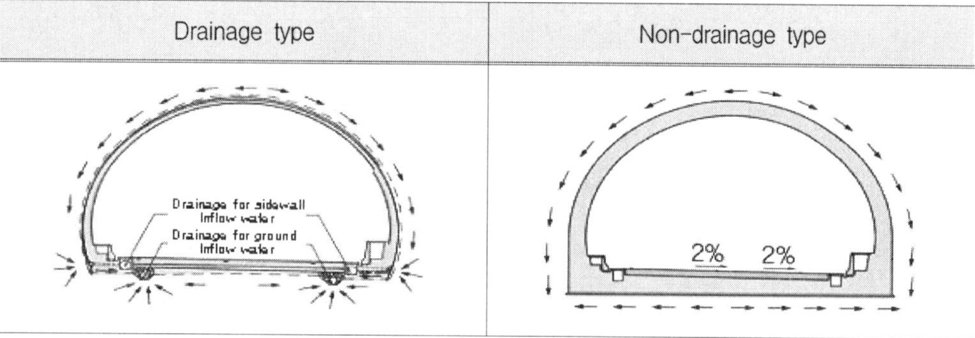

⑤ Drainage

- Install vertical drainage (HDPE polyethylene drain pipe ø100 mm) in the lower part of side wall for inflow water from surface of arch and wall sheet waterproof materials to induce the water to the drainage pipe.
- Induce cleaning water, emergency fire extinguish water and other water into drainage pipe (SPIRAL Seam DUCT pipe ø300 mm) through cross-sectional slope.
- Install water collection pipe (perforated drainage pipe, ø200 mm) to remove ground slab surface water pressure
- Connect P.V.C pipe for every C.T.C 50m to process water flowing into utility pipe conduit

【Side allowance, facility belt and drainage plan】

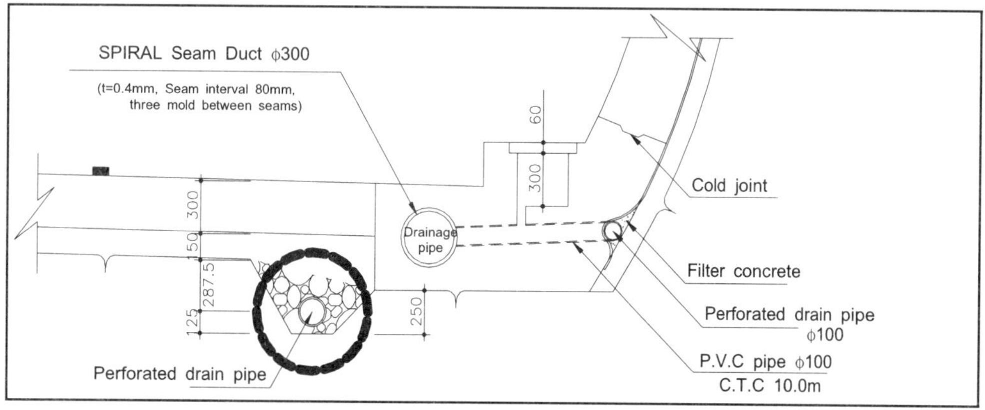

⑥ 공동구

- 공동구의 기능
  - 광통신 케이블 및 일반통신, 전원 케이블 설치공간 확보
  - 터널 내 조명, 환기, 교통관제 등을 위한 케이블 설치공간 확보
  - 각종 케이블의 추가설치, 교체, 점검 및 유지보수 공간 확보
- 규 모

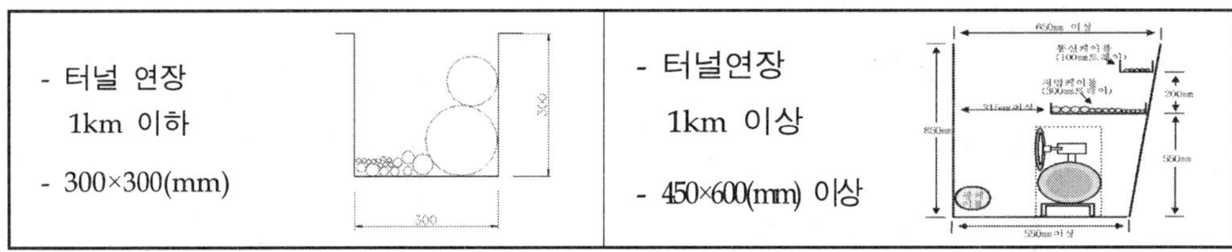

터널 연장 1km 이하	터널연장 1km 이상
- 터널 연장 1km 이하 - 300×300(mm)	- 터널연장 1km 이상 - 450×600(mm) 이상

1-2-3 터널 내공단면 종류

구 분	단 면 개 요	특 징
2차로 병 설	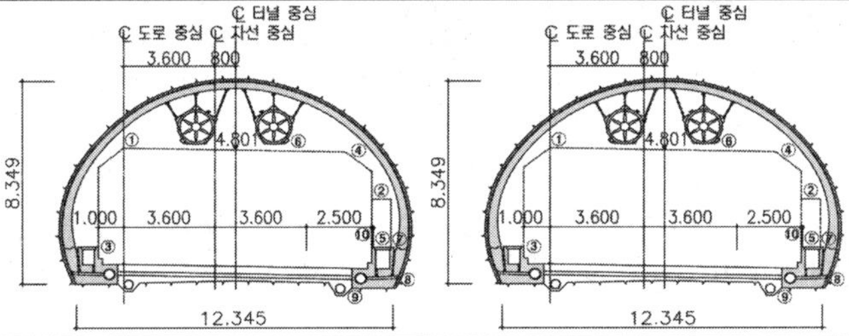	· 시공경험이 많아 시공성 우수 · 주변지반 이완 감소 · 편입용지 및 공사비 증가
4차로 대단면	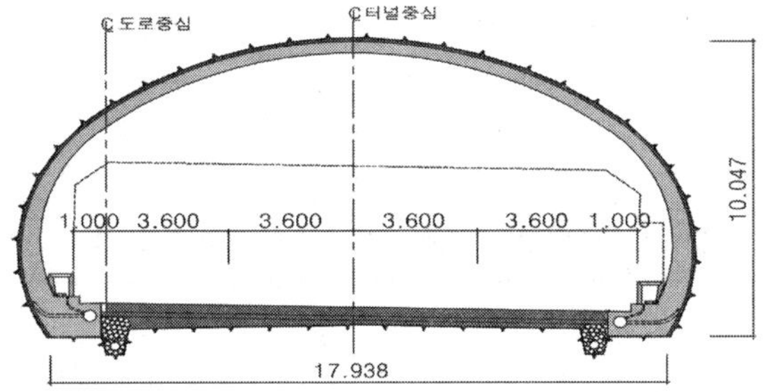	· 시공 공정 단순 · 공기단축 가능 · 편입용지 면적 감소 · 안정성 확보 불량 · 암질 불량 시 시공 곤란
2-ARCH	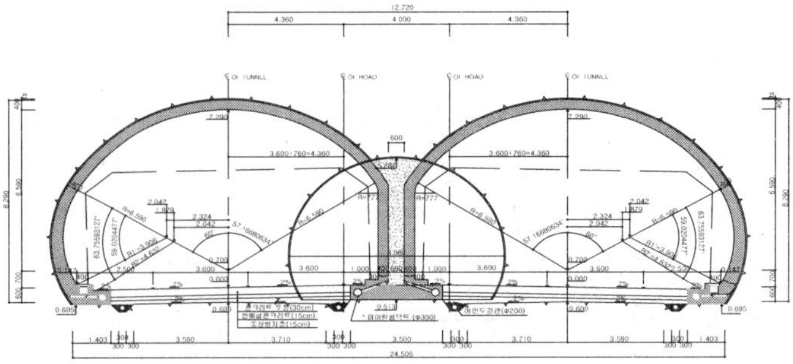	· 2차로 병설 터널에 비해 편입용지 면적 감소 · 시공 공정이 복잡 · Arch 중앙부 배수처리 불량

⑥ Utility pipe conduit

- Functionality
  - Secure installation space for fiber-optical / regular communication / power cable
  - Secure installation space for lighting, ventilation, traffic control cables.
  - Secure additional installation and maintenance space for various cables.

- Size

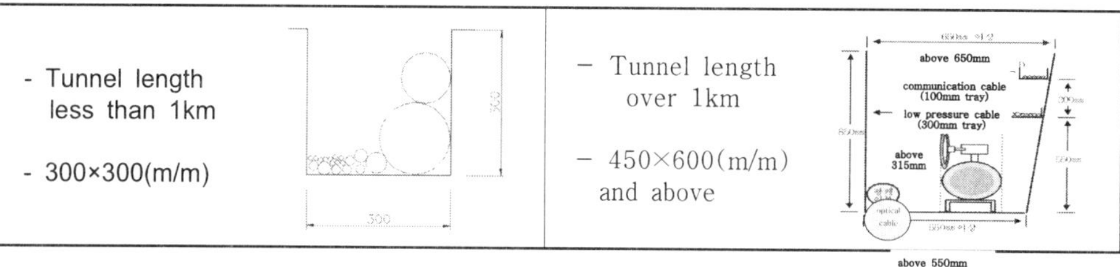

- Tunnel length less than 1km - 300×300(m/m)	- Tunnel length over 1km - 450×600(m/m) and above

1-2-3 Type of Tunnel Cross-section

Classification	Section overview	Characteristics
Two lane annex	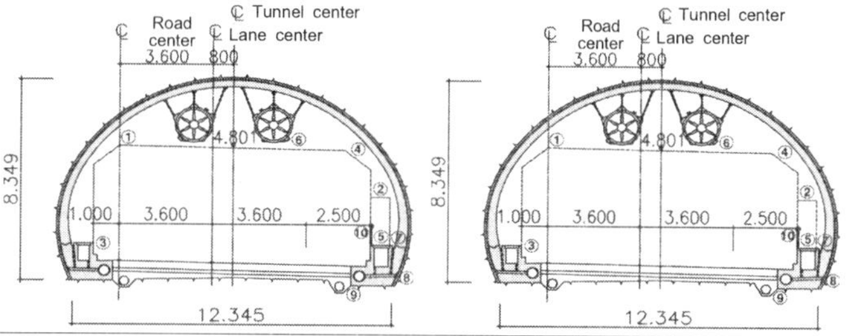	· Relatively easy to construct with plenty of previous construction experience · Minimum relaxation of nearby ground · Higher cost for construction and land purchasing
Four lane large section	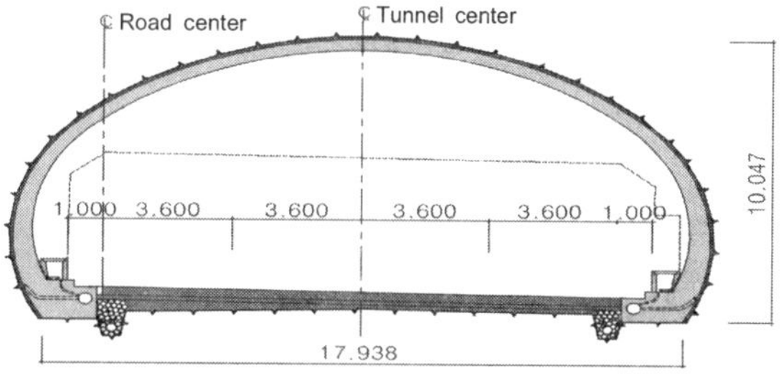	· Simple construction · Faster construction · Less land purchasing required · Low stability · Difficult to construct with low rock quality
2-ARCH	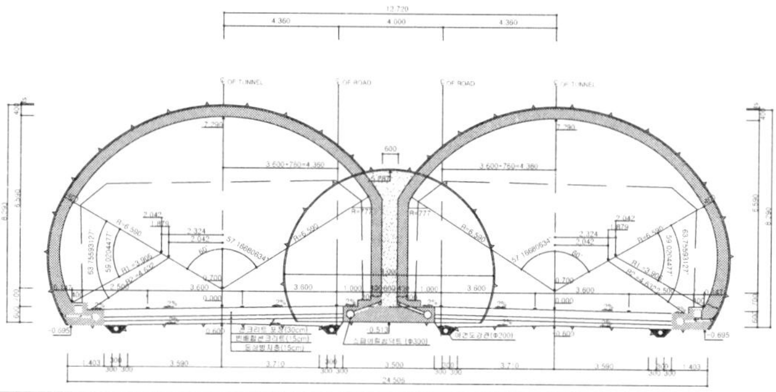	· Less land purchasing required compared to two lane annex · Complicated construction process · Arch center drainage is difficult

1-3 터널 환기 계획

1-3-1 환기 검토 흐름

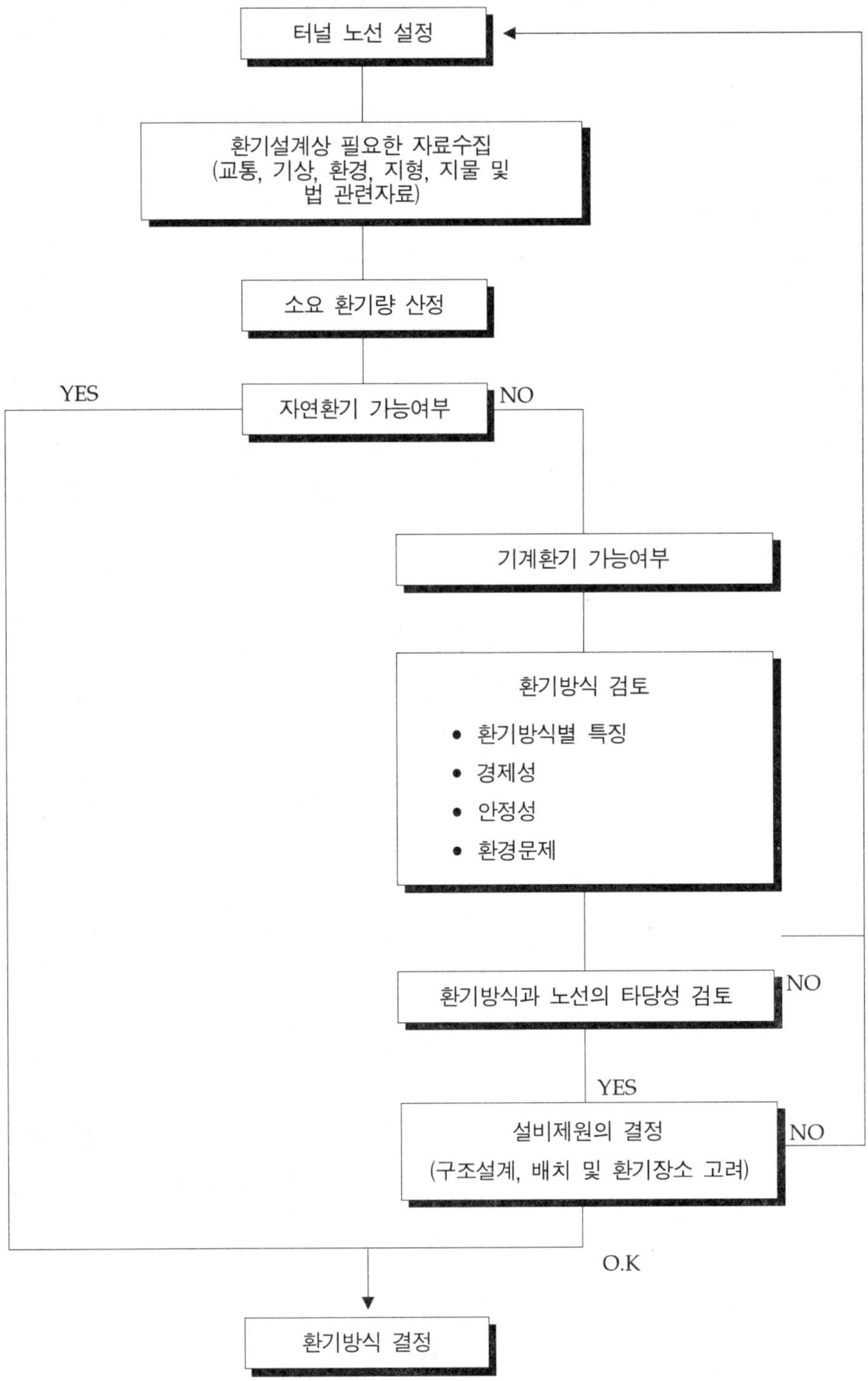

터널 노선 설정

환기설계상 필요한 자료수집
(교통, 기상, 환경, 지형, 지물 및
법 관련자료)

소요 환기량 산정

자연환기 가능여부

YES

NO

기계환기 가능여부

환기방식 검토
- 환기방식별 특징
- 경제성
- 안정성
- 환경문제

환기방식과 노선의 타당성 검토

NO

YES

설비제원의 결정
(구조설계, 배치 및 환기장소 고려)

NO

O.K

환기방식 결정

1-3 Tunnel Ventilation Planning

1-3-1 Flowchart for Ventilation Review

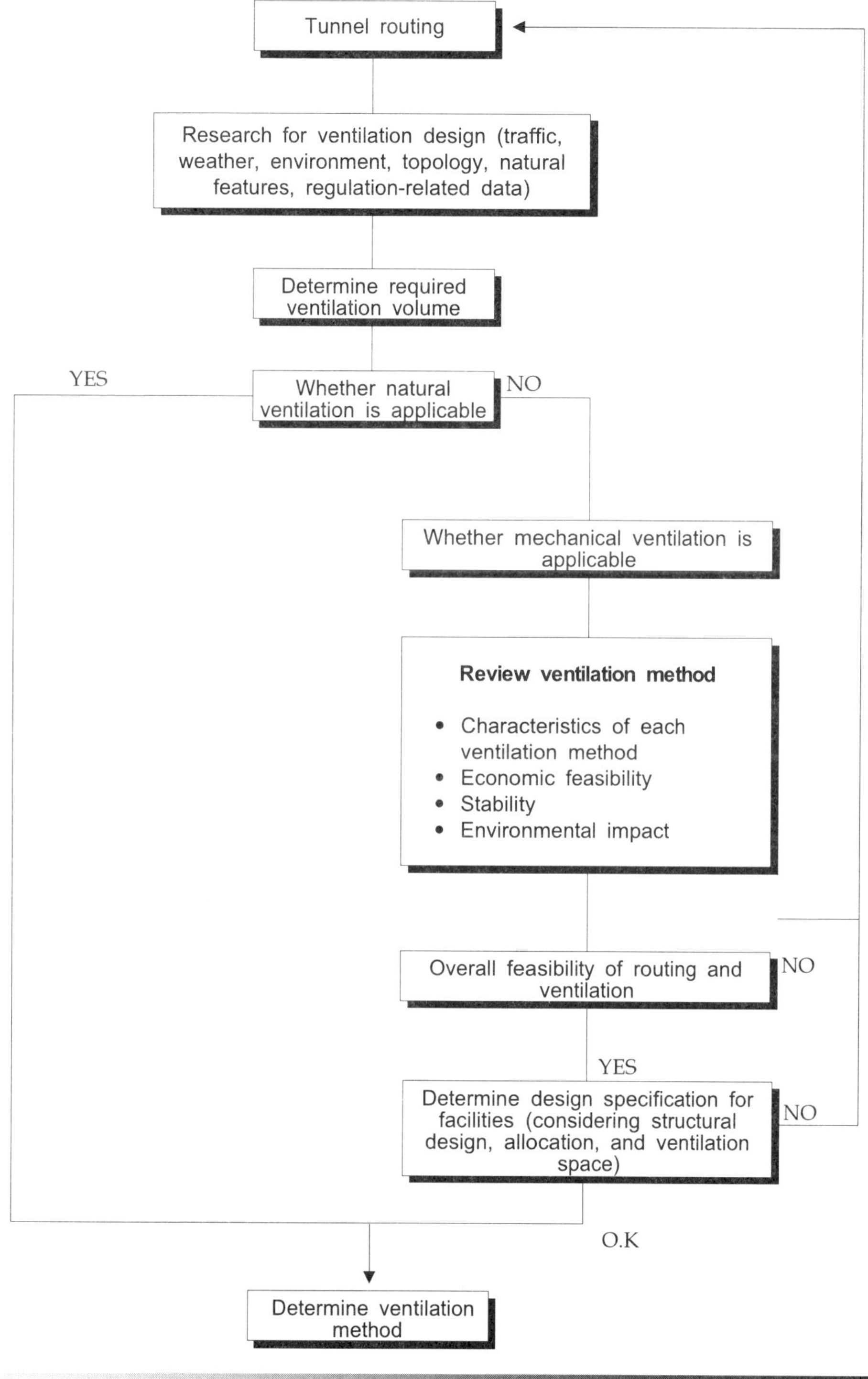

1-3-2 검토대상

① 환기 대상 물질

- 매연 : 터널 환기의 주된 대상물로서 운전자의 시야를 방해하는 요인
- CO : 인체에 생리적 악영향을 주며, 차량속도가 저속일 때 발생량 높음
- NOx : 인체에 악영향을 주는 요인이며, 차량속도가 고속일 때 발생량 높음

② 발생원인 및 영향

오염물질	발 생 원 인	직접적인 영향	간접적인 영향
매 연	일반적인 엔진배기와 불완전 연소	운전자의 시야감소	온실효과
CO	전 연소 및 각종첨가물의 연소	호흡기 및 순환기 질환유발	오존층 파괴 터널 주위의 생태계 파괴
NOx	고온 고압에 의한 산화가 주발생원인	폐기능 저하	스모그 현상 등 복합 오염유발

1-3-3 환기방식 검토

① 환기방식 종류

- 도로터널의 환기설비는 터널의 길이, 입지조건, 교통조건, 환경조건 등에 따라 환기 방식이 결정

- 유효하고 경제적 방법이 되도록 충분한 조사와 검토 후에 결정

- 환기방식은 크게 자연환기방식과 기계환기방식으로 대별

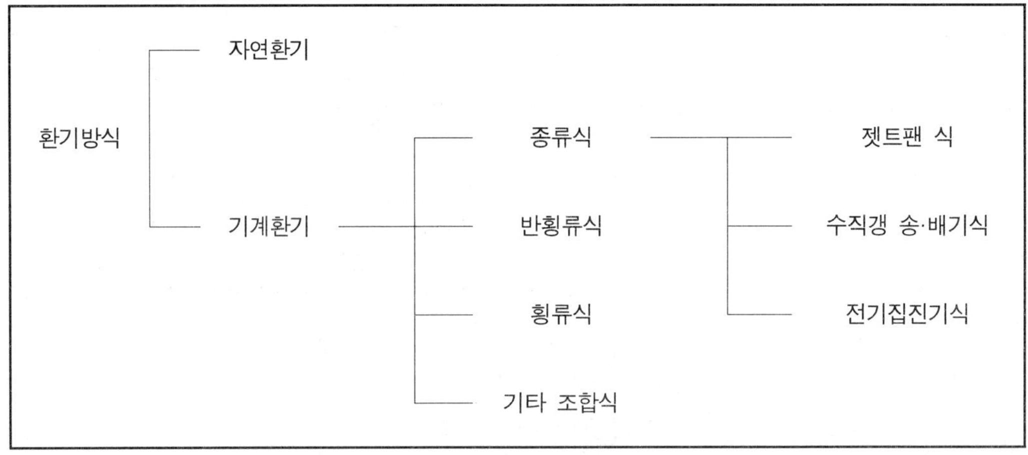

1-3-2 Subject for Review

① Subject required to be ventilated

- Smoke : main subject for tunnel ventilation as smoke disturb driver's field of vision
- CO : Bad effect to human body. Low speed yields more production
- NOx : Bad effect to human body. High speed yields more production

② Cause and effect of pollutant

Pollutant	Causes	Direct effect	Indirect effect
Smoke	Engine exhaust and imperfect combustion	Visual disturbance	Greenhouse effect
CO	Perfect combustion and combustion of various additives	respiratory and circulatory disease	Ozone layer destruction Ecosystem near the tunnel can be destroyed
NOx	Oxidation from high temperature and high pressure	lung deterioration	Complicated pollution such as smog

1-3-3 Ventilation Type Review

① Ventilation type

- Ventilation type for the tunnel is determined according to the length of tunnel, location, traffic, environmental consideration

- Determine with sufficient research and review to select effective and economic ventilation

- By and large, ventilation type can be divided into natural ventilation and mechanical ventilation

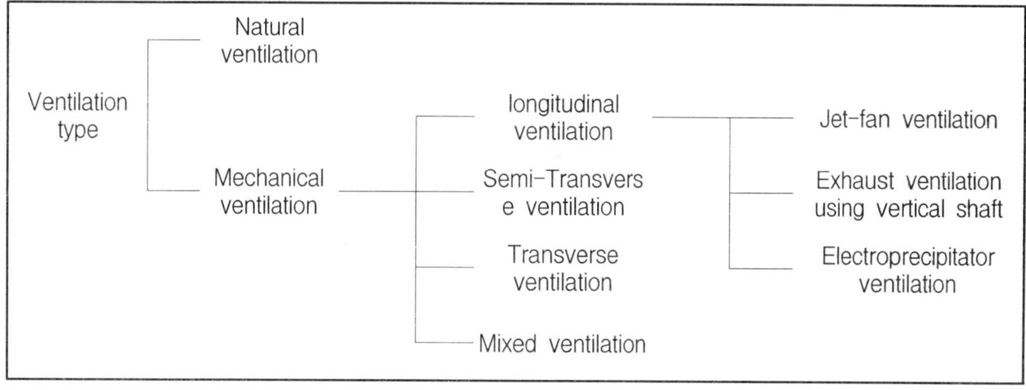

【터널 기계환기 방식의 종류】

제트팬 식	전기집진기식
	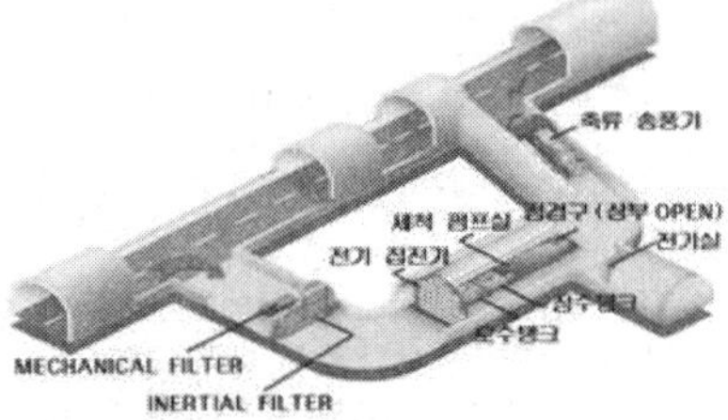 축류 송풍기 세척 펌프실 점검구(상부 OPEN) 전기실 전기 집진기 MECHANICAL FILTER INERTIAL FILTER
수직갱식	**집중배기식**
반횡류식	
	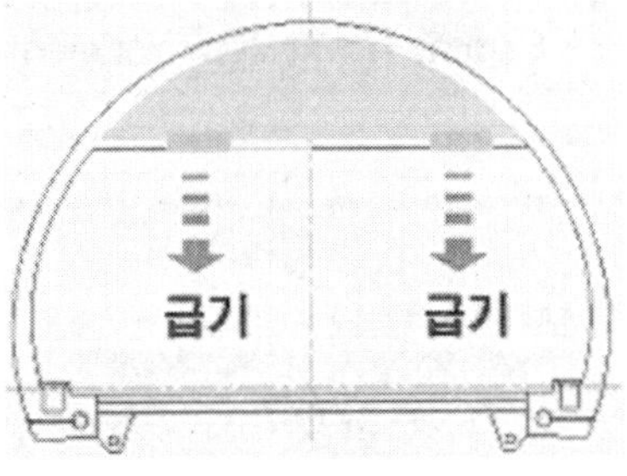 급기 급기
횡류식	
	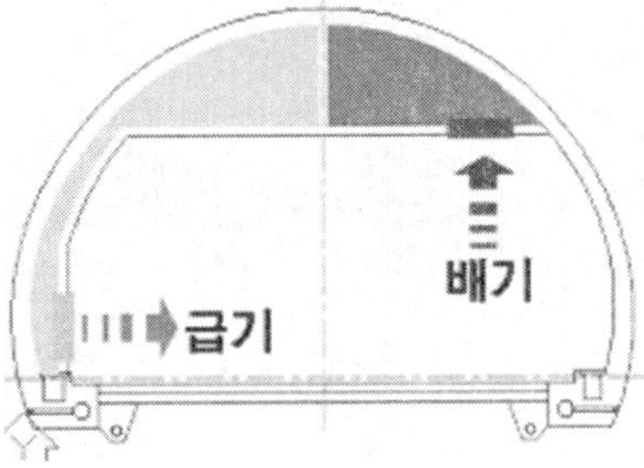 배기 급기

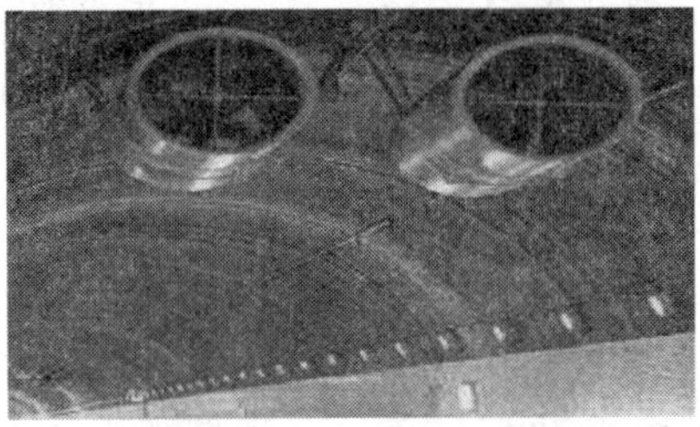

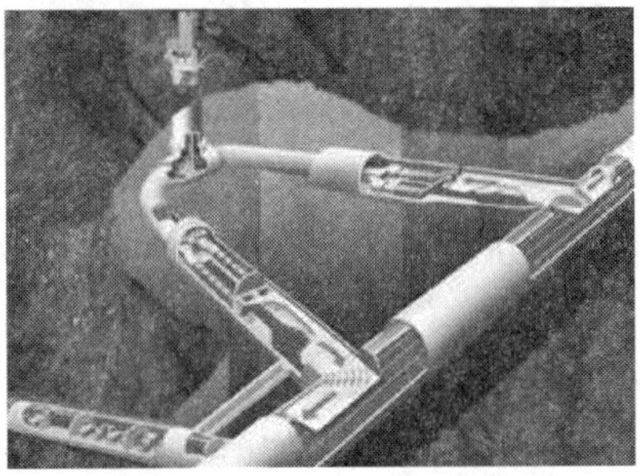

【Type of tunnel mechanical ventilation】

Jet-fan	Electroprecipitator 
	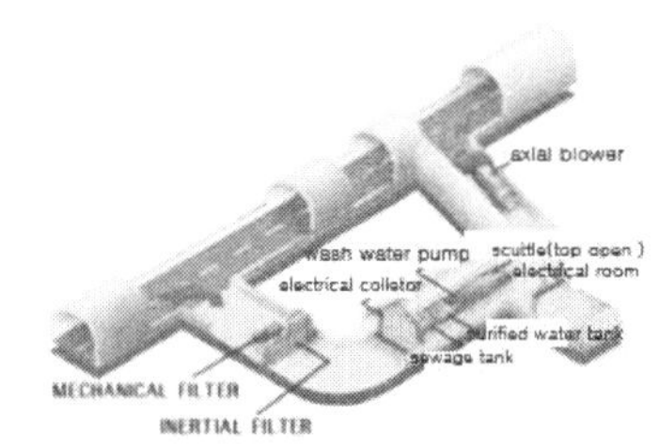 
Vertical shaft	Central exhaust
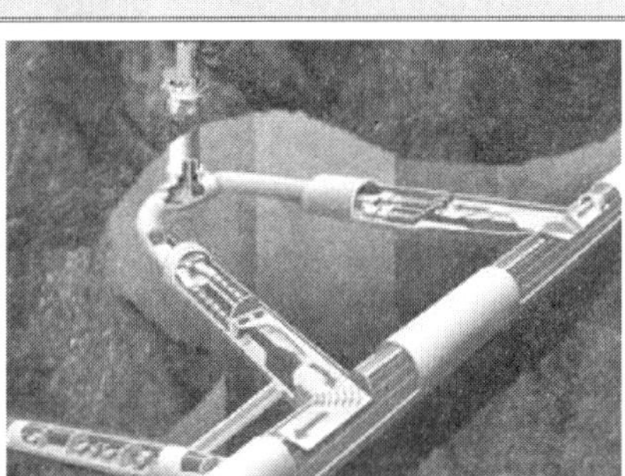	
Semi-Transverse	
	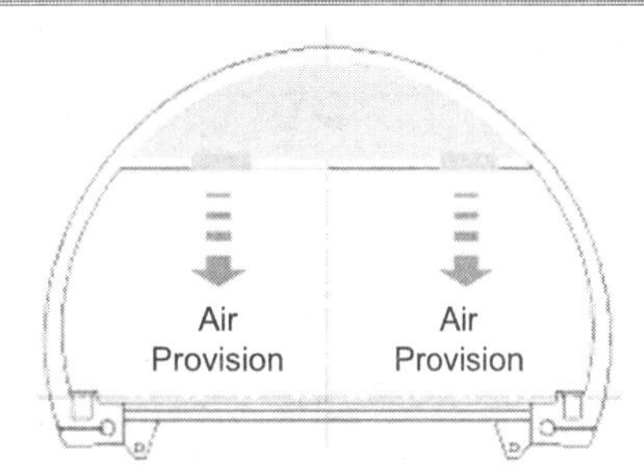
Transverse	
	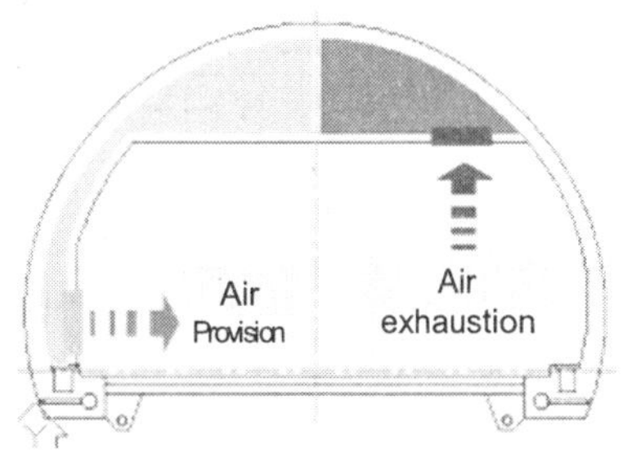

② 기계환기방식의 일반적인 적용 연장(일방향 교통기준)

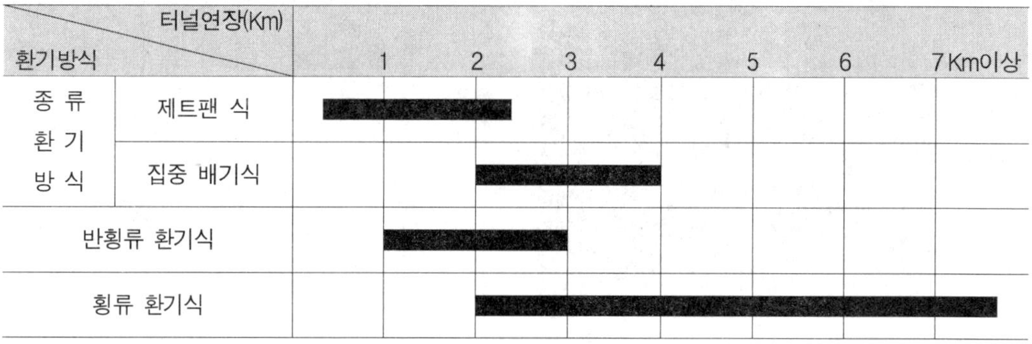

환기방식 \ 터널연장(Km)			1	2	3	4	5	6	7Km이상
종류 환기 방식	제트팬 식	■	■	■					
	집중 배기식			■	■				
반횡류 환기식			■	■					
횡류 환기식				■	■	■	■	■	■

③ 환기방식의 종류별 특징

방 식		개 념 도	특 징	적용상 제한
종류식	JET FAN 식	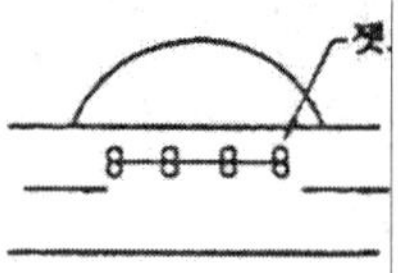	1. 설비비가 비교적 적게 든다. 2. 증설이 용이 3. 교통량의 대응에 비교적 수월 4. 터널단면을 비교적 적게 할 수 있다.	1. 차도 내 풍속은 안전 측면에서 최대 8m/s 이하로 유지 2. 설치개소 길이는 만족시켜야 한다.
	수직갱송·배기식	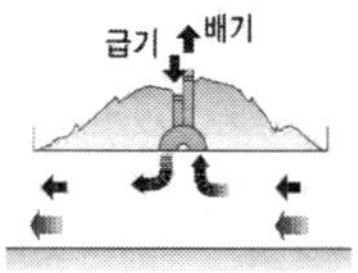 급기 배기	1. 갱구 부근의 환경은 양호 2. 상·하교통량의 변동에 의해 터널의 좌우구간에 환기 상태의 불균형이 발생 3. 덕트단면이 필요 없어 터널단면을 비교적 작게 할 수 있다.	1. 차도 내 풍속은 안전측면에서 최대 8m/s 이하로 유지 2. 위치선정에 신중 필요
반횡류식	송기 반횡류식	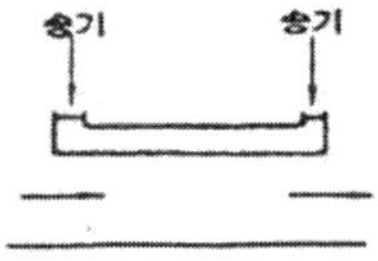 송기 송기	1. 자연풍의 영향이 적다 2. 덕트단면을 필요로 함 3. 차도 내 농도분포가 균일 4. 환기실 별도 설치	1. 차도 내 풍속은 안전측면에서 최대 8m/s 이하로 유지
	배기 반횡류식	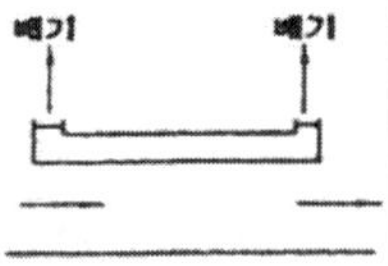 배기 배기	1. 차도 내 농도분포가 부분적으로 높아진다. 2. 덕트단면을 필요로 함 3. 갱구 부근의 환경이 양호 4. 환기실 별도 설치	1. 환기효율이 문제성과 적용 예가 없음
횡류식		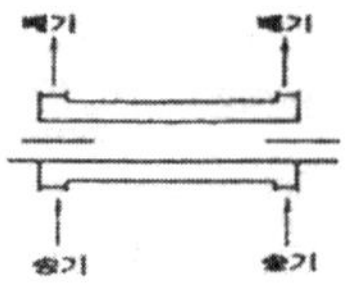 배기 배기 송기 송기	1. 2계통의 덕트면적이 필요하므로 시설비 및 유지비 고가 2. 화재 시 배연 효율이 좋다. 3. 환기실 별도 설치 4. 차도 내 농도분포가 일정	1. 제한이 없음 2. 경제성 검토 요망

② Applicable length of mechanical ventilation (based on unidirectional traffic)

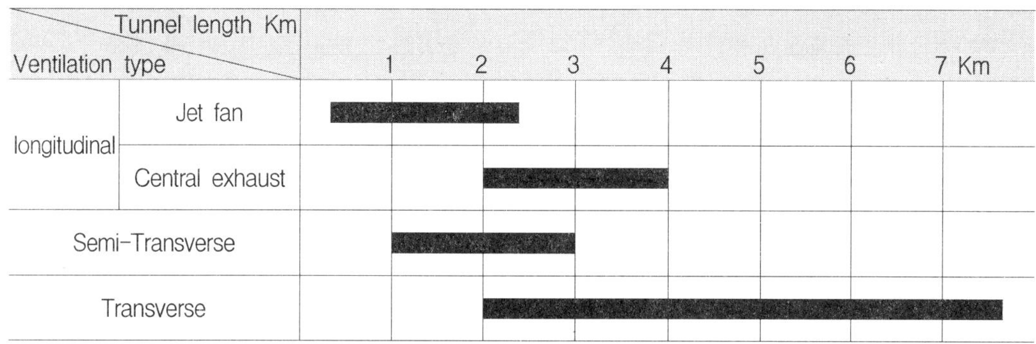

③ Characteristics by ventilation type

Type		Concept Drawing	Characteristics	Limitation
longitudinal	Jet fan		1. Low cost for equipment 2. Easy to extend 3. Easy to react to traffic volume 4. Smaller tunnel section	1. Wind speed within the lane should be under 8m/s to secure safety 2. Must meet installation interval
	Vertical shaft	Boost Emission	1. good for tunnel entrance 2. Unevenness of both ends of tunnel due to traffic volume change 3. Smaller tunnel section as no duct section is required.	1. Wind speed within the lane should be under 8m/s to secure safety 2. Carefully select location
Semi-Transverse	Air provision	Air-supply Air-supply	1. Low influence of natural wind 2. Duct section is required 3. Even density distribution within the lanes 4. Separate installation of ventilation room is required.	1. Wind speed within the lane should be under 8m/s to secure safety
	Air exhaustion	Emission Emission	1. Partially higher distribute within the lanes. 2. Duct section is required 3. good for tunnel entrance 4. Separate installation of ventilation room is required.	1. Issue on ventilation efficiency. no precious cases exist
Transverse		Emission Emission Air-supply Air-supply	1. High cost for facility and maintenance as two types of duct sections are required 2. Effective exhausting when there is fire 3. Separate installation of ventilation room is required. 4. Even density distribution within the lanes	1. No limitation 2. Consider economic feasibility

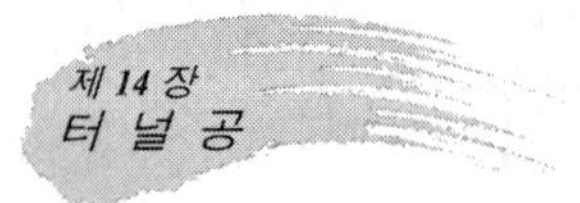

1-4 방재계획

- 방재설비는 방재와 관련된 각 제반규정을 충분히 검토하여 규정에 만족하도록 계획
- 방재설비 시설을 위한 공동구 및 구조물 내 매입설치방안 등 구조물에 영향을 미치는 부분에 대하여 터널 구조물 계획과 연계하여 계획
- 계획터널의 특성을 고려하여 방재설비의 운영을 위한 방안을 수립
- 기본적으로 터널연장을 기준으로 방재등급을 정하되, 교통량 등을 고려한 위험도 평가를 수행하여 위험도가 소정의 수준을 초과하는 경우에 방재등급을 상향 조정

【터널연장기준 방재등급 구분】

등 급	터널연장(L) 기준등급	위험도지수(X) 기준등급
1	3,000m 이상 (L ≧ 3,000m)	X > 29
2	1,000m 이상, 3,000m 미만 (1,000 ≦ L < 3,000m)	19 < X ≦ 29
3	500m 이상, 1,000m 미만 (500 ≦ L < 1,000m)	14 < X ≦ 19
4	연장 500m 미만 (L < 500)	X ≦ 14

【터널 위험도 평가기준】

세부평가항목			범 위	위험도지수
사고 확률	주행거리계 (교통량×연장) (Veh · km/tube · day)		8,000 미만	1.5
			8,000 이상~16,000 미만	2.5
			16,000 이상~32,000 미만	5.0
			32,000 이상~64,000 미만	7.5
			64,000 이상	10.0
터널 특성	표고차 및 경사도	입출구 표고차(m)	10 미만	0.5
			10 이상~20 미만	1.0
			20 이상~30 미만	1.5
			30 이상	2.0
		진입부 경사도(%)	3.0 미만	0.5
			3.0 이상	1.0
	터널높이(m)		7.5 이상	1.0
			5.0 이상~7.5 미만	2.0
			5.0 미만	3.0
	터널곡선반지름(m)		1,800m 이상	0.5
			1,800m 미만	1.0
대형 차량	위험물 수송 관련	대형차 혼입률 (%)	10 미만	0.5
			10 이상~17.5 미만	1.0
			17.5 이상~25 미만	1.5
			25 이상	2.0
		대형차 주행거리계 (대·km/tube·day)	500 미만	0.5
			500 이상~1,000 미만	1
			1,000 이상~2,500 미만	2
			2,500 이상~5,000 미만	4
			5,000 이상	6
		감시시스템	있음	0
			없음	1
		유도시스템	있음	0
			없음	1

1-4 Disaster Prevention Planning

- Plan disaster prevention facility to satisfy requirements of various relevant regulations
- Plan along with tunnel facility planning such as disaster prevention facilities in the utility pipe conduit or within other structure
- Prepare operation planning for disaster facility considering characteristics of planned tunnel
- Disaster prevention rate should be determined by tunnel length in principle. However, raise the rate when the risk of disaster excesses certain standard.

【Disaster Prevention Rate by Tunnel Length】

Rate	Tunnel length (L)	Risk index (X) Rate
1	Over 3,000m (L ≧3,000m)	X > 29
2	1,000m ~ 3,000m (1,000 ≦L < 3,000m)	19 < X ≦ 29
3	500m ~ 1,000m (500 ≦L < 1,000m)	14 < X ≦ 19
4	Below 500m (L < 500)	X ≦ 14

【Tunnel disaster risk evaluation standard】

Detailed evaluation items			Range	Risk index
Probability of accident	Sum of driving distance (Traffic volume ×length) (Veh · m/tube · ay)		Less than 8,000	1.5
			8,000 ~16,000	2.5
			16,000 ~32,000	5.0
			32,000 ~64,000	7.5
			Higher than 64,000	10.0
Tunnel characteristics	Altitude difference and slope	Altitude difference between entranceand exit (m)	Less than 10	0.5
			10 ~ 20	1.0
			20 ~ 30	1.5
			Higher than 30	2.0
		Slope of entrance (%)	Less than 3.0	0.5
			Higher than 3.0	1.0
	Tunnel height (m)		Higher than 7.5	1.0
			5.0 ~ 7.5	2.0
			Less than 5.0	3.0
	Radius of curvature of tunnel (m)		Higher than 1,800m	0.5
			Less than 1,800m	1.0
Heavy vehicles	Dangerous material transportation related	Ratio of heavy vehicles (%)	Less than 10	0.5
			10 ~ 17.5	1.0
			17.5 ~ 25	1.5
			Higher than 25	2.0
		Heavy vehicles total driving distance (number ·km/tube·day)	Less than 500	0.5
			500 ~ 1,000	1
			1,000 ~ 2,500	2
			2,500 ~ 5,000	4
			Higher than 5,000	6
		Monitoring system	Yes	0
			No	1
		Guidance system	Yes	0
			No	1

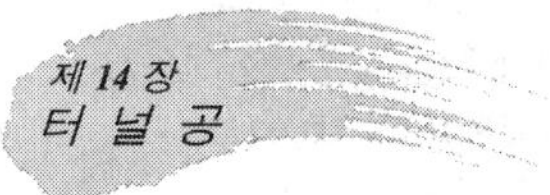

【고속도로 터널 방재시설 설치기준】

방재시설 \ 터널연장(m)			1등급 (3000 m이상)	2등급 (1000 m이상)	3등급 (500 m이상)	4등급 (500 m미만)	비 고
소화설비	소화기구		●	●	●	●	
	옥내소화전설비		●	●			
	물 분무설비		△				△ : 터널연장이 3,000m 이상으로 연평균일교통량과 터널연장의 곱이 90×103 대・km/day・tube 이상인 터널에 대해서 권장설비로 함
경보설비	비상경보설비		●	●	●		
	자동화재탐지설비		●	●			
	비상방송설비		●	●	●		
	비상전화		●	●	●		
	CCTV		●	●			
	라디오재방송설비		●	●	●	△	△ : 터널연장이 200 m 이상인 4등급 터널은 권장설비로 함.
	정보표시판		●	●			
피난설비	비상조명등		●	●	●	●(1)	(1) : 4등급 터널은 터널연장이 200m 이상인 경우에 설치함.
	유도표지판		●	●	●		
	피난대피시설	피난연락갱	●	●	●		
		비상주차대	●	●			
소화활동설비	제연설비		●	●			
	무선통신보조설비		●	●	●	△	△ : 4등급 터널의 경우, 라디오재방송설비가 설치되는 경우에 병용하여 설치함.
	연결송수관설비		●	●			
	비상콘센트설비		●	●	●		
비상전원설비	무정전전원설비		●	●	●	●(1)	(1) : 4등급 터널은 방재시설이 설치되는 경우에 시설별로 설치함.
	비상발전설비		●	●			

<범례> ● 기본설비 : 해당터널에 반듯이 설치하여야 하는 설비

△ 권장설비 : 설치의 필요성을 검토하여 필요시 설치하는 설비

【Highway tunnel disaster prevention facility installment standard】

Disaster prevention Facility \ Tunnel length(m)		First (3000m+)	Second (1000m+)	Third (500m+)	Fourth (500m-)	Remarks
Fire extinguishing facility	Fire extinguisher	●	●	●	●	
	Hydrant facility	●	●			
	Water sprinkler	△				△: Recommended facility for tunnel with length of 3,000m and more with yearly average daily traffic volume times tunnel length is more than 90x103 vehicles · km/day · tube
Alarm facility	Emergency alarm	●	●	●		
	Automatic fire detector	●	●			
	Emergency broadcasting	●	●	●		
	Emergency telephone	●	●	●		
	CCTV	●	●			
	Radio re-broadcasting facility	●	●	●	△	△: Recommended for fourth rate tunnel with length of more than 200m
	Information display	●	●			
Evacuation facility	Emergency light	●	●	●	●(1)	(1): for fourth rate tunnel, only for the tunnels with length of more than 200m
	Evacuation sign	●	●	●		
	Evacuation shelter – Evacuation communication shelter	●	●	●		
	Evacuation shelter – Emergency parking	●	●			
Fire extinguish activity facility	Smoke control	●	●			
	Wireless communication supplement facility	●	●	●	△	△: for fourth rate tunnel with radio re-broadcasting facility, use the facility jointly
	water pipe connection	●	●			
	Emergency consent	●	●	●		
Emergency power	UPS	●	●	●	●(1)	(1): for fourth rate tunnels, install UPS for each disaster prevention facility
	Emergency generator	●	●			

<Legend> ● Basic facility: Facility should be installed to the corresponding tunnels
△ Recommended facility: Facility can be installed if required

【방재시설별 설치위치 및 간격】

방 재 시 설			설치위치와 설치방법	설치간격
소화설비	소 화 기 구		주행차로 측벽에 격납상자를 설치하고 그 내부에 2개 1조로 하여 격납(대면통행 터널은 양 측벽)	50 m 이내
	옥내소화전 설비		주행차로 측벽설치	50 m 이내
	물분무설비		측벽설치(도로면 전체에 균일하게 방수되도록 한다)	물분무 헤드 간격 : 4~5m 방수구역 : 50 m 이내
경보설비	비상경보설비		소화기 또는 소화전함에 병설	
	화재탐지기		화재탐지기 최적성능을 확보할 수 있는 위치	환기방식별 필요인식 범위
	비상방송설비		측벽설치(피난연락갱, 피난대피소, 피난갱, 비상주차대 설치)	50 m 이내
	비상전화		피난대피시설(피난연락갱, 피난대피소, 피난갱, 비상주차대 설치)에 설치	250 m 이내
	CCTV		측벽설치(피난대피시설 및 터널 전 구간 감시가 가능하도록 설치함.)	200~400m 간격
	라디오재방송설비		터널 전 구간에서 청취 가능하도록 설치	
	정보표시판		터널입구 정보표지판 : 터널 전방 500m 이내	
			차로이용규제신호등	400~500 m 간격
피난설비	비상조명등		야간 점등회로를 이용하여 설치	
	유도표지판	A	피난대피시설 부근	
		B	피난대피시설이 설치된 측벽설치	피난대피시설간에 최소 4개소 이상
	피난대피시설	피난연락갱	병렬터널(차단문 설치)	250~300 m 이내
		피난갱	본선터널과 평행하게 설치하는 것을 원칙으로 함.	
		피난대피소	본선터널의 측벽이나 하부에 안전공간을 확보하여 설치	250~300 m 이내
		비상주차대	주행차로 길어깨, 대면통행 터널은 양 측벽	
소화활동설비	제연설비		환기설비와 병용	
	무선통신보조설비		라디오 재방송 설비와 병용	
	연결송수관설비		송수구 : 터널 입·출구부 방수구 : 옥내소화전설비와 병설	50 m 이내
	비상콘센트설비		소화전함에 병설	50 m 이내
비상전원설비	무정전 전원설비		시설별 설치	시설별
	비상발전설비		별도로 구획된 실내에 설치	

【Installing location and interval】

Disaster prevention facility			Installing location and method	Interval
Fire extinguishing facility	Fire extinguisher		Install housing box on the side wall and hang two extinguisher inside (Both wall for bidirectional tunnel)	Within 50m
	Hydrant facility		Install on the side wall	Within 50m
	Water sprinkler		Install on the side wall (water should be discharged uniformly into entire road surface)	Head interval : 4~5m Water discharging area : within 50m
Alarm facility	Emergency alarm		Install jointly with extinguisher or hydrant facility	
	Automatic fire detector		Optimum place to detect fire	Per recognizing perimeter
	Emergency broadcasting		Install on the side wall (Install in the evacuation shelters and emergency parking area)	Within 50m
	Emergency telephone		Install in the evacuation facility	Within 250m
	CCTV		Install on the side wall (All area including evacuation shelters should be monitored)	200~400m
	Radio re-broadcasting facility		Install to be heard from entire area inside tunnel	
	Information display		Tunnel entrance information display: within 500m from tunnel opening	
			Lane restriction light sign	400~500m
Evacuation facility	Emergency light		Install with automatic night lighting circuit	
	evacuation sign	A	Near evacuation facilities	
		B	Install on the side wall where evacuation facility is installed	Minimum four signs between evacuation facilities
	evacuation shelter	evacuation communication shelter	Twin tunnel (install isolation door)	250~300m
		evacuation shelter	install parallel to main lane tunnel	
		evacuation area	Secure safe area on the side wall or lower part of main lane tunnel	250~300m
		Emergency parking	road shoulder (both way for bidirectional tunnel)	
Fire extinguish activity facility	Smoke control		Install jointly with ventilation facility	
	Wireless communication supplement facility		Install jointly with radio re-broadcasting facility	
	water pipe connection		Water supply: Tunnel entrance Flood protection: Install jointly with hydrant facility	Within 50m
	Emergency consent		Install jointly with hydrant box	Within 50m
Emergency power	UPS		Install to each facility	per every facility
	Emergency generator		Install in the separate area	

1-5 터널 굴착 방법 계획

구 분	발파굴착 (Drill and Blast, NATM)	기계굴착 (TBM, Open)
단 면 및 시공도		
개 요	• 천공 및 발파에 의한 굴착	• 드럼에 나선형으로 부착된 원형 Pick 이 암반 절단
단면형상	• 자유조절 가능	• 원형
적용지반	• 충적 연약 지반에서 극경암 지반까지 거의 모든 지반에 적용 가능	• 연 · 경암에 효율적 • TBM 기계굴착으로 주변 암반 자체를 지보재로 활용
지보방법	• 숏크리트, 록볼트, 강지보	• 숏크리트, 록볼트, 강지보
장 점	• 굴착 중 지질변화에 대응 용이 • 대단면 굴착, 연약지반 시공가능 • 경사, 곡선반지름의 제약 없음 • 다양한 단면형상 굴착 가능 • 시공실적, 경험풍부 • 장비 및 초기투자비 저렴 • 동시다발적인 시공가능하며 굴착단면 최소화	• 발파공법에 비하여 지반의 이완 적음 • 작업원이 적고 시공상 안전 • 부분적인 경암지반이 분포한 경우는 발파공법을 병용 시공이 가능 • 여굴이 거의 없어 여굴률 3% 내외임 – 단면유지 정밀 • 굴진속도가 빠름
단 점	• 발파에 의한 소음·진동·비산대책 수립 및 안전시설 필요 • 발파에 의한 지반이완영역 증가 • 여굴 발생에 따른 공사비 증가 • 발파에 의한 분진, 가스 등 갱내 작업환경 불량 • 많은 인력의 필요에 따른 안전사고 유발 가능성	• 파쇄구간 및 암질 불량구간에서 불리함 • 분진의 발생량이 많아 환기 및 살수설비가 필요 • 극경암에서는 굴진속도 저하 • 용수가 있는 지반에서는 버력이 이토화되어 작업능률이 저하 • 전문 기술력 및 숙련된 기능공 소요

1-5 Tunnel Excavation Method Planning

Classification	Blasting excavation (Drill and Blast, NATM)	Mechine excavation (TBM, Open)
Working Drawing	 McDowell	
Overview	• Excavation through punching and blasting	• circular pick connected spirally into the drum cuts the rocks
Shape of section	• Freely adjustable	• Circular shape
Applicable ground	• Almost all type of ground	• effective for weak / strong rock • TBM machine excavation uses adjacent rock as support
Supporting method	• Shotcrete, rock bolt, steel support	• Shotcrete, rock bolt, steel support
Advantages	• Easy to cope with geological changes • large scale section excavation and weak ground construction are possible • No limitation in curvature and slope • Various shape of section is possible • Lots of previous experience • Low initial cost for equipment • Minimized excavation section • spontaneous construction is possible	• Lower relaxation of ground compared to blasting excavation • Few operator is required and safe to execute • For partial strong rock area, blasting excavation can be used jointly. • Almost none extra depth of dredging (about 3%) - Maintains precise section • Fast tunneling speed
Disadvantages	• Countermeasure for noise, vibration, scattering and safety measure required • Increased relaxation of ground • Extra depth of dredging increases cost • Working condition is poor due to gas and scattering from excavation • High risk of accident due to many operators	• Unfavorable in the crushed ground and poor rock conditioned area • Ventilation and water facility required due to heavy dust • Tunneling speed gets slower for super hard rock area • Lower efficiency for water containing ground • Technology and experienced experts are required

2. 지반조사 및 암반분류

2-1 터널설계를 위한 지반조사

2-1-1 단계별 조사 계획

예비조사	광역조사	상세조사	추가조사	시공중보완조사
• 지질도 분석 • 지형도 분석 • 원설계 분석 • 현장답사	• 인공위성 영상분석 (고해상도) • 물리탐사 • 지표지질조사	• 시추조사 • 정밀지표 지질조사 • 현장시험 • 실내시험	• 지질구조 확인 • 단층대 상세조사 –경사시추 –전기비저항탐사 • 지하수관련조사	• TSP 탐사 • 선진시추 • 감지공 천공 • 지하수위 관측공 설치

2-1-2 지반조사 항목 및 기준

조사항목	조사빈도	심 도	비 고
시추조사	• 터널 입출구부 각각 2개소 및 300m 마다 1개소	• 계획고하 2m까지	• 토사층 또는 연약대 등이 존재시 풍화암 또는 필요 심도까지
탄성파탐사	• 터널 개소당 입출구부 필요한 연장		• Cross Hole Test
전기비저항 및 전자탐사	• 터널 전 연장		• 터널계획심도 250m 이하구간 – 전기비저항탐사 • 터널계획심도 250m 이상구간 – 전자탐사
암반수압	• 터널 내 용출수 산정에 필요한 시추 위치		
화상정보	• 터널 개소당 입출구부 각각 2개소	• 암반구간	• 땅깎기부 기준 준용

2-2 터널구간의 암반분류

2-2-1 암반분류의 목적

① 암반분류는 터널의 설계 및 시공에 영향을 주는 지반의 여러 성질을 등급에 따라 구분

② 계획단계에서부터 조사, 설계, 시공에 이르는 전 과정에서 일관성 있게 적용될 수 있는 객관적인 지표로 사용하기 위해 실시

2-2-2 암반 분류 시 고려사항

① 암반 분류의 지표는 객관적인 평가가 가능하고, 터널 설계 및 시공과 관련된 암반특성을 포함하여 터널의 표준적인 굴착 및 지보패턴과 대비될 수 있는 요소들이어야 함

2. Ground Evaluation and Base Rock Classification

2-1 Ground Evaluation for Tunnel Designing

2-1-1 Evaluation Planning Stages

Preliminary research	Wide area research	Detailed research	Additional research	Supplement research during construction
• Geological map analysis • Topographical map analysis • Original design analysis • Site survey	• Satellite image analysis (high resolution) • Geophysical survey • Surface geological survey	• Boring research • Precise surface geological survey • Site test • Lab test	• Geological structure review • Fault detailed research - Directional drilling - Electrical resistivity survey • Underground water survey	• TSP survey • Probe drilling • Feeler hole drilling • Observation hole installation for underground water level

2-1-2 Ground Research Item and Standard

Research item	Survey frequency	Depth	Remarks
Boring	• Two for each entrance and exit of each tunnel and one per 300m	• 2m from designed level	• If soil layer or weak base exists, extend to required depth
Seismic survey	• Required length for each entrance and exit of each tunnel		• Cross Hole Test
Elextric resistivity and elextirc survey	• Entire tunnel length		• Whrer planned tunnel depth is less than 250m - Electirc resistivity survey • Whrer planned tunnel depth is higher than 250m - Electirc survey
Rock mass water pressure	• Boring location for effluent water calculation		
Image information	• Two for each entrance and exit of each tunnel	• rock base area	• apply standard of ground cutted area

2-2 Rock Mass Classification

2-2-1 Purpose

① Various ground characteristics affecting tunnel design and construction are classified by grade

② The survey is carried out to set up an unbiased index that can be applied from the planning stage to entire process including survey, design, and construction

2-2-2 Considerations when Classifying Rock Mass

① Factors used in rock mass classification should be used as objective comparison index for standard excavation and supporting pattern

【각종 암반 분류방법의 기준인자】

분 류 방 법	제 안 자	분 류 기 준 인 자	적 용
RSR (Rock Structure Rating)	Wickham 등, 1972	1. 암반 특성 2. 터널굴착방향과 불연속면 방향 3. 지하수 유입영향	Tunnel
RMR (Rock Mass Rating System)	Bieniawski, 1973	1. 일축압축강도 2. R.Q.D 3. 절리의 간격 4. 절리의 상태 5. 지하수 6. 절리의 방향	Tunnels, Mines, Slopes(SMR), Foundations
Q-System	Barton 등, 1974	1. R.Q.D 2. 절리군 수 3. 절리면의 거친 상태 4. 절리면의 변질정도 5. 지하수 6. 응력감소 인자 등 다수	Tunnels, Chambers

【Standard factor for rock mass classification】

Classification method	Proposer	Classification standard factor	Application
RSR (Rock Structure Rating)	Wickham et al, 1972	1. Rock base characteristics 2. Drirection of discontinuity to tunnel excavation 3. Underground water inflow direction	Tunnel
RMR (Rock Mass Rating System)	Bieniawski, 1973	1. Uniaxial compressive strength 2. R.Q.D 3. Joint spacing 4. Joint condition 5. Underground water 6. Joint direction	Tunnels, Mines, Slopes(SMR), Foundations
Q-System	Barton et al, 1974	1. R.Q.D 2. Number of joint sets 3. Roughness of joint surface 4. Alteration degree of joint surface 5. Underground water 6. Stress reducing factor, etc.	Tunnels, Chambers

3. 터널 상세 설계

3-1 터널 지보재

- 터널 굴착에 따라 원지반의 3축상태 응력이 2축상태로 변하게 되어 응력집중이 발생

- 이 때, 암반을 3축응력 상태로 유지하는 목적으로 지보재를 사용

3-1-1 지보공의 종류

① 지보공은 터널 주변의 원지반이 보유하고 있는 지보 기능을 최대한 활용할 수 있도록 설계하여야 함.

② 지보공 설계에 있어서는 지형, 지질, 지반의 역학적 특성, 토피의 대소, 용수의 유무, 굴착단면의 크기, 지표침하의 제약, 시공법 등 제반 설계조건을 종합적으로 고려하여 합리적인 지보공을 설계하여야 함.

[지보 단계별 지보재의 종류 및 개요]

구 분	종 류	개 념
1 차 지 보	숏크리트 와이어 메쉬 록볼트 강지보	• 터널굴착 후 주변지반의 지보기능을 발휘할 수 있도록 지반거동을 억제하여 안정상태에 이르도록 하기 위하여 설치 • 터널굴착에 의해 발생되는 지반의 거동을 조기에 억제시켜 지반이 보유하고 있는 지지력 및 강도를 최대한 이용할 수 있도록 하고 지반을 안정시킴으로써 안전하고 능률적인 갱내 작업이 가능토록 함.
2 차 지 보	내부라이닝	• 배수형 터널의 경우 모든 지반 하중을 1차 지보가 담당하고, 내부라이닝은 터널 내부 시설물의 보호·미관유지·쾌적한 공간유지 등의 부수적 기능 담당 • 방수형 터널의 경우 지하수 배출을 차단함으로써 발생하는 수압을 지지
보 조 공 법	훠폴링 파이프루프 그라우팅	• 지반이 불량한 경우 막장의 안정을 위해 설치하는 것으로 개념적으로는 1차 지보를 보강하는 것이다.

3. Tunnel Detailed Design

3-1 Tunnel Support

- Tunnel excavation yields stress concentration since the construction changes stress pattern for original ground from three axis stress to two axis stress.

- Support is used for ground to maintain three axis stress pattern

3-1-1 Types of Tunnel Support

① Support should be designed to maintain supporting function of original ground

② Consider dynamic characteristics of topography and geography, thickness of surface soil, water content, size of excavation section, limitation in surface settlement, construction methods, etc

【Type of support】

Classification	Type	Concept
Firsthand support	Shotcrete Wire mash rock bolt Steel support	• Install to suppress ground movement so that nearby ground can act as support • Maximize original bearing power and rigidity of ground through early suppressing to secure construction safety
Secondhand support	Inner lining	• For drainage type tunnel, almost all load is supported by firsthand support. Inner lining act as supplement with facility protection, aesthetic and clean area maintenance. • For waterproof type tunnel, prevent underground water from discharge and support resultant water pressure
supplement Construction	Forepoling Pipe loop Grouting	• Installed to secure safety of blind end when ground is not stable

3-1-2 보조 공법

- NATM공법에는 굴착 후 즉시 숏크리트(shotcrete)를 시공하는 것이 중요하며 그 사이 (보통 3~4시간) 막장이 자립하는 것이 안전시공의 전제가 됨
- 그러나, 강도가 작은 지반이나 지하수 유출이 많은 지반은 록볼트와 숏크리트 만으로 안정된 터널 시공이 어려우므로 적절한 보조공법을 검토하여야 함
- 따라서 주공법 외에 보조공법을 병행할 필요가 있음
- 보조(보강)공법은 그 사용 목적에 따라 다음과 같이 분류

【NATM 터널의 보강공법의 종류】

목 적	적용 공법	공 법 개 요
막장안정공법	Forepoling	- 경사각도 : 5° ~ 10° - 횡방향 간격 : 0.40~0.60m - 길이 : 2~4m - 재질 : 철근(D25), 강파이프(ø32mm) - 형식 : 모르타르 주입형 - 매 막장 또는 두 막장 마다 시공
	파이프루프 일단 그라우팅 다단 그라우팅	- 경사각도 : 15°~20° - 횡방향 간격 : 0.40~0.60m - 길이 : 6~16m - 재질 : 강관(ø50~100mm) - 중첩길이 : 3~4m • 시멘트 밀크를 전장에 걸쳐 1회 주입하여 단층 파쇄대 봉합 • Cement milk를 깊이별로 수회에 나누어 주입하여 단층파쇄대 적극 봉합(Packer 사용)
	수평고압분사 그라우팅	- 경사각도 : 6° - 횡방향 간격 : 0.40~0.60m - 길이 : 12m - 초 고압분사로 미고결 원지반 토사를 시멘트와 교반하여 콘크리트 구근 형성 - 중첩길이 : 4m정도
	수직고압분사 그라우팅	- 수직으로 시공 - 간격 : 0.6~1.00m - 천단부 상부 2.5~5.0m 두께로 주입
	Root Milk 시공	- 소정의 깊이까지 천공한 후 공내에 철근(D25)을 삽입하고 Packer, Casing 혹은 강관 등을 이용하여 시멘트 밀크를 압력 주입하여 철근과 지반이 일체로 작용하도록 하여 지반강도 증가 도모 - 공 간격 : 1.0m
	막장면 숏크리트	- 막장 지반 자립이 어려운 경우 숏크리트 타설로 보강 - 두께 : >30mm - 조건에 따라 록볼트와 함께 시공
차수공법	SGR 그라우팅 LW 그라우팅 MSG 그라우팅	- 터널 천단 상부에 2.52~5.0m의 차수층 형성 - 터널 좌우의 지중에 차수벽을 설치하여 지하수 유입을 차단하고 차수벽 바깥쪽에 지하수위 보존

3-1-2 Supplement Construction

- It is important for NATM method to apply shotcrete right after excavation. Safety of blind end depends on whether the blind end can self-support during the application of shotcrete (3~4 hours)
- Consider supplement construction for low rigid ground or ground with high underground water discharge as rock bolt and shotcrete cannot guarantee the safety of tunnel construction
- Therefore, use supplement construction method in addition to main construction method
- Supplement(reinforcement) construction is classified as below:

【Type of NATM tunnel supplement】

Purpose	Supplement construction	Overview
Blind end stabilizing method	Forepoling	- Slope angle: 5°~ 10° - Horizontal interval: 0.40~0.60m - Length: 2~4m - Material: Reinforce rod (D25), steel pipe (ø32mm) - Type: Mortar injection - Construct every one or two blind end
	Pipeloop Single layer grouting Multi layer grouting	- Slope angle: 15°~20° - Horizontal interval: 0.40~0.60m - Length: 6~16m - Material: Steel pipe (ø50~100mm) - Piled up length: 3~4m • Seal fault crash area by inject cement milk one time for entire length • Seal fault crash area completely by inject cement milk several time for each depth (Use Packer)
	Horizontal high-pressure grouting	- slope angle: 6° - Horizontal interval: 0.40~0.60m - Length: 12m - Create concrete bulb by mixing unconsolidated ground soil with cement - Piled up length: : about 4m
	Vertical high-pressure grouting	- vertical construction - Interval : 0.6~1.00m - inject to upper arch with 2.5~5.0m thickness
	Root Milk	- drill a hole to certain level to insert steel rod (D25). then inject cement milk with pressure using packer, casing or steel pipe so that the rod can act with original ground to reinforce ground rigidity - Drilling interval: 1.0m
	Blind-end shortcrete	- Cast shortcrete when blind-end self-suppot is not feasible - Thickness: >30mm - If needed, use rock bolt jointly
Water barrier method	SGR grouting LW grouting MSG grouting	- Form water barrier layer of 2.52~5.0m in the upper part of arch - Install water barrier wall to underground of left and right side of tunnel to prevent underground water from penetrating and maintain underground water level

3-2 표준지보패턴 설계

3-2-1 지보패턴 설계 방법

【지보패턴 설계 흐름도】

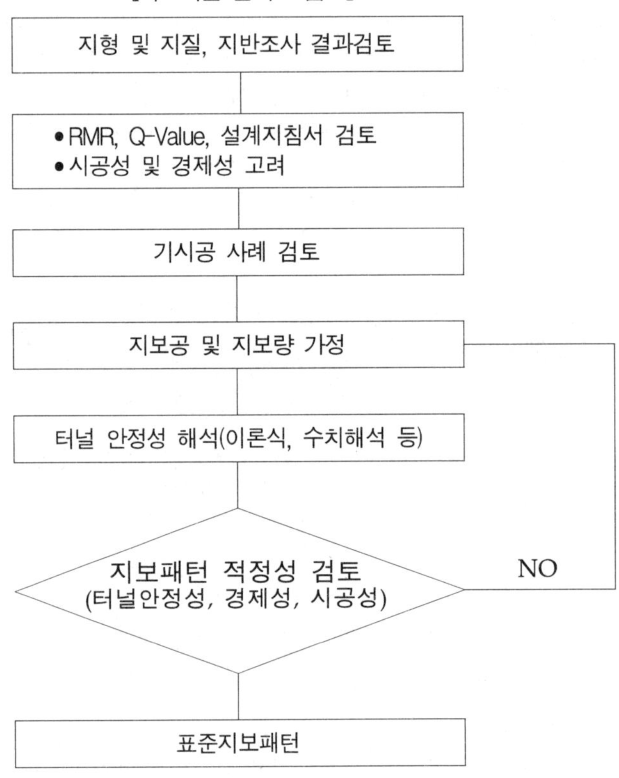

- 암반터널의 설계방법은 경험적 방법, 해석적 방법, 계측에 의한 방법 등으로 구분
- 표준패턴은 공학적 암반분류법을 이용한 경험적 방법에 의해 정한 후 구간별로 적용
- 암반분류기준은 설계단계에서 시공단계에 이르는 전 과정에서 일관성 필요
- 결정된 지보패턴에 대해 구간별로 적용한 후 지반조건을 고려하여 수치해석을 통해 안전성 검증 필요

경험적 방법	해석적 방법	계측에 의한 방법
•터널 시공사례를 통계적으로 분석 •암반조건과 지보량과의 관계 도출 •Q 방법	•터널 주변의 응력과 변형에 대한 해석 결과 이용	•불안정성의 탐지를 목적으로 굴착기간 동안 지반 거동 계측 •계측결과와 지반-지보 상호거동 분석결과 이용

3-2 Standard Support Pattern Design

3-2-1 Design Method

【Support pattern design flow】

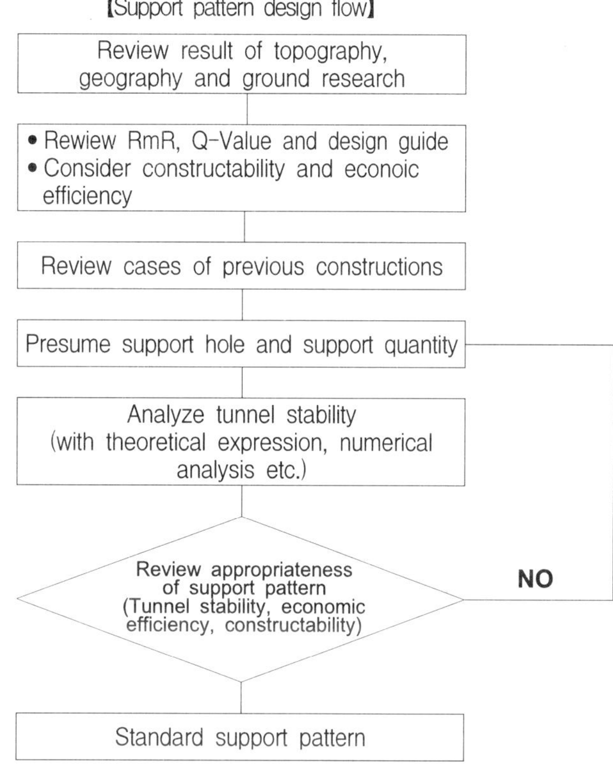

- Design method for rock base tunnel is divided into empirical method, analytical method and measurement method
- Standard pattern is determined by empirical method using engineering rock mass classification and applied for each section
- Rock mass classification should be consistent during the entire process
- Apply determined support pattern to each section. Then evaluate stability through numerical analysis

Empirical	Analytical	Measurement
• Statistically analyze previous tunnel construction experiences • Yield relation between rock base condition and support quantity • Q method	• Use analytical result of stress and deformation of tunnel area	• Measure ground movement to detect unstability during the excavation • Use measured result and analysis result of relation between ground-support interaction

3-2-2 암반분류법 적용

① 암반분류법 선정

- 지보패턴 결정을 위한 암반분류법은 설계단계에서 시공단계에 이르기까지 일관성 있게 적용될 수 있어야 함.

RMR System	Q - System
• 폭 10m인 터널에 대한 설계 지침 • 평가 Parameter가 다소 적음 (6가지) • 시추조사 및 실내시험으로 평가가 용이 • 국내 현장 기술자들에게 인지도가 높음	• 터널의 규모에 상관없이 지보량 결정 가능 • 평가 Parameter가 매우 세분화 되어 있음 • 지보 등급이 세분화 되어 있음(38등급) • Parameter평가는 시공 중 막장에서 용이

3-2-3 표준지보패턴 선정

① 굴진장 검토

- 무지보거리와 자립시간으로부터의 굴진장 산정

지보패턴	TYPE- I	TYPE- II	TYPE-III	TYPE-IV	TYPE- V	TYPE-VI
Q 값	> 40	10 ~ 40	4 ~ 10	1 ~ 4	< 1	
RMR	81 ~ 100	61 ~ 80	41 ~ 60	21 ~ 40	< 20	
무지보거리 (Q)	5.0m 스팬 / 10년	4.0m 스팬 / 6월	3.0m 스팬 / 1주일	1.5m 스팬 / 5시간	0.5m 스팬 / 10분	
	• Barton식에 의한 최대 무지보 연장 L = 2 · ESR · Q0.4 (도로 터널 ESR = 1.0)					
무지보거리 (RMR)	20m/2월 ~ 25m/20년	10m/8일 ~ 20m/3월	6m/13시간 ~ 10m/8일	2.0m/1시간 ~6m/13시간	자립불가 ~2.0m/1시간	

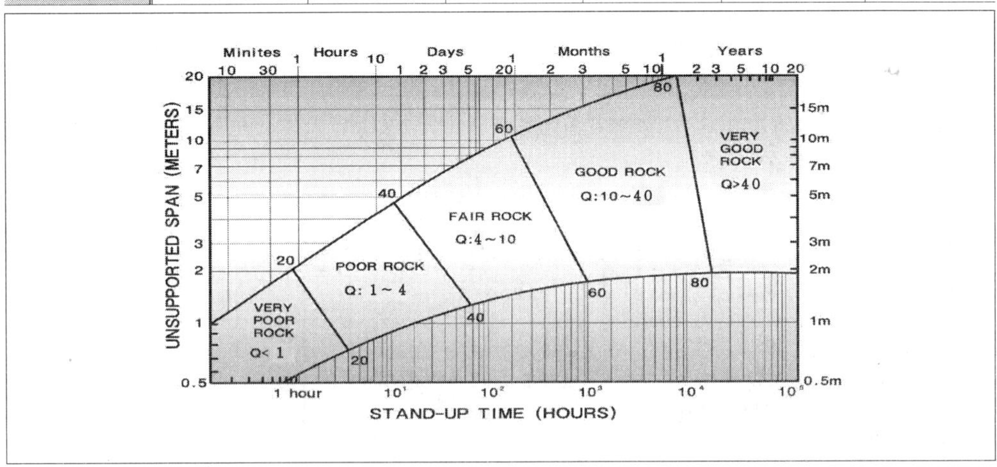

3-2-2 Rock Mass Classification

① Select rock mass classification method

- Rock mass classification should be consistent during the entire process

RMR System	Q - System
• Design guideline for tunnel with width of 10m • Relatively few evaluation Parameter (6) • Drilling survey and lab test is easy to perform • Well known to domestic site technicians	• Able to determine support quantity regardless of tunnel size • Large number of evaluation Parameter • Support grade is highly subdivided (38 grades) • Parameter evaluation is easy in blind-end

3-2-3 Standard Support Pattern Determination

① Unsupported tunnel length review

- Calculate unsupported tunnel length from non-support distance and Self-support time

Support pattern	TYPE-Ⅰ	TYPE-Ⅱ	TYPE-Ⅲ	TYPE-Ⅳ	TYPE-Ⅴ	TYPE-Ⅵ
Q value	> 40	10 ~ 40	4 ~ 10	1 ~ 4	< 1	
RMR	81 ~ 100	61 ~ 80	41 ~ 60	21 ~ 40	< 20	
Non-support distance (Q)	5.0m span / 10 year	4.0m span / 6 month	3.0m span / 1 week	1.5m span / 5 hour	0.5m span / 10 minute	
	• Barton equation maximum unsupported tunnel length L = 2 · ESR · Q0.4 (Road tunnel ESR = 1.0)					
Non-support distance (RMR)	20m/2 month ~ 25m/20 year	10m/8 day ~ 20m/3 month	6m/13 hour ~ 10m/8 day	2.0m/1 hour ~6m/13 hour	no self-support ~2.0m/1 hour	

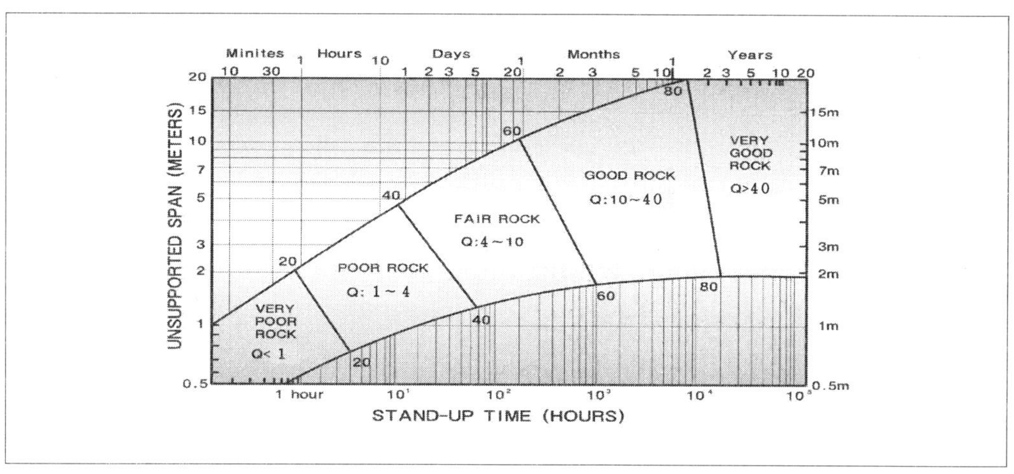

② 숏크리트 두께 검토

구 분	경 험 식	암 반 등 급				
		I	II	III	IV	V
Q - system (터널폭원:13.5m)	Q-Chart 적용	-	-	4~5	5~9	9 이상
RMR system	T=200-2.5×RMR≥50	-	5 이하	5~10	10~15	15 이상

③ 록볼트 길이 및 간격 검토

	구 분	경 험 식	록볼트 길이(m) (2차로, 터널폭원:13.5m)
경험식에 의한 록볼트 길이	Rabcewicz 식	$L \geq \omega/3 \sim \omega/5$ or $L \geq t$	2.7 ~ 4.5
	Q - system	$L = 2.0 + \frac{0.15B}{ESR}$	4.02
	RMR system	$L = 6 - 0.05 \times RMR \geq 3.0$	3.0 ~ 6.0
		$L = 0.5B - \frac{RMR}{200} \times B \geq 3.0$	3.0 ~ 6.75

	구 분	경 험 식	암 반 등 급				
			I	II	III	IV	V
암반등급에 따른 록볼트 횡간격	Rabcewicz 식	S<0.5×L or S<3×D	2.0~2.5	2.0~2.5	2.0~2.5	2.0~2.5	2.0~2.5
	Q - system	Q-Chart 적용	Random	2.3~2.5	2.1~2.3	1.7~2.1	1.7이하
	RMR system	S=0.5+0.025×RMR≥1.0	Random	2.0~2.5	1.5~2.0	1.0~1.5	1.0

④ 강지보 설치간격 검토

구 분	경 험 식	지 보 패 턴		
		TYPE-IV	TYPE-V	TYPE-VI
굴진장 (m)	-	1.5/3.0	1.2/1.2	1.0/1.0
RMR system (RMR < 40)	Sr(m) = 0.375 + 0.0375 × RMR	1.125~1.875	0.375~1.125	-
형 상	격자지보재		H-형강 지보재	

② Shotcrete thickness review

Classification	Empirical formula	Rock base grade				
		I	II	III	IV	V
Q - system (Tunnel width :13.5m)	Q-Chart	-	-	4~5	5~9	9+
RMR system	T=200-2.5×RMR≥50	-	5-	5~10	10~15	15+

③ Rock bolt length and interval review

	Classification	empirical formula	Rock bolt (m) (two lanes, tunnel width : 13.5m)
Rock bolt length by empirical formula	Rabcewicz equation	L ≥ $\omega/3 \sim \omega/5$ or L ≥ t	2.7 ~ 4.5
	Q - system	L=2.0+$\frac{0.15B}{ESR}$	4.02
	RMR system	L=6-0.05×RMR ≥ 3.0	3.0 ~ 6.0
		L=0.5B-$\frac{RMR}{200}$×B ≥ 3.0	3.0 ~ 6.75

	Classification	Empirical formula	Rock base grade				
			I	II	III	IV	V
Rock bolt horizontal interval by rock base grade	Rabcewicz formular	S<0.5×L or S<3×D	2.0~2.5	2.0~2.5	2.0~2.5	2.0~2.5	2.0~2.5
	Q - system	Q-Chart	Random	2.3~2.5	2.1~2.3	1.7~2.1	1.7-
	RMR system	S=0.5+0.025×RMR≥1.0	Random	2.0~2.5	1.5~2.0	1.0~1.5	1.0

④ Steel support installation interval review

Classification	Empirical formula	Support pattern		
		TYPE-Ⅳ	TYPE-Ⅴ	TYPE-Ⅵ
Unsupported tunnel length(m)	-	1.5/3.0	1.2/1.2	1.0/1.0
RMR system (RMR < 40)	Sr(m) = 0.375 + 0.0375 × RMR	1.125~1.875	0.375~1.125	-
Shape	Coffered support		H-shape steel support	

⑤ 지보패턴 사례

표준지보패턴			TYPE-Ⅰ	TYPE-Ⅱ	TYPE-Ⅲ	TYPE-Ⅳ	TYPE-Ⅴ	TYPE-Ⅵ	비 고
굴 착 공 법			전단면 굴 착	전단면 굴 착	전단면 굴 착	반단면 굴 착	반단면 굴 착	반단면 굴 착	
1회굴진(상/하)(m)			3.5	3.5	2.0	1.5/3.0	1.2/1.2	1.0/1.0	
숏크리트 두께(mm)			50 (일반)	50	80	120	160	160	강섬유
록볼트	길이(m)		3	3	4	4	4	4	SD35 (D25)
	간격 (m)	종	RANDOM	2.0	2.0	1.5	1.2	1.0	
		횡	RANDOM	2.0	1.5	1.5	1.5	1.5	
강지보 설치간격 (m)			-	-	-	1.5	1.2	1.0	격자 지보
내 부 라이닝	두께(mm)		300	300	300	300	300	300	
	철근		갱구부 등 중요구간에 설치						

⑥ 암반분류법 선정

- 구간별 표준지보패턴의 적용은 터널구간에서 실시한 지반조사 결과와 탄성파 탐사 및 전기비저항 탐사 등의 물리탐사결과를 종합하여 결정

⑤ Support pattern example

<table>
<tr><th colspan="3">standard support pattern</th><th>TYPE- I</th><th>TYPE-II</th><th>TYPE-III</th><th>TYPE-IV</th><th>TYPE-V</th><th>TYPE-VI</th><th>Remarks</th></tr>
<tr><td colspan="3">excavation method</td><td>Entire section</td><td>Entire section</td><td>Entire section</td><td>half section</td><td>half section</td><td>half section</td><td></td></tr>
<tr><td colspan="3">Tunneling (up/down)(m)</td><td>3.5</td><td>3.5</td><td>2.0</td><td>1.5/3.0</td><td>1.2/1.2</td><td>1.0/1.0</td><td></td></tr>
<tr><td colspan="3">Shotcrete thickness(mm)</td><td>50 (general)</td><td>50</td><td>80</td><td>120</td><td>160</td><td>160</td><td>steel fiber</td></tr>
<tr><td rowspan="3">Rock bolt</td><td colspan="2">Length(m)</td><td>3</td><td>3</td><td>4</td><td>4</td><td>4</td><td>4</td><td rowspan="3">SD35 (D25)</td></tr>
<tr><td rowspan="2">Interval (m)</td><td>V</td><td>RANDOM</td><td>2.0</td><td>2.0</td><td>1.5</td><td>1.2</td><td>1.0</td></tr>
<tr><td>H</td><td>RANDOM</td><td>2.0</td><td>1.5</td><td>1.5</td><td>1.5</td><td>1.5</td></tr>
<tr><td colspan="3">Steel support interval (m)</td><td>-</td><td>-</td><td>-</td><td>1.5</td><td>1.2</td><td>1.0</td><td>Coffered support</td></tr>
<tr><td rowspan="2">Inner lining</td><td colspan="2">T (mm)</td><td>300</td><td>300</td><td>300</td><td>300</td><td>300</td><td>300</td><td></td></tr>
<tr><td colspan="2">Reinforcement</td><td colspan="6">Install in critical areas such as tunnel opening</td><td></td></tr>
</table>

⑥ Rock mass classification determination

- Determine standard support pattern for each area through comprehensive review of ground survey, elastic wave survey and resistivity survey result

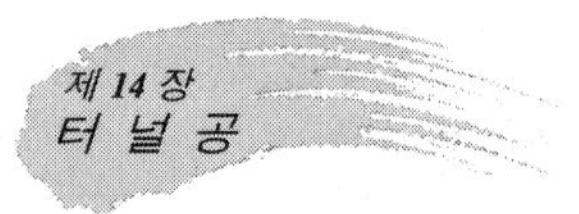

3-3 터널 굴착공법 검토

3-3-1 굴착공법의 종류 및 특징

구 분		적 용 조 건	주 의 사 항
전단면 공법		• 소단면에서 일반적 사용 • 지반조건이 극히 양호한 경우	• 터널 전체를 적용할 수는 없으므로 기타 방법으로의 전환도 검토해 놓을 필요가 있다.
수평분할굴착공법	Long Bench Cut 공법	• 전단면으로는 막장이 자립하지 않는 경우 • 다만, 지반조건이 비교적 안정하고 있어 시공단계 도중에 Invert 폐합이 필요 없는 경우	• 지반상황에 따라 시공단계 도중에 서Invert 폐합도 있을 수 있다.
	Short Bench Cut 공법	• 토사에서 팽창성 지반까지 가장 일반적인 공법 • 일반적으로 막장으로부터 30m 이내 또는 30일 이내에 Invert를 폐합한다.	• 변형, 침하가 현저한 경우에는 폐합시기를 빨리 하여야 할 필요가 있으므로 최단 Bench 장을 검토해 놓는다.
	Mutil Bench Cut 공법	• Short Bench Cut에서는 막장이 자립하지 않는 경우 • 지질에 따라 정설부를 지보하고 나서 상반을 굴착	• Bench를 다단으로 함에 따라 폐합시기가 늦어진다. 또 변형이 크게 된다.
	Mini Bench Cut 공법	• 도시터널에서 침하를 억제하는 경우 • 팽창성 지반에서 빠른 시기에 전 주변에 걸쳐 지보를 시공하는 경우 • 전단면 공법에서 지질악화에 대한 대책으로서 사용	• 막장의 안정성이 최대의 문제
	가 Invert 공법	• 도시터널에서 침하를 억제하는 경우 • Short Bench Cut에서 침하가 큰 경우의 대책으로서 사용	• 상반에서 가 Invert를 시공하므로 상반의 시공속도가 늦어져, 전체 시공속도도 저하된다.
선진도갱공법 (Pilot or Side Pilot)		• 도시터널에서 침하를 억제하는 경우 • 비교적 대단면에서 지반의 지지력이 부족한 경우	• Pilot의 단면 및 간격을 확보하기 위해 어느 정도의 단면이 필요 • 가축구조로 하기 어려우므로 팽창성이 큰 지반에서는 적용곤란
연직분할공법 (Center Diaphragm or Cross Diaphragm)		• 도시터널에서 침하를 극력 억지하여야 하는 경우 • 비교적 대단면으로 막장이 자립하지 않는 경우	• 중벽으로 분할하기 위하여 어느 정도의 단면 필요 • 중벽의 위치, 형상, 강성을 검토하여야 할 필요가 있다.
TBM 선진도갱공법 (TBM Pilot)		• 도시터널에서 발파진동의 영향을 최소화 하고자 하는 경우 • 지반조건이 비교적 안정한 경우	• 산악터널의 경우 연장이 긴 터널에서 적용하여야 하며, 시공 도중 파쇄대 등 지반조건이 불량한 경우 보강한 후 굴진하여야 한다.

3-3 Tunnel Excavation Method Review

3-3-1 Type and Characteristics of Excavation Method

Classification		Applicable condition	Attention
Entire section		• Generally used for small section • Ground condition should be very good	• Consider other alternatives as this method cannot be applied to entire tunnel area
Horizontal division drilling method	Long Bench Cut	• When Blind-end does not self-support when applying entire section method • Apply only when invert abolition during construction is not required as the ground condition is relatively fine	• Invert amalgamation can happen according to ground condition
	Short Bench Cut	• Most general method applicable to soil to expendable ground • Invert abolition: within 30m from blind end or within 30 days	• When there are serious deformation or subside, consider short bench cut to precipitate abolition
	Mutil Bench Cut	• Apply when Short Bench Cut do not make blind-end self-support • Support base part first and excavate upper part	• Abolition is delayed and ground deformation is increased as Bench becomes multi layer
	Mini Bench Cut	• Apply when suppressing subside for urban tunnel • Supports are constructed rapidly for entire area in expendable ground • Apply to cope with bad gound condition when using entire section method	• Major issue is safety of blind-end
	Temporary Invert	• Apply when suppressing subside for urban tunnel • Apply to cope with serious subside when using Short Bench Cut	• Construction may delayed due to temporary invert installment
Advanced pilot or side pilot method		• Apply when suppressing subside for urban tunnel • Apply when bearing power of ground is not sufficient for large section	• Requires certain section size to secure pilot section and interval • Hard to apply to expandable ground
Center Diaphragm or Cross Diaphragm		• Apply when suppressing subside excessively for urban tunnel • Apply when blind-end does not self-support where the section is relatively large	• Require certain section size to divide into center wall • Need to location, shape and rigidity of center wall
TBM Pilot		• Apply when suppressing vibration from excavation for urban tunnel • apply when ground condition is relatively fine	• Apply to lengthy mountain tunnel. If ground condition is not good enough, reinforce the ground before continue tunneling

【굴착공법의 종류】

굴 착 공 법	전 단 면 굴 착	상·하 반단면 굴착	링 컷 굴 착	중벽분할 굴 착	선진도갱 굴 착
단면도	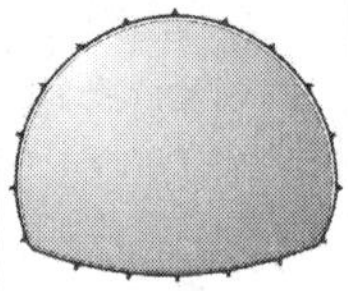	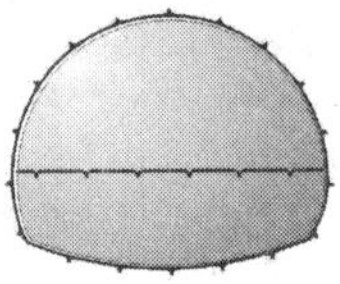	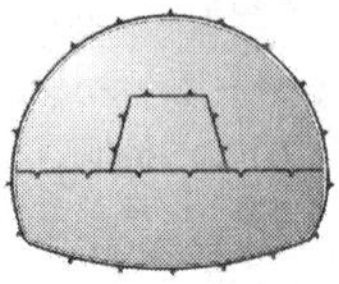	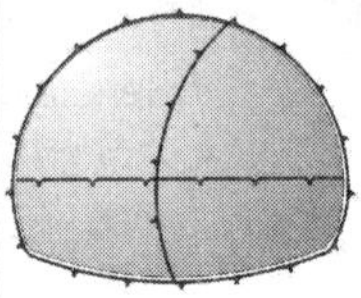	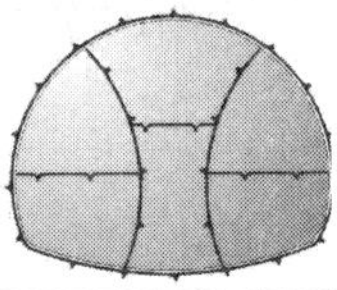

3-3-2 굴착공법 적용 사례

터 널 명	지보패턴별 굴착공법 적용 사례
우남터널	지보패턴 – 1 & 2 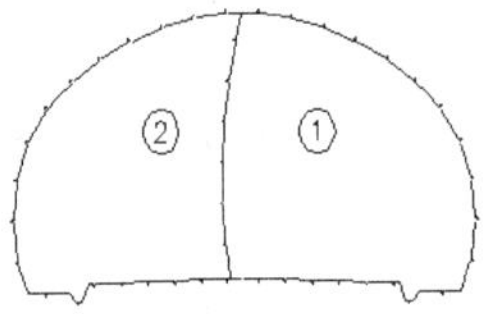 ② ① / 지보패턴 – 3 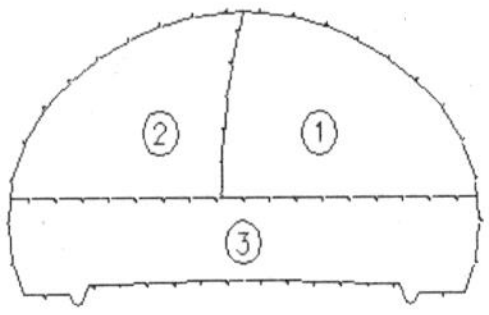 ② ① ③ / 지보패턴 – 4 & 5 ② ③ ① ④ ⑤ 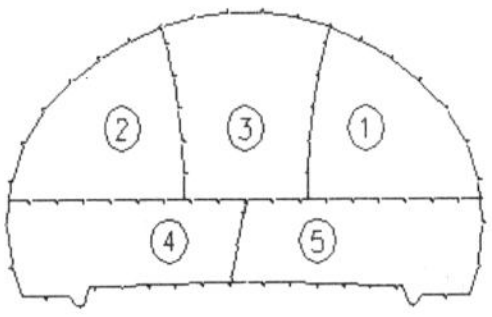
수암터널	지보패턴–1 ~ 3 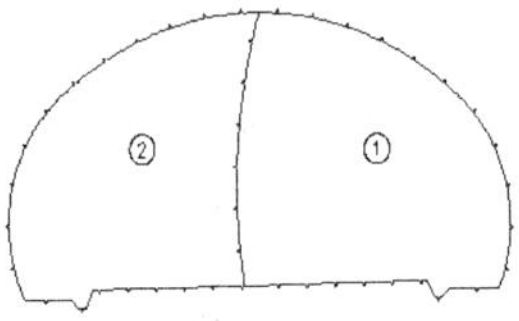 ② ① / 지보패턴–4 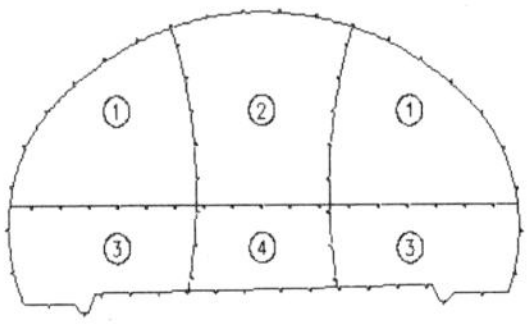 ① ② ① ③ ④ ③
사패산터널	지보패턴 – 1 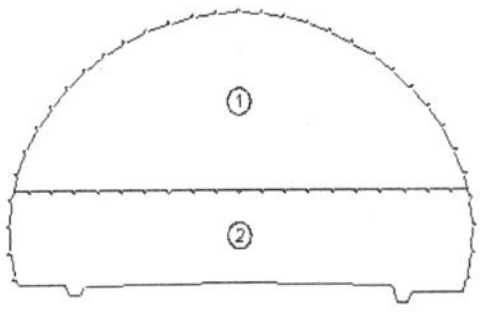 ① ② / 지보패턴 – 2 ~ 5 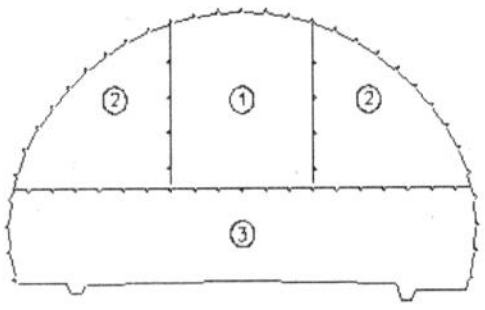 ② ① ② ③ / 지보패턴 – 6 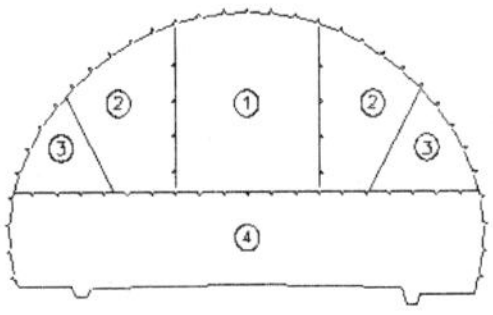 ③ ② ① ② ③ ④
청계터널	지보패턴 – 1~4 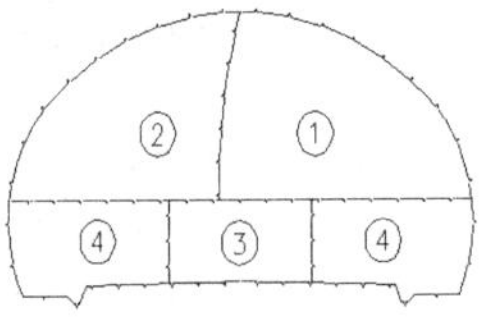 ② ① ④ ③ ④

【Type of excavation method】

Method	Entire Section	Up/Down half section	Ring cut	Center diaphragm	Pilot
Cross section	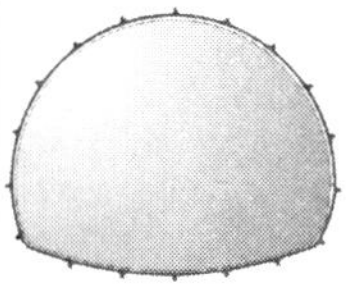	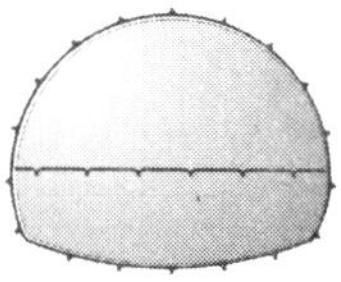	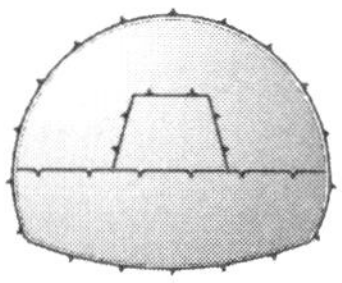	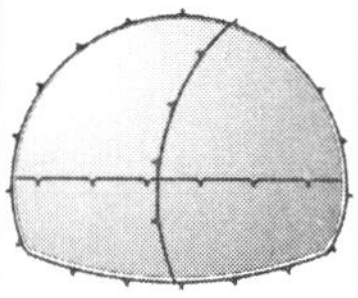	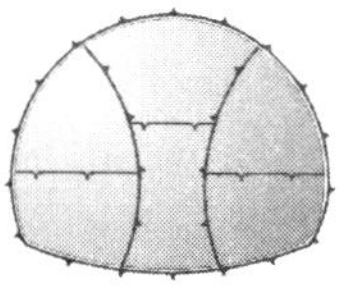

3-3-2 Application Example for Excavation Method

Name of tunnel	Application example for excavation method
U-nam tunnel	Support Pattern - 1 & 2 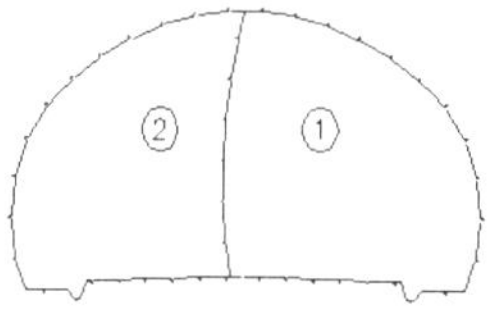 Support Pattern - 3 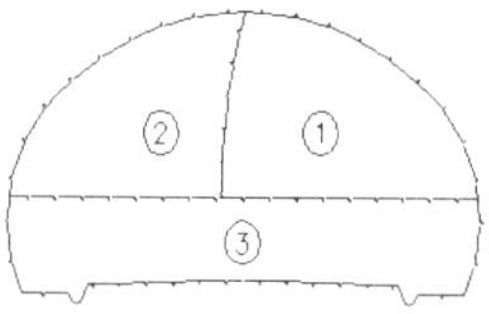 Support Pattern - 4 & 5 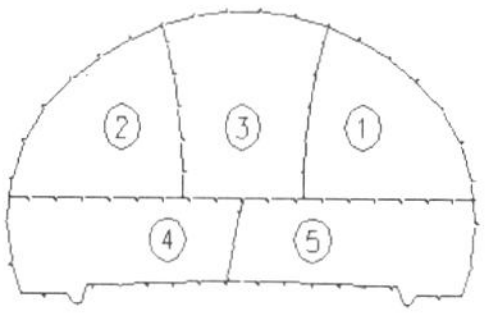
Suam tunnel	Support Pattern - 1 ~ 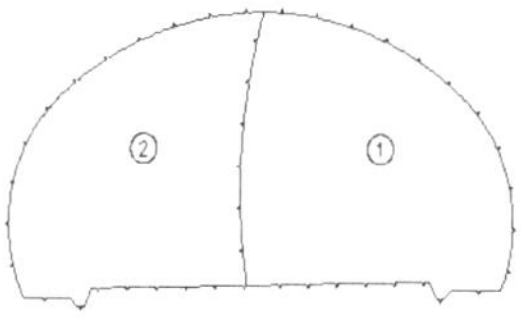 Support Pattern - 4 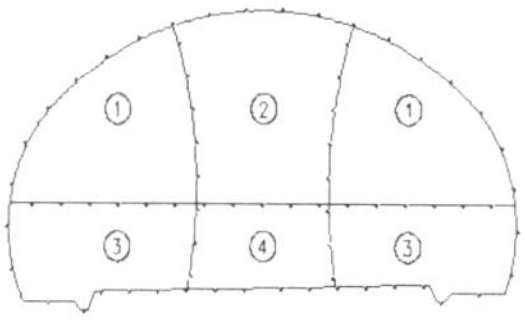
Sapesan tunnel	Support Pattern - 1 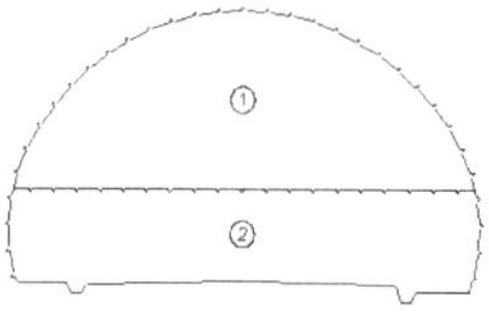 Support Pattern - 1~5 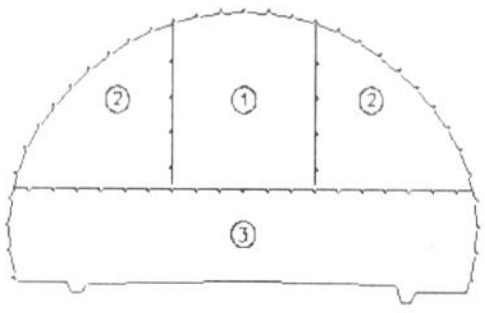 Support Pattern - 6 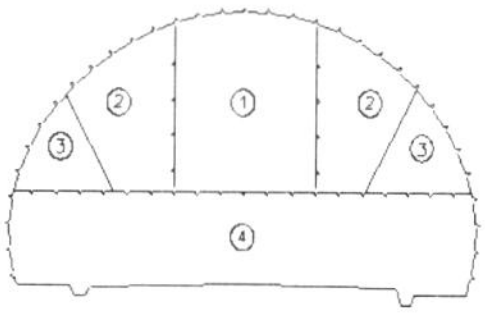
cheong-gye tunnel	Support Pattern - 1~4 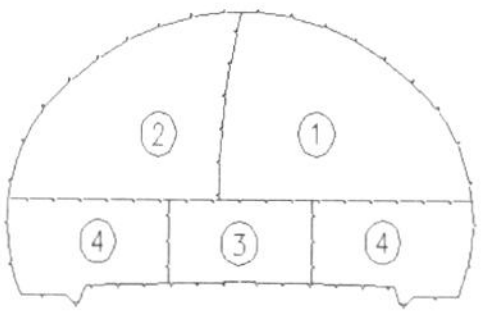

3-4 터널 안정성 해석

3-4-1 해석목적 및 사용프로그램

① 해석 목적

- 경험적 방법에 의한 지보설계의 적정성(안정성) 판단
- 해석결과를 계측 관리기준으로 활용

② 해석 항목

설계변수 결정을 위한 예비해석	표준지보패턴 안정성 검토	접속부 3차원 안정성 검토	위험구간 안정검토
• 사용프로그램 적정성평가 • 해석방법의 적정성평가 - 하중분담율 산정 - 벤치길이의 적정성 - 연약대의 영향 및 보강범위 선정 - 암반비탈면 해석방법	• 표준지보패턴 지보량의 적정성 평가 • 지보재의 작용하중을 설계기준과 비교 평가 - 연속체해석 - 불연속체해석	• 본선에 접속되는 부대갱의 영향 평가 - 본선-사갱부 - 급전구분소 구간 - 배전실 구간 - 신호기계실 구간 등	• 굴착 및 보강공법의 적정성 검증 - 저토피부 - 건축물 하부 통과부 - 단층부 - 계곡부

③ 사용 프로그램 및 해석분야

분 야	사용프로그램 및 구성 모델		해석 분야
지반 해석 분야	• FLAC 2D,3D • Pentagon2D,3D	• Linear & Non-Linear Elastic • M-C & H-B Elasto-plastic • Strain-Softening • Ubiquitous-Joint • Transversely isotropic	• 지보패턴별 해석 • 이방성/Creep 해석 • 접속부, 갱구부, 관통부 3차원 해석 • 침투류해석 • Seismic & Dynamic 해석
	• UDEC 2D	• Coulomb-Slip • Barton-Bandis	• 지보패턴별 2차원 불연속체 해석 • Key-Block 안정해석
동적해석 분야	• SIMQKE • ProShake	• Wave Propagating Theory • Dynamic Option • 직접법	• 인공지진파 생성 • 지진파에 의한 구조물 안정성
수리해석 분야	• SEEP/W • FRACMAN • Modflow-3D	• Volumetric Water Content Functions • Hydraulic Stochastic Functions • Hydraulic Conductivity Functions	• 터널 내 암반 유출수량 예측 • 불연속 암반 이방성 투수계수 산정 • 지하수 유동특성 검토

3-4 Tunnel Stability Analysis

3-4-1 Purpose of Analysis and Program to Use

① Analysis purpose

- Evaluate stability (feasibility) of support design done by empirical method
- Use the result of analysis as management standard

② Analysis items Preliminary analysis to determine design variables

Preliminary analysis to determine design variables	Stability review for standard support pattern	3-Dimensional stability of connection	Risk section stability review
• Appropriateness of program selection • Appropriateness of analysis method - Load distribution rate - Bench length - Effect of weak ground and area to be reinforced - Rock base slope analysis	• evaluate feasibility of support quantity of standard support pattern • Compare and evaluate applied load to support and design standard - continuum analysis - Non-continuum analysis	• Impact evaluation of subsidiary tunnel connected to main lane - Main lane-inclined shaft - electricity distribution area - signal machine room	• appropriateness of excavation and supporting method - Low soil surface - Passing under built structure - Fault - Valley

③ Program to use and model composition

Field	Program to use and model composition		Field of analysis
Ground analysis	• FLAC 2D,3D • Pentagon2D,3D	• Linear & Non-Linear Elastic • M-C & H-B Elasto-plastic • Strain-Softening • Ubiquitous-Joint • Transversely isotropic	• Support pattern • Anisotropy/Creep • Three demensional analysis of connection, opening and penetrating area • penetration • Seismic & Dynamic
	• UDEC 2D	• Coulomb-Slip • Barton-Bandis	• Two dimensional non-continuum per support pattern • Key-Block stability
Dynamic analysis	• SIMQKE • PROSHAKE	• Wave Propagating Theory • Dynamic Option • Direct method	• Artificial seismic wave • Structure stability from seismic wave
Numerical analysis	• SEEP/W • FRACMAN • Modflow-3D	• Volumetric Water Content Functions • Hydraulic Stochastic Functions • Hydraulic Conductivity Functions	• discharging water quantity from rock base into tunnel • Non-continuum rock base anisotropy permeability coefficient • Underground water liquidity review

④ 수치해석의 종류

【터널해석을 위한 수치해석법】

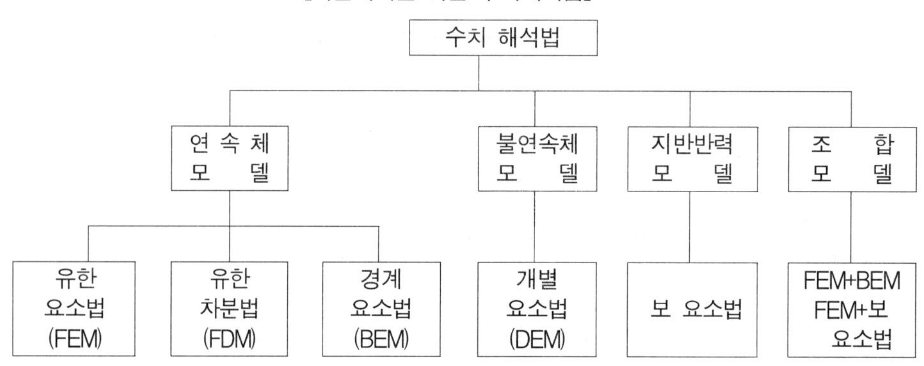

【수치해석법의 종류와 특징】

구 분	내 용	장 점	단 점
유 한 요소법 (F.E.M)	• 각 요소를 기하학적인 크기로 분할하고, 초기 지반 상태의 변화에 따라 응력·변형 해석을 실시하는 Implicit	• 고도의 복잡화된 지반 및 터널 특성의 해석이 가능 • 굴진에 따른 시간적인 영향 고려 가능 • 재료의 불균질성을 해석	• 대용량의 Computer 및 저장 용량 필요 • 경계요소 설정이 명확하지 않음
유 한 차분법 (F.D.M)	• 요소를 기하학적 크기로 분할하며, 미지요소 해석법은 Explicit. • 전체 행렬을 구성할 필요가 없으며 불균형력에 따라 미소변화	• 시간진행과 같은 거동해석 용이 • 행렬을 구성하지 않기 때문에 Computer용량이 적음 • 동적계산에 효과적	• 정적문제에 있어서 타 수치해석보다 소요시간이 길다
경 계 요소법 (B.E.M)	• 굴착 경계만으로 문제영역의 생성이 가능하고, 선형 방정식을 편미분방정식의 적분으로 해석하며, 굴착 경계는 외부경계로 적용	• 방정식 생성이 상대적으로 작다 • Computer 용량의 소형화 • 입·출력 과정이 용이 • 복잡한 건설공정이나 재료의 시간 의존성 고려 곤란	• 지반의 선형, 연속지반 거동에 국한 • 시간에 의한 효과를 Model화하기 곤란
개 별 요소법 (D.E.M)	• 연속체 Model화하지 않고, 개별적 Block으로 생성하고 지반체 변형이 강체 Block 사이의 절리를 따른 이동에 지배	• 절리도가 높은 Model에 대단히 효과적 • Computer 용량이 비교적 작음	• 절리의 위치, 방향의 입력이 계산에 필요하므로 이러한 Data 입수의 곤란
보 요소법	• 지반반력계수법으로 지칭 • Lining은 보 요소로 모형화 • 주변지반은 스프링 요소로 모형화	• 범용 구조계산 프로그램 사용 가능 • 해석과정, 연산시간, 컴퓨터 기억 용량 소모 적음	• 터널 및 지반상태의 단순화·지반 스프링 요소의 강성 및 외력 산정이 어려움

④ Type of numerical analysis

【Numerical analysis for tunnel analysis】

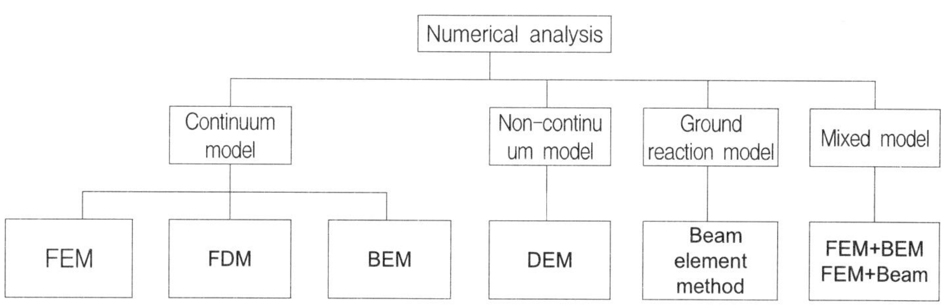

【Type and characteristics of numerical analysis】

Classification	Content	Advantage	Disadvantage
Finite element method (F.E.M)	• Divide each component to geometrical size to analyze stress and deformation according to initial ground status (Implicit)	• Possible to analyze complicated ground and tunnel characteristics • Time effect of tunneling can be considered • Analyze unevenness of material	• Requires large size of computer & memory capacity • Boundary setting is not quite clear
Finite Difference Method (F.D.M)	• Divide each component to geometrical size (explicit) • No need for entire matrix. slightly change according to uneven stress	• Easy to analyze movement such as time progress • Less computer capacity is required • Effective for dynamic calculation	• Requires more time on static modeling than other methods
Boundary element method (B.E.M)	• Creation of issued area is possible with excavation boundary. analyze linear equation with partial differential equation. Use excavation boundary as external boundary	• Relatively less creation of equations • Less computer capacity is required • Easy input / output • Can be used for simple construction with plain material	• Only for linear, continuous movement of ground • Hard to model time effect
Discrete element method (D.E.M)	• Create separate blocks rather than model continuum shape. movement along with joint between blocks governs ground deformation.	• Very effective for models with high rate of joint • Relatively less computer capacity is required.	• Hard to get data for joint location, direction
Beam Element method	• Ground reaction coefficient method • Model Lining as beam element and nearby ground as springs	• General-purpose structural calculation program can be used • Less process, time and computer memory is required	• Have to simplify tunnel and ground status. hard to calculate rigidity and external force of ground spring elements

⑤ 수치해석을 통한 터널안정성 검토

- 합리적인 설계를 위한 예비해석

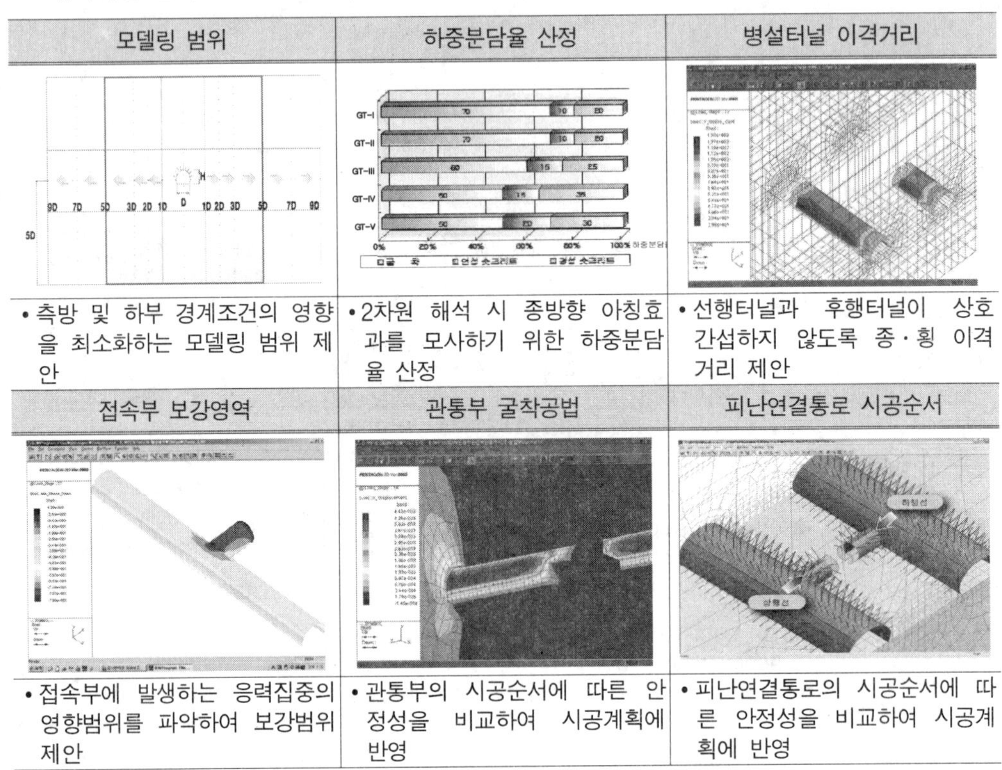

모델링 범위	하중분담율 산정	병설터널 이격거리
• 측방 및 하부 경계조건의 영향을 최소화하는 모델링 범위 제안	• 2차원 해석 시 종방향 아칭효과를 모사하기 위한 하중분담율 산정	• 선행터널과 후행터널이 상호 간섭하지 않도록 종·횡 이격거리 제안
접속부 보강영역	**관통부 굴착공법**	**피난연결통로 시공순서**
• 접속부에 발생하는 응력집중의 영향범위를 파악하여 보강범위 제안	• 관통부의 시공순서에 따른 안정성을 비교하여 시공계획에 반영	• 피난연결통로의 시공순서에 따른 안정성을 비교하여 시공계획에 반영

- 적용지보패턴 적정성 검토

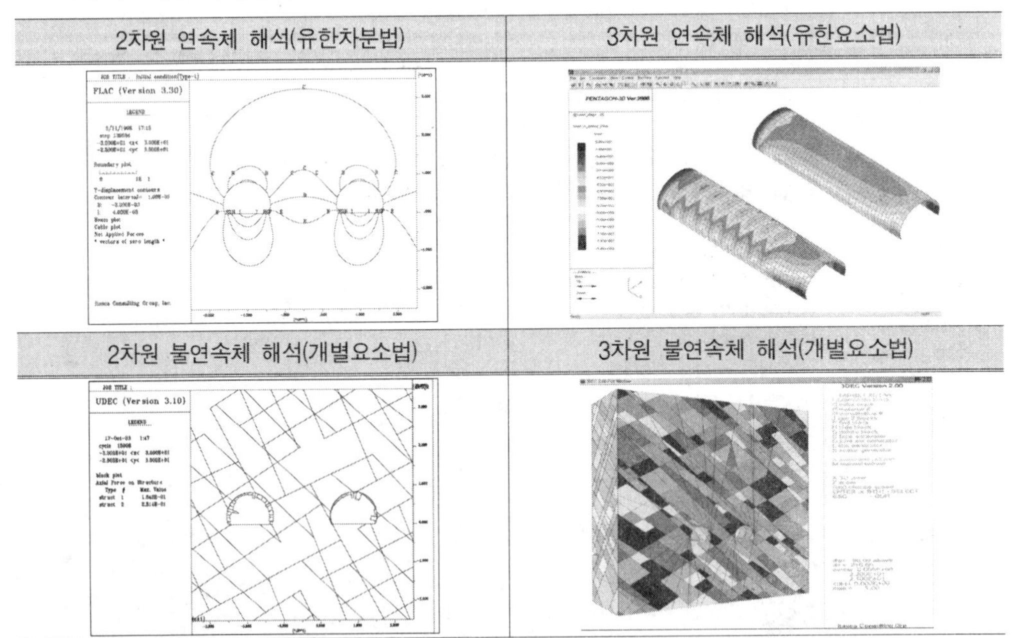

2차원 연속체 해석(유한차분법)	3차원 연속체 해석(유한요소법)
2차원 불연속체 해석(개별요소법)	**3차원 불연속체 해석(개별요소법)**

⑤ Tunnel stability review through numerical analysis

- Preliminary analysis for logical design

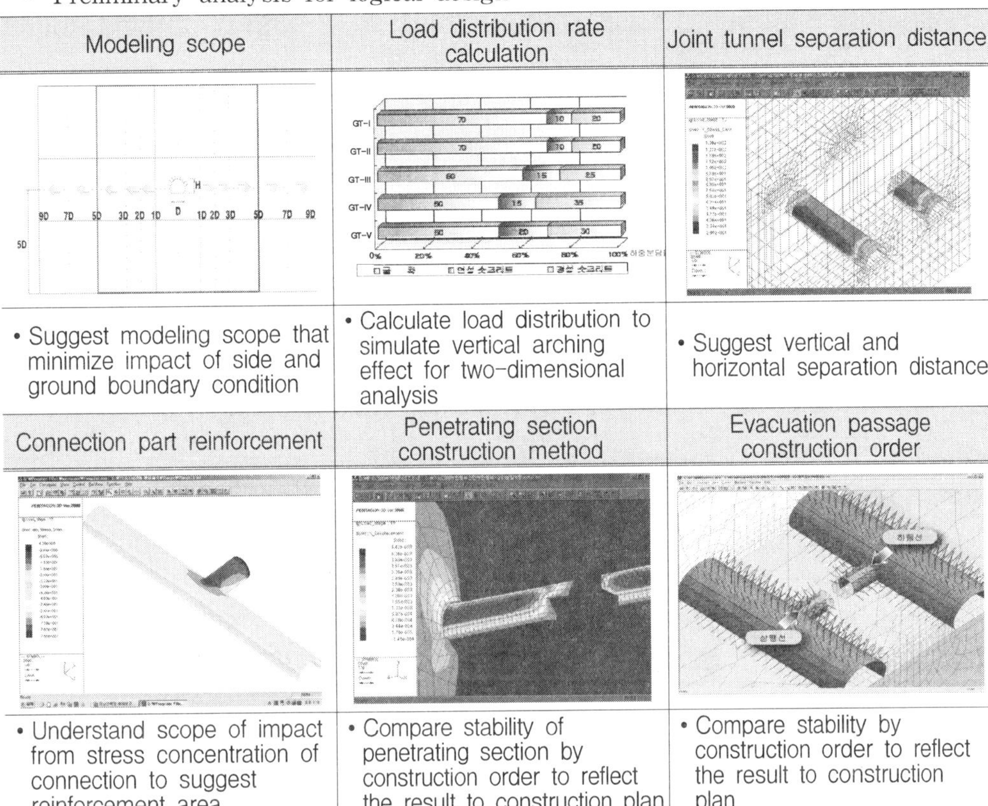

Modeling scope	Load distribution rate calculation	Joint tunnel separation distance
• Suggest modeling scope that minimize impact of side and ground boundary condition	• Calculate load distribution to simulate vertical arching effect for two-dimensional analysis	• Suggest vertical and horizontal separation distance
Connection part reinforcement	**Penetrating section construction method**	**Evacuation passage construction order**
• Understand scope of impact from stress concentration of connection to suggest reinforcement area	• Compare stability of penetrating section by construction order to reflect the result to construction plan	• Compare stability by construction order to reflect the result to construction plan

- Appropriateness of applied support pattern

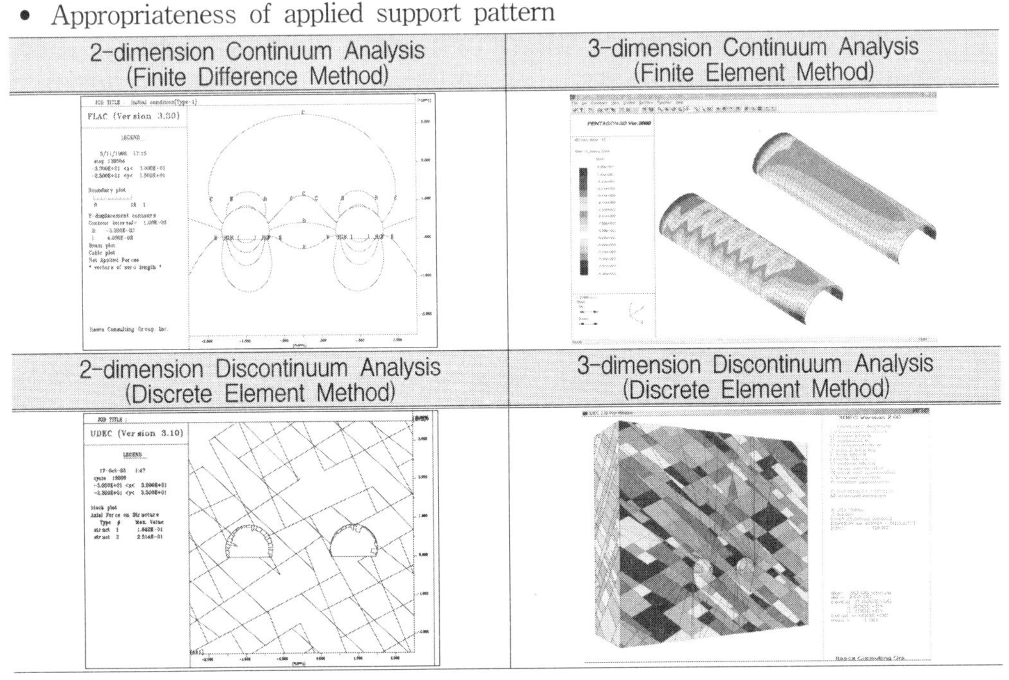

2-dimension Continuum Analysis (Finite Difference Method)	3-dimension Continuum Analysis (Finite Element Method)
2-dimension Discontinuum Analysis (Discrete Element Method)	**3-dimension Discontinuum Analysis (Discrete Element Method)**

- 주요 구간 안정성 검토

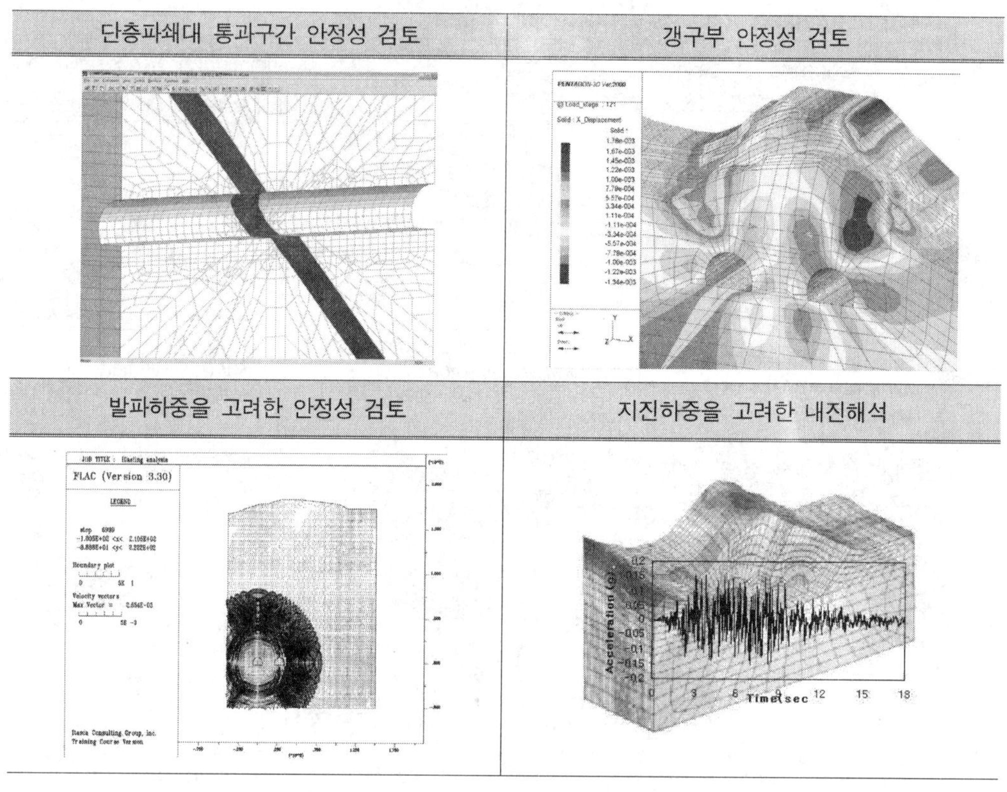

- 콘크리트 라이닝 안정성 검토
  - 지반 및 하중조건에 따른 콘크리트 라이닝의 두께 및 철근배근의 적정성 평가

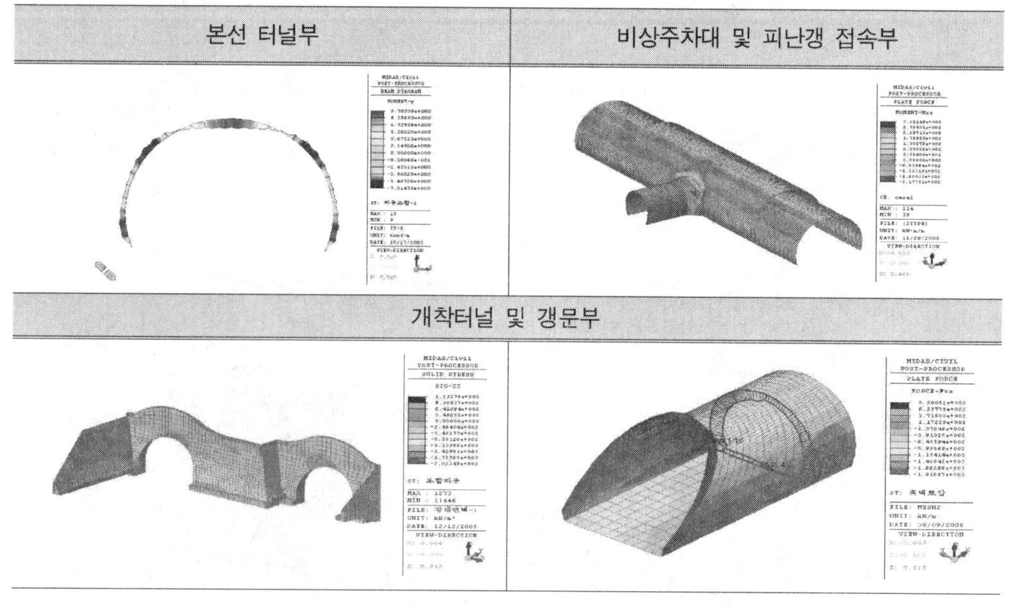

- Stability review for major area

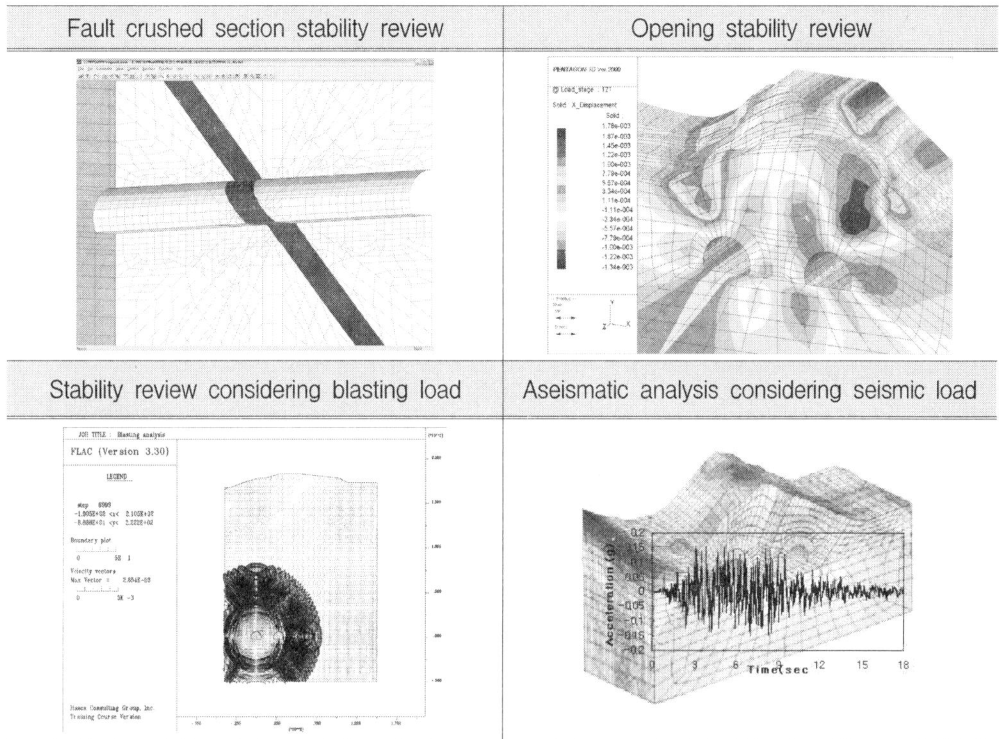

- Concrete lining stability review
  - Review appropriateness of concrete lining thickness and reinforcing rod distribution

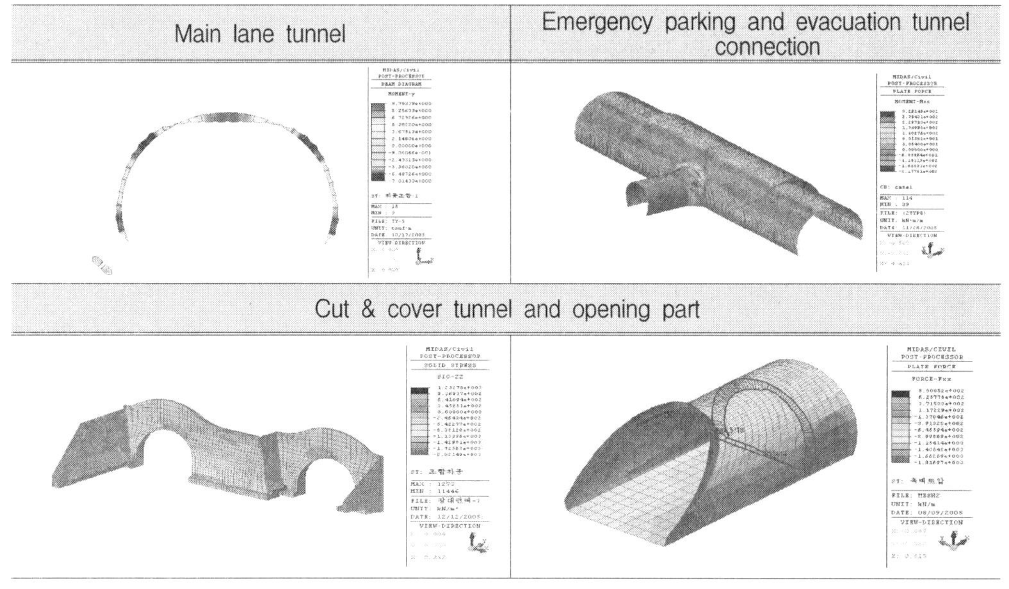

3-5 발파 설계

3-5-1 발파 설계 시 고려사항

발파 설계 시 고려사항	설계반영 사항
• 지반상태를 고려한 발파패턴 설계	• 지반종류 및 암반 등급을 고려한 발파패턴
• 굴진장 및 주변 환경에 적합한 심발 공법	• 갱구부 소음진동 감소 • 장공발파
• 적재 및 운반에 용이한 버력 크기 고려	• 공 간격 및 최소 저항선 조정
• 발파에 의한 허용 기준치 설정	• 구조물에 따른 허용 진동속도 설정
• 여굴 및 주변 암반의 이완 최소화	• 외곽공에 제어발파

3-5-2 발파 설계순서

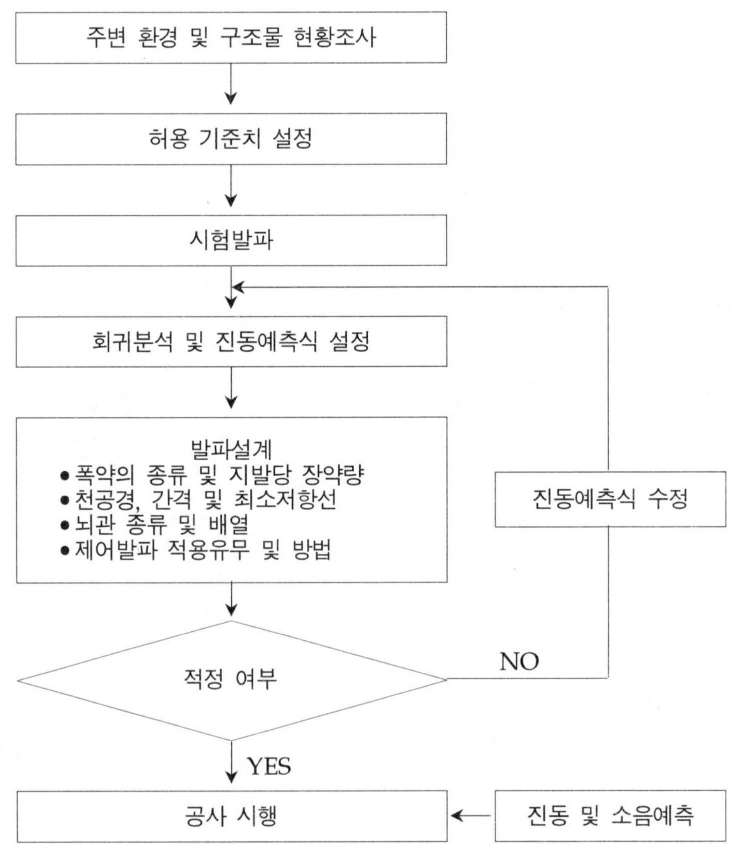

3-5 Blasting Design

3-5-1 Considerations for Blasting

Considerations when designing blasting	Design application
• Ground condition	• Blasting pattern with consideration to ground type and rock base grade
• Cut method appropriate to tunnel unsupported length and nearby environment	• Reduction of noise and vibration in tunnel entrance • Long hole blast
• Size of waste ore appropriate to load and transfer	• Adjusting hole interval and minimum burden
• Allowance standard for blasting	• Setting allowable vibration speed depending on the sturcture
• Minimize over-break and relaxation of nearby rock base	• Controlled blast on outer wall hole

3-5-2 Blasting Design Process

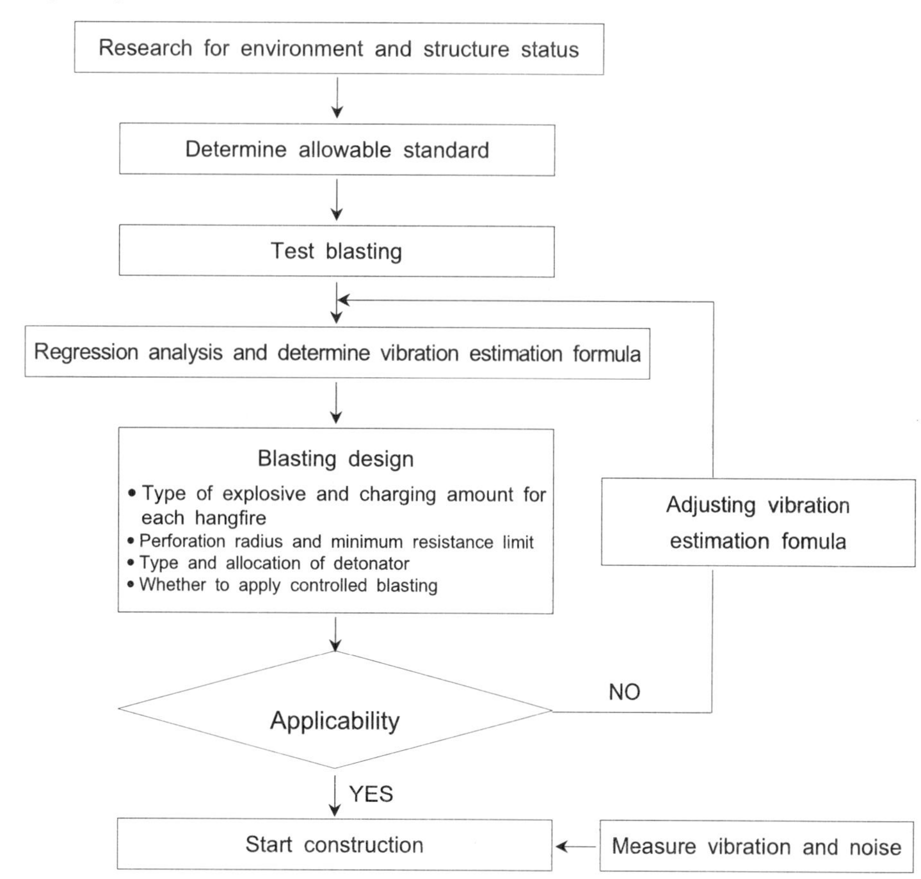

3-5-3 발파 용품

① 터널 발파용품 선정기준

구분	폭 약	기폭 시스템 (뇌관류)
선정 기준	• 물리적 외력에 안전한 폭약 • 제조 · 운반 및 사용에 안전한 폭약 • 발파진동 및 소음이 작은 폭약 • 지하수 유입에 대한 내수성이 우수한 폭약 • 여굴 및 주변 암반의 이완을 최소화할 수 있는 폭약	• 취급하기 쉽고 경제성이 우수한 뇌관 • 안전성 및 불발확인이 용이한 뇌관 • 수공이나 열악한 환경에서도 시공성이 우수한 뇌관 • 지발당 약량 제한에 따른 단차 확보가 가능한 뇌관 • 단차에 제한 없는 기폭 시스템 선정 • 발파잔류물에 의한 환경오염 최소화

② 기폭 뇌관

구 분	전기식 뇌 관	비전기식 뇌관 (NONEL)
개요도	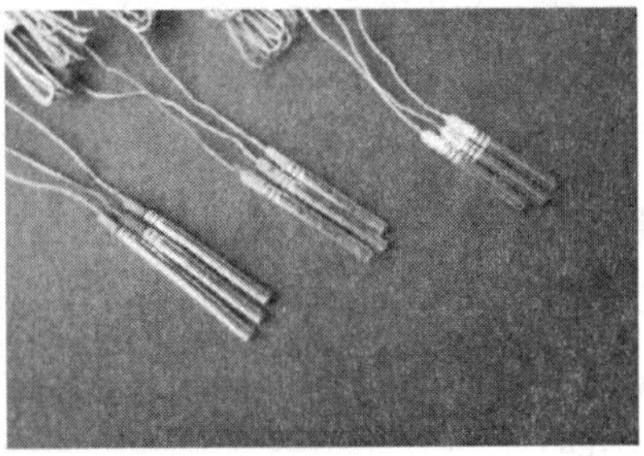	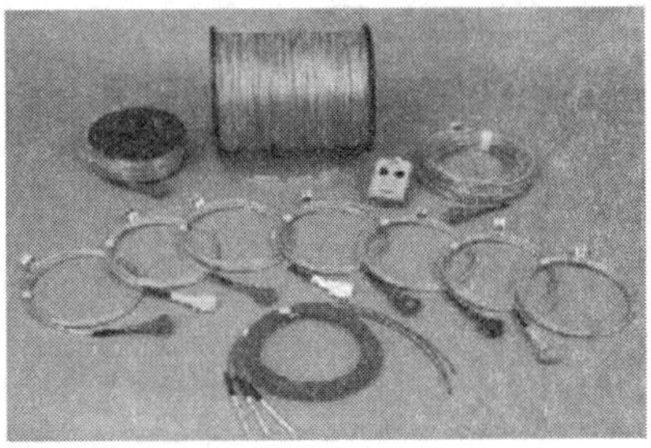
특 징	• 대중화 및 취급이 용이 • 터널 발파에 시공성이 우수함 • 결선 여부 및 저항 확인이 용이하여 내정전 뇌관사용으로 대전이 우려가 없음 • 전기적 성질을 이용함으로서 발파실패 사전예방 • 표면뇌관이 터지는 소음이 없음 • MS 및 LP뇌관 조합 시 41단까지 지발이 가능하고 다단 발파 시 무한단차 가능 • 시공성 및 경제성 우수 • 플라스틱 튜브관이 없어 환경오염 문제를 원천적으로 해결	• 물리적 외력, 미주전류, 정전기, 천둥, 전파 등에 안전 • 단차가 무한대이며, 도심 미진동 발파 가능 • 발파실패를 사전에 예방하기 어려움 • 표면뇌관이 터지는 소음발생 • 전기뇌관에 비하여 다소 고가 • 플라스틱 튜브의 미 회수에 의한 환경오염 발생

※최근에는 전자뇌관이 개발되었음

3-5-3 Blasting Supplies

① Criteria for blasting supply selection

Classification	Explosive	Detonating system (detonator)
selection standard	• Stable from physical force from outside • Safe to produce, transfer and apply • Low vibration and noise • High water-resistance (against underground water discharge) • Minimize over-break and relaxation of nearby rock base	• Easy to operate, economic • Safe and easy to check misfire • Easy to install in poor conditions • Able to secure level difference through limiting amount of explosive in each bangfire • No limit in level difference • Minimize residue from blasting for environmental preservation

② Detonator

Classification	Electiric	Non-electric (NONEL)
concept	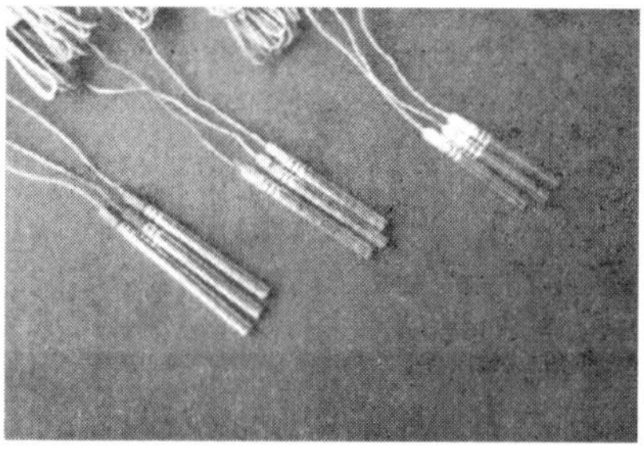	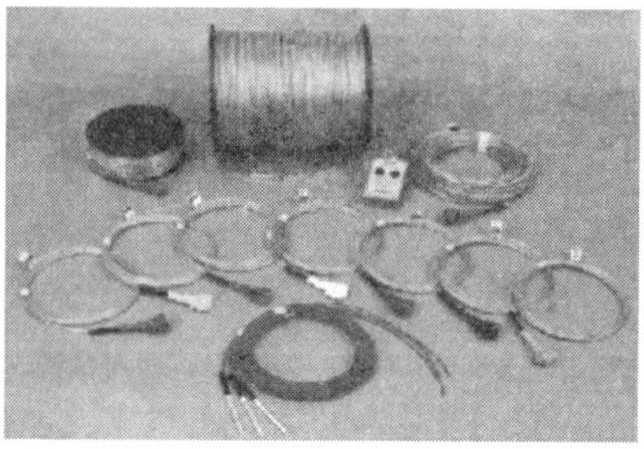
Characteristics	• Widely known, easy to use • Easy to apply for tunnel blasting • Easy to check wiring and resistance. Anti-static electric detonator is used. • Prevent blasting failure by using electric behavior • No noise for surface detonator explosion • 41 step delay is possible with combination of MS and LP detonator. unlimited stage difference for blasting • Easy and economic to use • No environmental pollution from plastic tube	• Safe from external forces, stray current, static electricity, lighting, radio wave • Unlimited stage difference. no-vibration blasting for urban area • Hard to prevent blasting failure • Noise for surface detonator explosion • Relatively expensive than electric detonator • Environmental pollution from unrecovered plastic tube

※ Electronic detonator is developed recently

3-5-4 발파공법의 검토

① 제어발파 방법

구 분	Smooth Blasting	Cushion Blasting	Pre-Splitting	Line Drilling
개요도	S, S, W $\frac{S}{W} \leq 0.5 \sim 0.8$	1차발파 2차발파 자유면 파단면 Cushion Blasting공 발파 직전 자유면	자유면 파단면	W 0.5~0.75W 자유면 W 0.75W 무장약공열(발파후 파단면) 1열 2열 3열
개 요	• 최종열에 좁은 공간격, 최소저항선, 소구경 화약을 공중앙에 배치하여 약장약으로 주변에 미치는 영향 최소화	• 소량의 분상장약과 전색재, 공기에 의한 완충효과 활용 • 장약을 자유면 쪽에 붙여 수행	• 발파 전에 선행 발파를 실시하여 인위적 파단면을 형성함으로서 진동의 전파경로 차단	• 보안물 방향에 천공경의 2~4배 간격으로 천공하여 인위적으로 진동 전파경로 차단
천공 간격	• S:W=(0.5~0.8):1 • S : 공간격 • W : 최소저항선	• 넓은 공간격	• 공경에 비례 • SB보다 좁은 공간격	• 공경의 2~4배 • 전열 최소저항선의 0.5~0.75배
장 점 및 단 점	• 주발파와 동시 시행 • 정밀폭약 이용으로 장약 용이 • 주변 암반이완 손상 적음 • 여굴이 적고 평활한 굴착면 확보	• Line-Drilling에 비해 천공 수 적음 • 불량한 암질에서도 효과적 • 부분 장약 • 주발파 완료 후 시행 • 장약 시 주의	• 균질한 암반에서 효과적 • 암반균열 최소화 • 선행 발파 시 소음, 진동 우려 • 숙련공 필요	• 진동 경감 • 암반균열 최소화 • 무장약 • 과도한 천공비 • 숙련공 필요
용 도	• 노천 · 터널발파	• 노천발파	• 노천발파	• 노천 · 터널발파

3-5-4 Blasting Method Review

① Controlled blasting method

Classification	Smooth Blasting	Cushion Blasting	Pre-Splitting	Line Drilling
concept	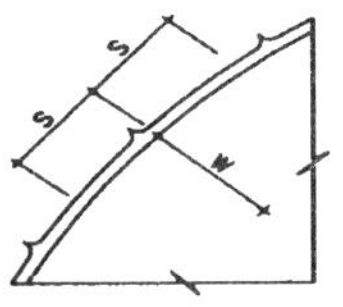 $\frac{S}{W} \leq 0.5\sim0.8$	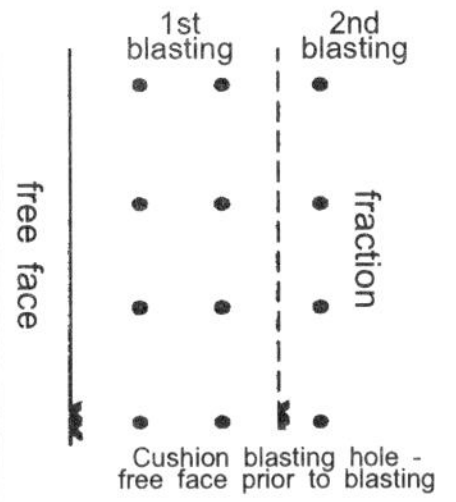 1st blasting, 2nd blasting, free face, fraction, Cushion blasting hole - free face prior to blasting	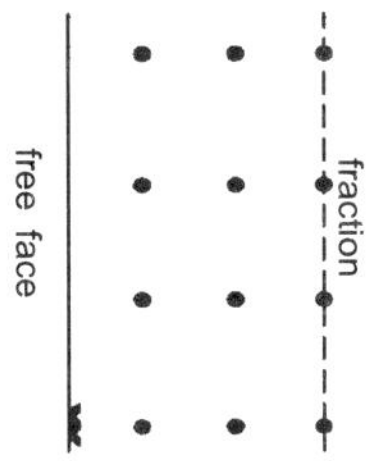 free face, fraction	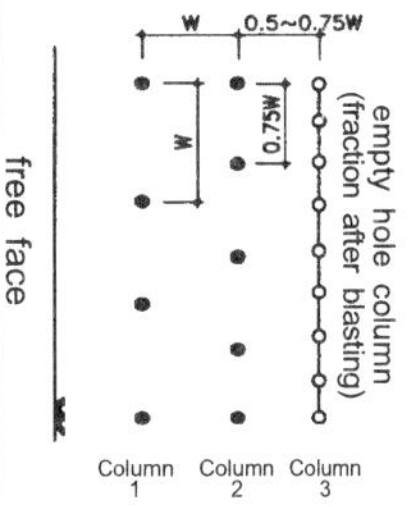 W, 0.5~0.75W, free face, empty hole column (fraction after blasting), Column 1, Column 2, Column 3
overview	• Minimize influences to nearby area through allocating minimum burden, small-caliber explosive with interval narrowest in the last row at the center of the hole	• Use absorbing effect by small amount of powdered charger, sealant, air • attach explosives to the free-face side	• Perform pilot blasting before main one to cut off the radio wave movement path	• Cut off the radio wave movement path through perforate with interval of 2 to 4 times perforation radius
hole interval	• S:W=(0.5~0.8):1 • S : hole interval • W : minimum burden	• Wide hole interval	• Proportionate to hole radius • Smaller than SB	• 2 to 4 times hole radius • 0.5 to 0.75 times front row minimum burden
advantage and disadvantage	• Simultaneous to main blasting • Easy to charge FINEX • Less rock base relaxation • Less over-break and even blasting face	• Less number of holes compared to Line-Drilling • Effective even with poor rock quality • Partial charging • Conduct after main blasting • Safety concern on charging	• Effective with even rock quality • Minimize rock cleavage • Noise & vibration concern on pilot blasting • Requires expert engineer	• Less vibration • Minimize rock cleavage • No charge • Excessive perforating cost • Requires expert engineer
usage	• Open air / tunnel	• Open air	• Open air	• Open air / tunnel

③ 발파진동 영향 검토

- 발파진동 허용치

【민원에 대한 발파진동 허용치(환경분쟁 조정사례)】

건 물 구 분	I	II	III	IV
건 물 종 류	문화재, 컴퓨터시설물 주변	주택, 아파트	상가	R.C 건물, 공장
허용 진동치	0.2	0.3	1.0	1.0 ~ 4.0

【구조물에 대한 발파진동 허용치】

[단위:Kine(cm/sec)]

등 급	I	II	III	IV	V
구 분	진동예민 구조물	조적식(벽돌,석재 등) 벽체와 목재로 된 천정을 가진 구조물	지하기초와 콘크리트 슬래브를 갖는 조적식 건물	철근 콘크리트 골조 및 슬래브를 갖는 중·소형 건축물	철근 콘크리트 또는 철골구조 및 슬래브를 갖는 대형건축물
	문화재 등	재래가옥 저층 일반가옥 등	저층 양옥, 연립주택 등	중·저층 아파트, 중소상가 및 공장	내진구조물 (고층아파트 대형건물 등)
허용진동치	0.3	1.0	2.0	3.0	5.0

- 발파진동 추정방법
  - 최대입자속도 추정식

구	분	추 정 식
서 울 지 하 철	화 강 암	$V = K \cdot W^{0.57} \cdot D^{-1.7}$
	편 마 암	$V = K \cdot W^{0.5} \cdot D^{-1.5}$
미 국 광 무 국		$V = 160 \cdot (D / W^{1/2})^{-1.6}$

  - 발파진동 적용식

발 파 위 치	적 용 식	적 용 발 파 진 동 상 수
심 발 발 파	$V = K \cdot W^{0.57} \cdot D^{-1.7}$	$K = Ei(Ri \cdot Sc + 60)$
확 대 발 파		$K = Ei(Ri \cdot Sc + 30)$

주) Sc : 화강암의 일축 압축강도

③ Blasting vibration impact review

- Blasting vibration allowance

【Basting vibration allowance (examples of environmental dispute mediation】

Structure classification	I	II	III	IV
Structure type	Cultural assets, computer facilities	Housing, apartments	Retail shops	R.C structure, factories
Allowed vibration	0.2	0.3	1.0	1.0 ~ 4.0

【Blasting vibration allowance for structures】

[Unit: Kine(cm/sec)]

Grade	I	II	III	IV	V
Classification	sensitive to vibration	Masonry (brick, stone, etc) wall and wood ceiling structure	basement foundation and concrete slab masonry structure	reinforced concrete frame and slab small and mid sized structure	reinforced concrete frame and slab large sized structure
	cultural asset	conventional housing, low story ordinary housing	low story western-style housing, row houses	low and mid size apartment, small and medium retail stores and factories	quake-proof structure (high-rise apartment / large scale buildings)
Allowed vibration	0.3	1.0	2.0	3.0	5.0

- Blasting vibration estimation method
  - Maximum particle velocity estimation formula

Classification		presumed formula
Seoul Metro	granite	$V = K \cdot W^{0.57} \cdot D^{-1.7}$
	gneiss	$V = K \cdot W^{0.5} \cdot D^{-1.5}$
US Bureau of mining		$V = 160 \cdot (D / W^{1/2})^{-1.6}$

  - Blasting vibration application equation

blasting location	equation	applied blast vibration constant
Cut	$V = K \cdot W^{0.57} \cdot D^{-1.7}$	$K = Ei(Ri \cdot Sc + 60)$
Extended		$K = Ei(Ri \cdot Sc + 30)$

Note) Sc : single axile compression intensity of granite

- 진동상수값(K) 계산식(도심 지하철 발파진동식)

구	분			내 용
발 파 진 동 식	진 동 식			$V = K \cdot (D / W^{1/3})^n$
	진 동 상 수 식			$K = Ei\ (Ri \cdot Sc + Qi)$
n 값	화 강 암			−1.7
	편 마 암			−1.5
폭 약 보 정 치 (Ei)	다 이 너 마 이 트			1.0
	에 멀 젼 폭 약			0.8
	초안(초유) 폭약			0.65
암종에 따른 보정치 상 수 (Ri)	화 강 암			0.0371
	편 마 암			0.0206
발파형식에 따른 보정치 (Qi)	화강암	개착식	바닥발파	80
			계단발파	50
		터널식	심발발파	60
			확대발파	30
	편마암	개착식	바닥발파	60
			계단발파	30
		터널식	심발발파	40
			확대발파	10

주) Sc : 암석의 일축압축강도

④ 발파소음 영향 검토

- 발파소음의 특성
  - 화약이 자유면으로부터 비교적 얕은 곳에서 폭발했을 때 폭발에너지는 탄성파나 발파음으로 주위에 전파되어 소비
  - 폭원에서 어느 정도 떨어진 곳에서는 음파로 전파
  - 발파음은 지속시간이 매우 짧고, 주요 주파수는 50~150Hz 범위

- Vibration constant value (K) calculation formula (metro subway blasting vibration equation)

classification				content
blast vibration equation	vibration equation			$V = K \cdot (D / W^{1/3})^n$
	vibration constant equation			$K = Ei (Ri \cdot Sc + Qi)$
n value	granite			-1.7
	gneiss			-1.5
explosive adjustment (Ei)	dynamite			1.0
	emulsion			0.8
	ammonium nitrate (AN-FO) explosive			0.65
rock type adjustment constant (Ri)	granite			0.0371
	gneiss			0.0206
blasting type adjustment (Qi)	granite	open air	ground	80
			stairway	50
		tunnel	cut	60
			extended	30
	gneiss	open air	ground	60
			stairway	30
		tunnel	cut	40
			extended	10

Note) Sc : single axile compression intensity of the rock

④ Blast vibration impact review

- Characteristics of blast vibration
  - explosive energy is transferred as elastic wave or blasting noise to nearby area when explosive explodes from relatively swallow location from free-face
  - transferred as sound wave to remote areas
  - blast noise lasts very shortly and main frequency is between 50~150Hz

- 음압이 인체나 구조물에 미치는 영향

음 압	인 체	구 조 물
180dB(A)	-	건조물 피해
170dB(A)	-	대부분의 창유리 파손
150dB(A)	고막 손상	창유리 일부 파손
140dB(A)	-	창유리 파손 없음
120dB(A)	고통 받음	-
100dB(A)	장시간 난청	-
80dB(A)	안 전	-

- 건설소음 규제기준

시 간 별 / 대 상 지 역	조 석 (05:00~08:00) (18:00~22:00)	주 간 (08:00~18:00)	심 야 (22:00~05:00)
주거 · 녹지 · 준도시 지역 중 취락지구 및 운동 · 휴양지구, 자연환경보전지역, 학교 · 병원 · 공공도서관 부지경계선으로부터 50m 이내 지역	65 이하	70 이하	55 이하
상업 · 공업 · 농림지역, 준농림 지역 및 준도시 지역 중 취락지구 외의 지구, 미고시 지역	70 이하	75 이하	55 이하
공사장 소음 규제 (주간)	소음발생 시간이 1일 2시간 미만은 + 10dB, 2~4시간일 때는 + 5dB로 보정		
적 용	70dB(A)		

- 발파소음 추정식

적 용 식	비 고
dB(A) = 20 · log10(P/Po)	• dB : 음압 Level(dB) • 음압실효치 : $P = 82 \cdot (D/W^{1/3})^{-1.2}$ - W : 장약량(kg), D : 이격거리(m) • Po : 기준 음압 실효치(= 2×10^{-4} μbar)

- Influence of sound pressure to human body or built structure

Sound pressure	Human body	Structure
180dB(A)	–	Structural damage
170dB(A)	–	Break all window glasses
150dB(A)	eardrum damage	Break some window glass
140dB(A)	–	Break no window glass
120dB(A)	Pain	–
100dB(A)	Long term hearing difficulty	–
80dB(A)	Safe	–

- Construction noise restriction standard

Time / Area	Morning and night (05:00~08:00) (18:00~22:00)	daytime (08:00~18:00)	Midnight (22:00~05:00)
housing, green area, dwelling area and recreation area of semi-urban area, environmental preservation area, within 50m from school, hospital, public library	65 or less	70 or less	55 or less
commercial, agricultural, industrial area, semi-agricultural area, area other than dwellings in semi-urban area	70 or less	75 or less	55 or less
construction sit noise restriction (daily)	less than 2 hours daily: + 10dB, between 2~4 hours daily: + 5dB		
application	70dB(A)		

- Blasting noise estimation formula

Formula	Remarks
dB(A) = 20 · log10(P/Po)	• dB : sound pressure Level(dB) • effective value of sound pressure: $P = 82 \cdot (D/W^{1/3})^{-1.2}$ - W : charged explosive (kg), D : separation distance (m) • Po : standard sound pressure effective value (= 2×10^{-4} μbar)

⑤ 발파 작업 시 주의사항

구 분	주 의 사 항
누설전류	• 장약 시 누설전류에 의한 폭약의 폭발로 인명 피해가 예상되기 때문에 누설전류의 발생이 되지 않도록 하여야 함. → 비전기식 뇌관 사용
용수개소	• 용수개소에서의 터널용 폭약 사용은 불발의 원인이 되기 때문에 가급적 지양하고, 장약 후 장시간 보존시키지 않도록 시간개념을 도입하여야 함.
발파제어	• 발파진동은 현장여건(암의 강도, 절리, 토피층의 토사의 두께 등)에 따라 다르게 측정된다. 따라서, 시공 중 발파진동치 추정 시 발파진동상수는 시험발파를 실시하여 95% 이상의 신뢰도가 있는 값을 선정, 적용하여야 한다. • 실측 진동치가 0.3kine이 상회하는 경우 발파순차를 지연시켜 지발당 장약량을 최소화함으로써 진동을 감소시켜야 함. V995%0=2409,65(SD)-2,34 V(50%)=1172,02(SD)-2,34 PPV(cm/sec) 1 0,1 10 10 Scaled Distance(m/kg1/2)
발파소음	• 발파진동뿐 아니라 발파소음은 도로터널에서 직접적인 민원으로 진전되기 때문에 소음 억제를 위해 갱구입구에 방음차단막 등의 방음시설을 설치하여 소음을 억제하여야 함.
방음문 및 방호벽	

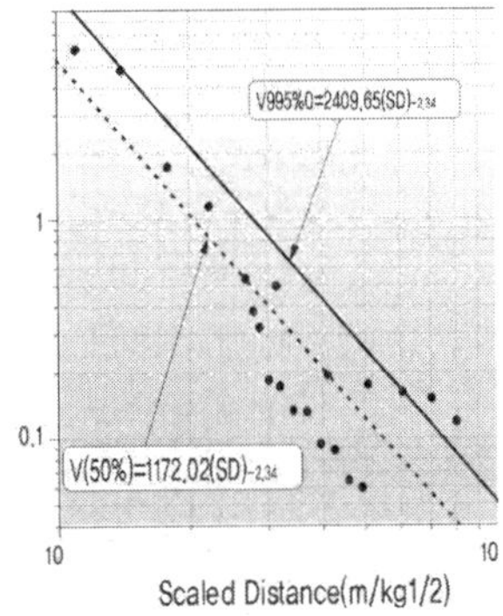

⑤ Considerations on blasting work

<table>
<tr><th>Classification</th><th colspan="2">Considerations</th></tr>
<tr><td>Current leakage</td><td colspan="2">• Prevent explosives from ignite with leaked electricity as this may lead to serious life damages → use non-electric detonator</td></tr>
<tr><td>Water content area</td><td colspan="2">• Refrain from using explosive in the area of water content. Apply time concept to avoid leaving explosive charged for long period of time</td></tr>
<tr><td>Blasting control</td><td>• Blasting vibration is measured differently due to site conditions such as rock rigidity, joint, soil thickness, etc. As a result, conduct pilot blasting to make sure vibration estimation has 95% of reliability

• When measured vibration exceeds 0.3kine, delay blasting order to minimize vibration</td><td>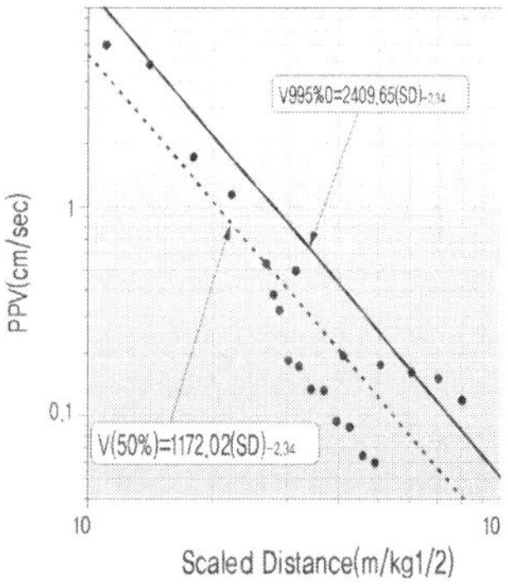
</td></tr>
<tr><td>Blasting noise</td><td colspan="2">• Install sound-proof facilities such as noise barrier at the tunnel openning since blasting noise may cause public dispute</td></tr>
<tr><td>Soundproof door and protective wall</td><td></td><td></td></tr>
</table>

3-6 방수 · 배수 설계

3-6-1 방수형식의 적용

① 방수형식

- 도로터널의 경우 대부분 산악지에 위치하여 지질이 양호한 경우가 많으므로 대부분 배수형 부분방수 형식을 채택하여 왔으며, 이 방법이 경제적으로 유리

【비배수형 완전방수와 배수형 부분방수의 비교】

구 분	배수형(부분방수) 방수	비배수형(완전방수) 방수
적 용 조 건	• 지하수 유출이 비교적 적은 구간 • 터널 심도가 깊고(50m 이상) 지하수위가 높아 2차 라이닝에 수압이 크게 작용하는 경우	• 하천 통과부 등 지하수유출이 과다하여 유지관리상 문제가 있는 구간 • 지하수 강하 시 터널주변 지표침하로 피해가 예상되는 경우
터널구조형태	• 마제형 또는 난형	• 원형 또는 원형에 가까운 난형
설계 하중	• 라이닝 콘크리트 자중	• 수압 • 라이닝 콘크리트 자중
공 사 비	• 원형단면에 비해 경제적	• 원형에 가까운 단면으로 굴착량이 많아 비경제적임.
유 지 관 리	• 유지관리비가 비교적 많이 든다. • 누수 시 보수가 가능함.	• 청결하여 유지관리가 용이하다. • 유지관리비가 적게 든다. • 누수 시 보수가 어려움.
주변 환경에 미치는 영향	• 지하수위 저하로 지하수 이용에 영향을 미치며 얕은 심도의 터널에서 주변 지반침하 발생.	• 지하수위를 계속 유지할 수 있으므로 주변 환경에 거의 영향을 주지 않는다.
시 공 성	• 대단면 및 변단면에 시공이 가능함. • 수압을 고려하지 않으므로 구조적으로 안전하며 라이닝 부재의 두께가 상대적으로 얇다. • 누수 시 보수가 가능하다.	• 대단면 및 변단면에서의 시공이 어려움. • 라이닝의 두께가 커지게 되며, 철근 보강이 필요하다. • 방수재에 결함이나 손상이 있는 경우 압력수가 유출하게 되어 복구가 곤란하다.

3-6 Waterproof / Drainage Design

3-6-1 Waterproof Type Application

① Waterproof type

- Drainage partial waterproof method has been used for road tunnel as the structure is located in mountainous area mostly with stable topological condition.(economically more feasible as well)

【comparison between non-drainage complete waterproof and drainage partial waterproof】

classification	drainage partial waterproof	non-drainage complete waterproof
applying condition	• Area with less underground water discharge • Deep tunnel depth (50m and above), higher water level so that water pressure is influencing heavily to the second lining	• Area with excessive underground water discharge • Damage is expected with surface settlement when water level decreases
tunnel structure shape	• Horse shoe or egg	• Crcular or egg close to circle
design weight	• Lning concrete weight	• Water pressure • Lining concrete weight
cost	• Economic compared to circular share	• Expensive due to large amount of excavation
maintenance	• Maintenance cost is relatively high • Able to repair when there is leakage	• Easy, clean and unexpensive to maintain • Hard to repair when there is leakage
influence to nearby environment	• Decreased underground water level • Ground sinking for low depth tunnel	• Almost no influence to nearby environment as the water level remains the same.
construction easiness	• Able to construct in large scale section or changing section • Structurally stable as there is no need to consider water pressure. relatively thin lining • Able to repair when there is leakage	• Hard to construct in large scale section or changing section • Lining gets thicker and reinforce rod is required • When there is defect or damage in waterproof material, it becomes very hard to repair the leakage of underground water

② 방수재료

- 터널 방수설계에 있어서는 터널 주변 시설물에 대한 영향뿐만 아니라 경제성 및 현재 적용되고 있는 방수공법 · 기술수준 · 사용자재 · 시공성 등에 대한 검토가 선행되어야 하며, 방수공법으로는 구체방수공법 · 도막방수공법 · Sheet 방수공법 등이 있으며, 각각의 특징 및 장단점은 다음과 같다.

【터널구간 방수공법 비교검토 및 선정】

공 법	구체방수	도막방수	Sheet 방수공법	
			Asphalt	ECB Sheet
재료구성 성 분	• AE제 • 혼합재	• Asphalt 합성수지	• 고무화아스팔트 (콤파운드)	• ECB Sheet • EVA Sheet • VE Sheet
장 점	• 수밀성이 큼 • 구조물 자체가 방수 역할 • 시공성 양호 • 경제적	• 제품이 다양 • 시공실적이 많음 • 경제적 • 시공성 양호	• 공법 일반화 • 시공 경험 풍부	• 확실한 방수효과 • 온도변화에 안정 • 공기 단축
단 점	• 국내 사용실적 없음 • 완전한 방수효과기대가 어려움	• 역청재의 경화 시간 과다 소요 • 내구성이 없음	• 습기에 약함 • 이음부, 우각부 누수 발생 • 하자보수 어려움	• 이음부 처리 어려움 • 터널 아치부 처짐

【터널 방수 상세도】

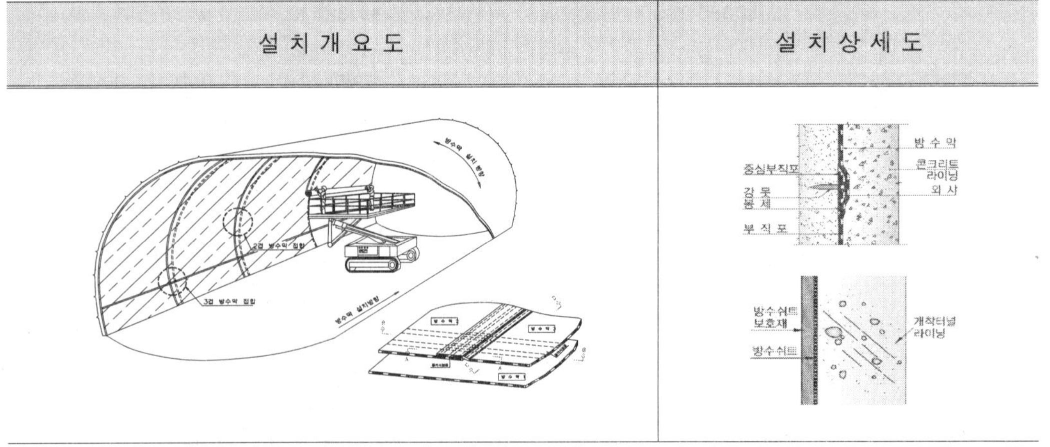

② Waterproof material

- Consider influence to facilities near the tunnel, economic feasibility, waterproof engineering trend and technique, material, construction easiness. There are body type, membranous type, sheet type waterproof. Characteristics, advantage and disadvantage of each method are shown below :

【comparison among tunnel waterproof method】

Method	Body	Membranous	Sheet	
			Asphalt	ECB Sheet
material	• AE agent • Admixture	• Asphalt Compound	• Rubber asphalt (compound)	• ECB Sheet • EVA Sheet • VE Sheet
advantage	• High water tightness • Structure body itself performs waterproof function • Easy to construct • Economic	• Various products • Lots of experiences • Economic • Easy to construct	• Generalized method • Lots of experiences	• Very effective waterproof • Stable for temperature change • Faster construction
disadvantage	• No domestic experience • Cannot expect complete waterproof	• Hardening time of bituminous agent is too long • Less durability	• Weak to humidity • Leakage in connection and arch sector • Hard to maintain	• Hard to finish connection part • Sag in arch sector

【tunnel waterproof detailed drawing】

concept	details

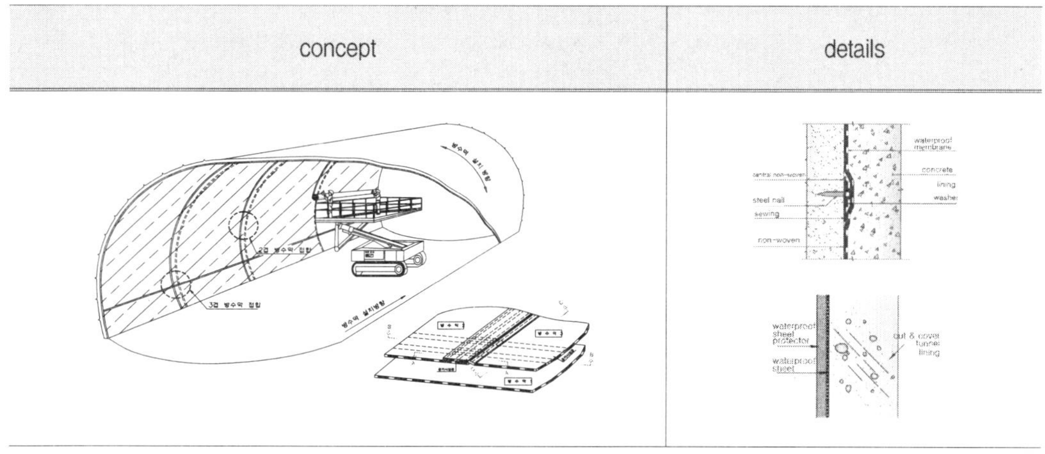

3-6-2 터널 내 용수처리

① 터널에서 처리되어야 할 용수

- 터널 측벽 숏크리트 및 지하배수공으로 유입되는 지하수
- 청소 및 기타 노면으로 유입되는 노면수

② 배수 파이프는 겨울철 동해방지를 위해 지역적 조건, 동결심도 및 용수량 상태 등을 충분히 고려하여 철저히 시공

③ 배수관로는 집수정까지 반드시 연속된 기울기를 갖도록 시공하고, 서로 다른 지보패턴이 연결되는 부분에서는 단차발생에 주의

【용수처리의 흐름도(오 · 폐수 비분리 시)】

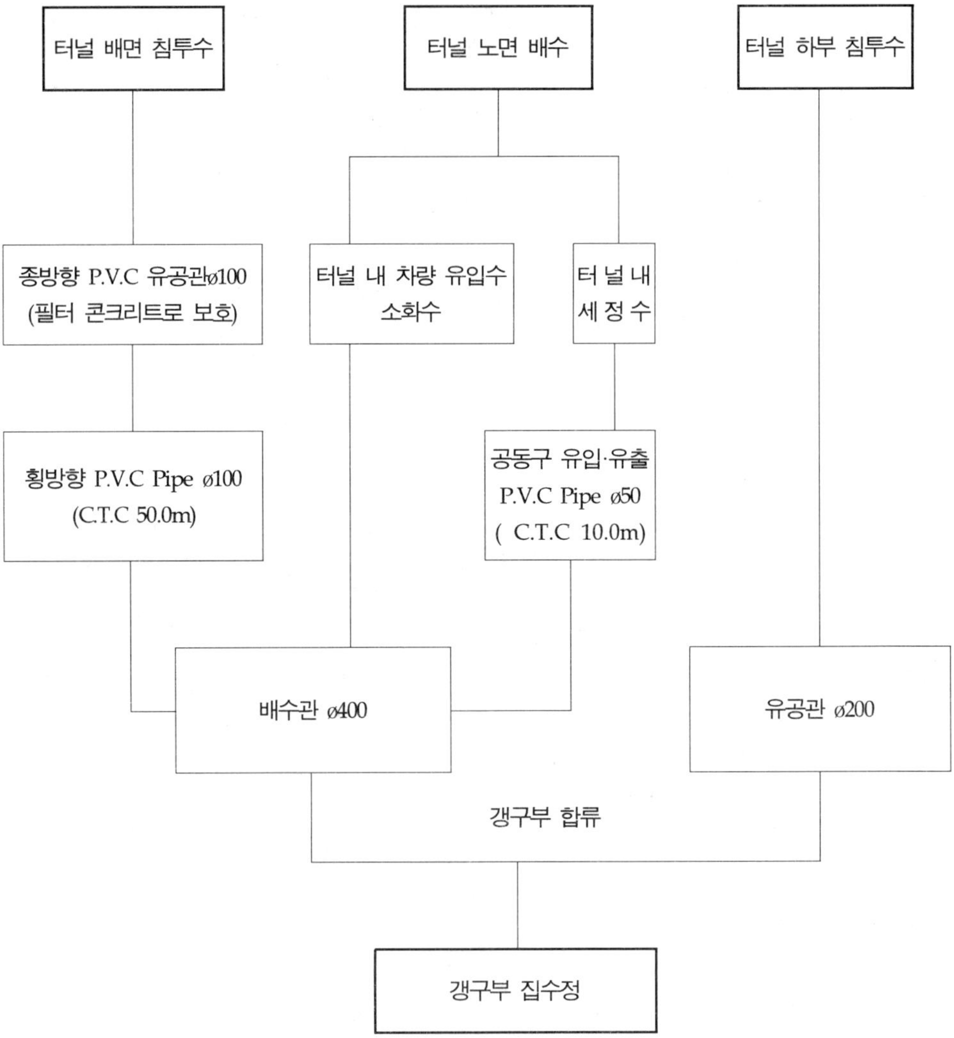

터널 배면 침투수

터널 노면 배수

터널 하부 침투수

종방향 P.V.C 유공관ø100
(필터 콘크리트로 보호)

터널 내 차량 유입수
소화수

터 널 내
세 정 수

횡방향 P.V.C Pipe ø100
(C.T.C 50.0m)

공동구 유입·유출
P.V.C Pipe ø50
(C.T.C 10.0m)

배수관 ø400

유공관 ø200

갱구부 합류

갱구부 집수정

3-6-2 Water Management in Tunnel

① Type of water required to be managed in tunnel

- Underground water discharged to tunnel sidewall shortcrete and underground drainage
- cleaning water and other surface water

② Drainage pipe should consider regional factors, freezing penetration, and water amount to avoid winder frost

③ Drainage pipe line should have continuous slope to collector well. attention to the connection of different support pattern to avoid stage difference

【Water management flow (polluted and waste water are not separated)】

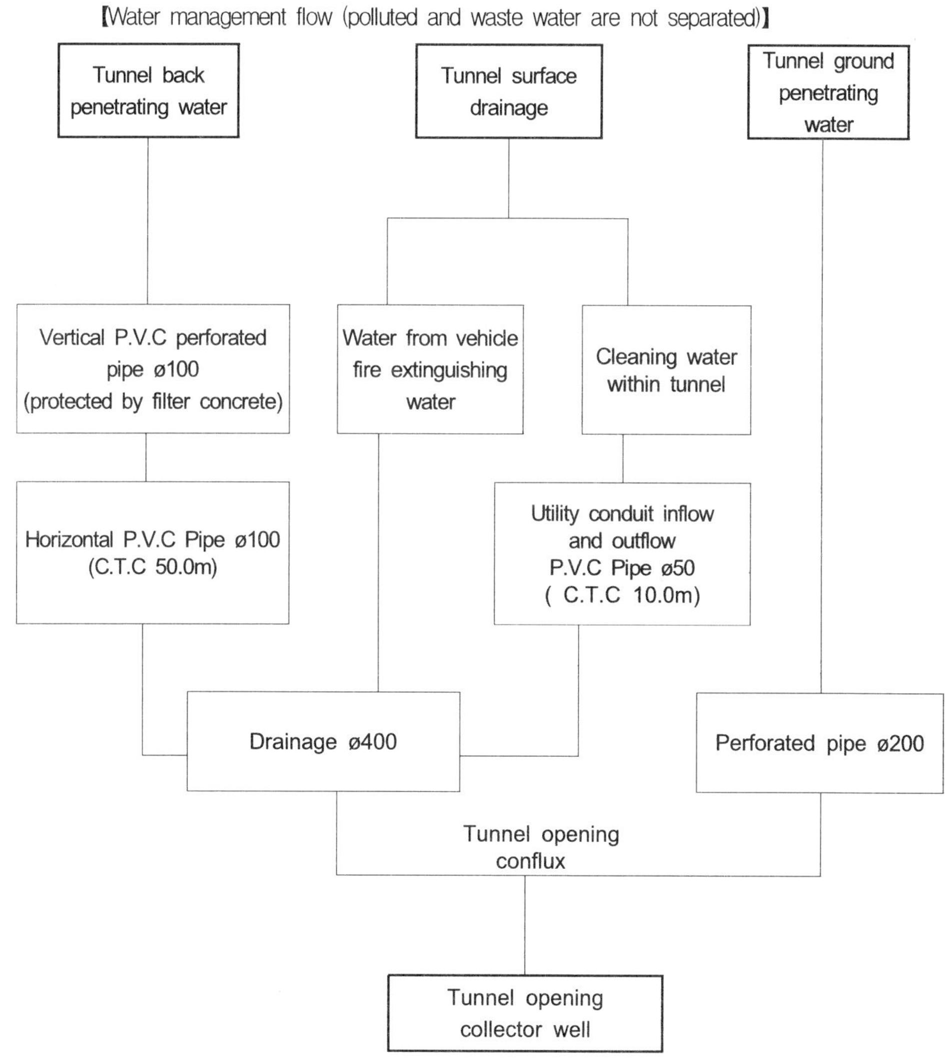

3-7 갱구부 및 갱문 설계

3-7-1 개 요

① 갱문은 지형 및 지반조건, 기상조건, 주변환경, 유지관리상의 편의성 등을 고려한 위치 및 형식으로 결정되어야함

② 터널의 갱구부는 터널 구간 중 피복두께가 가장 얇은 부분이므로 암석이 풍화되었거나 절리가 발달되어 있는 경우가 대부분

③ 특히, 지표수에 의하여 영향을 받기 쉬울 뿐만 아니라 급경사지와 접하는 경우에는 편토압이 작용하기도 함

3-7-2 갱구위치 선정 시 주요 고려사항

① 안정성 측면

- 배면의 토압, 낙석 등에 대해 안정한 구조물로 계획
- 지형의 조건이 횡단면상 가능한 한 대칭적 구조가 되도록 하여 편토압을 받지 않는 위치 선정
- 늪이나 계곡과는 교차를 피하도록 선정
- 교량과 근접되는 경우는 갱문 기초의 지반반력 분포 범위와 교대의 굴착과의 관계를 충분히 검토하여 선정

② 시공성 관련

- 시공이 용이하고 제약조건이 없도록 계획
- 근접하는 교대 등 구조물과 영향을 충분히 고려한 위치 및 구조 계획
- 갱구부근에 계획된 터널관련 구조물, 유지관리 시설 등의 배치에 대해 고려

③ 미관 측면

- 터널의 갱문은 주변지형과 자연스럽게 접하는 구조
- 콘크리트면이 크면 조명상 야외조도가 크게 되어 완화조명의 수준 적용에 영향이 있으므로 가능한 한 콘크리트면이 작은 설계가 미관 측면에 유리

④ 기상 조건

- 갱문은 산사태, 이상용수, 적설지에 따른 눈의 유입, 눈사태, 제설작업 등 고려

3-7 Tunnel Opening and Gate Design

3-7-1 Overview

① Tunnel gate should be located and selected with considerations on topology and ground condition, weather, nearby environment, maintenance easiness

② In most cases, rocks are weathered and joints are developed around tunnel opening since the covering is thinnest

③ Especially, opening are easily influenced by surface water. If the opening is on stiff slope, opening can be influenced by declining earth pressure as well

3-7-2 Major Consideration for Selecting Tunnel Gate

① Stability

- Stable structure to the backside soil pressure and falling rock
- Design soil condition to be horizontally symmetrical so that there will be no declining earth pressure
- Avoid crossing to wetland and valley area
- When located near bridge, consider relation between ground reaction distribution scope and bridge excavation

② Construction easiness

- Easy to construct with less limitation as possible
- Plan location and structure with consideration on impact from nearby structures such as bridges
- Consider allocation of tunnel related structures and maintenance facilities

③ Aesthetic considerations

- Connect to nearby topology condition smoothly
- Expose concrete face as small as possible since concrete face decreases effectiveness of dimmed lighting with high level of reflection

④ Weather condition

- Consider landslide, irregular water content, snow inflow, snow-slide, snow removal

3-7-3 갱문 형식

① 갱문 형식

구 분	Bell Mouth 변형	Bird Beak형	Arch 면벽형
측면도	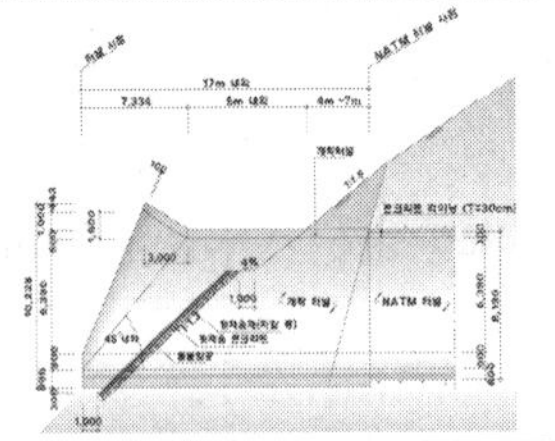	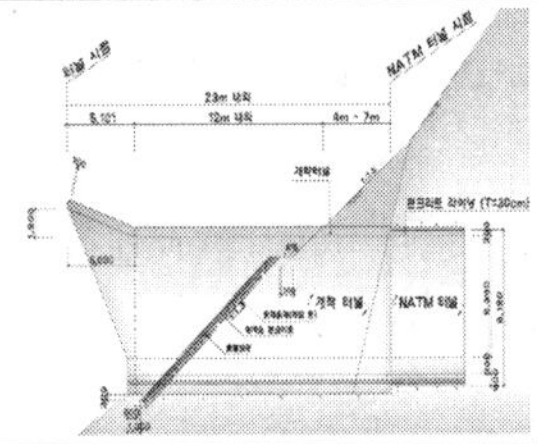	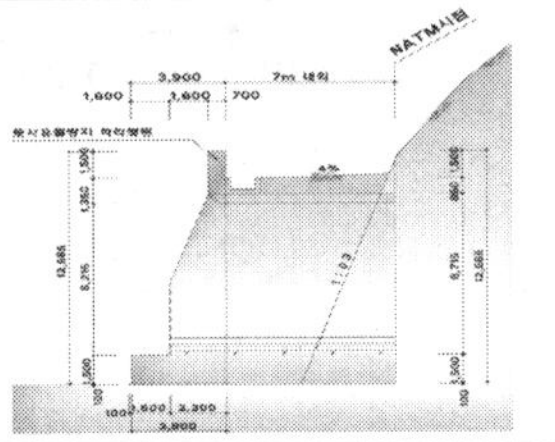
정면도	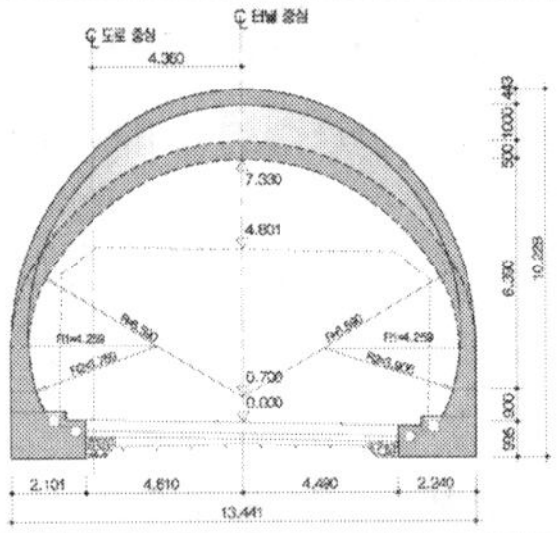	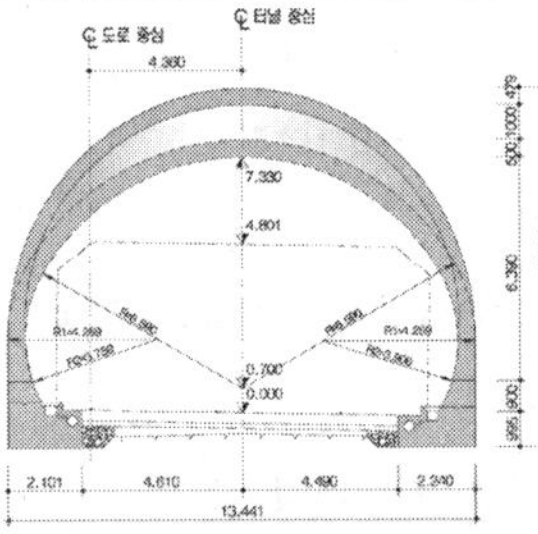	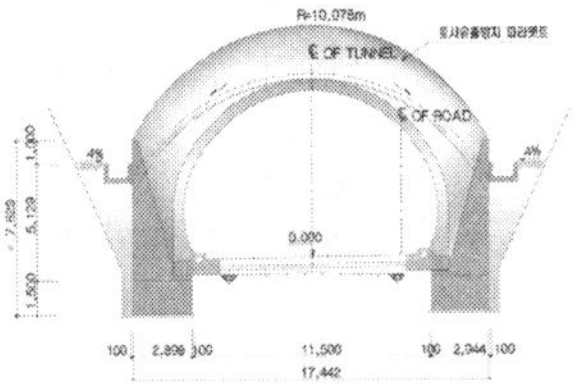
조감도		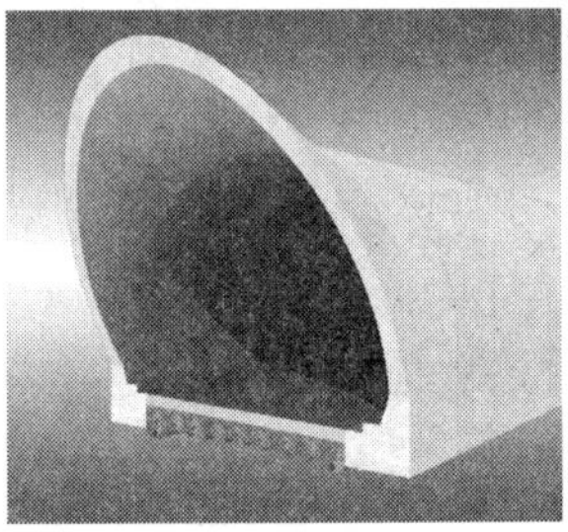	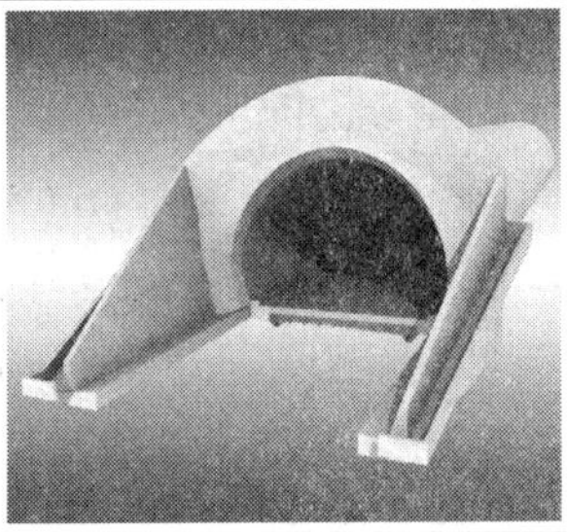
적 용 지 형	• 갱구상단 자연경사면 : 완사면(30° 미만) • 강우 시 유속 영향작음 • 낙석·산사태 우려적음 • 갱구 NATM 위치 선정 : 토피고 3~5m	• 갱구상단 자연경사면 : 급사면(30° 이상) • 강우 시 유속 영향이 큼 • 낙석 · 산사태 우려 많음 • 갱구 NATM 위치 선정 : 토피고 3~5m	• 갱구상단 자연경사면 : 급사면(30° 이상) • 편토압 예상지역 • 토류 옹벽식 구조물이 유리한 경우 • 배수처리 필요 지형
장단점	• 도로와 자연스럽게 접속하여 유도되므로 운전자에게 안정감을 준다. • 자연비탈면이 완만한 지형으로 조화를 이루어 미관이 수려함 • 개착터널 상부에 인위적인 흙쌓기가 필요	• 도로와 자연스럽게 접속하여 유도되므로 운전자에게 안정감을 준다. • 자연비탈면이 급하여 낙석 및 산사태로 인한 도로유실을 방지한다. • 주변지형과 조화를 이루어 미관이 수려하다 • 개착터널 상부에 인위적인 흙쌓기가 필요	• 터널 상부에 인위적인 되메우기량이 적다. • 터널 진입시 운전자에게 위압감을 준다 • 인위적인 구조물 설치로 주변환경과 조화가 부자연스럽다.

3-7-3 Type of Tunnel Gate

① Type of tunnel gate

Classification	Modified Bell Mouth	Bird Beak	Arch facing wall
Side view	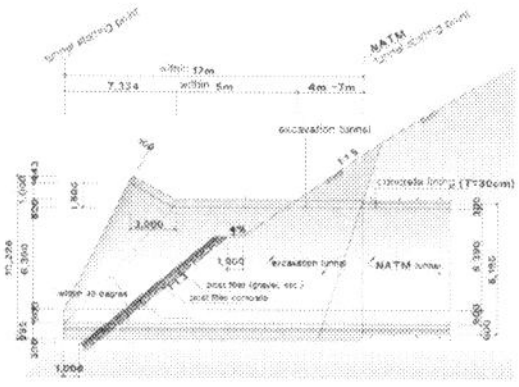	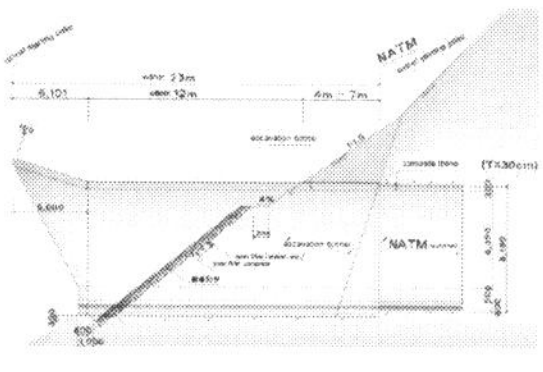	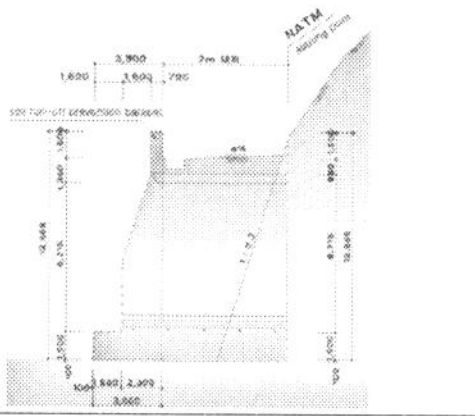
Front view	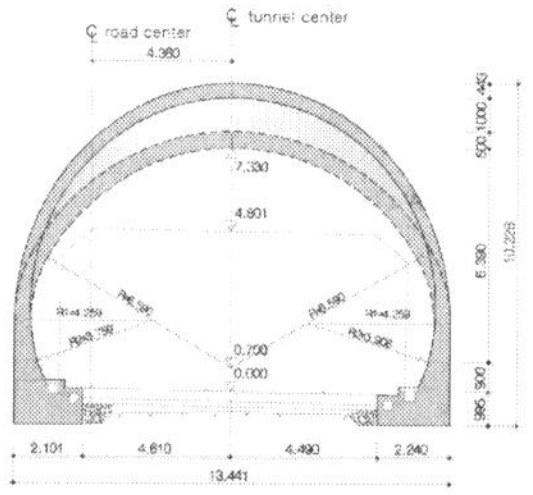	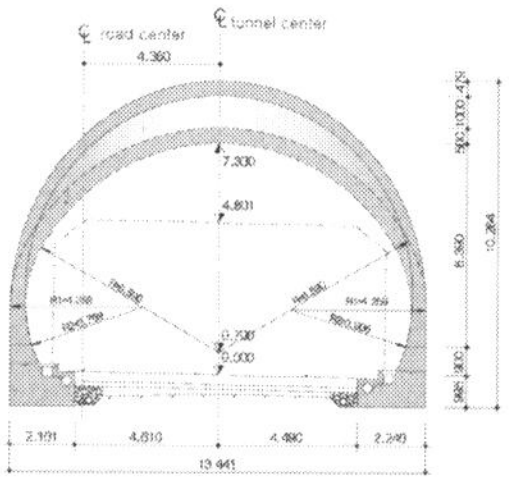	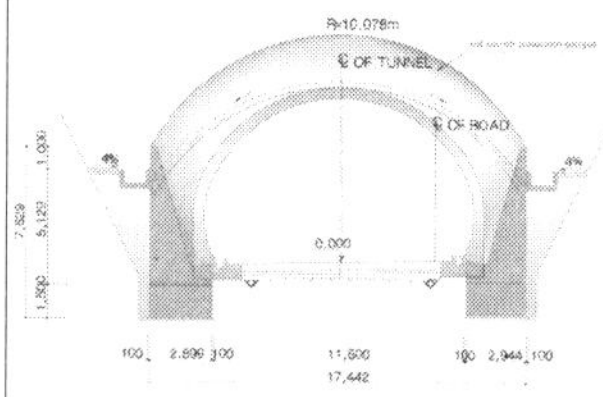
Bird"s eye view			
Applicable topology	• Natural slope on top of the gate : Gentle slope (less than 30°) • Less effects of water stream during rainfall • Less probability of falling rocks or landslide • Selecting the location of gate NATM : Soil surface height 3~5m	• Natural slope on top of the gate : Stiff slope (30° or higher) • Bigger effects of water stream during rainfall • Higher probability of falling rocks or landslide • Selecting the location of gate NATM : Soil surface height 3~5m	• Natural slope on top of the gate : Stiff slope (30° or higher) • Area where declining earth pressure is expected • Area where retaining wall structure is appropriate • Area where drainage is required
Advantage and disadvantage	• Driver feels comfortable as the gate smoothly connects to the road • Natural slopes effortlessly blends in with the surrounding to provide beautiful environment. • Artificial landfill is required on the upper side of tunnel	• Driver feels comfortable as it smoothly connects to the road • Stiff slope can trigger falling rocks or landslide. • Great aesthetics by complementing the surrounding environment • Artificial landfill is required on the upper side of tunnel	• Very little backfill is required on the upper side of tunnel • Driver feels uncomfortable when entering the tunnel • Unnatural look due to artificial structures that are installed

3-8 터널내장 및 내부도장

3-8-1 내장의 목적

- 터널의 내장은 아래와 같은 목적으로 설치
  - 조명효과 증대, 터널 내 시각 환경을 양호하게 유지
  - 터널 벽면을 일원화하여 미관 유지 및 터널 내 소음 흡수

3-8-2 내장재료로서의 선정조건

구 분	선 정 조 건
내구성	장기간에 걸친 환경조건에서 재료의 성질 유지
내오염성	터널 내 분진 오염에 강한 재료
반사성	터널 내 조명효과를 최대한 확보할 수 있는 재료
유지보수	부분적 파손, 균열에 용이하게 유지 보수할 수 있는 재료
청소	오염물질 제거에 용이하고 방수성이 있는 재료
색상	시각적 즐거움, 인지성을 줄 수 있는 색상 선택의 가능성
시공성	단순한 작업에 의한 시공 가능성
흡음성	차량 소음을 감소시킬 수 있는 재료

3-8-3 내장재료의 종류 및 비교검토

구 분	타 일	도 장	Concrete PANEL	PERLITE BLOCK (PANEL)
장 점	• 미려한 색상 • 내구성 우수 • 내식성 양호 • 내후성 양호 • 불연제 • 반사성 양호	• 다양한 색상 (도안 가능) • 내구성 양호 • 건식공법 가능	• 내식·내후성 양호 • 내구성 우수 • 불연재	• 색상 조정 • 내식·내후성 양호 • 방습성 양호 • 반사율 우수 • 흡음성 양호
단 점	• 줄눈의 오염성 높다 • 숙련된 면처리 필요 (탈락 가능성)	• 방수·흡음성 불량 • 반사성 높다 • CON'C 면정리 필요	• 흡음성 불량 • 조명이 난반사함	• 판넬면의 난반사 우려
시공성	• 세심한 시공 필요	• 시공 간단	• 시공 간단	• 시공성 불량 • 세심한 시공 필요
유지보수	• 반영구적 • 보수 가능 • 세척 양호	• 정기적 재도장 • 결함부 미관 불량 • 유지보수·청소 용이	• 유지보수·청소용이	• 반영구적 • 파손 시 보수곤란 • 세척 곤란
공사비	• 다소 고가	• 저렴	• 고가	• 고가
적용 예	• 국내 터널 다수	• 마감도장	• 외국에서 적용	• 남산 1호 터널 • 외국에서 적용

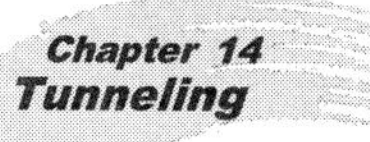

3-8 Tunnel Interior and Painting

3-8-1 Purpose of Interior

- Installed to meet the following purposes:
  - Increased lighting effect, clean visual environment
  - Aesthetical uniformity of the wall and noise absorption

3-8-2 Interior Material Selection Criteria

Classification	criteria
Durability	Maintain material characteristics for long period of time
Pollution resistance	Highly resistant to dust pollution inside tunnel
Reflection	Maximize lighting effectiveness
Maintenance	Easy to cope with partial damage and cleavage
Cleaning	Water-resistant, easy to remove taints
Color	Color of visually delight and easy to recognize
Construction easiness	Simple construction
Noise absorption	Material to reduce vehicle noise

3-8-3 Type of Interior Materials and Comparison

Classification	Tile	Paint	Concrete PANEL	PERLITE BLOCK (PANEL)
Pros	• Excellent color • High durability • Good corrosion resistance • Good decay • Resistance • Nonflammable • Good reflection	• Various color (design) • Good durability • Dry wall support system	• Good corrosion/ decay resistance • High durability • Nonflammable	• Adjustable color • Good corrosion / decay resistance • Good humid resistance • Good reflection • Good noise absorption
Cons	• Joints are easy to be tainted • Experienced finish required (not suitable)	• Poor water resistance/ noise absorption • Too much reflection • CON'C face arrangement required	• Poor noise absorption • Diffused reflection of lighting	• Diffused reflection of panel
Construction easiness	• Delicate finish	• Easy to construct	• Easy to construct	• Hard to construct • Delicate finish
Maintenance	• Semi-permanent • Easy to repair • Good to clean	• Periodical re-paint • Defected area is aesthetically poor • Easy to repair and clean	• Easy to repair and clean	• Semi-permanent • Hard to repair and clean
Cost	• Relatively expensive	• Cheap	• Expensive	• Expensive
Examples	• Many domestic tunnels	• Finish painting	• Many foreign tunnels	• Namsan 1st tunnel • Applicable to foreign tunnels

3-8-4 내장의 설치

◎ 국내 도로터널의 내장은 좌·우 측벽에만 설치

구 분	터 널 연 장	설치 높이
I	300m 미만	미 설 치
II	300m 이상 ~ 1,000m 미만	2.0m
III	1,000m 이상	3.0m

주) 1. 단, 연장 300m 미만의 터널이라도 종단경사 3% 이상, 평면곡선반지름 500m 이하, 대형차 혼입율이 높을 것으로 예상되는 경우, 터널이 연속되어 설치된 구간에서는 내장재를 설치할 수 있다.

2. 검사원 통로 설치 시 검사원 통로 하단 0.25m, 상단 0.20m와 일반구간의 공동구 상단 0.20m는 터널 청소차량의 청소가능범위를 고려하여 타일을 부착하지 않는다.

【내장타일의 설치 범위(예)】

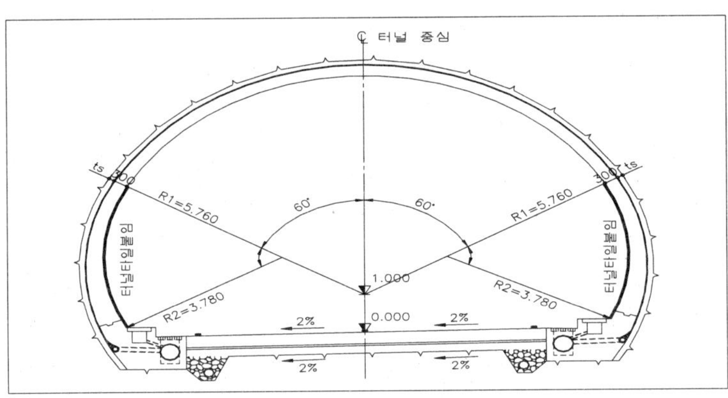

내장타일 설치	내오염도장 설치

3-8-4 Interior Installation

◯ For domestic road tunnel, install only on right and left side of the wall

Classification	Tunnel length	Installing height
I	less than 300m	none
II	300m ~ 1,000m	2.0m
III	1,000m and longer	3.0m

Note) 1. Even for tunnel with length of less than 300m, interior can be installed when vertical slope is above 3%, plane curvature radius is less than 500m, and ratio of heavy vehicle is expected to be high

2. When inspector passage is installed, do not attach tile to lower 0.25m and upper 0.20m of the passage and upper 0.20m of utility conduit considering cleaning capacity of tunnel cleaning vehicle

【example of interior tile installation scope】

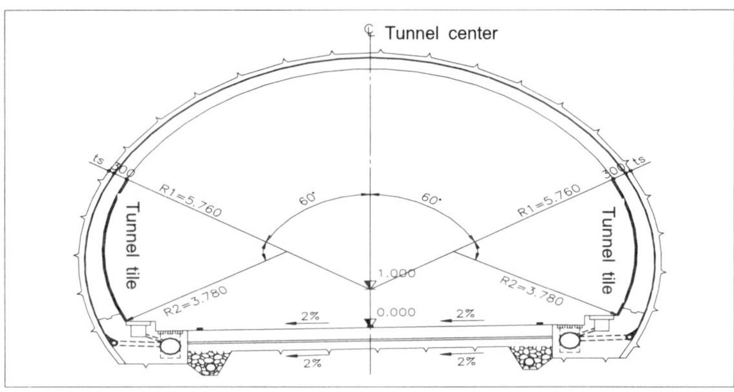

Interior tile	Pollution-resistance paint

3-9 계　　측

3-9-1 계측의 목적 및 계획

① 계측 목적

- 터널 굴착 시 지반의 거동 분석과 각 지보재의 지보 효과를 측정 및 분석하여 터널 주변에 미치는 영향과 터널 구조물의 안정성을 확인하여 시공 시 제반 안정성을 부여하고 장래 계획 시 자료로써 활용하는데 있음

- 시공되는 터널은 입체 구조물이므로 제한된 사전 지질조사에서 얻는 정보의 한계와 지반의 굴착방법 및 시공순서, 시공의 정확성 등에 따른 응력의 발생량, 지반의 강도 등이 불확실함

- 계측수단을 통하여 정확하게 파악하여 대처방안을 강구

- 구체적인 계측의 목적을 요약하면 다음과 같다.
  - 지반 거동의 관리
  - 지보공 효과의 확인
  - 안정상태의 확인
  - 근접 구조물의 안전성 확인
  - 장래 공사 설계의 자료 확보

② 계측계획

- 단계별 계측 설정은 다음과 같다.
  - 계측항목 설정 : 산악터널, 지하철
  - 계측단면 결정 : 지질조사, 주계측 및 보조계측 단면의 배치
  - 계측항목의 결정 : 계측목적에 부합, 터널규모 고려
  - 관리기준 설정 : 변위기준결정, 근접구조물 허용변형
  - 계측기 사양 결정 : 계기의 용량, 정밀도, 검출기 형식
  - 설치위치 결정 : 계기의 배치
  - 계측간격, 회수 결정

3-9 Measurement

3-9-1 Purpose and Planning of Measurement

① Purpose

- To verify tunnel structure safety and impact to nearby area through analyzing ground movement and supporting effect of each support so that the construction safety is secured as well as the measurement result can be used as future planning basis

- As the constructed tunnel is three dimensional structure, there will be limitation and uncertainty from preliminary topology analysis data such as stress induced from construction precision and order, strength of ground

- Use measurement method to understand those data precisely and plan proper counter measurement

- To summarize the purpose of measurement:
  - management of ground movement
  - check the effect of support hole
  - check the stability
  - check the safety of nearby structures
  - secure data for future planning and construction

② Measurement plan

- Ataged measurement plan is as below:
  - select measurement subject : mountainous tunnel, subway
  - select measurement section : topology survey, allocation of main measurement and sub measurement section
  - select measurement items : meet measurement purposes and consider tunnel scale
  - determine management standard : changing standard, allowable deformation change of nearby structure
  - determine measurement device specification : capacity, preciseness, detector type
  - select installing location : allocation of devices
  - determine interval and time

③ 계측계획

- 지반조건에 따른 계측항목의 중요도

항목 / 지반조건	일상관리계측 (계측A)					대표위치계측 (계측B)				
	막장 관찰 조사	내공 변위 측정	천단 침하 측정	숏크리트 응력 계측	록볼트 인발 시험	지중 변위 측정	록볼트 축력 측정	라이닝 응력 측정	원지반 시료 시험	갱내탄성 파속도 측정
경암지반 (단층, 파쇄대 제외)	◎	◎	◎	◎	△	△	△	△	△	△
연암지반 (큰 소성 발생 없음)	◎	◎	◎	◎	△	△	△	△	△	△
연암지반 (큰 소성 발생)	◎	◎	◎	◎	◎	◎	◎	○	○	△
토 사	◎	◎	◎	◎	◎	○	◎	△	○	△

(주) ◎ : 반드시 실시하는 항목, ○ : 실시하여야 할 항목, △ : 필요 시 실시하는 항목

【일상관리계측 단면도】

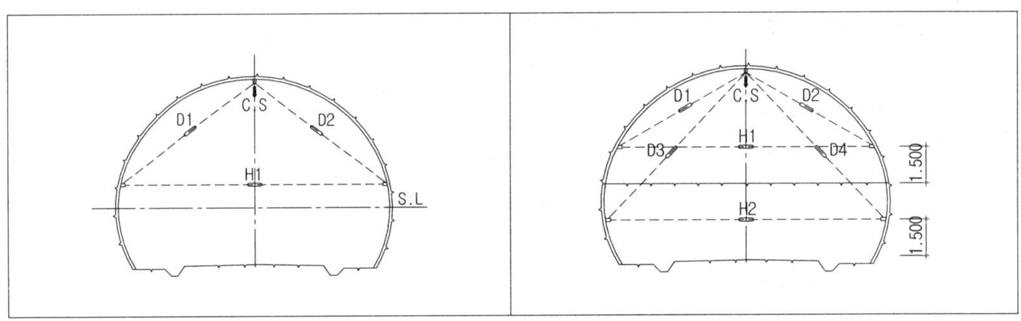

【대표위치계측 단면도】

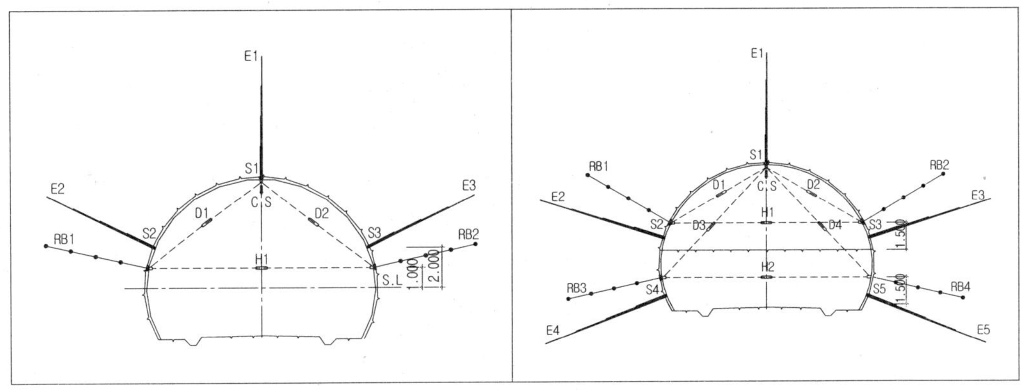

③ Measurement plan

- Importance of measurement item according to ground condition

Item / Ground condition	Daily management measurement (measurement A)				Representative location measurement (measurement B)					
	Blind-end survey	Interior change	Shotcrete stress measurement	Rock bolt pull-out test	Ground change	Rock bolt axial force	Lining stress	Rock bolt pull test	Original ground sampling test	Elastic wave velocity
Hard-rock (exclude staged, cracked area)	◎	◎	◎	◎	△	△	△	△	△	△
Soft-rock (no large plasticity)	◎	◎	◎	◎	△	△	△	△	△	△
Soft-rock (large plasticity)	◎	◎	◎	◎	◎	◎	◎	○	○	△
Soil	◎	◎	◎	◎	◎	○	◎	△	○	△

(Note) ◎ : highly obligatory, ○ : obligatory, △ : recommended if needed

【Cross-section for daily management measurement】

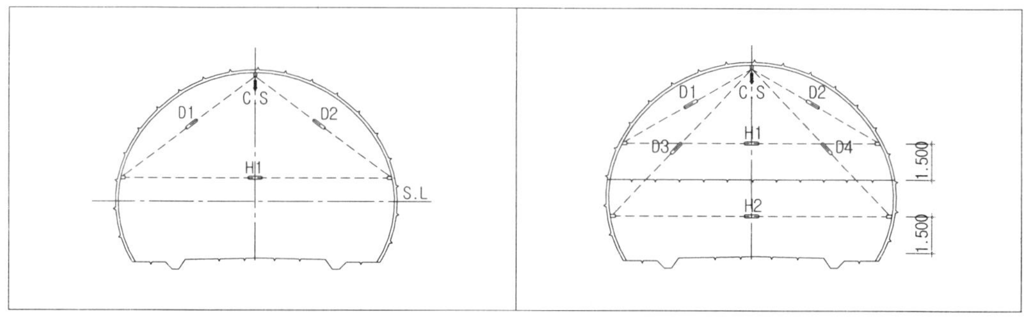

【Cross-section for Representative location measurement】

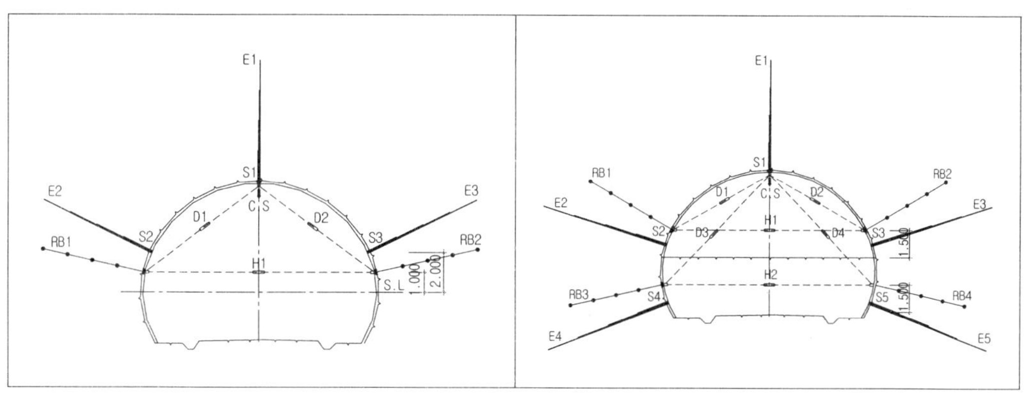

④ 계측평가 사항

구 분	항 목	계측에 의한 평가사항
일상 관리 계측 (계측 A)	막장 관찰 조사	발파 후 막장 상태를 관찰하여 안정성 분석 ① 암반의 풍화도, 등급 ② 절리의 상태 ③ 지하수 상태
	내 공 변 위	터널 측벽의 변위량, 변위속도, 수렴상태 등 측정 ① 주변지반의 안정성 ② 1차 지보재 설계 시공의 타당성 ③ 2차 라이닝 타설시기 판단
	천 단 침 하	터널 천단의 침하량, 침하속도, 수렴상태 등 측정 ① 터널 상부 기반의 안정성 ② 1차 지보재의 타당성 ③ 2차 라이닝 타설시기 판단
대표 위치 계측 (계측 B)	지 중 변 위	터널 굴착면 주변의 소성 영역 측정 ① 록볼트 길이의 타당성 ② 소성영역의 확대성 여부 ③ 지보재의 안정성
	숏크리트 응력	① 숏크리트의 축력 크기 및 수렴상태 ② 숏크리트에 의한 작용토압의 안정성
	록볼트 축력	① 록볼트에 생긴 변형을 측정하여 축력 효과 확인 ② 록볼트 길이 및 개수 증가 기본자료 ③ 소성영역 확인

【유지관리 계측 단면】

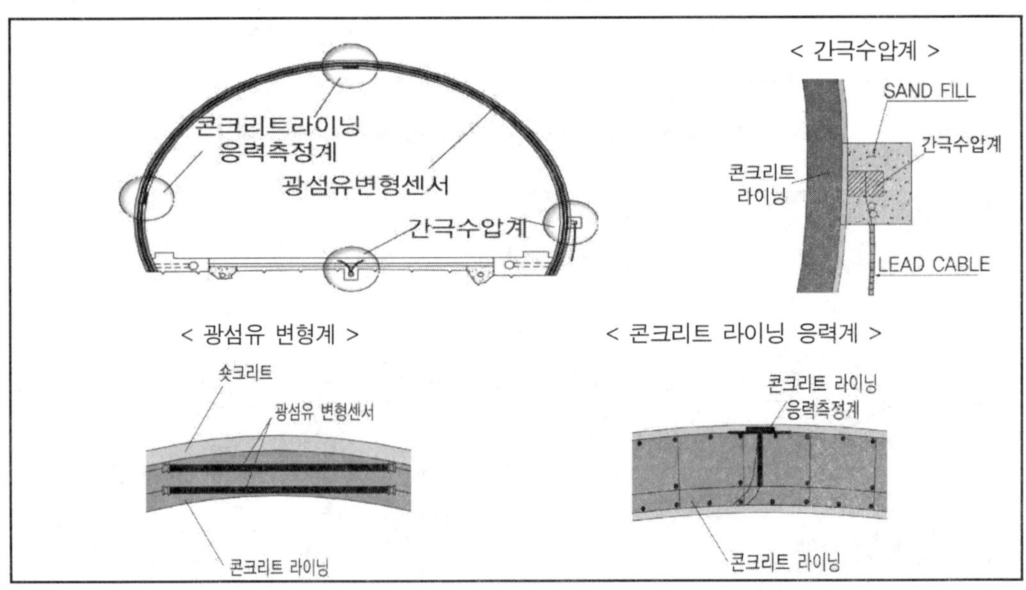

④ Measurement evaluation items

Classification	Item	Items to be evaluated through measurement
Daily management measurement (measurement A)	Blind-end survey	analyze stability through observing blind-end after blasting ① rock base weathering, grade ② joint status ③ underground water status
	Interior change	measure change amount, change speed, convergence status ① stability of nearby ground ② feasibility of firsthand support design and construction ③ determine when to cast second lining
	Arch settlement	measure tunnel arch settlement amount, speed, convergence status ① stability of upper ground of tunnel ② feasibility of firsthand support design and construction ③ determine when to cast second lining
Representative location measurement (Measurement B)	ground change	measure plasticity area nearby excavation ① appropriateness of rock bolt length ② whether plasticity area is extending or not ③ stability of support
	Shotcrete stress	① shotcrete stress amount and convergence status ② stability of soil pressure from shotcrete
	Rock bolt axial force	① measure rock bolt deformation to check the effect of axial force ② basic data for rock bolt length and amount increase ③ check plasticity area

【Cross-section of maintenance measurement】

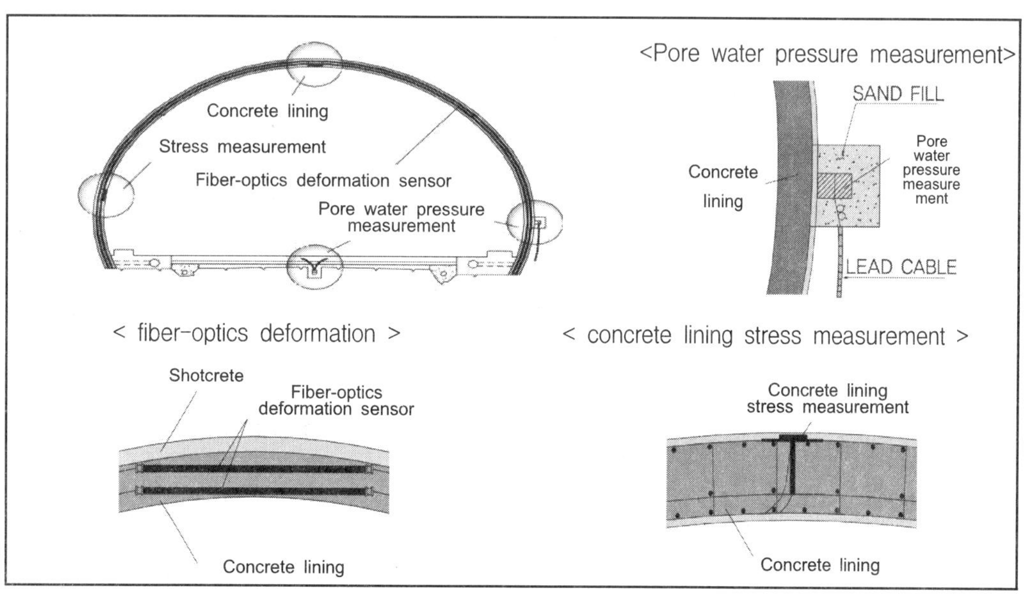

3-10 공사 중 환기

3-10-1 유해물질의 허용기준

- 환기를 함에 있어서 지질·입지조건 및 작업방법 등을 충분히 고려한 뒤, 발생 원인이 되는 유해물질의 허용기준과 그 특성 등을 확인하는 것이 중요

【터널 공사 시 유해가스의 허용농도】

유해가스	허용농도
CO	100 ppm
NO_x	25 ppm
CO_2	5,000 ppm

【유해가스의 허용기준】

종 류	비중 (공기 1.0)	발파범위 (Vol %)	허용농도 (ppm)		산업안전 보건법 규제치	비 고
			일본	미국		
황화수소 (H_2S)	1.199	-	10	10	10 ppm	-
메 탄 (CH_4)	0.55	5.3~14.0	-	-	1.5%	-

【분진의 허용농도 (일본 산업위생학회 자료)】

구 분	분진의 종류	허용농도 (㎎/㎥)	
		흡수성 분진	총 분진
제1종 분진	골석, 납, 알루미늄, 알루미나, 규조토, 유화광, 유화소광, 벤토나이트, 칼리나이트, 활성탄, 흑연	0.5	2
제2종 분진	유리규사 10% 미만의 광물성 분진, 고화철 카본블랙, 석탄, 고화아연, 이산화치탄, 포틀랜트 시멘트, 석탄석, 대리석, 선향 재료분진, 곡분, 선진, 목분, 초분, 콜크분, 베크라이트	1	4
제3종 분진	기타 무기 및 유기분진	2	8

3-10 Ventilation during Construction

3-10-1 Allowance Standard of Hazardous Material

- It is important for ventilation to check allowance standard and characteristics of hazardous material after sufficient consideration on topology, location and construction methodologies

【Allowable density of hazardous gas during tunnel construction】

hazardous gas	allowable density
CO	100 ppm
NO_x	25 ppm
CO_2	5,000 ppm

【Allowance standard of hazardous gas】

Type	Gravity (air 1.0)	Blasting scope (Vol %)	Allowable density (ppm)		Regulation from industrial safety and health law	Remarks
			JPN	US		
Hydrogen sulfide (H_2S)	1.199	-	10	10	10 ppm	-
Methane (CH_4)	0.55	5.3~14.0	-	-	1.5%	-

【Allowable density of dust (industrial sanitation society data)】

Classification	Type of dust	Allowable density (mg/m³)	
		Absorptive	Total
tier 1 dust	talc ore, lead, aluminum, diatomite, sulfide mineral, bentonite, kalinite, activated carbon, black lead	0.5	2
tier 2 dust	mineral dust with glass silica less than 10%, pit steel carbon black, coal, pit zinc, dioxide chitan, portland cement, marble, incense, corn flour, wood powder, cork powder, bakelite	1	4
tier 3 dust	other organic or non-organic dust	2	8

3-10-2 공사 중 환기방식의 종류

① 터널 관통 시는 자연풍에 의한 환기력 이용

② 굴착공사 중에는 NATM 터널 연장이 200m 이상인 경우 기계적인 강제환기 방식을 적용

③ 강제환기 방식은 굴착연장이 1000m 미만인 경우 송기식 적용

【강제환기 방식별 특성비교】

구 분	송 기 식	배 기 식	송·배기식
개략도			
장·단점	• 외부 신선공기를 작업면에 공급함으로써 오염된 공기가 터널을 통과하여 외부로 배기되므로 터널 전체가 오염될 수 있다. • 닥트가 작업면에서 30m 이상 떨어져 있을 시는 환기효율이 저하되므로 로칼팬 설치가 예상된다. • 비닐 닥트를 사용할 수 있지만 누풍율을 고려하여야 한다.	• 발파가스를 곧바로 터널외부로 배기시킴으로써 터널내 다른 지역이 오염되지 않는다. • 닥트가 작업면에서 30m 이상 떨어져 있을 시는 환기효율이 저하되므로 로칼팬 설치가 예상된다. • 비닐 닥트를 사용할 때 닥트가 찌그러지거나 마찰손실이 증가하므로 스틸닥트 사용이 바람직하다. • 작업이 진척됨에 따라 닥트와 팬을 추가 설치할 수 있다.	• 송기식과 배기식의 단점을 서로 보완하여 환기효율을 높였다. • 배기닥트는 비닐닥트를 사용할 수 있다. • 설비비 및 운전비가 다소 고가이나 환기효과가 우수하다.

3-10-2 Type of Ventilation during Construction

① Use natural ventilation when tunnel penetrates

② During excavation, apply machine forced ventilation for NATM tunnel with length of more than 200m

③ Apply air supply type for tunnel with length of less than 1,000m

【Comparison among forced ventilation types】

Classification	Air supply	Air exhaustion	Mixture of supply and exhaustion
Concept	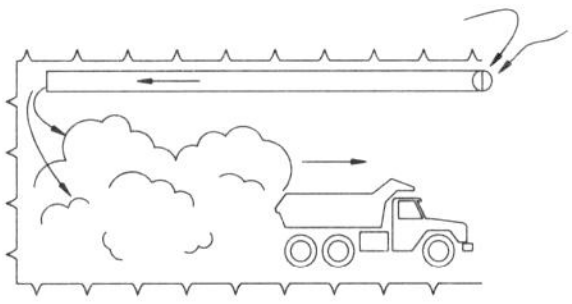	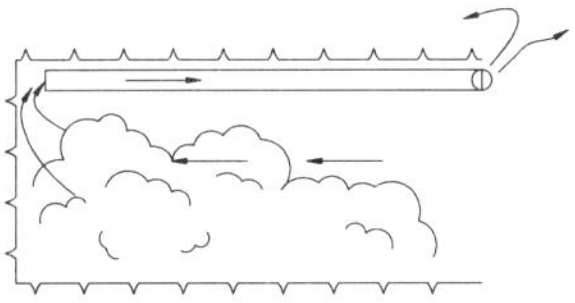	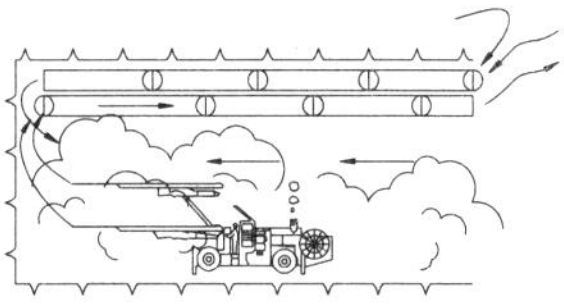
Pros and cons	• As fresh air is supplied to operational area while polluted air is passing through tunnel, entire tunnel can be polluted • When duct is separated form working area for more than 30m, local fan is required to maintain ventilation efficiency • If needed, use vinyl duct with considerations on wind leakage	• Blasting gas is directly exhausted to outside so that entire tunnel pollution is prevented. • When duct is separated from working area for more than 30m, local fan is required to maintain ventilation efficiency • Steel duct is recommended as vinyl duct may be battered • May add more duct and fan	• Mutual supplementation of supply type and exhaustion type to increase ventilation efficiency • May use vinyl for supply duct • Facility and operation is relatively expensive while ventilation effectiveness is excellent

3-10-3 소요 환기량 산정 시 고려 오염원

① 화약발파 후 가스

- 터널 내에서 발파·굴착하였을 경우, 환기계산의 산정기준으로 하는 유해가스의 종류와 폭약 1kg 당 표준발생량은 다음과 같음

구 분	배출된 CO 가스의 합계	NO_x 가스
에멀전계	3.5×10^{-3}	'0.75×10^{-3}
정밀폭약	9×10^{-3}	17×10^{-3}

② 디젤엔진의 배기가스

- 갱내에서 사용하는 디젤기관에 의해 발생되는 유해가스에 대한 환기는 그 디젤기관 PS 당 발생을 기준치로 해서 산출
- 갱내에서는 배기가스 처리장치가 없는 기관을 사용해서는 안 되며, 표준발생량은 다음과 같음

디젤엔진의 종류	Air cleamer 가 있는 경우	Air cleamer 가 없는 경우	가동율
셔블로우더계	1.08 m^3/min.ps	2.16 m^3/min.ps	60%
덤프트럭계	0.67 m^3/min.ps	0.84 m^3/min.ps	60%
기타 기계	0.65 m^3/min.ps	0.82 m^3/min.ps	50%

③ 작업자의 호흡 산소량

- 작업원의 호흡에 의한 갱내의 공기의 오염은 CO_2 임.
- 호흡에 의한 CO_2 가스의 발생은 발파 후 가스 및 디젤기관의 배기가스 등에 의한 경우와 비교해서는 극히 적은 양이지만, 장시간 작업 시 증가
- 특히, 많은 사람이 작업하는 경우는 무시할 수 없으며, 작업원의 호흡에 의한 환기량은 다음과 같음

$$Q \geq \frac{Y}{\alpha - \beta} = \frac{1.2\times10^{-3}}{5,000\times10^{-6} - 0.03\times10^{-2}} = 0.3\text{m}^3/\text{min}\cdot\text{人}$$

여기서, α : CO_2 허용농도 (5,000 ppm)

β : 대기중의 CO_2 농도 (0.03%)

γ : 작업자가 토출하는 CO_2량 (1.2×10^{-3} m^3/min·人)

- 상기 계산식에서 얻은 0.3m^3/min에 환경 악화요인을 고려하여 안전율(10배)을 고려하여 설계 산정 시에는 3m^3/min ~ 10m^3/min을 적용

3-10-3 Pollution Source Consideration for Required Ventilation Amount

① Gas from explosive blasting

- When blasting and excavating within tunnel, standard gas amount of hazardous gas used for ventilation calculation is shown below:

Classification	sum of CO gas	NO_x gas
Emulsion	3.5×10^{-3}	'0.75×10^{-3}
FINEX	9×10^{-3}	17×10^{-3}

② Exhausting gas of diesel engine

- Calculate ventilation of hazardous gas generated from diesel engine by using gas generation standard per diesel engine PS
- Do not use engine without exhausting gas process device in the tunnel. standard generation amount is shown below:

Type of diesel engine	Air cleaner exists	Air cleaner not exist	Working ratio
Shovel loader	1.08 m^3/min.ps	2.16 m^3/min.ps	60%
Dump truck	0.67 m^3/min.ps	0.84 m^3/min.ps	60%
Others	0.65 m^3/min.ps	0.82 m^3/min.ps	50%

③ Breathe of workers

- Breathing of workers generates CO2 in the tunnel
- The amount of CO2 generated from breathe is very small compared to blasting gas or diesel engine gas. However, if the work continues for a longer time, the amount increases
- Especially in the site with large number of operator, breathe of workers should not be ignored. Ventilation required by breathe of workers can be obtained from the following equation:

$$Q \geq \frac{Y}{\alpha-\beta} = \frac{1.2\times10^{-3}}{5,000\times10^{-6}-0.03\times10^{-2}} = 0.3\text{m}^3/\text{min}\cdot\text{人}$$

where, α : CO_2 allowable density (5,000 ppm)

β : CO_2 density in atmosphere (0.03%)

γ : CO_2 from workers (1.2×10^{-3} m^3/min · 人)

- The result from the equation 0.3m^3/min is multiplied by safety ratio (10) to yield design application value of 3m^3/min ~ 10m^3/min

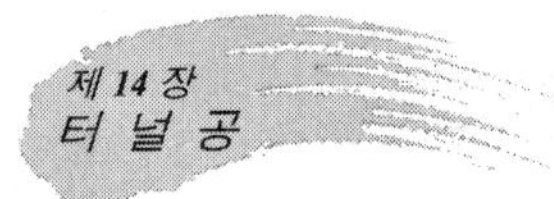

④ 산소결핍 공기량

- 터널 내의 공기의 산소농도가 항상 18% 이상이 되도록 유지하여야 함

⑤ 숏크리트 시공 시 분진량

- 숏크리트는 제 2종 분진이며, 환경기준상 환기보다는 집진설계에 충분한 검토가 필요
- 분진 발생량의 국내 실측데이터가 없기 때문에 일본 오누끼 터널 실측량 22.01mg/m^3 을 이용하여 국내에서는 25mg/m^3 을 적용
- 단, 습식일 경우는 제외

⑥ 자연가스 발생량

- 터널 공사에서 발생될 수 있는 가스로는 메탄(CH_4), 황화수소(H_2S) 등이 있음
- 자연발생 가스는 자연조건에 따라 그 발생량이 변동되므로 일률적으로 정하는 것은 곤란하므로 각 터널의 사전조사 등을 통하여 결정

④ Hypoxia air

- Oxygen density of the air in the tunnel should be remained as 18% or above

⑤ Dust from shotcrete cast

- Shocrete is one of second tier dust. dust collection design is more important factor to review than ventilation
- As there is no domestic measured data, apply 25mg/m^3 using onuki tunnel measurement value of 22.01mg/m^3
- Exclude when wet wall support system

⑥ Natural gas

- Natural gas from tunnel construction can be methene(CH_4), hydrogen sulfide (H_2S), etc.
- It is very hard to estimate amount of natural gas since the amount varies by natural condition. Determine the value by preliminary survey.

Chapter 15

Effect Assessment

1. 교통영향분석·개선 대책

1-1 개 요

- 교통영향분석 · 개선 대책은 도로의 신설 또는 확장으로 인해 대량의 통 수요를 유발하거나 유발할 우려가 있는 사업을 시행하거나 시설을 설치할 경우, 발생하는 교통량·교통흐름의 변화 및 교통안전에 미치는 영향을 조사·예측·분석하고 그와 관련된 각종 문제점을 최소화하기 위하여 수립하는 대책을 말한다.

1-2 교통영향분석·개선 대책의 내용

- 교통영향분석 · 개선대책 수립 대상은 「도시교통정비 촉진법」 제15조 제1항 및 동법 시행령 제13조의2 제3항의 별표1에 따라 다음과 같이 구분된다.

<표 6-15> 교통영향분석 및 개선 수립 대상

구 분		대상사업의 범위		시행 시기
		중앙교통영향심의	지방교통영향심의	
「도로법」 제8조에 따른 도로 건설	고속국도 · 일반국도	총연장 30km 이상인 신설 노선 중 인터체인지, 분기점, 교차부분 및 다른 간선도로와의 접속부	총연장 5km 이상인 신설 노선 중 인터체인지, 분기점, 교차부분 및 다른 간선도로와의 접속부	도로구역 결정전
	지방도	-	총연장 5km 이상인 신설 노선 중 인터체인지, 교차로 및 다른 간선도로와의 접속부	도로구역 결정전
「국토의 계획 및 이용에 관한 법률」 제2조 제6호에 따른 기반 시설		-	총연장 5km 이상 신설 노선 중 인터체인지, 교차부분 및 다른 간선도로와의 접속부	실시계획 인가전

1-2-1 교통영향분석·개선대책의 내용

- 교통영향평가분석 · 개선대책 수립 시 포함되어야 할 내용은 다음과 같다.
  - 대상사업의 시행으로 교통에 미치는 영향의 시간적 · 공간적 범위
  - 대상사업별 교통의 문제점에 대한 교통개선대책에 관한 사항
  - 교통개선대책의 수립사항을 반영한 사업계획의 내용

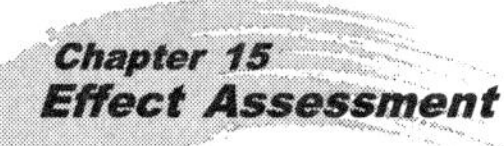

1. Transportation Influence Assessment and Reform Measures

1-1 Abstract

- The Transportation Influence Assessment and Reform Measures aims to counteract and ease many problems concerning the various aspects of transportation, such as traffic congestion, changes in traffic flow, and road safety that emerge during or after road construction (be it new or refurbishment of roads) by examining, forecasting, and analysing.

1-2 Contents of Transportation Influence Assessment and Reform Measures

- The subjects of Transportation Influence Assessment and Reform Measures are classified under clause 1 of Article 15 and clause 3 of Article 13-2 of the Urban Traffic Improvement Promotion Act.

<fig 6-15> Subjects of Transportation Influence Assessment and Reform

Type		Project Scale		Effective
		Deliberation of Highway Intersection Traffic Impact	Deliberation of Provincial Roads	
Road Projects concerning The Article 8 of the 「Road Act」	national highways & general national roads	Interchange, junction, crossway and joint with other trunk roads among new routes over 30km in length	Interchange, junction, crossway and joint with other trunk roads among new routes over 5km in length	Prior to the designation
	provincial roads	-	Interchange, crossway and joint with other trunk roads among new routes over 5km in length	Prior to the designation
Infrastructures of clause 6 of Article 2 in the 「National Land Planning and Utilization Act」		-	Interchange, crossway and joint with other trunk roads among new routes over 5km in length	Prior to approval of execution plan

1-2-1 Details of Transportation Influence Assessment and Reform Measures

- Transportation Influence Assessment and Reform Measures should include the following;
  - duration and spatial range of influences on a given construction.
  - solutions for road problems concerning the given construction.
  - a construction plan reflecting the transportation reform

1-2-2 교통현황 조사

- 장래 교통수요 예측 및 사업 시행에 따른 문제점 및 개선방안 도출을 위한 교통현황 조사 항목은 다음과 같다.
  - 주변의 토지이용 현황 조사
  - 사회 · 경제지표 현황 조사
  - 개발계획 현황 조사
  - 가로망 현황 조사
  - 교통시설 이용 현황 조사
  - 교통안전시설 및 교통사고 발생 현황 조사
  - 교차로 기하구조 및 신호 운영체계 현황 조사
  - 교통량 현황 조사

1-2-3 교통수요 예측

- 교통수요예측은 평가 대상사업에 대한 분석 목표연도를 기준으로 사업을 시행하였을 경우와 사업을 시행하지 않았을 경우를 비교하여 어느 정도의 영향을 미치게 되는가를 파악하는데 쓰인다. 따라서 사업을 시행하지 않는 경우와 시행할 경우의 두 가지로 나누어 수요를 예측한다.

1-2-4 교통영향분석·개선대책의 변경 심의

- 교통영향분석 · 개선 대책의 심의 후 공사 내용 및 규모 등의 변경 시 재 심의를 받아야 하며 변경 심의 여부에 대한 기준은 다음과 같다.
  - 도로건설사업의 경우 그 도로(차도와 보도를 포함한다. 이하 같다)의 건설구간 내 인터체인지, 분기점 및 교차 부분의 도로면적 합계가 기존 교차로의 인터체인지 · 분기점 및 교차 부분의 도로면적 합계의 30퍼센트를 초과하여 증가하는 경우(여러 번 확장하는 경우에는 이를 모두 합산한다). 다만, 도로폭이 12m 미만의 도로만으로 접속하여 교차로가 생기는 경우에는 인터체인지 · 분기점 및 교차부분의 도로 면적에 합산하지 않는다.
  - 공사가 7년 이상 중지된 후 다시 공사를 시작하는 경우
  - 토지의 이용을 변경하거나 건축물의 배치를 변경하여 개선 필요사항 등에 포함된 교통개선대책의 실효성이 현저하게 감소된 경우와 사업지구의 외부 교통개선대책의 이행이 불가능한 경우
  - 「도시교통정비 촉진법」 제13조의2 제5항 본문에 따른 국토해양부장관의 고시에서 정하는 교통개선대책의 변경 허용 인정 범위를 초과하여 교통개선대책을 변경하는 경우

1-2-2 Traffic State Report

- The following is a list of survey clauses about traffic conditions to be used for devising a solution program to ease possible problems and potential traffic demand

① property(land) occupancy of surrounding area

② index of society & economy

③ development plan targeted at the area

④ condition of street network

⑤ statistics of public transportation demand and usage

⑥ transportation safety facilities and traffic accidents in the proximity

⑦ intersection structure and traffic light operating systems

⑧ traffic amount

1-2-3 Predicting the Amount of Traffic

- The prediction of traffic amount measures out possible influences a given road construction has by cross analysing potential traffic in the proximity without construction and with construction according to the goal year in which the construction is estimated to reach completion. In another words, the numbers are drawn from the two hypothetical calculations.

1-2-4 Deliberation of Changes in Transportation Influence Assessment and Reform Measures

- After the initial Transportation Influence Assessment and Reform Measures has been reviewed, it must go through a re-deliberation should there be any changes to the scale or any other matters in the content, and the re-deliberation standard are as follows;

① The given construction includes over 30% increase in the total area of the initial crossroad, junction, and or intersection (including sidewalks) - any further increases are combined in the total as well. The widths of all four roads meeting at an intersection that don't exceed 12m are not to be included.

② Any discontinuation in the construction for over 7 years.

③ Utilization of the land is being differed, effectiveness of traffic reformation is being diminished tremendously by the changes in the orientation of buildings despite the requirements issued for the reformation, and the project district's external traffic reformation could no longer be carried out.

④ Permissible range of modifiable size of roads as notified by the Ministry of Land, Transport and Maritime Affairs in accordance with the clause 5 of Article 13-2 of the Urban Traffic Readjustment Promotion Act is exceeded, and no longer considered suitable.

2. 환경영향평가

2-1 개 요

- 환경영향평가는 대상사업의 사업계획을 수립함에 있어 당해 사업의 시행으로 인하여 환경에 미치는 영향을 미리 예측하고 분석하여 환경영향을 줄일 수 있는 방안을 강구하는 것이다.

- 환경영향평가제도는 환경에 미치는 오염을 예방하는 제도로써 각종 사업의 수립 및 시행에 있어 당해 사업의 경제성·기술성 뿐만아니라 환경적 요인도 비교·검토하여 최적의 사업계획안을 모색하는 과정이다. 따라서 환경영향평가서의 작성 목적은 각종 도로 사업으로 인한 환경오염문제를 사전에 파악하고 그 해결책을 제시함으로써 환경적으로 건전하고 지속 가능한 개발이 되도록 하는 것이다.

2-2 환경영향평가의 내용

- 환경영향평가의 대상이 되는 도로는「도로법」제2조 제1호, 제7조 및「국토의 계획 및 이용에 관한 법률」제2조 제13호에 따른 도로로서 대상 규모는 다음과 같다.

① 4km 이상의 신설도로

- 「국토의 계획 및 이용에 관한 법률」제6조제1호에 따른 도시지역에서는 폭 25m 이상의 도로인 경우에만 해당한다.
- 「도로법」제8조제1호에 따른 고속국도와「국토의 계획 및 이용에 관한 법률 시행령」제2조제2항제1호 나목·바목에 따른 자동차전용도로 또는 지하도로의 경우에는 해당하지 않는다.

② 2차로 이상으로서 10km 이상의 확장도로

③ 신설과 확장을 함께하는 경우로서 다음 계산식에 따라 산출한 수치의 합이 1 이상 도로

$$\frac{\text{신설구간 길이의 합}}{4km} + \frac{\text{확장구간 길이의 합}}{10km}$$

④ 도로의 신설로서 도시지역과 비도시지역에 걸쳐 있는 경우에는 다음 계산식에 따라 산출한 수치의 합이 1 이상인 도로(4차로는 폭 25m 이상으로 본다.)

$$\frac{\text{비도시구간 길이의 합}}{4km} + \frac{\text{도시구간 길이의 합}}{4km}$$

2. Environmental Assessment

2-1 Abstract

- The environmental assessment is carried out prior to the construction of roads to assist in establishing project plans and it provides solutions in reducing environmental effects the construction has beforehand.
- Assessment of Environmental Effects(AEE) is a system instituted to prevent pollution, and is also a procedure to search for the environmentally optimal construction plans by analysing not only the economical and technical capabilities of the constructors, but also the environmental factors concerning the project. Thus, the goal of completing the assessment is to forewarn of any environmental impact the construction might have on its surroundings and finding preventive measures before and during the operation itself to allow for an environmentally-friendly development.

2-2 Contents of the Assessment of Environmental Effects

- All roads that fall under the clause 1 of Article 2 of Road Act, the Article 7 of Act on Planning and Use of National Territory, and the clause 13 of Article 2 are to be objects of AEE and interpreted as;

① A new road exceeding 4km in length

- Only applicable to roads surpassing 25m in width within urban zones under the clause 1 of Article 6 in the 「Act on Planning and Use of National Territory」
- The following are not object of AEE; the clause 1 of Article 8 in the 「Road Act」, highways built in accordance with the clause 1 of Article 8 in the 「Act on Planning and Use of National Territory」, and motorways or underpasses built under the clause 2-1 B and F of Article 2 of the same Act

② Expanded 2-lane highways exceeding 10km length

③ Integrated construction of roads consisting of both new installation and expansion of existing roads with calculated results exceed '1' under the following formula;

$$\frac{length\,of\ installation}{4km} + \frac{length\,of\,\text{expansion}}{10km}$$

④ The result exceeding 1 in the following equation for a new installation of road in non-urban areas connecting the urban zones (4-lane highways are considered to have 25m width)

$$\frac{length\,of\,non-urban\,roads}{4km} + \frac{length\,of\,urban\,roads}{4km}$$

2-3 환경영향조사

2-3-1 환경영향평가 대상지역의 설정

- 도로건설 사업에 따른 환경영향평가를 시행함에 있어서 먼저 환경영향이 우려되는 대상지역을 설정한다.

- 대상지역의 설정은 도로건설사업을 시행함에 있어, 사업시행으로 인한 환경적인 영향이 미칠 것으로 예상되는 계획노선 주변지역의 자연환경과 생활환경, 사회·경제환경 등 제반환경 상에 미치는 영향을 객관적이고 과학적인 방법에 의거 정량분석을 실시하고, 정량분석이 곤란한 분야에 대해서는 기존 자료를 토대로 객관성 있는 정성분석을 통해 시행한다.

<표 6-18> 환경영향평가 대상

구 분		평 가 대 상 기 준	비 고
도로의 건 설	신 설	• 연장 4km 이상	- 도로법 제2조, 제7조 - 국토의 계획 및 이용에 관한 법률 제2조 13호
	확 장	• 2차로 이상으로서 10km 이상	
	신설+확장	$\frac{\text{신설구간 길이의 합}}{4km} + \frac{\text{확장구간 길이의 합}}{10km} \geq 1$	
	도시+비도시	$\frac{\text{비도시구간 길이의 합}}{4km} + \frac{\text{도시구간 길이의 합}}{4km} \geq 1$	
재 협의		• 협의내용이 통과된 날부터 5년 이내 착공하지 않은 사업 • 공사가 7년 이상 중지되었다가 다시 착공하는 사업 • 변경규모가 30%이상 증가 확대된 사업	
변경 협의		• 재협의 대상에 해당되지 않으며 변경규모가 10%이상 증가 확대된 사업	

주) 「자연환경보전법」에 의거 자연경관영향의 협의내용 포함

- 환경영향평가 대상지역 설정을 위한 주요 항목은 다음과 같다.

  ① 사업의 입지적, 공간적 범위

  ② 사업의 시간적 범위

2-3-2 환경영향요소 추출

- 도로건설사업 시행에 따른 환경영향요소의 추출은 도로건설 사업계획 및 현장조사, 사업특성, 지역특성을 감안 공사단계와 운영단계로 구분하여 추출한다.

2-3 Examination of Environmental Effects

2-3-1 Selecting Areas for AEE

- Areas with environmental effects concerned become primary subjects for AEE for any examination performed on road construction.

- Target areas for AEE are chosen according to environmental impact the construction would have on the proximity, taking into consideration any natural surroundings, life environment, and socio-economic environment. The process is carried out scientifically as well as objectively, and existing database is utilized should there be any complications in quantitative analysis.

<fig. 6-18> AEE Applicable Subjects

Category		Assessment Standards	
Road Constructi on	Installation	• Over 4km in length	- Clause 7 of Article 2 of the Road Act - Clause 13 of Article 2 of the Act on Planning and Use of National Territory
	Expansion	• Over 10km in length (2-lane highways)	
	Installation + Expansion	$\frac{\text{length of installation}}{4km} + \frac{\text{length of expansion}}{10km} \geq 1$	
	Urban + non-Urban	$\frac{\text{length of non-Urban}}{4km} + \frac{\text{length of Urban roads}}{4km} \geq 1$	
Re-deliberation		• Projects that are 5 years overdue • Continuation of projects that are over 7 years or more of discontinuation • Projects that are increased in scale by 30% or more	
Alternative Deliberation		• Re-deliberation not applicable and the scale of the project being increased by 10% or more	

Environmental Scenery Effects Agreement included in accordance with 「Natural Environment Conservation Act」

- The decisive factors in selecting areas applicable with AEE are as follows;

① Locational and spatial range of the project

② Time margin of the project

2-3-2 Extracting the Factors of Environmental Effects

- Extracting the factors of environmental effects in road construction is done with the locality of the site, construction plan, on-location analysis, and project feature into consideration, and is divided into construction phase and management phase.

① 공사단계 환경요소 추출 항목

- 자연형질의 변경 : 산림벌채, 하천개수 및 변경, 육상지형의 변형
- 자재채취 및 운반 : 골재채취 및 운반, 토사 운반, 기자재 운반(육상)
- 시설공사(건설공사) : 절·성토, 굴삭, 항타, 콘크리트공사, 배수공사, 공사 인부 투입

② 운영단계 환경요소 추출 항목

- 구조물의 이용 : 도로, 교량, 박스
- 운행시 : 차량운행에 따른 소음·진동 발생, 배기가스 발생

2-3-3 환경영향 평가항목의 설정

- 도로건설사업의 환경영향평가에 대한 평가항목의 설정은 환경부고시 제2008호-223호「환경영향평가서 작성 등에 관한 규정」에 제시된 도로건설사업의 주요 평가항목과 환경영향요소를 비교·검토하여 지형·지질, 대기질, 동·식물상, 소음·진동, 교통 등 중점평가항목을 설정하여 분석한다.

- 또한, 중점평가 항목 이외의 항목에 대해서는 도로건설 사업의 시행에 따른 영향 정도를 파악하여 평가의 효율성을 고려한 현황조사항목과 제외항목으로 구분하여 설정한다.

<표 6-19> 평가 대상 항목

대기환경	수환경	토지환경	자연생태환경	생활환경	사회·경제
기상 대기질 악취 온실가스	수질 (지표·지하) 수리·수문 해양환경	토지환경 토양 지형·지질	동·식물상 자연환경자산	친환경적 자원순환 소음·진동 위락·경관 위생·공중보건 전파장해 일조장해	인구 주거 산업

2-3-4 환경영향 평가항목의 조사방법

- 환경조사는 예측 및 평가의 시행여부 결정을 위한 자료 수집과 예측 및 평가가 시행될 경우 기본 데이터를 정리하기 위하여 시행하는 것이다.

- 원칙적으로 국가나 지방 공공단체 등 공공기관이 공표한 자료를 수집·정리하고 자료수집과 병행하여 현지조사를 통해 보완한다.

① Construction Phase : Extracting Factors of Environmental Effects
- Form and quality changes : felling, river improvement and alteration, geographical alterations to the area
- Extraction and transportation of resources : extraction and/or transportation of aggregate, sand and soil, and tools and equipments(by land)
- Facility construction : ground cutting, excavation, pile driving, concretion, drainage works, and manning the construction

② Management Phase : Extracting Factors of Environmental Effects
- Fabric utilization : roads, bridges, and boxes
- Operating vehicles : generated noise, tremor, and exhaust gas

2-3-3 Establishing Clauses for Evaluation of Environmental Effects

- The clauses by which the environmental effects are evaluated are determined considering the geographical, geological, zoological, and botanical factors, along with noise pollution, tremor, and traffic factors that arise in a construction, which are all indicated as environmental effects in the notification of the Ministry of Environment.
- Also, additional clauses besides the main factors for evaluation are categorized into two : the site evaluation clauses about the effects of construction and additional clauses.

<fig 6-19> List of Evaluation Subjects

Atmospheric Environment	Water Environment	Terrain & Soil	Ecosystems	Life Environment	Socioeconomics
weather atmosphere odor greenhouse - gas	water quality (surface & subterranean) water supply and floodgate marine - environment	terrain environment pedalogy topography geology	animals and flora natural environmental resources	environmentally -friendly resource circulation noise & tremor leisure and scenic quality public health & sanitation electric obstacle sunilght blockage	population residency industry

2-3-4 Method to Examine Evaluation Clauses of Environmental Effects

- Environmental check up is collecting data for a viable forecasting and assessment. During forecasting and assessment, the check up is instituted to arrange initial data.
- Fundamentally, the government or public bodies collect and organize official data and they are reinforced through local research.

2-4 간이평가절차

2-4-1 주요 내용

- 평가서의 작성 및 협의절차가 해당사업의 환경에 미치는 영향에 관계없이 일률적인 절차에 따라 진행됨에 따른 평가기간 및 작성비용의 절감을 위해 단순 반복되는 사업으로서 환경영향이 미미한 사업은 평가절차를 간소화하는 간이평가절차를 운영하도록 함
- 간이평가절차의 대상은 평가규모, 사업지역 특성(폐쇄성수역, 오염물질 체류지역, 집단주거지역 등)을 고려하여 설정하고, 평가계획서 심의위원회에서 대상여부를 결정하도록 함

2-4-2 간이평가절차

- 평가서작성 초기단계에서 사업자가 간이평가절차 해당여부를 평가계획서 심의위원회에 심의요청
- 사업자는 평가서를 작성하여 협의기관 협의, 관계기관 및 주민의견수렴 병행실시
  (일반) 초안작성→관계기관 · 주민 의견수렴→평가서작성→평가협의→사후관리
  (간이) 간이평가서→관계기관 · 주민 의견수렴 및 평가협의→사후관리
- 의견수렴내용 등을 반영한 최종평가서의 제출 및 확인절차를 규정하여 주민, 관계기관, 협의기관의 의견이 반영되었는지 확인할 수 있도록 함

2-4-3 사전환경성검토와의 관계

- 환경정책기본법에 따른 사전환경성검토를 실시한 경우에는 환경영향평가를 하면서 사전환경성검토서의 내용을 활용할 수 있도록 함
- 사전환경성검토과정에서 평가항목 · 범위 등에 대해 구체적으로 협의된 경우 스코핑절차를 생략할 수 있도록 함
- 사전환경성검토협의시 의견수렴이 환경영향평가법에 따른 의견수렴을 대체할 수 있다고 판단되는 경우에는 평가계획서 심의위원회의 심의를 거쳐 평가서초안에 대한 의견수렴 및 절차를 생략할 수 있도록 함
- 이 경우 사전환경성검토서의 내용, 의견수렴 절차의 적정성 등을 종합적으로 고려하여 평가계획서 심의위원회에서 결정하도록 함

2-4 Simplified Evaluation Procedure

2-4-1 Main Content

- Filling out of evaluation sheets and or negotiation process of road projects that are of minimal environmental effects are simplified to reduce repetitiveness, time, and cost : a simplified version is provided for such road projects.

- Eligibility for simplified evaluation is decided by a project's scale, locality (densely populated areas, water bodies of dense eutrophication, etc.), and the decision will be reviewed by the Committee of Evaluation.

2-4-2 Simplified Evaluation Procedure

- The evaluation at an early stage allows for the contractor to ask for a Simplified Evaluation through consulting the Committee.

- The contractor must fill out an assessment form including approval from eligible council, government bodies and residents of the site.
  (Regular Form) a rough draft → collection of views from gov't bodies and local residents → complete evaluation form → negotiation → post-management
  (Simplified Form) Simplified Form → collection of views from gov't bodies and local residents + negotiation → post-management

- The final version of evaluation forms including the collection of various approvals must include feasible verification of those bodies, be it local residents or government bodies.

2-4-3 Pre-evaluation of Environment and Its Bearing

- If an environmental evaluation is done before a construction (under the Framework Act on Environmental Policy), the data may be used for later environmental effect evaluations.

- In such pre-evaluations, the 'scoping' procedure may be skipped, if the pre-evaluation is fully detailed with valid assessment clauses and complete negotiations.

- During the negotiation procedure of a pre-evaluation of environment, it is possible to omit the 'collection of views' process from the rough draft of the form only if the negotiations are considered to overrule Assessment of Environmental Impact Act, and only after a thorough review by the Committee of Evaluation.

- In such cases, the Committee of Evaluation decides on alternative procedure of evaluation and assessment through a careful consideration of reasoning, and post environmental effects of the construction.

2-4-4 간이평가절차 대상 기준

<표 6-20> 간이평가절차 대상 기준

구 분	기 준
사업규모	● 평가대상 최소규모의 2.0배 이하
지역특성	● 사업지역에 환경적·생태적으로 보전가치가 높은 다음의 지역이 포함되지 아니한 사업 -「자연환경보전법」 제34조에 따른 생태·자연도 1등급 권역 -「습지보전법」 제8조에 따른 습지보호지역 및 습지주변관리지역 -「자연공원법」 제2조제1호에 따른 자연공원 -「야생동·식물보호법」 제27조 및 제33조에 따른 야생동·식물특별보호구역 및 야생동·식물보호구역 -「문화재보호법」 제2조제3항에 따른 보호구역 -「금강수계 물관리 및 주민지원 등에 관한 법률」 제4조에 따른 수변구역 -「낙동강수계 물관리 및 주민지원 등에 관한 법률」 제4조에 따른 수변구역 -「영산강·섬진강수계 물관리 및 주민지원 등에 관한 법률」 제4조에 따른 수변구역 -「한강수계 상수원수질개선 및 주민지원 등에 관한 법률」 제4조에 따른 수변구역

2-4-4 Simplified Evaluation Standard

<Table 6-20> Simplified Evaluation Standard

Type	Standard
Scale of Project	• Less than twice the size of standard scale for evaluation
Locality	• Areas excluding naturally and ecologically well-preserved regions such as follows; - Article 34 of the 「Natural Environment Conservation Act」 - Wet Land Conservation Zones and proximities in the Article 8 of the 「Conservation of Wetlands Act」 - Natural Parks in the clause 1 of Article 2 of the 「Natural Parks Act」 - Special Wildlife Conservation Zones and Wildlife Conservation Zones of the Articles 27 and 33 of the 「Protection of Wild Fauna and Flora Act」 - Conservation Zones of clause 3 of Article 2 of the 「Protection of Cultural Properties Act」 - Water sides in Article 4 of the 「Act on the Management of Water and Support for Residents of the Kum-kang Water Systems」 - Water sides in Article 4 of the 「Act on the Management of Water and Support for Residents of the Nakdong River System」 - Water sides in Article 4 of the 「Act on the Management of Water and Support for Resident of the Young-san River and Sum-jin Water Systems」 - Water sides in Article 4 of the 「Act on the Improvement of Water Quality and Support for Residents of the Riverhead of the Han River System」

Chapter 16
Road Area Determination

도로구역결정 흐름도

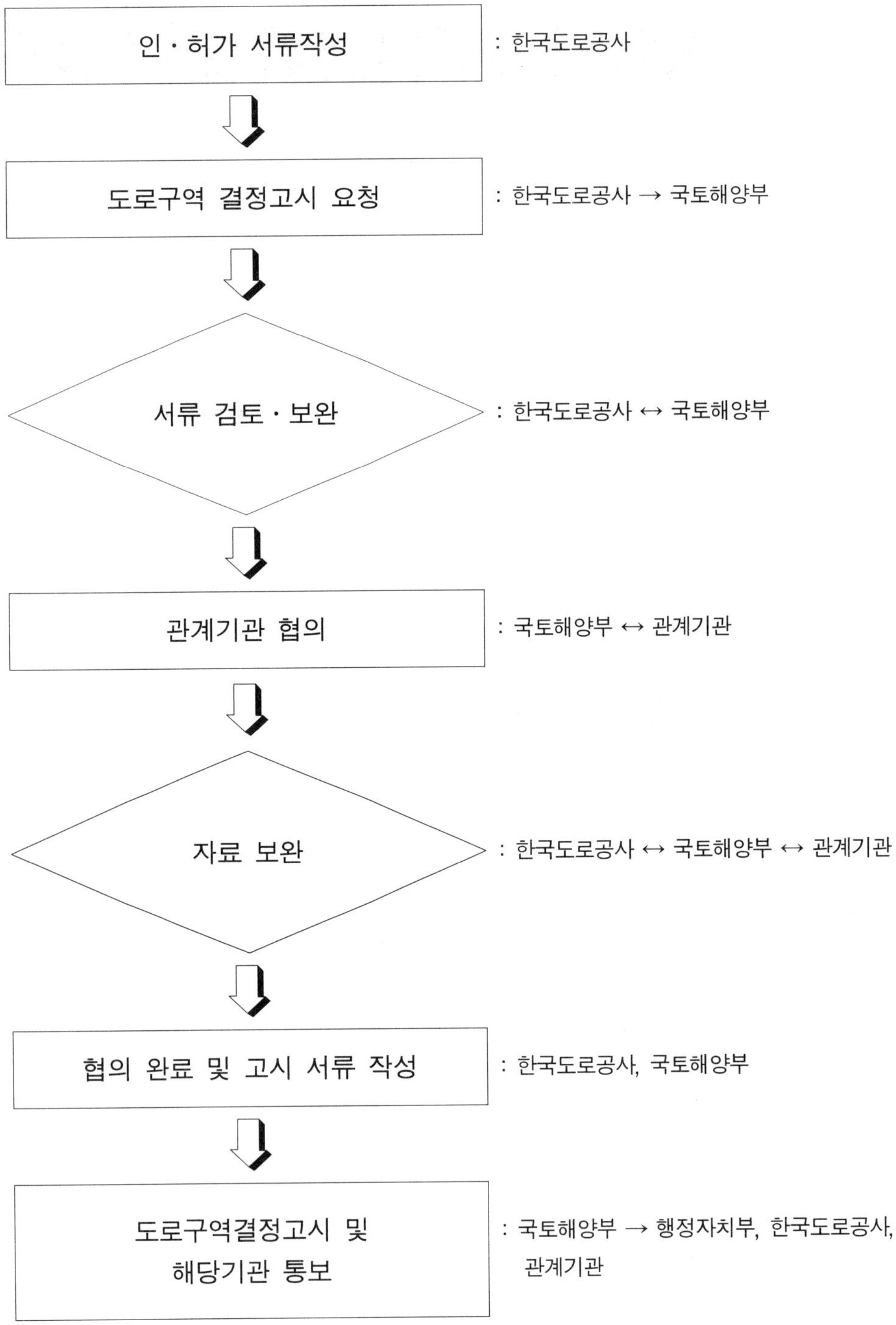

인·허가 서류작성	: 한국도로공사
⇩	
도로구역 결정고시 요청	: 한국도로공사 → 국토해양부
⇩	
서류 검토·보완	: 한국도로공사 ↔ 국토해양부
⇩	
관계기관 협의	: 국토해양부 ↔ 관계기관
⇩	
자료 보완	: 한국도로공사 ↔ 국토해양부 ↔ 관계기관
⇩	
협의 완료 및 고시 서류 작성	: 한국도로공사, 국토해양부
⇩	
도로구역결정고시 및 해당기관 통보	: 국토해양부 → 행정자치부, 한국도로공사, 관계기관

Flowchart for Road Area Determination

Documentation for approval	: Korean Expressway Corporation

Request for official announcement for road area determination	: Korean Expressway Corporation → Ministry of Land, Transport and Maritime Affaires

Documentation review and complement	: Korean Expressway Corporation ↔ Ministry of Land, Transport and Maritime Affaires

Consultation with related agencies	: Ministry of Land, Transport and Maritime Affaires↔Related Agencies

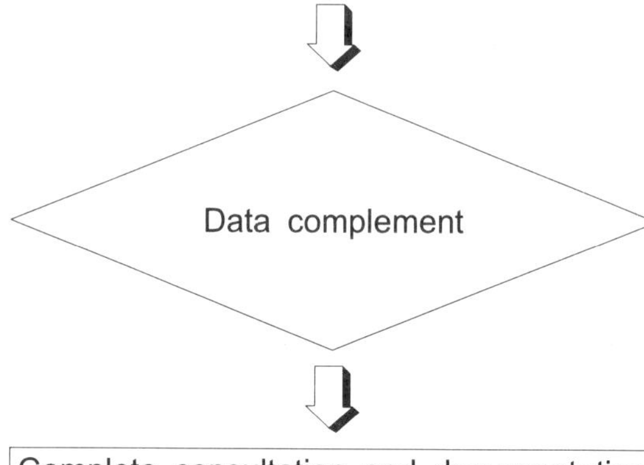

: Korean Expressway Corporation ↔ Ministry of Land, Transport and Maritime Affaires ↔ Related Agencies

Complete consultation and documentation for official announcement	: Korean Expressway Corporation, Ministry of Land, Transport and Maritime Affaires

Official announcement for road area determination and report to corresponding agency	: Ministry of Land, Transport and Maritime Affaires → Ministry of Public Administration and Security, Korean Expressway Corporation, related agencies

1. 협의서류 작성

1-1 용지경계 확정

- 평면 및 종단면도, 횡단면도를 검토하여 용지경계 기준에 따라 구간별로 용지 경계를 확정한다.

- 용지경계 기준은 도로공사 자체 방침에 따라 다음과 같이 적용한다.

구 간		용지경계	비 고
토공구간	흙쌓기부	측구 및 비탈면 끝으로부터 1m	
	땅깎기부	측구 및 비탈면 끝으로부터 2m	
	측도	측구 및 비탈면 끝	
교량구간	교고 20m 이상	교량 끝으로부터 10m	
	교고 20m 미만	교량 끝으로부터 5m	

1-2 편입면적 산출

- 관할 행정기관에 요청하여 실시설계 결과 확정된 용지경계에 포함되는 지적도를 경계로부터 여유 있게 발급받는다.

- 또한, 도로구역 협의 서류에 포함될 편입토지의 토지대장, 지적도등본, 토지이용계획확인원도 해당기관에 방문 또는 인터넷을 통하여 발급받는다.

- 용지경계에 포함된 토지의 지번을 발췌하고, 각 지번의 편입면적을 산출하여 행정구역별로 집계한다.

- 농지, 산지 등 협의 시 필요한 토지의 종류별 면적을 산출하고, 농업보호구역·농업진흥구역 등 특정 용도 토지에 대한 면적도 산출한다.

1-3 용지도 작성

- 지적평면도상에 도로구역 경계선을 굵은 실선으로 표기한다.
- 확장공사의 경우 당초 도로구역과 변경되는 도로구역이 명확히 구분될 수 있도록 색상 등을 표기하여 작성한다.

- 기타 세부사항은 국토해양부 도로구역결정고시 업무추진 매뉴얼에 따른다.

1. Documentation for Consultation

1-1 Parcel Boundary Determination

- Review the drawings of ground, vertical section and cross section to determine parcel boundary per each sector in accordance to site boundary standard.

- Korean Expressway Corporation applies the following parcel boundary standard:

Sector		Parcel boundary	Remarks
Land Fill	Filling	1m from the end of gutter or slope	
	Section	2m from the end of gutter or slope	
	Measure	The end of gutter or slope	
Bridge	20m and higher	10m from the end of the bridge	
	Below 20m	5m from the end of the bridge	

1-2 Incorporated Area Determination

- Request the land registration map for the area contiguous to the parcel boundary determined by detailed design from related administrative agency.

- Issue land-book, certified copy of land registration map, land-use planning certification map of the incorporated parcels from the administrative agency or through Internet.

- Extract parcel numbers of the parcels included in the boundary. Identify the size of incorporated area of each parcel number and sum them up by respective administrative jurisdiction.

- Identify the size of the incorporated parcels by their types (farmlands, mountainous districts, etc) Special land-use area should be identified as well (protective farmland, promotive area for farming, etc)

1-3 Land Use Map Preparation

- Identify land area boundary with thick solid line in the land registration map.

- For extension work, identify changed road area from original road area with different color scheme.

- Refer to work process manual for official announcement of road area determination from Ministry of Land, Transport and Maritime Affaires for other details.

1-4 협의서류 작성

1-4-1 서류 작성 시 유의사항

- 모든 협의서류의 위치도는 칼라도면 사용을 원칙으로 한다.
- 편입용지가 누락되지 않도록 철저히 확인한다.
- 사업명은 해당노선, 구간, 신설, 확장, 시설명 등이 명확히 나타나도록 표기하고, "실시설계" 발주명을 기재해서는 안 된다.
- 사업목적에는 해당 노선 또는 시설물의 설치목적을 기재해야 하며, 실시설계를 수행하는 목적을 기재해서는 안 된다.
- 사업계획서에는 휴게소, 인터체인지, 영업소, 지사 등 부대시설 현황을 반드시 기재하여야 한다.
- 작성부수는 통상 5부 내외로서, 서류 종류별 협의대상 기관 수에 따르므로 협 의 요청 전 필요 부수를 파악하여 작성한다.

1-4-2 협의서류 종류

- 도로구역 협의 시 필요한 서류는 "도로법 제25조의2 (다른 법률에 의한 인·허가 등의 의제)"에 해당하는 협의 중 해당 법률에서 요구하는 서류이다.
- 다만, 상기법에 제시된 협의사항 중 당해 설계와 관련 있는 협의서류만 작성한다.
(예를 들어, 편입예정 용지경계에 초지가 미포함이면 해당 서류는 불필요)

1-4-3 기 타

- 상기 외 도로구역 결정협의에 관한 세부사항은 국토해양부 도로구역결정고시 업무추진 매뉴얼에 따른다.
- 협의서류 작성 및 협의요청 전 반드시 관련 법률 개정여부를 확인하여 개정사 항을 반영하여 협의를 실시하도록 한다.

2. 협의실시

2-1 협의요청

- 도로구역 결정은 국토해양부장관이 승인하도록 되어있으며, 도로구역 결정 협의를 진행하기 위해서는 먼저 한국도로공사에서 국토해양부 도로구역 결정 협의요청을 하여야 한다.
- 국토해양부에서는 협의서류 검토 후 협의기관에 도로구역 결정에 관한 협의를 요청한다.

1-4 Documentation for Consultation

1-4-1 Notices on Document Preparation

- Use color map for all consultation documents.
- Check comprehensively not to omit any incorporated parcels.
- The name of the project should include respective route, section, new road or extension, related facilities. Do not put the name of "execution design" order.
- In the purpose of the project, indicate the purpose of respective route or facility. Do not put the purpose of the execution design instead.
- Document of the project plan should indicate facility status such as resting area, interchange, business office, branch office.
- Usually, consultation requires about five hard copies of the documentation. However, check how many hard copies are required as this depends on the types of documentation and number of agencies participated in consultation.

1-4-2 Types of Consultation Documents

- The documents required for road area determination are described in "the Article 25-2 of Road Act (admission by other laws)"
- Prepare only the documents required and related to the current design. (For example, no documents required for grassland if there is no such land-use in the incorporated area.)

1-4-3 Others

- Refer to work process manual for official announcement of road area determination from Ministry of Land, Transport and Maritime Affaires for other details.
- Check the updates for relevant laws and regulation before preparing the documentation.

2. Consultation Process

2-1 Request for Consultation

- Road area determination is subject to be approved by Minister of Land, Transport and Maritime Affaires. To proceed with road area determination consultation, Korea Expressway Corporation requests consultation to Ministry of Land, Transport and Maritime Affaires.
- Ministry of Land, Transport and Maritime Affaires, after reviewing the consultation documents, requests consultation to relevant administrative agencies.

- 협의기관에서는 국토해양부의 협의 요청을 받은 날로부터 후 30일 이내에 회신하는 것이 원칙(도로법 제25조의2)이나, 부득이한 경우 검토기간이 길어지는 경우도 있다.

- 도로공사 담당자는 조속한 협의 완료를 위해 설계용역사와 협력하여 협의기관의 요구에 적극 협조하도록 한다.

2-2 협의사항 보완

- 관계기관이 협의서류 검토 중 서류상 보완이 필요하면 국토해양부를 통해 협의 서류 보완을 요청할 수 있다.

- 국토해양부로부터 관계기관의 협의서류 보완요청 사항이 통보되면, 협의 자료 보완 또는 방문 협의를 실시한다.

2-3 협의의견 회신 및 조치계획 제출

- 모든 관련기관의 검토의견이 국토해양부로 회신되면 국토해양부에서는 의견들을 정리하여 한국도로공사에 의견을 송부하고, 조치계획 제출을 요청한다.

- 요청에 따라 조치계획을 작성하고, 도로구역 결정서류 및 기타 도로구역 결정고시에 필요한 서류도 국토해양부에서 정하는 양식에 따라 작성하여 제출한다.

- 도로구역 결정서류는 국토해양부에서 해당 사업의 도로구역 결정 시 관련기관에 송부하는 서류로서, 사업개요 · 편입용지조서 · 용지도 등이 수록된다.

- The agency participated in the consultation should reply within 30 days from the request (the Article 25-2 of Road Act). If needed, this may delay.

- The person in charge in Korea Expressway Corporation, in collaboration with detailed design engineering company, deals with any requests from agencies participated in the consultation to precipitate the consultation process.

2-2 Supplementation for Consultation Documents

- The agencies participated in the consultation may request additional documents through Ministry of Land, Transport and Maritime Affaires.

- If additional documents are requested by Ministry of Land, Transport and Maritime Affairs, prepare the documents or set up the discussion visit.

2-3 Consultation Result Report and Management Plan Submittal

- When all the agencies complete consultation and send their opinion to Ministry of Land, Transport and Maritime Affaires, the Ministry sends them to Korea Expressway Corporation and requests for management plan.

- The Corporation prepares the management plan as well as other documents required for road area determination.

- The documents are sent to relevant agencies when road area is officially determined (the summary of the project, incorporated parcel transcript, land-use map, etc.)

Chapter 17

Total Cost

1. 개 요

- 총사업비란 건설사업에 소요되는 모든 경비로서, 공사비 · 보상비 · 시설부대경비로 구성된다.

- 총사업비 관리 대상사업은 국가예산의 지원을 받는 사업 중 사업기간이 2년 이상으로서 총사업비가 500억 이상인 토목사업이 해당된다. 통상적으로 고속도로 신설 또는 확장 사업은 km당 공사비가 200억원을 상회하므로 총사업비 관리 대상사업에 거의 해당된다고 볼 수 있다.

- 총사업비 협의는 계획 및 설계단계의 각 단계가 끝나고 실시하며, 공사 중에는 설계변경 규모가 일정 규모 이상일 경우 실시한다.

- 본 매뉴얼에서는 실시설계 완료 후 실시하는 총사업비 협의에 대해서만 다루도록 하며, 여기에서 다루지 않은 내용에 대해서는 기획재정부 총사업비 관리지침에 따른다.

- 총사업비 관리지침은 기획재정부에서 매년 1회 발행한다.

2. 총사업비 협의 절차

- 총사업비 협의는 한국도로공사가 국토해양부를 통해 기획재정부와 협의를 실시하는 것이며, 다음과 같은 순서로 실시한다.
  - <한국도로공사 ↔ 조달청> 단가 적정성 의뢰 및 결과 반영
  - <한국도로공사 → 국토해양부> 총사업비 협의 요청
  - <국토해양부> 총사업비 사전 심사 및 협의서류 보완
  - <국토해양부 → 기획재정부> 총사업비 협의 요청
  - <기획재정부 → 국토해양부 → 한국도로공사> 총사업비 조정결과 통보
    (필요 시 타당성 재조사 실시 후 총사업비 조정결과 통보)

1. Overview

- Total cost means all expenses occurred during the construction work including construction cost, compensation cost and all other attached facility costs.

- The subject of total cost management is the civil engineering and construction work subsidized by the government with total cost of more than 50 billion won and the term of work of more than 2 years. In general, new highway construction works and highway extension works are the subject of total cost management as the construction cost is more than 20 billion per kilometer.

- Total cost consultation is held on every end of planning and designing stage. During the construction, the consultation is held only when the scope of design change reaches certain level.

- This manual only focus on total cost consultation held after finishing detailed design. For more information not described in here, refer to the total cost management guideline of Ministry of Strategy and Finance.

- Total cost management guideline is published by Ministry of Strategy and Finance every year.

2. Consultation Process for Total Cost

- During the total cost consultation, the Korea Expressway Corporation, through Ministry of Land, Transport and Maritime Affaires, discusses total cost with Ministry of Strategy and Finance The consultation is proceeded with the following order
  - <Korea Expressway Corporation ↔ Public Procurement Service>
    Request for unit cost appropriateness review and reflect the result
  - <Korea Expressway Corporation → Ministry of Land, Transport and Maritime Affaires> Request for total cost consultation
  - <Ministry of Land, Transport and Maritime Affaires>
    Preliminary Review on total cost and complement the consultation documents
  - <Ministry of Land, Transport and Maritime Affaires → Ministry of Strategy and Finance> Request for total cost consultation
  - <Ministry of Strategy and Finance → Ministry of Land, Transport and Maritime Affaires → Korea Expressway Corporation> Report the result of total cost adjustment
    (If needed, the result will be reported after conducting feasibility reexamination)

3. 조달청 사전검토 실시

- 총사업비 관리대상사업은 총사업비 관리지침에 따라 총사업비 협의 전 조달청에 의뢰하여 설계결과에 대한 단가 적정성 검토를 실시하여야 한다.

- 요청서류는 공구별 설계예산서와 단가산출서이며, 협의요청 전 조달청 담당자에게 문의하여 기타 필요한 설계도서를 준비한다.

- 요청 후 회신까지 소요되는 기간은 1개월이며, 자료보완 등 부득이한 사유 발생 시 1개월을 초과할 수 있다.

- 조달청에서 협의자료의 보완이 필요하다고 판단하는 경우 한국도로공사에 보완요청을 할 수 있으며, 이 경우 보완자료를 신속히 제출하여 협의가 지연되는 일이 없도록 한다.

- 조달청 검토가 완료되면 검토결과에 따른 단가 파일(EBS)이 송부되며, 그 단가 파일의 분석을 통해 검토내용의 이상 유무·공구별 통일성 등을 확인하여 필요 시 자체 보완을 실시하고, 근거를 보관하도록 한다.

- 조달청에서는 단가 적정성 검토결과 송부 시 검토에 따른 수수료 납부를 한국도로공사에 요청한다. 수수료는 「조달사업에 관한 법률」에 따라 조달청장이 별도로 정하며 검토 요청 공사비의 0.04%이다.

4. 총사업비 협의

- 총사업비 협의를 위해 선행되어야 하는 절차는 다음과 같으며, 해당 절차가 완료되었다는 증빙자료를 총사업비 협의자료에 첨부하여야 한다.
  - 환경영향평가 및 교통영향분석 및 개선대책 협의결과 반영
  - 관계기관(지자체) 등 협의결과 반영
  - 건설기술관리법시행령 제38조의13에 따른 설계의 경제성 검토(VE) 실시
  - 조달청 설계단가 적정성 검토

3. Preliminary Review of Public Procurement Service

- The construction work subject of total cost management should, in accordance with total cost management guideline, request the unit cost appropriateness review to Public Procurement Service on the design results before total cost consultation.

- The required documents are design budget for each section of work and bill of unit price. Discuss with person in charge in Public Procurement Service to prepare additional design drawings required.

- The result of the review will be reported within one month from the request. If there are unavoidable causes (ex. additional information required), the report can be delayed.

- When Public Procurement Service requires additional information for the consultation, Korean Expressway Corporation can request the information. In this case, the information should be submitted as soon as possible to avoid any delays.

- When the review is finished, the resultant EBS is sent. EBS should be reviewed to make sure if there are any issues. If needed, supplementation should be conducted and the accompanying information should be stored.

- Public Procurement Service requests the service charge when sending the unit price appropriateness review result to Korean Expressway Corporation. The service charge is based on 'the Government Procurement Act' and decided by the head of Public Procurement Service. (0.04% of the requested construction cost)

4. Total Cost Consultation

- The process preceding total cost consultation is shown as below. The documentary evidence that the following process has been finished should be included in the total cost consultation documents.
  - Environmental impact assessment and traffic impact assessment consultation result and any reflection
  - Consultation result of discussion with related organization (municipalities, etc.) and any reflection
  - Value Evaluation (VE) on the design based on article 38-13 of the Enforcement Decree of the Construction Technology Management Act.
  - Design unit price appropriateness review from Public Procurement Service

- 총사업비 협의를 위한 서류는 기본적으로 다음과 같으며, 요청 전 담당자와 협의하여 추가로 필요한 자료를 준비하여야 한다.
  - 총사업비 조정요구서
  - 실시설계용역 요약보고서 및 결과보고서(성과품과는 별도로 정리된 현황자료)
  - 실시설계 결과에 대한 조달청장의 검토의견서
  - 실시설계 반영된 환경영향평가 결과, 교통영향분석 및 개선대책, 지자체 협의결과

- 모든 서류가 준비되면 국토해양부로 한국도로공사 예산팀을 통해 총사업비 조정 요청 공문을 발송하고, 국토해양부 담당자에게 총사업비 조정내역에 관해 설명 후 필요 시 서류를 보완토록 한다.

- 총사업비 내역에 대한 검토가 끝나면 국토해양부에서 기획재정부로 총사업비 조정 요청을 하며, 필요 시 기획재정부 담당자에게 총사업비 내역에 대한 설명을 하도록 한다.

- 기획재정부에서 요청서류 검토 후 총사업비 조정결과를 통보하면 총사업비 협의가 완료되며, 협의기간은 공문 접수일로부터 약 30~60일 가량 소요된다.

5. 타당성 재조사

- 타당성 재조사는 다음의 경우에 실시한다.
  - 실시설계 완료단계에서 총사업비가 예비타당성조사 대상규모에 해당되나 예비타당성조사를 실시하지 않은 사업
  - 물가상승분 및 손실보상비 증가분을 제외한 총사업비가 기 확정된 총사업비 대비 20% 이상 증가한 사업
  - 교통수요 예측치가 이전 단계 대비 30% 이상 감소한 사업
  - 예산 등 기타 사유로 기획재정부장관 또는 중앙관서의 장이 필요하다고 인정하는 경우

- 타당성 재조사 용역은 한국개발연구원(KDI)이 대학, 연구소, 엔지니어링 회사 등을 참여시켜 실시하며, 용역 수행기간은 6개월을 원칙으로 하여 탄력적으로 조정한다.

- The document required for total cost consultation is shown as below. Additional information should be prepared through discussion with person in charge prior to the consultation.
  - Total cost adjustment request
  - Summary report and result report of execution design service (Status report separated from the deliverable)
  - Review report of the head of Public Procurement Service on execution design result
  - The result of environmental / traffic impact assessment and municipality consultation reflected to the execution design

- When all the documentation is ready, official document for total cost adjustment request is sent to Ministry of Land, Transport and Maritime Affaires through Korean Expressway Corporation. The total cost adjustment should be explained to the person in charge at the Ministry so that additional documentation should be supplemented if required.

- When the review on total cost is finished, Ministry of Land, Transport and Maritime Affaires requests total cost adjustment to Ministry of Strategy and Finance. Total cost may be explained to the person in charge in the Ministry of Strategy and Finance if needed.

- Total cost consultation is finished when Ministry of Strategy and Finance reports the total cost adjustment result. 30 to 60 business days are required for the consultation.

5. Feasibility Reexamination

- Feasibility reexamination is conducted in the following cases:
  - Preliminary feasibility study is not conducted at the end of execution design while the cost of construction work is subject to preliminary feasibility study
  - Total cost is increased by more than 20% from the initial total cost excluding inflation and loss compensation cost.
  - Traffic demand estimation is decreased by more than 30% from the preceding stage.
  - All other circumstances that the head of Ministry of Strategy and Finance other Ministers decide to be needed.

- Feasibility Reexamination is conducted by KDI with participation of universities, research institutions and engineering companies. The official term of the reexamination is 6 months but can be adjusted if needed.

- 타당성 재조사 시 수행되는 내용은 사업의 개요 및 재조사의 쟁점 파악, 사업계획의 적절성 검토, 경제성 분석, 정책적 분석 등을 실시하여 이를 종합하여 사업의 타당성을 판단한다.

- 타당성 재조사가 완료되면 기획재정부에서는 타당성 재조사 결과를 반영한 총사업비 조정결과를 통보하며, 사업의 타당성이 없다고 판단되는 사업에 대해서는 사업규모·적정투자시기 조정 등 사업에 대한 추진전략을 제시한다.

- The contents of the reexamination includes identifying the issues of the work, reviewing the appropriateness of the business plan, analysing the economic feasibility and political feasibility. The feasibility should be examined with consideration of all forementioned aspects.

- When feasibility reexamination is completed, Ministry of Strategy and Finance reports the total cost adjustment result which reflects the result of feasibility reexamination. When the work is not feasible enough, adjusted business plan is suggested with adjustment of work scope and investment timing, etc.

Expressway Design Manual

인　　쇄 : 2010년 5월 10일
발　　행 : 2010년 5월 20일
발 행 처 : 한국도로공사
보 급 처 : 사단법인 한국도로교통협회
주　　소 : 서울시 강남구 대치3동 987-14
T E L : 02) 3490-1000
F A X : 02) 552-5879

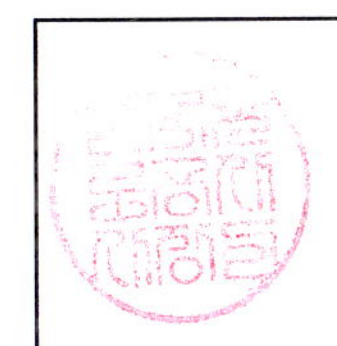

판매처 : 도서출판 건설정보사
주　소 : 서울시 용산구 갈월동 70-9
T E L : 02) 717-3396~7
F A X : 02) 717-3398
등　록 : 1998년 12월 24일 제 3-1122호
http://www.gunsulbook.co.kr

ISBN 978-89-6295-064-9　　93530　　정가 30,000원